Asynchronous Machines

Valentin Asanbayev

Asynchronous Machines

Basic Calculation Elements from Field Equations

Springer

Valentin Asanbayev
Culver City, CA, USA

ISBN 978-3-030-92283-2 ISBN 978-3-030-92284-9 (eBook)
https://doi.org/10.1007/978-3-030-92284-9

This Springer imprint is published by the registered company Springer Nature Switzerland AG
The registered company address is: Gewerbestrasse 11, 6330 Cham, Switzerland

To
daughter Anna and son Oleg

Preface

The current stage in the development of the methods for research and analysis of the electromagnetic processes occurring in the electric machines is characterized by the fairly widespread use of the theory of the electromagnetic field described by Maxwell's equations. The field calculations are very laborious and are usually characterized by the excess of information, which creates serious difficulties in generalizing the results of an analysis. In this regard, it seems appropriate to carry out the research aimed at increasing the efficiency of the use of the field calculations. In this work, as one of the possible ways to solve this problem, it is considered an approach that implies the development of the method that makes it possible to perform the electromagnetic calculations of the asynchronous machine based on the direct use of the field theory.

Electromagnetic and electromechanical processes occurring in the electric machine are due to the effect of the resulting magnetic field created in it by the currents flowing in the stator and rotor windings of the machine. In this regard, the use for the calculation purposes of the relations expressed in terms of the corresponding quantities representing the resulting magnetic field of the machine should be considered natural. However, as practice has shown, the calculated expressions obtained in this case have the rather complex structure, are difficult to the physical interpretation, and thus do not meet the requirements of the engineering practice.

In the practice of engineering calculations, the resulting magnetic field of the machine is represented as a result of the superposition of the less complex in structure "partial" magnetic fields generated by the currents of the stator and rotor windings of the machine separately. In this regard, the modern theory of calculating the electric machine is based on the consideration of these "partial" magnetic fields, the use of which simplifies the calculations, as well as guarantees the convenience and clarity of an analysis. Therefore, the calculation methods resulting from the use of the "partial" magnetic fields have become widespread in practice.

In the technical literature, there are two ways to select the "partial" magnetic fields of the electric machine. According to the first method, the resulting magnetic field of the machine is represented as a result of the superposition of the total self-induction magnetic fields and the mutual induction magnetic fields created by the currents of the stator and rotor windings separately (the “inductively coupled circuits” concept). In the second method, the resulting magnetic field of the electric machine is represented as a result of the superposition of the machine magnetizing field existing in the machine in the no-load mode and the stator and rotor leakage magnetic fields that occur in the machine in the under-load mode (the “no-load and under-load” concept). In the technical literature, the decomposition of the resulting magnetic field of the electric machine into the “partial” magnetic fields is usually carried out on the basis of physical considerations, which indicates the need to develop the sufficiently rigorous techniques and procedures that allow the resulting magnetic field of the machine to be represented as decomposed into the “partial” magnetic fields. The solution to this issue forms the basis of the content of this work.

The issues discussed in this work are considered on the basis of the use of the equivalent circuits arising from the analysis of the resulting magnetic field that takes place in the electric machine. Moreover, such equivalent circuits are built here for the resulting magnetic field, determined in the two ways. According to the first method, the resulting magnetic field of the machine is set as a result of the superposition of the machine magnetizing field that exists in the machine in the no-load mode and the stator and rotor leakage magnetic fields that occur in the machine in the under-load mode (the “no-load and under-load” concept). By the second method, the resulting magnetic field of the machine is determined by Maxwell’s equations (the “field” concept). In this work, we first consider equivalent circuits corresponding to the resulting magnetic field determined by the superposition method (the “no-load and under-load” concept). As a generalization of the results obtained from considering such machine equivalent circuits, the general provisions are formulated and the specific techniques and procedures are recommended. The provisions and also the techniques and procedures created in this way are focused on the construction of the equivalent circuits arising from the consideration of the resulting magnetic field determined using Maxwell’s equations. Note that the techniques and procedures proposed in this work are based on ideas that support the “no-lad and under-load” concept. The use of the equivalent circuits allows the results of the field studies to be presented quite easily and visually through the terms and concepts adopted in the theory for calculating the electric machines, as a result of which the new opportunities are provided for further improving the methods for calculating the machine.

In this work, the author did not seek to propose the certain clarifications in the methods of calculating the electric machine, but mainly tried to indicate the ways to achieve the further possible refinements.

This work is focused on the determination of the basic calculation elements of the asynchronous machine (the circuit loops, parameters, equivalent circuits, and voltage equations). For this purpose, the equivalent circuits obtained on the basis of the analysis of the resulting magnetic field set in the machine using Maxwell’s equations are used. In this case, the basic calculation elements of the electric machine are determined as a result of the conversion of the equivalent circuits constructed in this way. For this, in this work, the corresponding techniques and procedures that can be used when converting the equivalent circuits have been developed. In this way, it is provided the determination of the calculation elements of the machine directly from field equations.

The modern methods for calculating the electric machines are based on the use of the values of the voltages *U* and currents *I* of the machine windings (phase system of units). The field calculations are performed using the vectors of the electromagnetic field intensities *E* and *H* (specific system of units). It follows from this that the development of the methods with the new promising capabilities is associated with the use of the specific system of units. Consequently, the application of the specific system of units (*E*, *H*) is the necessary condition for the successful solution of the tasks of this work.

The practical implementation of the provisions developed in this work is demonstrated by the example of considering the basic calculation elements of the asynchronous machine with the wound and squirrel-cage rotor.

The materials presented in this work are devoted to the discussion of the general issues of the theory for electromagnetic calculation of asynchronous machines. In this sense, this work can be seen as the introduction to the author's book, published in 2015 and entitled as "Alternating Current Multi-Circuit Electric Machines: A New Approach to the Steady-State Parameter Determination." Thus, the specified author's book can be considered as the logical continuation of this work.

The proposed book can be useful to the engineers and scientists involved in the calculation and analysis of the electric machines, as well as the teachers and students of the technical universities.

Valentin Asanbayev

Contents

Chapter 1
Introduction: Basic Calculation Elements of Asynchronous Machine

1.1 Basic Calculation Elements: Problems of Determination

The electromagnetic calculation of the asynchronous machine is a rather complex system of techniques and calculation procedures. Therefore, the consideration in the full scope of the issues of the theory of the electromagnetic calculation of the asynchronous machine requires a large amount of research and analysis. In this regard, in this work, the main attention is paid to the issues that form the basis of the theory of the electromagnetic calculation of the asynchronous machine. The modern methods for calculating the asynchronous machine are based on the use of the voltage equations (Kirchhoff's equations), equivalent circuits, as well as on the introduction of the circuit loops and parameters of the electric machine [1–21]. In this work, the voltage equations, equivalent circuits, circuit loops, and parameters are considered as the basic calculation elements of the asynchronous machine. These four most important calculation elements of the electric machine are interconnected, however, in this work, they are whenever possible, considered as the "separate and independent" elements. Below, we will briefly describe the current state of the problems associated with the determination of the calculation elements of the asynchronous machine and also analyze the main trends in the further improvement of the methods for their determination.

1.1.1 *Circuit Loops*

In the theory for calculating the electric machines, the concept of the winding circuit loop is central, but this concept is currently receiving little attention. There is still no clear and precise formulation of this concept. This state of affairs does not contribute to the further successful development of the theory for calculating the electric machine.

V. Asanbayev, *Asynchronous Machines*,
https://doi.org/10.1007/978-3-030-92284-9_1

In the circuit theory, under the circuit loop is understood as any closed path formed during the passage of the branches connected at the nodes of the circuit under the consideration. Therefore, the circuit loops follow, in this case, from the consideration of the accepted circuit and are determined based on the analysis of the real structure of this circuit.

In the theory of the electric machines, the circuit loops are determined in relation to the values of the voltages set on the terminals of the stator and rotor windings [1–21]. These voltages are balanced by the voltage drops across the ohmic resistances of the stator and rotor windings as well as by the values of the emf induced in the stator and rotor windings by the resulting magnetic field taking place in the electric machine. Therefore, the configurations of the circuit loops and the elements used in them depend on the nature of the assignment of the machine resulting magnetic field. In the technical literature, the machine resulting magnetic field is determined in terms of the use of the "inductively coupled circuit" concept and "no-load and under-load" concept. In this regard, depending on the nature of the assignment of the machine resulting magnetic field, two types of circuit loops representing the stator and rotor windings can be considered. When using the "inductively coupled circuit" concept, the machine resulting magnetic field is represented as a result of the superposition of the total self-induction magnetic fields and mutual induction magnetic fields caused by the currents of the stator and rotor windings, individually. In this case, the machine windings are represented as the inductively coupled circuit loops, for example [1, 8, 13]. In this case, the ohmic resistances and the total self-and mutual induction reactances of the stator and rotor windings are considered as the elements used in these circuit loops.

When using the "no-load and under-load" concept, the machine resulting magnetic field is represented as a result of the superposition of the machine magnetizing field taken place in the machine in the no-load mode and the stator and rotor leakage magnetic fields existing in the machine in the under-load mode. In this case, the machine windings are represented as the conditionally electrically connected circuit loops, for example [13]. As the elements used in these circuit loops are considered the ohmic resistances and the leakage reactances of the stator and rotor windings and also the machine magnetizing reactance.

The introduction of the winding circuit loops allows us to build the equivalent circuit and to draw up the voltage equations for the electric machine. Thus, the voltage equations and equivalent circuits can be considered as tools for the visualization of the stator and rotor winding circuit loops.

In the theory of the electric machines, the procedure for forming the winding circuit loops is described for the simple case when there is the single winding on the stator and the single winding on the rotor, but this procedure is not usually specifically specified, for example [1, 2]. To some extent, because of this, in particular, the issues related to the formation of the machine circuit loops, especially the circuit loops of the multi-winding electric machines, have not yet received due attention from the specialists.

Above, we were talking about the circuit loops representing the real physical windings used in the electric machines. In practice, the electric machines the windings of which are implicit are also used, for example, electric machines with the solid rotor, etc. The eddy currents induced in the solid rotor are distributed over the cross section of the rotor. In this regard, it becomes necessary to form the circuit loops of the eddy currents induced in the rotor taking into account their spatial distribution associated with the phenomenon of the current displacement. In [13], the procedure for the formation of the circuit loops of the eddy currents induced in the machine solid rotor is described. In this work, we will adhere to the ideas used in [13].

In the theory of the electric machines, the voltage equations are usually first compiled, i.e., the circuit loops of the windings are established. Then, using these voltage equations, the equivalent circuit of the machine is obtained. This sequence of the winding circuit-loops formation is used when the machine resulting magnetic field appears to be decomposed into the "partial" magnetic fields (for which the "inductively coupled circuits" concept or "no-load and under-load" concept can be used). When the machine resulting magnetic field is determined by Maxwell's equations, this sequence of the winding circuit-loop formation can be disrupted. In [13], it is shown that in this case, it is more rational to proceed from the algorithm which assumes first the formation of the winding circuit loops and the visualization them by the equivalent circuit, and then using the equivalent circuit obtained in this way, the voltage equations can be drawn up. Thus, according to the first method when the machine resulting magnetic field is represented as the result of a superposition of the "partial" magnetic fields the machine circuit loops are formed in the process of drawing up the voltage equations. According to the second method, when the machine resulting magnetic field is determined by Maxwell's equations, the machine circuit loops are formed and visualized in the process of building the equivalent circuit of the electric machine.

In the theory of electric machines, the voltage equations are written in relation to the values of the voltages set on the terminals of the stator and rotor windings. In this regard, the winding circuit loops are formed, in this case, in relation to the voltage values of the machine windings. As shown in [13], when using the resulting magnetic field defined by Maxwell's equations, the machine circuit loops are formed in relation to the air gap emf (emf corresponding to the surface passing through the middle of the radial length of the air gap). In relation to the air gap emf, the three machine circuit loops can be identified: the magnetizing circuit loop of the machine, the stator winding circuit loop, and the rotor winding circuit loop. The circuit loops of the stator and rotor windings include the ohmic resistances. In this work, the ohmic resistances of the stator and rotor windings do not get considered. Therefore, the circuit loops of the stator and rotor windings are considered, here, for convenience, as the stator and rotor circuit loops. In this work, we use the circuit loops formed in relation to the machine air gap emf. The use of the circuit loops obtained in this way simplifies the procedures for constructing the equivalent circuit of the electric machine [13].

1.1.2 Parameters

The voltage equations and equivalent circuits obtained using the two concepts (the "inductively coupled circuits" concept and "no-load and under-load" concept) contain the ohmic resistances and reactances. These ohmic resistances and reactances are considered as the parameters of the electric machine. The parameter determination process can be considered as consisting of the following two stages. The first stage involves drawing up the machine voltage equations, which means the formation of the circuit loops and the construction of the machine equivalent circuit. The second stage is associated with the calculation of the machine magnetizing circuit and the stator and rotor leakage magnetic circuits, as well as the determination of the ohmic resistances and reactances used in the voltage equations and machine equivalent circuit. (In this work, as noted above, the ohmic resistances of the windings do not consider and, in addition, we restrict ourselves here to consider the slot leakage reactances of the stator and rotor windings.)

In the theory of electric machines, the question of determining the parameters has been always considered as a separate independent problem. If we consider the inductive parameters of the machine, then the expressions for them are usually obtained from the calculation of the "partial" magnetic fields which together represent the resulting magnetic field of the machine (the total self-induction magnetic fields and the mutual induction magnetic fields of the stator and rotor windings in the concept of the "inductively coupled circuits" and the machine magnetizing field and the slot leakage magnetic fields of the stator and rotor windings in the "no-load and under-load" concept). In this case, we can note the following sequence of the actions the implementation of which leads to the determination of the inductive parameters of the electric machine. First, the magnetic fluxes are determined and the relationships between the magnetic fluxes and the mmf of the machine windings are established. When determining the magnetic fluxes, the magnetic circuit of the machine magnetizing field and the stator and rotor slot leakage magnetic circuits are represented as decomposed into the regions (sections) in such a way as to ensure the possibility of using the total current law. Then, the procedure for determining the machine magnetizing flux and the stator and rotor slot leakage magnetic fluxes is greatly simplified. Then, the emf induced in the machine windings by these magnetic fluxes is determined. Further, the winding emf obtained in this way is represented through the corresponding reactances, the values of which are determined as the ratio of the emf induced in the winding to the current flowing in it (in the case of determining the mutual induction reactance, the emf of the winding is divided by the current flowing in the other winding). As a result of the application of these procedures, the inductive parameters of the machine are determined. It should be noted that the important feature of determining the inductive parameters is the decomposition of the machine magnetic circuits into the regions (sections). In this work, this method of decomposing the machine magnetic circuits into the regions is widely used and it forms the basis for the "fragmentation" concept.

Using the parameters determined in such a way ensures reliability of the calculation and analysis of the processes in the machine with the relatively simple design. However, until now, the problem of determining the machine parameters continues to be quite relevant. This problem is especially typical for the machines with the high level of the use of the active materials, for example, for such machines as the powerful high-speed synchronous turbo-motors, turbogenerators, machines with the solid rotor, etc., for example [5, 13, 19, 20].

The problem of determining the parameters of the electric machine has its own features. The methods used in drawing up the voltage equations (and, consequently, in the formation of the winding circuit loops and in the construction of the machine equivalent circuit), and the methods used in obtaining the analytical expressions for the machine parameters do not represent a system of the organically interrelated techniques and procedures. This provision follows from the fact that within the framework of the concepts considered above, the compilation of the voltage equations, and the construction of the equivalent circuit, as well as the determination of the machine parameters, in general case, is difficult to carry out based on an unified understanding the nature of the machine magnetic field distribution. In the two winding electric machine (the single winding on the stator and the single winding on the rotor), it is more or less possible to adhere to the same picture of the magnetic field distribution both when drawing up the voltage equations and constructing the equivalent circuit and when determining the parameters used in them. However, for example, the picture of the magnetic field distribution in the multi-winding electric machine becomes much more complicated. Therefore, in this case, serious difficulties arise in providing an unified understanding of the nature of the magnetic field distribution in the machine, both when drawing up the voltage equations and constructing the equivalent circuit and when determining the parameters applied in them. As shown in [13], in the area of the weak current displacement, the problems associated with drawing up the voltage equations and constructing the equivalent circuit, as well as determining the parameters used in them can be solved more or less successfully. However, for the area of the strong current displacement, this problem becomes practically insoluble due to the overly complex picture of the magnetic field distribution in the electric machine.

A provision described above creates certain difficulties in ensuring the required level of the methodology and consistency in the presentation of the theory for electromagnetic calculating the asynchronous machine. According to [13], the difficulties noted above can be eliminated in the case of using the distribution pattern for the resulting electromagnetic field determined by Maxwell's equations. With this approach, the determination of the machine calculation elements is provided in the single process of considering the machine resulting magnetic field [13]. Thus, in this case, the main disadvantage of the method used in the technical literature to determine machine parameters is eliminated.

1.1.3 Equivalent Circuits

The machine equivalent circuit, as the tool for performing the electromagnetic calculations, provides a high level of clarity of analysis. Therefore, the equivalent circuits are widespread. In the system of the concepts accepted above, the equivalent circuit of the electric machine is usually obtained on the basis of using voltage equations. Therefore, the voltage equations are drawn up and then, using them, the equivalent circuit of the electric machine is built. Thus, the equivalent circuit can be considered as the tool for visualizing the machine voltage equations.

In the equivalent circuit based on the "inductively coupled circuits" concept, the stator and rotor windings are represented as the inductively coupled circuit loops. The equivalent circuit obtained in this way contains the inductively coupled reactances which create serious difficulties in the way of the practical application of such the equivalent circuit.

As noted above, the use of the "no-load and under-load" concept is associated with the introduction of the referred values of the emf and currents of the rotor winding. Under these conditions, in the equivalent circuit built on the basis of the use of the "no-load and under-load" concept, the stator and rotor windings are represented as the conditionally electrically connected circuit loops [1]. Such equivalent circuit of the electric machine is widespread.

When drawing up the voltage equations and constructing the equivalent circuit, the operating mode of the asynchronous machine is reduced to the conventional operating mode of the machine with the standing rotor, for example [1, 6]. This means that in the voltage equations and the equivalent circuit of the electric machine, the voltages (emf) and the currents used in them have the frequency of the power supply network.

Currently, the equivalent circuits are not the main means of calculating and analyzing the electromagnetic processes in the electric machine. This role is played by the voltage equations obtained for the machine windings. However, as shown in [13], when passing to the specific quantities (E, H) and determining the resulting magnetic field of the electric machine using Maxwell's equations, the equivalent circuit of the machine is first established, and then, using this equivalent circuit, the voltage equations of the machine windings are compiled. Thus, when using the specific quantities (E, H), one should expect a slight increase in the role of the equivalent circuits in the theory for calculating the electric machines.

1.1.4 Voltage Equations

In the modern theory of electric machines, the voltage equations for the machine windings follow from the use of the two concepts: the "inductively coupled circuits" concept and the "no-load and under-load" concept [1–14]. These concepts are based on the ideas of the nature of the formation of the machine resulting magnetic field. In the "inductively coupled circuits" concept, the machine resulting magnetic field is created by the currents flowing in the inductively coupled stator and rotor windings. In this case, as noted above, the machine resulting magnetic field is the result of the superposition of the total self-induction magnetic fields and the mutual induction magnetic fields created by the currents of the stator and rotor winding separately. In the "no-load and under-load" concept, the machine resulting magnetic field is considered as the result of the superposition of the magnetizing field existing in the machine in the no-load mode and the leakage magnetic fields taking place in the electric machine in the under-load mode.

When drawing up the voltage equations, one usually proceeds from the second Kirchhoff's law. According to this law, the sum of the voltage applied to the winding terminals and the emf induced in it is equal to the voltage drop across the ohmic resistance of this winding. In the "inductively coupled circuits" concept, this definition is applied to the stator and rotor windings flowing round by the currents. In the "no-load and under-load" concept, this definition is usually applied to the stator winding when the machine is in the no-load mode and to the rotor winding when the machine is in the under-load mode. In the "inductively coupled circuit" concept, the voltages applied to the winding terminals are balanced by the voltage drops across the ohmic resistances of the windings, as well as by the voltage drops across the total self-and mutual induction reactances of the stator and rotor windings. In the "no-load and under-load" concept, the voltages applied to the winding terminals are balanced by the voltage drops across the ohmic resistances of the windings, and by the voltage drops across the leakage reactances of the stator and rotor windings, and also by the voltage drop across the magnetizing reactance of the electric machine.

In the voltage equations drawn up for the rotor winding using the "no-load and under-load" concept, the referred to the stator winding rotor quantities are usually used. In practice, the referred to the stator winding rotor quantities can be also applied in the voltage equations obtained by the use of the "inductively coupled circuits" concept which is done to eliminate the difficulties associated with determining the parameters used in these voltage equations [1]. In this regard, the voltage equations containing the referred rotor quantities have gained practical application. Despite the different initial provisions, the voltage equations obtained on the basis of the two concepts provide the same results for the values of the voltages and the currents determined on the terminals of the electric machine windings [14].

As follows from the above, the voltage equations for the machine windings do not directly follow from the electromagnetic field theory described by Maxwell's equations. This provision is confirmed by the fact that when drawing up the voltage equations, the machine resulting magnetic field created by the joint action of the currents of the windings is decomposed into the "partial" magnetic fields (the total self-induction magnetic fields and mutual induction magnetic fields of the stator and rotor windings in the "inductively coupled circuits" concept and also the magnetizing field of the machine and the leakage magnetic fields of the stator and rotor windings in the "no-load and under-load" concept), the total action of which provides, with the certain approximation, the resulting magnetic field of the machine [14]. It should be noted that the difficulties of extending the methods used above for determining the machine resulting magnetic field to the case of defining the resulting magnetic field by Maxwell's equations are the main obstacle for a long time to the successful application of the electromagnetic field theory in the development of the calculation methods with the new qualitative capabilities.

1.2 Basic Calculation Elements from "Field" Concept: Features of Determination

Above, the basic calculation elements of the electric machine were considered on the basis of the decomposition of the machine resulting in a magnetic field into the "partial" fields. In the technical literature, the decomposition of the resulting magnetic field into the "partial" magnetic fields is carried out on the basis of physical considerations. In this case, the phase quantities are usually applied (Э, I).

We can proceed from the assumption that the further development of the methods for determining the basic calculation elements of the electric machine (the circuit loops, parameters, equivalent circuits, and voltage equations) is associated with the use of the resulting magnetic field determined by Maxwell's equations ("field" concept). In modern conditions, the various field methods of an analysis based on the use of Maxwell's equations are rapidly developing. In this regard, we can assume that at present there is the very real basis for the further development of the theory for calculating the electric machines based on the use of the "field" concept.

The "field" concept is based on the use of the resulting field determined using Maxwell's equations. The specific quantities (E, H) are used in Maxwell's equations, with help of which, in this case, the resulting magnetic field of the electric machine is described. Consequently, the specific quantities (E, H) are applied for describing the resulting magnetic field of the machine. In [13], the approach is developed that allows one to determine the calculation elements of the machine based on the use of the resulting magnetic field specified using Maxwell's equations. By the approach developed in [13], the calculation elements of the multi-winding electric machines of the various design are considered. It is shown that the machine calculation elements obtained using the approach developed in [13] are in the agreement with the calculation elements used in the theory of electric machines. Nevertheless, this approach continues to be in the isolated position from the theory of the calculation of the electric machines. This work is focused on eliminating this isolation.

The calculation elements of the electric machines can be expressed through the phase (Э, I) and specific (E, H) quantities. The phase quantities (Э, I) represent the volume power released in the winding conductors. In this regard, the results obtained using the calculation methods developed applying the phase quantities (Э, I) have a fairly high level of physical interpretation and visualization. Therefore, these methods are better suited to accommodate the assumptions made. Besides, the phase quantities (Э, I) can be relatively easily measured. Therefore, the results obtained by the methods developed with the use of the phase quantities ($\mathcal{E}$, I) can be relatively easily verified experimentally.

When using the specific quantities (E, H), the amount of power released in the conductor is defined as the difference in a power entering one surface and leaving the other surface of the conductor. In this regard, the specific quantities (E, H) characterize the transfer of energy, the value of which can be determined by the well-known Poynting vector. The calculation methods developed using the specific quantities (E, H) are more formal in nature, therefore, the results obtained by these methods are more difficult to interpret physically. However, on the other hand, the specific quantities (E, H) directly operate with the electromagnetic field theory described by Maxwell's equations. Thus, the specific quantities (E, H) have a powerful potential for developing the methods with a higher level of quality characteristics.

As practice shows, the methods with a relatively high level of physical interpretation and visualization are more preferable for using them in engineering practice. In this regard, the aforementioned drawback of the method obtained using the specific quantities (E, H) can be eliminated by converting it into the system of the phase quantities ($\mathcal{E}$, I). As noted above, in the system of the phase quantities ($\mathcal{E}$, I), it is easier to take into account the assumptions, and the results of the calculations can be relatively easily verified experimentally. Therefore, the phase system of units ($\mathcal{E}$, I) is more preferable for making the final calculations. Thus, we can proceed from the idea that the results obtained applying the specific quantities (E, H) can be used to give the methods developed by the phase quantities ($\mathcal{E}$, I) the higher quality level. At the same time, such expectation may turn out to be very significant and fundamental, since the calculation methods based on the use of the specific quantities (E, H) directly operate with the results of solving Maxwell's equations.

1.3 Work Structure

The modern theory used to carry out the practical calculations of the electric machine assumes, in accordance with the "no-load and under-load" concept, the decomposition of the machine resulting magnetic field into the "partial" magnetic fields. When using the "field" concept, the development of the calculation method is carried out on the basis of considering the machine resulting magnetic field arising from the solution of Maxwell's equations. Thus, the modern theory for calculating the electric machine is based on the use of the "partial" magnetic fields into which the machine resulting magnetic field is decomposed and the calculation theory developed using the "field" concept is based on the application of the machine resulting magnetic field determined by Maxwell's equations. Thus, the methods used the "no-load and under-load" concept and the methods obtained on the basis of the application of the "field" concept have different starting provisions. This state of affairs creates quite serious difficulties in the way of further improving the methods for calculating the electric machine.

In this work, this problem is solved by bringing to the unified initial basis the methods created on the basis of using the "no-load and under-load" concept and developed on the basis of applying the "field" concept. For this purpose, the method developed on the basis of the "no-load and under-load" concept is generalized to the case of using the machine resulting magnetic field arising from the implementation of the superposition principle. As a result of this generalization, the provisions have been formulated and corresponding techniques and procedures have been developed. Now, these provisions and techniques can be used in defining the calculation elements of the electric machine when the machine resulting magnetic field is determined by Maxwell's equations. In this case, the unified approach to the determination of the calculation elements of the electric machine is provided both when using the "field" concept and when applying the "no-load and under-load" concept. Thus, the methods developed using the "no-load and under-load" concept and the "field" concept acquire the single initial basis which in this case creates the more favorable conditions for the preparation of the more extensive recommendations for improving the methods of calculating the electric machine. In this work, the determination of the basic calculation elements of the asynchronous machine is achieved using the field equations. Below, we will demonstrate the implementation of this idea.

In accordance with the above, this work can be conditionally considered as consisting of the five parts. In the first part (Chap. 1), the description of the current state of the problems associated with the determination of the calculation elements of the electric machine is given, as well as a brief analysis of the main trends in the further improvement of the calculation methods is demonstrated. In the second part (Chaps. 2–4), the general issues of an analysis of the steady-state conditions of the asynchronous machine are considered. In the third part (Chaps. 5–10), analysis and the systematization of the main provisions of the calculation method developed using the phase quantities ($Э$, I) are proposed. In the fourth part (Chaps. 11–13), the description of the approach supposing the derivation of the calculation elements based on the consideration of the machine resulting magnetic field determined in accordance with the "no-load and under-load" concept as the result of the superposition of the "partial" magnetic fields is shown, and also in this part of the work, the formulated provisions are given and the corresponding techniques and procedures allowing us to find the calculation elements of the electric machine on the basis of using the resulting magnetic field determined by Maxwell's equations are established. In the fifth part of the work (Chaps. 14–16), the general approach for determining the machine calculation elements based on the use of the resulting magnetic field defined by Maxwell's equations ("field concept") is demonstrated. In this part, as the example, the description of the schemes for determining the calculation elements of the asynchronous machines with the wound and single-cage rotor is given.

Appendix A provides a brief description of the main provisions on which the theory for calculating the asynchronous machine is based. In addition, the various techniques and procedures used in determining the calculation elements of the machine are considered. It will be shown below that the development of the theory for calculating the electric machine based on the application of the field equations will be carried out using the provisions as well as the techniques and procedures given in Appendix A.

1.4 Summary

1. This work is devoted to the development of the method that makes it possible to establish the calculation elements of the asynchronous machine on the basis of the consideration of the resulting magnetic field determined by solving Maxwell's equations. In this work, the circuit loops, parameters, equivalent circuits, and voltage equations are considered as such calculation elements. The direct use of the results of the field researches in the development of the method for determining the calculation elements of the electric machine will open new opportunities for its further improvement.
2. In the technical literature, the "inductively coupled circuits" concept and the "no-load and under-load" concept are used to set the resulting magnetic field of the electric machine. In the "inductively coupled circuits" concept, the resulting magnetic field is represented as a result of the superposition of the "partial" magnetic fields: the total self-induction magnetic fields and the mutual induction magnetic fields of the stator and rotor windings. When using the "no-load and under-load" concept, the resulting magnetic field is represented as a result of the superposition of the "partial" magnetic fields: the magnetizing field of the electric machine (no-load mode) and leakage magnetic fields of the stator and rotor windings (under-load mode). The use of the "partial" magnetic fields greatly simplifies the procedure for defining the calculation elements of the electric machine.
3. The practical approach for determining the calculation elements of the electric machine is based on the use of the "no-load and under-load" concept, which is due to the fact that the "no-load and under-load" concept can be relatively easily implemented both by the calculation methods and experimentally. Therefore, the results received on the basis of the "no-load and under-load" concept are relatively easy to the physical interpretation, which contributes to the increase in the level of reliability of analysis.
4. In practice, the "no-load and under-load" concept is quite well developed for the electric machines of the relatively simple design (for the machines with single winding on the laminated stator and the single winding on the laminated rotor). The application of the "no-load and under-load" concept for the electric machines of the complex design (e.g., for powerful turbine generators, machines with solid rotor, etc.) causes serious difficulties, and in some cases, its use in the form accepted in practice becomes impossible. In this regard, the acute issue arises related to the need to develop the method that allows one to find the calculation elements of the electric machine on the basis of considering the resulting magnetic field determined by Maxwell's equations ("field" concept).

5. In the "no-load and under-load" concept, the machine resulting magnetic field is decomposed into the "partial" magnetic fields. Therefore, the determination of the calculation elements of the electric machine, in this case, is carried out using these "partial" magnetic fields. In the "field" concept, the machine resulting magnetic field is determined using Maxwell's equations. In this case, the machine resulting magnetic field is directly applied to define the calculation elements of the electric machine. In this regard, the methods arising from the application of these two concepts have different starting provisions. Consequently, the results obtained in this way turn out to be incomparable which will complicate their use in practice.
6. This problem can be solved by bringing to the single baseline the methods arising from the application of the "no-load and under-load" concept and the "field" concept. To achieve this goal, the method can be first developed to determine the calculation elements of the electric machine based on the use of the resulting magnetic field defined by the application of the superposition principle and the "no-load and under-load" concept. Second, based on the analysis of the method obtained in this way, the provisions can be formulated and the techniques and procedures that can be used to determine the calculation elements of the electric machine based on the resulting magnetic field defined by Maxwell's equations can be developed. In this case, the methods developed on the basis of the application of the "no-load and under-load" concept and the "field" concept acquire the single initial basis which opens up new opportunities for improving the methods of calculating the electric machine. This work demonstrates the implementation of this provision.
7. In the technical literature, the voltage equations, as the calculation elements of the electric machine, are determined based on the two concepts: the "no-load and under-load" concept and the "inductively coupled circuits" concept. The voltage equations arising from the use of the "inductively coupled circuits" concept contain the reactances, the fairly accurate determination of which is associated with certain difficulties. To eliminate this problem, these voltage equations are represented in terms of the referred rotor quantities. The second type of voltage equations used in practice is associated with the use of the "no-load and under-load" concept. A feature of the voltage equations obtained in this way is the use of the referred to the stator winding rotor quantities. Consequently, the voltage equations used in practice are expressed in terms of the referred rotor quantities.
8. The equivalent circuit, as the calculation element of the electric machine, usually follows directly from the voltage equations of the machine. In practice, the equivalent circuits obtained on the basis of the "no-load and under-load" concept have become widespread. At present, the equivalent circuits are not the main means for the calculation and analysis. This role is played by the voltage equations of the electric machine. However, when using the "field" concept and moving to the specific quantities (E, H), first, the equivalent circuit of the machine is constructed and then, using this equivalent circuit, the voltage equations of the machine are compiled. Thus, when using the "field" concept one should expect a slight increase in the role of the equivalent circuits in the theory for calculating the electric machine.

9. The voltage equations and the equivalent circuits can be considered as tools for visualizing the circuit loops of the electric machine. The machine circuit loops are usually distinguished in relation to the voltage values determined on the terminals of the stator and rotor windings. Depending on the method of setting, the resulting magnetic field, the two types of machine circuit loops are distinguished. When using the "inductively coupled circuits" concept, the stator and rotor windings are presented as the inductively coupled circuit loops. For practical use, these circuit loops are expressed through the referred rotor quantities. When using the "no-load and under-load" concept, the stator and rotor windings are considered as the conditionally electrically connected circuit loops. Such, the machine circuit loops contain the referred stator winding rotor parameters which help to simplify calculations.
10. In the technical literature, the concept of the "winding circuit loop," the issues related to the selection of the circuit loops, etc., are usually not specifically specified, which indicates the lack of proper attention to these problems. On the transition to the specific quantities (E, H) and on the determination of the resulting magnetic field by Maxwell's equations, the need to develop the special techniques and procedures focused on establishing the machine circuit loops is becoming relevant.
11. The voltage equations, equivalent circuits, and circuit loops contain the ohmic resistances and reactances. These ohmic resistances and reactances represent the parameters of the electric machine. In practice, the parameters arising from the use of the "no-load and under-load" concept have become widespread. In the technical literature, the issues of determining the parameters are usually considered as separate independent tasks. This is due to the fact that the methods used in drawing up the voltage equations and building the equivalent circuits and the methods applied in determining the expressions for the machine parameters do not represent a system of organically interrelated techniques and procedures. All this is the consequence of the fact that within the framework of the concepts used in the theory of the electric machines, it is very difficult to adhere to the single picture of the distribution of the magnetic field both when drawing up the voltage equations and building the equivalent circuits as well as when determining the expressions for the parameters used in them. This problem is especially common for multi-winding electric machines. This state of affairs does not contribute to the effective development of the theory for calculating electric machines.
12. The calculation elements of the electric machine can be determined from the "field" concept. With this approach, the determination of the parameters is carried out in the process of forming the circuit loops and building the equivalent circuit of the electric machine. In other words, the formation of the circuit loops, the construction of the equivalent circuit, as well as the determination of the parameters are achieved in the single process of considering the machine resulting magnetic field found by Maxwell's equation. Thus, in this case, the main disadvantage of the method used in the technical literature to determine the parameters is eliminated.

References

1. Voldek. A.: Electric Machines, Energiya, Leningrad (1974)
2. Ivanov-Smolensky, A.: Electric Machines, Energiya, Moscow (1980)
3. Vazhnov, A.: Electric Machines, Energiya, Leningrad (1968)
4. Alger, P.: Induction Machines. Their Behavior and Uses, Gordon and Beach, New York (1970)
5. Livshits, A.: Electric Machines, vol. 1-3, United Scientific and Technical Publishing, Moscow (1936)
6. Richter, R.: Electric Machines, vol. 1-5, United Scientific and Technical Publishing (Gosenergoizdat), Moscow (1934-19397).
7. Kopylov, I.: Mathematic Modeling of Electric Machines, Vysshaya Shkola, Moscow (1987)
8. Bragstad, O.: Theory of AC Machines with an Introduction to the Theory of Steady-State Alternating Currents (Translated from German), Izdatelstvo Lensoveta i Oblispolkoma, Leningrad (1933)
9. Kostenko, M., Piotrovsky, L.: Electric Machines, Energiya, Leningrad (1973)
10. Adkins, B.: The General Theory of Electrical Machines, Chapman and Hall, London (1959)
11. Kimbrak, E.: Power System Stability, vol. 3, Synchronous Machines, Gosudarstvennoe Energetiheskoe Izdatelstvo (1949)
12. Kostenko, M.: Electric Machines. Special Part, Gosenergoizdat, Moscow (1949)
13. Asanbayev. V.: Alternating Current Multi-Circuit Electric Machines. A New Approach to Steady-State Parameter Determination, Springer Charm Heidelberg, New York (2015)
14. Seely, S.: Electromechanical Energy Conversion, McGraw-Hill Book Company, Inc., (1962)
15. Heg, B.: Electromagnetic Calculations (Translated from English), United Scientific and Technical Publishing, Moscow (1934)
16. Boldea, I., Nazar, S.: The Induction Machine Design Handbook, 2nd ed., CRC Press/Taylor and Francis, Boca Raton (2010)
17. Sipailov, G., Kononenko, E., Kharkov, K.: Electric Machines. Special Course, Vysshaya Shkola, Moscow (1987)
18. Treshchev, I.: Electromechanical Processes in AC Electric Machines, Energiya, Leningrad (1980)
19. Kazovsky, E.: Transient Processes in AC Electric Machines, Publishing of Academy of Science USSR, Moscow-Leningrad (1962)
20. Danilevich, Y., Dombrovsky, V., Kazovsky, E.: Parameters of AC Electric Machines, Publishing Nauka, Moscow (1965)
21. White, D., Woodson, H.: Electromechanical Energy Conversion, John Wiley and Sons, Inc., New York (1064)

Chapter 2
Electromechanical Interactions in Asynchronous Machines: Basic Physical Phenomena

2.1 Electromechanical Energy Conversion

In asynchronous machines, the electromechanical interaction accompanied by the electromechanical energy conversion is carried out by the rotating magnetic field created by the currents flowing in the stator and rotor windings [1–16]. The magnetic field created by the three-phase stator winding has a rotational speed equal to n_1 and the rotor speed is n. In this case, the frequency of rotation of the magnetic field created by the currents of the rotor winding is equal to the difference between the frequency of the rotation of the stator field n_1 and the frequency of the rotation of the rotor n, that is, $n_2 = n_1 - n$. The frequency of rotation of the rotor magnetic field relative to the stator is equal to the sum of the rotation frequency of the rotor magnetic field n_2 and the rotor speed n, that is, $n_2 + n = n_1 - n + n = n_1$. It follows from this that the magnetic field created by the currents of the rotor winding has a rotation frequency equal to the rotation frequency of the stator magnetic field. Consequently, in the operating mode, the magnetic fields created by the currents flowing in the windings of the stator and rotor of the asynchronous machine are stationary relative to each other. As a result, the single rotating resulting magnetic field is created in the electric machine. This field has a constant force effect on the current-carrying rotor windings. The existence of this constant force causes the electromagnetic torque of the electric machine [1–16].

The electromechanical energy conversion taking place in the electric machine can be considered from the point of view of the "circuit" or from the "energy" point of view [3–17]. In the first case, the electromechanical conversion of energy in the electric machine is associated with the induction of the emf in its windings due to the change in their relative spatial arrangement caused by the rotation of the rotor. In this case, the electromechanical energy conversion in the electric machine can be considered as a result of the interaction of the magnetic field of the air gap with the currents of the rotor winding. The presence of such the interaction in the electric machine, leading to the appearance of the electromagnetic torque acting on the rotor,

V. Asanbayev, *Asynchronous Machines*,
https://doi.org/10.1007/978-3-030-92284-9_2

can also be characterized by the change in the mutual inductance between the stator and rotor windings, for example, [2, 3, 8, 12, 13, 15].

When connecting the terminals of the stator winding to the power supply network, the current occurs in it. In accordance with the total current law, this stator current creates the magnetic field in the electric machine. According to the electromagnetic induction law, the change in the time of this magnetic field induces the emf in the stator and rotor windings. Under the influence of this emf, the current arises in the rotor winding. The interaction of the current-carrying rotor winding with the air gap magnetic field creates the electromagnetic force and consequently, the electromagnetic torque which leads the machine rotor to rotate. Thus, the electromagnetic energy entering the stator winding from the power supply network is converted into the mechanical energy of the rotor rotation.

When using the "energy" point of view, the electromagnetic energy conversion in the machine can be considered as a result of the transferring (e.g., from the stator surface to the rotor surface in the case of the motor) an electromagnetic power $P_{эм}$ and converting it into a mechanical power $P_{мх}$ which leads the rotor into the rotation [17–20]. The transfer of the power $P_{эм}$ by the electromagnetic field is described by the known Poynting vector representing through the vectors of the electric and the magnetic field intensities [21]. Therefore, the stator winding electromagnetic power $P_{эм}$ can be represented through Poynting vector. Then, this power can be considered flowing out from the stator bore surface as the Poynting vector, for example, [3]. Thus, the stator winding electromagnetic power $P_{эм}$ is transferred to the machine air gap. Furthermore, from the air gap, the stator winding electromagnetic power $P_{эм}$ as Poynting vector enters the rotating rotor.

The total angular frequency ω_1 of the vectors of the electric and the magnetic fields, representing the Poynting vector, consists of the angular frequency of the rotor rotation ω and the angular frequency of the vectors of the electric and the magnetic fields of the rotor ω_2, i.e., $\omega_1 = \omega + \omega_2$. Therefore, in the rotating rotor, the electromagnetic power $P_{эм}$ representing the Poynting vector can be considered as the sum of the two components. The first component of the electromagnetic power $P_{эм}$ reflects the power losses P_2 dissipated in the rotor winding. The second component of the electromagnetic power $P_{эм}$ represents the mechanical power $P_{мх}$ causing the electromagnetic torque indicating the phenomenon of the force effect of the air gap magnetic field on the current-carrying rotor winding.

Thus, we can note that the sufficiently complex process of the electromechanical interaction taking place in the electric machine in the operation modes is accompanied by various physical phenomena. As noted earlier, such physical phenomena, in this work, are considered the phenomenon of the magnetic field formation, the phenomenon of the emf induction, the phenomenon of the force effect of the magnetic field, and the phenomenon of energy transferring by the electromagnetic field. These listed phenomena allow us to describe the features of the rather complex processes of the electromechanical energy conversion in the electric machine.

2.2 Phenomenon of Magnetic Field Formation: Total Current Equation

This phenomenon was first discovered experimentally by the Danish physicist H Oersted. He found that the current in the conductor causes the magnetic field in the space surrounding the conductor. The French scientists Biot and Savard in 1920 and Ampere in the same year established the laws that allow us to determine the magnetic field caused by the current.

Integral form of the total current law. In practice, the equation known as the total current law is widely used to describe the phenomenon of magnetic field formation. For this, the following equation, we have

$$\oint Hdl = I \tag{2.1a}$$

where H is the magnetic field intensity, I is the current flowing in the conductor, dl is the length of the element of the integration contour.

In Eq. (2.1a), the positive direction of the elementary length dl of the integration contour and the positive direction of the current I are determined by the right-handed screw rule (Fig. 2.1). The relation (2.1a) asserts that in the magnetic field the linear integral of the magnetic field intensity taken along any closed circuit loop is equal to the current penetrating the surface bounded by this closed circuit loop. In the general case, if the surface is pierced by several current-carrying conductors (Fig. 2.1b), then, in the expression (2.1a) instead of the current I the total current $\sum I$ determined by the algebraic addition of these currents is used. From here, we have

$$\oint Hdl = \sum I \tag{2.1b}$$

The windings of the electric machines can be considered as a system of conductors distributed on the stator and rotor surfaces. In this regard, the total current $\sum I$ represents, in this case, the sum of the currents flowing in the conductors of the winding. In the technical literature, the Expression (2.1a) is known as the integral

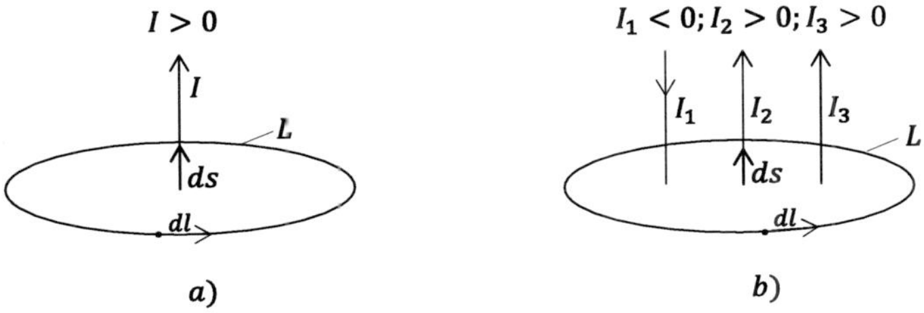

Fig. 2.1 To the application of the total current law: (**a**) circuit loop embracing the single current-carrying conductor; (**b**) circuit loop embracing the several current-carrying conductors

form of the total current law. The total current law confirmed experimentally can be used to determine the magnetic field created by the current flowing in the closed conductor. The use of this law is more practical when the magnetic field is characterized by symmetry and the magnetic field lines are known. It should be noted that, in this case, the application of the total current law greatly simplifies the determination of the magnetic field intensity. In the case of the magnetic field of the complex configuration, when it is not possible preliminarily to identify the field lines with the same field intensities, the use of the total current law causes certain difficulties because in (2.1a) the vector H cannot be removed from the sign of the integral. However, it should be noted that the total current law represented in the integral form allows us, in some cases, to solve the issue of determining the magnetic field intensity vector. Therefore, expression (2.1b) is widely used in the theory of electric machines.

In addition to the field intensity vector H, the magnetic field is characterized by the magnetic flux density vector B. The second vector of the magnetic flux density B is determined through the vector of the magnetic field intensity H as

$$B = \mu_0 \mu_r H = \mu H \tag{2.2}$$

Here, B is the vector of the magnetic flux density, μ_r and μ_0 are the relative and absolute magnetic permeabilities, respectively; μ_0 is the magnetic constant of the vacuum; $\mu = \mu_0 \mu_r$. Expression (2.2) is valid for the case when the vectors H and B have the same direction, and their values are linearly proportional.

In the theory of the electric machines, expression (2.1a) is usually used to determine the magnetic flux density in the machine air gap and also the leakage magnetic flux densities in the stator and rotor slots of the electric machine.

The flux density vector B through some surface S forms the magnetic flux Φ which is determined, for the general case, by the following expression

$$\Phi = \int BdS \tag{2.3}$$

If the magnetic flux density vector B is directed perpendicular to the surface S, then, the expression (2.3) is simplified and takes the form $\Phi = BS$. In electric machines, distributed multi-turn windings are usually used. Therefore, the magnetic flux Φ forms with the multi-turn winding the flux linkage Ψ determined as $\Psi = wk_w\Phi$, where wk_w is the effective number of the turns of the distributed winding. Here, w means the real number of the turns of the distributed winding and k_w is known as the winding factor.

The flux linkage of the winding Ψ is measured in the same units as the magnetic flux Φ since the number of the turns w is the dimensionless quantity. In the theory of the electric machines, the emf induced in the winding is considered as a result of changing the flux linkage Ψ over time. Thus, the use of the total current equation represented in Eq. (2.1a) allows us to define the vector of the magnetic field intensity H, the vector of the magnetic flux density B, the magnetic flux Φ and the flux linkage of the winding Ψ.

Subsequently, we will consider another integral form of writing the total current law. In the theory of the electric machines, the contour L representing the machine magnetic circuit and embracing the currents of the windings is divided into the regions (sections) in such a way that in the regions obtained in this way the values of the H do not change, i.e. $H = const$ for the fixed magnetic circuit section, and, in this case, the magnetic field in them becomes usually one-dimensional. Then, for the left part of the Eq. (2.1b), we can obtain

$$\oint Hdl = \sum_{n=1}^{m} \frac{B_n}{\mu_n} l_n = \sum_{n=1}^{m} \frac{\Phi_n l_n}{\mu_n S_n} = \sum_{n=1}^{m} \Phi_n R_n$$

where $\Phi_n = B_n S_n$ is the magnetic flux of the nth region having the cross-section area equal to S_n and the length equal to l_n; $R_n = l_n / \mu_n S_n$ is the reluctance of the nth region.

The right part of the Eq. (2.1b) represents the mmf acting in the integration contour L and is denoted by the letter F. Now, we can obtain

$$\sum_{n=1}^{m} \Phi_n R_n = F \tag{2.4}$$

Expression (2.4) represents the second Kirchhoff's law for the magnetic circuit and reflects the analogy between the magnetic and the electric circuits. The total current law in the form shown in Eq. (2.4) is convenient to use for the magnetic circuit consisting of the regions with the different but within the one secluded region the constant magnetic permeabilities. This provision and expression (2.4) are often used when calculating the machine magnetic circuit, for example [1–7].

It is often assumed that the machine's magnetic field is mainly localized in the air gap [1–8]. In this regard, the magnetic field is often neglected in the machine stator and rotor steel and it is believed that $\mu_c = \infty$. The air gap length δ for the asynchronous machine design is relatively small compared to the pole pitch τ. Under these conditions, we can assume that the magnetic flux density in the air gap has only the radial component, that is, the magnetic flux density lines cross the air gap radially, and along the length of the air gap, the magnetic field intensity $\dot{H}$ is constant in magnitude. Then, for the single length of the air gap, Eq. (2.4) can be represented as $\dot{B}\delta = \mu_0 \dot{F}$, where δ is the length of the air gap. From here, it arises for the magnetic flux density in the air gap the following expression

$$B = \frac{\dot{\mu}_0}{\delta} \dot{F} \tag{2.5}$$

Here, the $\dot{F}$ represents the mmf required to conduct the magnetic flux through the single air gap [1–19]. In the theory of the electric machines, the mmf concept arising from the total current Eq. (2.1b) under the condition of the one-dimensionality of the magnetic field in the air gap is very important. The total current equation in the form shown in Eq. (2.5) is often used to determine the magnetic flux density in the machine air gap.

Differential form of the total current law. This form of the total current law connects the value of the current density δ at the given point in a space with the magnetic field intensity H. To obtain such the dependence, we use the known from the vector analysis the Stokes theorem which for some vector A is expressed as $\int Adl = \iint rotAdS$. By this expression and Eq. (2.1a), and also bearing in mind that $\sum I = \iint \delta dS$, we can obtain

$$\int Hdl = \iint rotHdS = \iint \delta dS$$

From here, it is followed that

$$rotH = \delta \tag{2.6}$$

In Eq. (2.6), the vector δ represents the current density determined as $\delta = \sigma E$. Here, σ is the specific electrical conductivity. The quantity of E reflects the vector of the electrical field intensity.

Expression (2.6) is known as the total current law in the differential form, for example [22]. Eq. (2.6) is called the first Maxwell's equation. The physical meaning of this equation is reduced to the fact that the current-carrying conductor excites the magnetic field in the vicinity of the conductor. In practice, in addition, the conductor can be flowed around by the current from the external source. The concept of the external source is used as the means of the simplifying task. So, for example, when considering the magnetic field in the electric machine, the excitation system is represented as the external source which limits the region of the existence of the magnetic field, and thus, the considering task is greatly simplified. The external sources excite the electromagnetic field, but they do not depend on this field since the action of the external sources is based on the use of the physical phenomena extraneous to the field under the consideration. Therefore, when determining the magnetic field, the external sources are represented as the known functions of the external current density δ_{cm} or of the external electric field intensity E_{cm}. The intensity of the external electric field E_{cm} causes, in a certain volume, the external current density δ_{cm} which is determined by the formula $\delta_{cm} = \sigma E_{cm}$. To introduce the current density of the external source δ_{cm} into the field equations, it is necessary to use the condition of the form

$$\delta = \sigma(E + E_{\text{ст}}) \tag{2.7}$$

In the presence of the external sources, the first Maxwell's equation is represented from Eqs. (2.6) and (2.7) as

$$\text{rot}H = \sigma E + \delta_{\text{ст}} \tag{2.8}$$

From here, we can conclude that both the conduction current and the external current excite the vortex magnetic field. Equations (2.6) and (2.7) are applied when the magnetic field intensity vector has several components. In electric machines, these equations are often used to determine the two-dimensional magnetic field.

2.3 Phenomenon of emf Induction: Electromagnetic Induction Equation

This phenomenon was discovered by the English physicist M Faraday in 1831, the essence of which is explained by the fact that with any change in the magnetic field crossing the closed conductor, the emf is induced in it. Empirically it was found that the value of the emf is directly proportional to the rate of the change of the magnetic flux crossing the closed conductor.

An integral form of the electromagnetic induction law. The formula to determine the value of the induced emf was proposed by Faraday. This formula looks like

$$Э = -\frac{d\Phi}{dt} \tag{2.9}$$

The positive direction of the induced emf is determined using the right-hand screw rule: if the rotation of the right-hand screw ensures the movement of its tip in the direction of the magnetic field, then, the direction of rotation of the head of this screw indicates the positive direction for the induced emf (Fig. 2.2). The accepted positive direction of the induced emf corresponds to the right-hand system, while the real direction of the induced emf corresponds to the left-hand system (Fig. 2.2). In Eq. (2.9), the minus sign is used to reconcile the induced emf direction obtained from the experience with the emf direction taken as positive. Based on Eq. (2.9), we can conclude that the change in time of the magnetic flux Φ leads to the emf induction in the closed conductor. This induced emf indicates that the induced electric field appears around the conductor along its entire length. In other words, the emf is induced along the entire length of the conductor. Therefore, the value of the induced

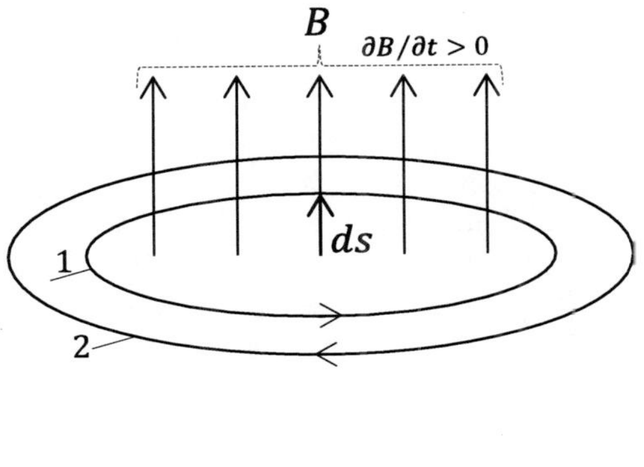

Fig. 2.2 To the application of the electromagnetic induction law: (1) positive direction of the induced emf; (2) real direction of the induced emf

emf can be obtained by summing the electric field intensity along the entire length of the conductor, i.e.

$$Э = \int Edl \tag{2.10}$$

Now, based on Expressions (2.9) and (2.10), and also replacing the total time derivative by the partial derivative, we can get

$$\int Edl = -\frac{\partial \Phi}{\partial t} \tag{2.11}$$

Here, the partial derivative means that the emf can be induced not only by changing the magnetic flux over a time but also by moving or deforming the conductor.

By the condition (2.4), the expression (2.11) can be represented in the more general form

$$\int Edl = -\frac{\partial}{\partial t}\iint BdS \tag{2.12}$$

where $\Phi = \iint BdS$.

The left part of the formula (2.12) represents the circulation of the vector E along the contour of the integration L and reflects the emf of the conductor, i.e. $Э = \int Edl$. The presence of the emf causes in the closed conductor the current that flows in the direction of the vector E and creates the magnetic flux of this closed conductor. The right part of the formula (2.12) contains the time derivative of the magnetic flux crossing the surface bounded by the closed current-carrying conductor. Equation (2.12) can be used to describe and analyze the wider range of tasks associated with this physical phenomenon.

In the electric machines, multi-turn windings are used. As already noted earlier, the machine magnetic flux $\varPhi$, interlinking with the turns of the winding, forms the flux linkage $\varPsi$ of the winding. Therefore in Eq. (2.11) instead of the magnetic flux $\varPhi$ the flux linkage $\varPsi$ can be used, i.e.

$$\int Edl = -\frac{\partial \varPsi}{\partial t} \tag{2.13}$$

In expressions (2.11) and (2.13), the length of the integration contour takes different values. In Eq. (2.11), the length of the integration contour is the length of the considered closed conductor. In Eq. (2.13), the length of the integration contour represents the total length of the multi-turn winding.

We will consider the magnitude of the emf occurring at the ends of the linear conductor of the length L. Suppose that this conductor moves with the speed υ in the magnetic field with the flux density B. In this case, we can proceed from the formula of the form

$$Э = \int_a^b [vB]dl \tag{2.14}$$

The magnitude of the emf measured at the ends of the conductor of the length L is the sum of the emf induced along the entire length of the conductor. In this case, the conductor ends marked with the letters a and b represent the starting and ending points of the integration path. The direction of the emf induced in the conductor is determined by the right-hand rule. If the magnetic flux enters the palm of the right hand, and the ben thumb indicates the direction of the conductor's speed, then, the direction of the emf induced in the conductor and the direction of the four outstretched fingers of the right hand coincide.

With the perpendicular spatial orientation of the vectors B and v, as well as the conductor L, we can obtain from Eq. (2.14) for the emf induced along the entire length of the conductor

$$Э = BLv \tag{2.15}$$

When the vectors B and v as well as the conductor of the length L are mutually orthogonal, the emf induced in the conductor is maximum. This condition is ensured in the electric machines. Therefore, expression (2.15) is often used to determine the emf induced in the windings of the electric machines. The windings of the electric machines usually represent a system of series-connected distributed conductors. Therefore, when using the relationship (Eq. 2.15) to determine the emf of the winding, this expression is multiplied by the effective numbers of the conductors found as $2wk_w$, where the magnitude of w represents the real number of the series-connected conductors of the winding, k_w is winding factor.

In the theory of the electric machines, the value of the emf induced in the winding is taken into account using the reactance x, i.e. $Э = xI$, where I is the current flowing in the winding, for example [1–8]. Due to this representation of the emf, Kirchhoff's equations can be expressed through the currents flowing in the windings. This procedure greatly simplifies the calculations of the currents. In this regard, the condition $Э = xI$ has become widespread in the formation of Kirchhoff's voltage equations used in the research and an analysis of the machine operating modes. Now, the reactance x arises from $x = Э/I$. From here and taking into account Eq. (2.10), we can obtain

$$x = \frac{Э}{I} = \frac{\int Edl}{I} \tag{2.16}$$

This expression is valid for determining the reactances of the geometrically linear conductors the cross-section dimensions of which are small in the comparison with their length. The spatial orientation of the geometrically linear conductor uniquely determines the direction of the current density vector δ and therefore the direction of the electric field intensity vector E. Based on the foregoing, we can assert that the geometric linearity of the conductor is the main condition necessary to determine the reactance of the conductor.

Differential form of the electromagnetic induction law. Earlier, we have given the equations of the electromagnetic induction law represented in the integral form. Now, we will consider the differential form of the electromagnetic induction law. For this, by the Stokes' theorem of the form $\oint Edl = \iint rotEdS$, the left part of Eq. (2.12) can be shown as

$$\oint Edl = \iint rotEdS = -\frac{\partial}{\partial t}\iint BdS$$

Considering that the sub-integral functions are equal, we have

$$rotE = -\frac{\partial B}{\partial t} \tag{2.17}$$

This equation is called the second Maxwell's equation and represents the differential form of the electromagnetic induction law, for example [22]. This law states that the time-varying magnetic field excites the electric field.

2.4 Phenomenon of Force Effect of Magnetic Field

The magnetic field has force effect on the current-carrying conductor. The law establishing the force effect of the magnetic field on the current-carrying conductor was discovered by the French scientist A. Ampere in 1920. In practice, the force effect between the two parallel current-carrying conductors is determined as

$$F = BLI \tag{2.18}$$

If the angle between the magnetic flux density vector B and the direction of the current-carrying conductor of the length equal to L is α, then, the Ampere's law can be represented as

$$F = (B\sin\alpha)LI \tag{2.19}$$

In Eq. (2.19), the magnitude B sin $\alpha = B_{\perp}$ represents the component of the magnetic flux density vector perpendicular to the current-carrying conductor. Therefore, it follows that the value of the electromagnetic force F depends on the

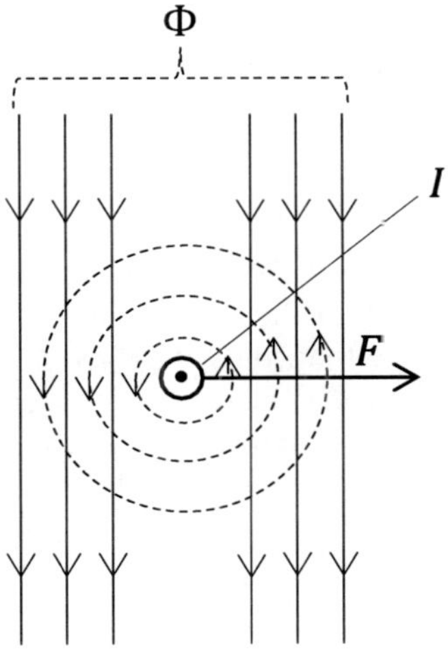

Fig. 2.3 Demonstration of the conventional mechanism of the force effect of the magnetic field on the current-carrying conductor

component of the magnetic flux density vector B perpendicular to the current-carrying conductor. The direction of the force F is perpendicular to the surface formed by the magnetic flux density vector B and the current-carrying conductor and determined by the right-hand rule. The direction of the force F can also be determined based on the concept of the deformation of the magnetic field lines (Fig. 2.3). As shown in Fig. 2.3, on the left of the current-carrying conductor, the magnetic field lines are directed in accordance with lines of the external magnetic field, and on the right of the conductor, the magnetic field lines are inserted into them. Therefore on the left of the current-carrying conductor, the common field lines are thickened and on the right, they are sparse. This means that the field lines exert the force on the current-carrying conductor from the left to the right.

The condition (2.19) can also be obtained by Lawrence's equation having the form $F_L = q(E + \upsilon \times B)$, where υ is the velocity of the charge. Here, the first component $F_e = qE$ reflects the force effect on the charge located in the electric field, and the second component $F = q(\upsilon \times B)$ describes the force effect on the charge moving in the magnetic field. The second component of Lawrence's force can be used to determine the electromagnetic force arising in the electric machines [17–20]. It can be shown that the condition $F = q(\upsilon \times B)$ can be represented as the Eq. (2.19).

In many technical solutions, including the electric machines, the angle α used in Eq. (2.19) is equal to 90°. In this case, the directions of the magnetic flux density vector B and of the current-carrying conductor of the length equal to L are mutually orthogonal, and the electromagnetic force effect on the current-carrying conductor acquires the maximum value and is determined by the expression obtained in Eq. (2.19).

Electromagnetic torque. With the rotational movement of the current-carrying conductor, the force F causes the electromagnetic torque the magnitude of which is determined by the expression of the form

$$M = FR \tag{2.20}$$

where R is the radius of the rotation of the current-carrying conductor.

With the translational movement of the current-carrying conductor, the work S performed by the force F at the distance dx can be described by the equation of kind $dS = Fdx$. With the rotational movement of the current-carrying conductor, this expression for the dS acquires the known form $dS = FRd\alpha$, since $dx = Rd\alpha$. Now, the work performing by the force F in terms of the expression (2.20) can be represented as

$$dS = Md\alpha \tag{2.21}$$

An electromagnetic power determining as $P_{эм} = dS/dt$ can be represented from Eq. (2.21) as

$$P_{эм} = \frac{dS}{dt} = M\frac{d\alpha}{dt} = \omega M \tag{2.22}$$

where $\omega = d\alpha/dt$ is the angular velocity of the rotation of the current-carrying conductor in the geometric radians per second.

From Eq. (2.22), we have for the electromagnetic torque

$$M = \frac{P_{эм}}{\omega} \tag{2.23}$$

As follows from this expression, the electromagnetic torque is calculated using the very simple formula. The expressions (2.18) and (2.23) shown earlier can be considered as the initial to calculate the electromagnetic force and the electromagnetic torque taking place in the electric machine in the operating modes, for example [1–19].

2.5 Phenomenon of Energy Transfer by Electromagnetic Field: Poynting Vector

Electromagnetic field energy is distributed in a space that is consequent of energy transferring by the electromagnetic field. An analysis of the phenomenon of energy transfer by the electromagnetic field can be used the Poynting vector [21].

In the theory of the alternating currents, to determine the power $\dot{P}_i$ dissipated in the circuit elements, the voltage and the current complexes ($\dot{U}_{im}$ and $\dot{I}_{im}$) are used

[3, 8, 10, 12, 15, 22]. To do this, the voltage complex value $\dot{U}_{im}$ can be multiplied by the conjugate value of the current complex $\dot{I}_{im}$. Then, for the power $\dot{P}_i$, we can have the known expression of the form $\dot{P}_i = 0.5(\dot{U}_{im}\dot{I}^*_{im})$.

The Poynting vector is expressed through the electric and the magnetic field intensities $(\dot{E}, \dot{H})$. According to the electromagnetic induction law, the voltage complex $\dot{U}_{im}$ corresponds to the vector complex $\dot{E}$ and the current complex $\dot{I}^*_{im}$ proceeding from the total current law corresponds to the vector complex $\dot{H}$. Here, the vectors $\dot{E}$ and $\dot{H}$ are represented through the amplitude values. Now, by an analogy with the expression $\dot{P}_i = 0.5(\dot{U}_{im}\dot{I}^*_{im})$, we can introduce the Poynting vector determined as

$$\dot{\Pi} = 0.5\left[\dot{E}\dot{H}^*\right] \tag{2.24}$$

The vector $\dot{\Pi}$ obtained in Eqs. (2.24) can be used to analyze the phenomenon of energy transferring in the electric machine by the electromagnetic field [3, 17, 19–21].

In electric machines, an electromagnetic power is converted into mechanical power (in the case of the motor) and mechanical power is converted into electromagnetic power (in the case of the generator). In this regard, to analyze the electromagnetic processes associated with the electromechanical energy conversion, the Umov-Poynting vectors can be, in the general case, used [3]. This work is dedicated to the consideration of the basic calculation elements of asynchronous machines. Therefore, we use here the Poynting vector.

2.6 Summary

1. The electromechanical interaction in the asynchronous machine can be represented as a result of the force effect of the air gap magnetic field on the current-carrying rotor winding and of the electromagnetic torque appearance in the consequence of which the energy conversion process is implemented in the electric machine. The electromechanical processes taking place in the electric machine during its operation modes are accompanied by various physical phenomena. In this work, the phenomenon of the magnetic field formation (total current law, Ampere's law), the phenomenon of the emf induction (electromagnetic induction law, Faraday's law), the phenomenon of the force effect of the magnetic field (Ampere's law), and the phenomenon of the energy transfer by the electromagnetic field (Poynting vector) are considered as the basic physical phenomena.
2. The electromechanical processes taking place in the electric machine are described by the equations representing the basic physical phenomena. These equations are represented in integral and differential forms. In the modern theory of electric machines, the equations described in the integral form have been

received the preferential position. This state of affairs is mainly due to the fact that historically the calculations and analysis of the processes taking place in the electric machine are usually implemented using the quantities represented in the phase (integral) system of units ($\dot{Э}, \dot{I}$).

3. In the theory of the electric machines, the total current law is used to calculate the magnetic circuit divided into separate regions containing usually the one-dimensional magnetic field and having the constant value of the magnetic field intensity. Under these conditions, the equation of the total current law can be represented in the form of the second Kirchhoff's law for the magnetic circuit, and in this form, it reflects the analogy between the magnetic and the electric circuits. In this regard, the calculations are greatly simplified.
4. The electromagnetic induction law is used to determine the emf induced in the machine windings. In the theory of the electric machines, the winding emf is taken into account by the corresponding reactances. This technique allows Kirchhoff's equations to express through the currents flowing in the machine windings, and the equations obtained in this way allow one to solve them in relation to these currents. Therefore, Kirchhoff's equations representing the winding currents are widely used in the practice of engineering calculations.
5. The electromagnetic forces can be considered as the results of the interaction of the air gap magnetic field and the current-carrying rotor winding. These forces cause the electromagnetic torque determining the electromechanical properties of the electric machine. The formula of kind $M = P_{эм}/\omega_1$, where $P_{эм}$ represents an electromagnetic power of the machine and ω_1 reflects the angular frequency of the stator magnetic field rotation, can be used to calculate the electromagnetic torque of the electric machine. This expression is the starting point for the analysis of the electromagnetic torque taken place in the electric machine it the operating modes.
6. The electromechanical interaction taking place in the electric machine is characterized by energy conversion. In this case, in the electric machine, an electromagnetic energy can be converted into mechanical energy (motor) or mechanical energy can be converted into electromagnetic energy (generator). In the general case, the Umov-Poynting vectors can be used to analyze the electromechanical processes in the electric machine. However, this approach to the analysis of the electromechanical processes in the electric machine has not yet received the attention it deserves. This work is limited by considering the issues associated with the electromagnetic processes described by the Poynting vector.

References

1. Kostenko, M., Piotrovsky, L.: Electric Machines, Publishing Energiya, Leningrad (1973)
2. Voldek, A.: Electric Machines, Publishing Energiya, Leningrad (1974)
3. Ivanov-Smolensky, A.: Electric Machines, Publishing Energiya, Moscow (1980)
4. Vazhnov, A.: Electric Machines, Publishing Energiya, Leningrad (1968)

5. Richter, R.: Electric Machines, vol., 1-5, United Scientific and Technical Publishing (Gosenergoizdat), Moscow (1934-1939).
6. Livshits, A.: Electric Machines, vol., 1-3, United Scientific and Technical Publishing (Gosenergoizdat), Moscow (1936)
7. Alger, P.: Induction Machines. Their Behavior and Uses, Gordon and Beach, New York (1970)
8. Del Toro, T.: Electric Machines and Power Systems, Prentice-Hall, Inc., Englewood Cliffs, New Jersey (1985)
9. Boldea, I., Nazar, S.: The Induction Machine Design Handbook, 2nd ed., CRC Press/Taylor and Francis, Boca Raton (2010)
10. Say, M.: Alternating Current Machines, John Wiley and Sons, Inc., New York (1983)
11. McPherson, R., Laramore, R.: Electric Machines and Transformers, 2nd ed., John Wiley and Sons, Inc., New York (1990)
12. Fitzgerald, A., Kingsley, C., Umans, S.: Electric Machinery, McGraw-Hill, 5th ed., New York (1990)
13. Krause, P.: Analysis of Electric Machinery, McGraw-Hill, New York (1986)
14. Chapman, S.: Electric Machinery Fundamentals, 4th ed., McGraw-Hill, Inc., New York (2005)
15. Sarema, M.: Electric Machines. Steady-State Theory and Dynamic Performance, West Publishing Company, St. Raul, New York (1986)
16. Gross, C.: Electric Machines, CRC Press, Taylor and Francis Group, New York (2007)
17. Seely, S.: Electromechanical Energy Conversion, John Wiley and Sons, Inc., New York (1959)
18. White, D., Woodson, H.: Electromechanical Energy Conversion, John Wiley and Sons, Inc., New York (1959)
19. Levi, D., Panzer, M.: Electromechanical Power Conversion, McGraw-Hill Book Company, New York (1969)
20. Turovskiy, YA.: Technical Electrodynamics, (Translated from Polish), Publishing Energiya (1974)
21. Poynting, T.: On the Transfer of Energy in the Electromagnetic Field, Trans. of Roy. Soc., pt.2, (1884)
22. Shimoni, K.: Theoretical Electrical Engineering (Translation from German), Publishing Mir, Moscow (1973)

Chapter 3
Interconnections of Basic Physical Phenomena: Kirchhoff's Equations

3.1 Voltage Equations: Phase and Specific System of Units

An analysis of the electromagnetic processes in the electric machines is carried out by the voltage equations compiling for the circuit loops reflecting the stator and rotor windings. The voltage equations obtained in this way correspond to the condition when the phase windings are performed from the geometrically linear conductors the cross-section dimensions of which are relatively small compared to the length of the winding conductors. Below, the voltage equations are represented in the phase and specific system of units.

3.1.1 Phase System of Units

Secluded circuit loop. We will consider the linear circuit loop with the number of turns equal to w. Consequently, it is accepted that the turns of the circuit loop are made of the linear conductors. We assume that this circuit loop embraces the uniformly distributed magnetic flux $\dot{\Phi}$. Because of the uniformly distributed magnetic flux $\dot{\Phi}$, the value of the electric field intensity $\dot{E}$ does not change along the length l of the circuit loop, i.e. $\dot{E} = \text{const}$. Which the use of the condition (2.10) and (2.11) arising from the electromagnetic induction law, for the emf induced in the secluded circuit loop consisting of the series-connected w conductors and having the length equal to l, we can have

$$\dot{Э} = \dot{E}wl \qquad \dot{Э} = -j\omega\dot{\Psi} \tag{3.1}$$

where $\dot{\Psi} = w\dot{\Phi}$

By the equation for Ohm's law represented in (2.7) as $\dot{\delta} = \frac{1}{\rho}\left(\dot{E} + \dot{E}_{cm}\right)$, the electric field intensity $\dot{E}$ is determined in the following form

V. Asanbayev, *Asynchronous Machines*,
https://doi.org/10.1007/978-3-030-92284-9_3

$$\dot{E} = \rho\dot{\delta} - \dot{E}_{cm} \tag{3.2}$$

Here, it is assumed that the current density $\dot{\delta}$ is uniformly distributed over the cross sections of the conductors forming the circuit loop. Taking into account Eq. (3.2), the first equation used in (3.1) can be shown as

$$\dot{Э} = \rho wl\dot{\delta} - wl\dot{E}_{cm} \tag{3.3}$$

In (3.3), we introduce the following expressions

$$r = \rho\frac{w}{S}l \quad \dot{I} = \dot{\delta}S \quad \dot{U} = wl\dot{E}_{cm} \tag{3.4}$$

where r is the circuit-loop ohmic resistance in the phase system of units; $\dot{U}$ is the voltage of the external source used in the circuit loop; $\dot{I}$ is the current flowing in the circuit loop; and S is the cross section of the circuit-loop conductors.

Now, Eq. (3.3) acquires the form

$$\dot{U} = r\dot{I} + (-\dot{Э}) \tag{3.5}$$

Equation (3.5) can be rewritten as $\dot{U} + \dot{Э} = r\dot{I}$. From here, it follows that the sum of the voltage $\dot{U}$ applied to the terminals of the secluded circuit loop and the emf $\dot{Э}$ induced in this circuit loop is equal to the voltage drop across the ohmic resistance of the circuit loop. Thus, Eq. (3.5) represents Kirchhoff's equation expressed for the secluded multi-turn circuit loop. Equation (3.5) corresponds to the phase system of units.

In the theory of electric machines, the emf $\dot{Э}$ induced in the circuit loop is taken into account with the use of the reactance x representing the considered circuit loop. Then, for the emf $\dot{Э}$ used in (3.5), we can have that

$$(-\dot{Э}) = jx\dot{I} \tag{3.6}$$

where x is the reactance of the secluded multi-turn circuit loop in the phase system of units. Now, the Eq. (3.5) taking into account (3.6) acquires the form

$$\dot{U} = r\dot{I} + jx\dot{I} \tag{3.7}$$

In accordance with (3.6), the reactance x used in (3.7) is determined from the condition of the form $x = \dot{Э}/\dot{I}$.

The Eq. (3.7) allows us to determine the value of the current flowing in the circuit loop. Really, at the given values of the voltage of the external source $\dot{U}$ as well as the ohmic resistance r and the reactance x, for the circuit-loop current $\dot{I}$, we can have that $\dot{I} = \dot{U}/(r + jx)$. From this, it follows that the condition of kind $\dot{Э} = x\dot{I}$ shown in (3.6) is introduced to give to the voltage equation the form that makes it possible to simplify the procedure for determining the value of the current flowing in the circuit loop. This technique is widely used in the theory of calculating electric machines.

In Kirchhoff's equation given in the form (3.7), the reactance x is used. Below, we will briefly consider how the reactance x can be determined. For this purpose, the flux linkage of the circuit loop $\dot{\Psi}$ should be found in accordance with the second equation shown in (3.1). Using the equation of the total current law, the flux linkage $\dot{\Psi}$ is determined for the given value of the current $\dot{I}$ flowing in the circuit loop. Now, by the equation of the electromagnetic induction law, the value of the emf $\dot{Э}$ induced in the circuit loop arises as a result of the change in a time of the flux linkage $\dot{\Psi}$. With the known values of the circuit-loop current $\dot{I}$ and the emf $\dot{Э}$ induced in the circuit loop, the reactance x follows as the ratio of the emf $\dot{Э}$ to the current $\dot{I}$ flowing in the circuit loop. Thus, we can argue that Kirchhoff's equation shown in Eq. (3.7) represents the result of using Ohm's law, Ampere's law, and Faraday's law.

Inductively coupled circuit loops. Equation (3.5) is valid for the secluded circuit loop. Equation (3.5) can be used to determine the voltage equations for the n inductively coupled circuit loops.

Performing the same transformations as for the secluded circuit loop, it can be shown that the voltage Eq. (3.5) keeps its form for each n inductively coupled circuit loops. In this regard, Eq. (3.5) can be represented for the kth circuit loop of the n inductively coupled circuit-loop system as

$$\dot{U}_k = r_k\dot{I}_k + (-\dot{Э}_k) \tag{3.8}$$

where $k = 1, 2, 3, \ldots, n$.

In this case, the emf $\dot{Э}_k$ induced in the kth circuit loop is determined by the currents flowing in the n inductively coupled circuit loops. Therefore, for this case, the emf equation (3.6) acquires the form for the kth circuit loop

$$(-\dot{Э}_k) = \sum_{s=1}^{n} jx_{sk}\dot{I}_s \tag{3.9}$$

where $s = 1, 2, 3, \ldots, n$ and $k = 1, 2, 3, \ldots, n$.

In (3.9), the emf of the kth circuit loop $\dot{Э}_k$, where k takes the values from $k = 1$ to $k = n$, can be represented as the sum of the total self-induction emf of the kth circuit loop $\dot{Э}_{kk}$ and the mutual induction emf of the kth circuit loop $\dot{Э}_{sk}$, where s takes the values from $s = 1$ to $s = n$ (with the exception $s = k$). Consequently, the emf of the kth circuit loop $\dot{Э}_k$ can be represented as $\dot{Э}_k = \dot{Э}_{kk} + \dot{Э}_{sk}$. Here, the emf $\dot{Э}_{kk}$ can be taken into account by using the total self-induction reactance x_{kk} and the emf $\dot{Э}_{sk}$ by applying the mutual induction reactances x_{sk}.

Now, from (3.9), the total self-induction emf $\dot{Э}_{kk}$ and the mutual induction emf $\dot{Э}_{sk}$ of the kth circuit loop can be expressed as

$$\begin{gathered}(-\dot{Э}_{kk}) = jx_{kk}\dot{I}_k \\ (-\dot{Э}_{sk}) = \sum_{s=1}^{n} jx_{sk}\dot{I}_s\end{gathered} \tag{3.10}$$

where $s \neq k$.

Here, the value $x_{kk}\dot{I}_k$ reflects the total self-induction emf of the kth circuit loop $\dot{Э}_{kk}$, where x_{kk} represents the total self-induction reactance of the kth circuit loop in the phase system of units, and $\dot{I}_k$ is the current flowing in the kth circuit loop. When the magnitude of s obtains the values from $s = 1$ to $s = n$ (with the exception $s = k$), the value $x_{sk}\dot{I}_s$ reflects the mutual induction emf of the kth circuit loop $\dot{Э}_{sk}$, where x_{sk} represents the mutual induction reactance of the kth circuit loop in the phase system of units, and $\dot{I}_s$ is the current flowing in the sth circuit loop.

Now, for the kth circuit loop, the voltage Eq. (3.8) can be reduced to the following form taking into account the emf equation shown in (3.9)

$$\dot{U}_k = r_k\dot{I}_k + jx_{kk}\dot{I}_k + \sum_{s=1}^{n} jx_{sk}\dot{I}_s \tag{3.11}$$

where $s \neq k$.

Using the Eq. (3.11), the voltage equations can be written for all n circuit loops. The equations represented in (3.11) can be considered as the initial voltage equations to obtain Kirchhoff's equations for the circuit loops reflecting the multi-winding electric machines [10].

When compiling the voltage equations of kind (3.11), the "inductively coupled circuits" concept (superposition principle) is used, that is, it is assumed that the emf $\dot{Э}_k$ induced in the kth circuit loop is determined by the current $\dot{I}_k$ flowing in the kth circuit loop and by the currents off the other inductively coupled circuit loops $\dot{I}_s$. As shown, for example, in [1], the equations obtained in the form of (3.11) cannot be recommended for the engineering practice, since the sufficiently accurate determination of the reactances used in them is associated with certain difficulties. To eliminate this problem, the Eq. (3.11) are subjected to the appropriate transformation. The essence of these transformations is based on the fact that the total self-induction reactances of the circuit loops are represented as a result of the sum of the leakage reactances and the self-induction reactances of the circuit loops as well as in accordance with the "no-load and under-load" concept, the reduction technique is applied in the voltage equations shown in (3.11). In this case, the considered circuit loops can be represented as interlinked with the one common magnetic flux (magnetizing flux) created by the current flowing in the previously accepted k^{th} circuit loop (in the technical literature, the stator winding is usually accepted as such circuit loop). In this case, all circuit loops form their own leakage fluxes interlinking with the particular circuit loop. In these conditions, the quantities used in the Eq. (3.11) will have the values referred to the number of the turns of the kth circuit loop, for example, [1–9]. In addition, the frequencies of the currents of the circuit loops are referred to as the frequency of the current flowing in the previously accepted k^{th} circuit loop. The voltage equations obtained in this way can be used to analyze the transient and steady-state behaviors of the electric machine.

3.1.2 Specific System of Units

Secluded circuit loop. The voltage equations obtained in (3.5) and (3.8) have been represented in the phase system of units. In this part of this work, these equations are expressed through the specific quantities representing the vectors of the electromagnetic field intensities $(\dot{E}, \dot{H})$.

We proceed from Eq. (3.2) shown as $\dot{E} = \rho\dot{\delta} - \dot{E}_{cm}$. In this equation, we consider the value of the current density $\dot{\delta}$ under the assumption that the circuit loop has been performed from the round linear conductor with the uniform distribution of the current density $\dot{\delta}$ and it (circuit loop) contains the w turns. Then, the vector of the current density $\dot{\delta}$ used in the Eq. (3.2) can be determined using the equation for total current law written for the w round conductors as $\dot{H}'(2\pi R') = \dot{\delta}(\pi R^2)w$. Here, the value $\dot{H}'$ corresponds to the surface determined by some conditional radius R', and R is the radius of the circuit-loop conductor. Now, for the current density $\dot{\delta}$, we can have that $\dot{\delta} = \dot{H}'\frac{2R'}{R^2 w}$. Taking into account the condition $\dot{\delta} = \dot{H}'\frac{2R'}{R^2 w}$, the Eq. (3.2) can be expressed as

$$\dot{E} = \rho\frac{2R'}{R^2 w}\dot{H}' - \dot{E}_{cm} \tag{3.12}$$

In (3.12), we can introduce the ohmic resistance of the circuit loop r^* and the calculated value of the magnetic field intensity $\dot{H}$ for which arise

$$r^* = \frac{\rho}{R/2} \quad \dot{H} = \frac{\dot{H}'}{(R/R')w} \tag{3.13}$$

Taking into account (3.13), Eq. (3.12) can be expressed as

$$\dot{E} = r^*\dot{H} - \dot{E}_{cm} \tag{3.14}$$

Using in this equation the condition of the form $\dot{U}^* = \dot{E}_{cm}$, for the circuit loop having the unit length, we obtain

$$\dot{U}^* = r^*\dot{H} + (-\dot{E}) \tag{3.15}$$

where $\dot{U}^*$ is the voltage applied to the terminals of the circuit loop and represented in the specific system of units.

From (3.15), it follows that Kirchhoff's equation shown in the specific system of units is expressed through the electric field intensity $\dot{E}$ caused by the induced in the circuit-loop emf, and also through the calculated value of the magnetic field intensity $\dot{H}$ created by the currents flowing in the turns of the circuit loop. For the electric field intensity $\dot{E}$ used in (3.15), we can have similar to (3.6)

$$\left(-\dot{E}\right) = jx^*\dot{H} \tag{3.16}$$

Now, Eq. (3.16) can be expressed as

$$\dot{U}^* = r^*\dot{H} + jx^*\dot{H} \tag{3.17}$$

In (3.17), the quantities r^* and x^* represent the ohmic resistance and reactance of the circuit loop, respectively. They are expressed in specific system of units. The value of the reactance x^* can be found using the reasoning applied earlier.

For given values of voltage $\dot{U}^*$ as well as the parameters $r^*$and x^*, Eq. (3.17) can be used to determine $\dot{H}$ of the circuit loop.

Inductively coupled circuit loops. Equations (3.15) and (3.17) are valid for the secluded circuit loop. In the case of considering the multi-circuit-loop system, the equation for the kth circuit loop can be shown by an analogy with the equation given in (3.15)

$$\dot{U}_k^* = r_k^*\dot{H}_k + \left(-\dot{E}_k\right) \tag{3.18}$$

where $\left(-\dot{E}_k\right) = \sum_{s=1}^{n} jx_{sk}^*\dot{H}_s$.

In accordance with (3.18), the electric field intensity $\dot{E}_k$, where k takes the values from $k = 1$ to $k = n$, can be represented as the sum of the electric field intensity $\dot{E}_{kk}$ caused in the kth circuit loop by the induced total self-induction emf and the electric field intensity $\dot{E}_{sk}$ caused in the kth circuit loop by the induced mutual induction emf, i.e. $\dot{E}_k = \dot{E}_{kk} + \sum_{s=1}^{n} \dot{E}_{sk}$, where s takes the values from $s = 1$ to $s = n$ (with the exception $s \neq k$). Here, the electric field intensity $\dot{E}_{kk}$ can be taken into account by the total self-induction reactance x_{kk}^* and the electric field intensity $\dot{E}_{sk}$ can be taken into account by the mutual induction reactance x_{sk}^*. The reactances x_{kk}^* and x_{sk}^* are represented in the specific system of units. Then, for the electric field intensity $\dot{E}_{kk}$ caused in the kth circuit loop by the induced total self-induction emf and for the electric field intensity $\dot{E}_{sk}$ caused in the kth circuit loop by the induced mutual induction emf, we can have

$$\begin{aligned}(-\dot{E}_{kk}) &= jx_{kk}^*\dot{H}_k \\ (-\dot{E}_{sk}) &= \sum_{s=1}^{n} jx_{sk}^*\dot{H}_s\end{aligned} \tag{3.19}$$

where $s \neq k$.

In (3.19), the quantity $x_{kk}^*\dot{H}_k$ reflects the value of the electric field intensity $\dot{E}_{kk}$ caused in the kth circuit loop by the induced total self-induction emf, where $\dot{H}_k$ is the magnetic field intensity created by the current flowing in the kth circuit loop. When s takes the values from $s = 1$ to $s = n$ (with the exception $s \neq k$), the quantity $\sum_{s=1}^{n} jx_{sk}^*\dot{H}_s$ reflects the value of the sum of the electric field intensities $\sum_{s=1}^{n} \dot{E}_{sk}$ caused in the kth circuit loop by the induced mutual induction emf, where $\dot{H}_s$ is the magnetic field intensity created by the current flowing in the sth circuit loop. Now, Eq. (3.18) in terms of the conditions shown in (3.19) can be expressed as

$$\dot{U}_k^* = r_k^*\dot{H}_k + jx_{kk}^*\dot{H}_k + \sum_{s=1}^{n} jx_{sk}^*\dot{H}_s \tag{3.20}$$

where $s \neq k$.

In (3.20), r_k^* is the ohmic resistance of the kth circuit loop in the specific system of units.

To give the form that is convenient for practical use, the equations shown in (3.20) are subjected to the transformation for which the techniques discussed above can be used. It should be noted that the transformation of the Eq. (3.20) is simplified, since in the specific system of units, the need to reduce the circuit loops by the number of the winding turns disappears, for example [10].

3.2 Voltage Equations: Some Features

We will consider some features of the voltage equations obtained in (3.12) and (3.20) for the n inductively coupled circuit loops and represented in the phase and specific system of units. As it is noted earlier, these equations have been written for the system of the linear circuit loops. In this case, the space orientation of the circuit loop uniquely determines the direction of the current density in the circuit loop. In this regard, the linear conductor is the circuit loop of the conductor at the same time. From the Eqs. (3.12) and (3.20), it arises that when using the linear conductors, the electric and magnetic fields associated with the circuit loops can be considered individually which simplifies an analysis.

In electric machines, the windings made from the wires of the limited cross section are often used. In these conditions, the direct use of the equations of the form (3.12) and (3.20) may cause difficulties, since the provisions noted earlier for the system of the linear conductors are not, in the general case, applicable for the conductors of the limited cross section. According to [10], the independent consideration of the electric and magnetic fields is not, in the general case, correct for the conductor with the limited cross section. In this regard, compiling Eqs. (3.12) and (3.20) for the circuit loops representing the conductors with the limited cross sections presents a certain difficulty. To solve this problem, the conductor with the limited cross section can be represented as the set of the conditional linear circuit loops [10]. However, this approach leads to the excessive complication of the calculation process.

In asynchronous machines, the rotor winding is often performed by the use of conductors with limited cross sections. In this regard, in the rotor winding, the current displacement phenomenon takes place in the general case. In practice, it is believed that the current density in the rotor winding is directed along the machine axis, and its value does not change along the machine axis. In this case, the value of the current density of the rotor winding can be changed in the radial and tangential directions. Therefore, in the general case, the compiling process of the voltage equations for the rotor winding becomes extremely difficult. In [10], the voltage equations for the rotor winding are proposed for the case, when the rotor winding current density is changed only in the radial direction. In this work, we consider the case, when the current density of the rotor winding can change only in the radial direction.

The voltage equations shown in (3.12) and (3.20) are represented in the phase and specific system of units. In the phase system of units, equation (3.12) is expressed through the voltages and currents of the circuit loops. The machine circuit-loop voltages and currents are relatively easy to control during the operating mode of the electric machine. Consequently, the use of the voltage equations represented in the phase system of units simplifies the process of comparing the calculated data with the data obtained as a result of the experiment which helps to increase the reliability of analysis.

In (3.20), the voltage equations have been expressed through the components of the electric and magnetic field intensities ($\dot{E}, \dot{H}$). The vectors $\dot{E}$ and $\dot{H}$ are used in the field equations represented in the differential form (Maxwell's equations). Therefore, the vectors $\dot{E}$ and $\dot{H}$ can be used to obtain the voltage equations by applying Maxwell's equations. This feature of using the vectors $\dot{E}$ и $\dot{H}$ is very important because, in this case, new opportunities for the further improvement of the machine calculation methods are opened up. The features of the practical implementation of these new opportunities have been demonstrated in [10].

3.3 Currents and emf of Winding: Phase and Specific Systems of Units

As follows from (3.6) and (3.12), the voltage equations expressed through the phase quantities can be considered in relation to the voltages and the currents $(\dot{U}, \dot{I})$ or in relation to the emf and the currents $(\dot{Э}, \dot{I})$. Here, the value of $\dot{U}$ reflects the voltage of the external source. The voltage $\dot{U}$ creates the current in the circuit loop, but the voltage value does not depend on the current flowing in the circuit loop as we can accept. In this regard, in relation to the considered circuit loop, the voltage $\dot{U}$ can be represented as the emf of the external source $\dot{Э}_{cm}$ for which we can have the condition of the form $\dot{Э}_{cm} = \dot{U}$. The emf of the external source $\dot{Э}_{cm}$ is the result of the action of other physical phenomena (for example, mechanical, chemical, etc.); therefore, the use of the emf $\dot{Э}_{cm}$ (the voltage $\dot{U}$) is not entirely rational. The electromagnetic processes taking place in the electric machine are directly associated with the emf and the currents $(\dot{Э}, \dot{I})$ arising as a result of the action of the electromagnetic induction law and total current law. In this regard, and in accordance with the objectives of this work, we will use the quantities $(\dot{Э}, \dot{I})$ represented the circuit-loop emf and the currents in the phase system of units.

In the specific system of units, the circuit-loop emf and the current $(\dot{Э}, \dot{I})$ correspond to the electric field intensity $\dot{E}$ caused by the induced in the circuit-loop emf $\dot{Э}$ and to the magnetic field intensity $\dot{H}$ created by the current $\dot{I}$ flowing in the circuit loop. To consider the features of the connections between the emf and the currents $(\dot{Э}, \dot{I})$ and the electric and magnetic field intensities $(\dot{E}, \dot{H})$, we will use the following conditions

$$\dot{Э} = \oint \dot{E}dl \qquad \dot{I} = \oint \dot{H}dl \tag{3.21}$$

The equations given in (3.21) can be used to transfer the emf and the currents of the winding from the phase system of units to the specific system of units and vice versa. Below, we demonstrate this.

We will consider the features of the application of the equations shown in (3.21) taking into account the conditions in the electric machine. In the second equation of kind $\dot{I} = \oint \dot{H}dl$, the value $\dot{I}$ represents the current flowing in the secluded circuit loop. In the asynchronous machine, the currents flow in the distributed multi-turn windings. In this regard, in the theory of the electric machines, the distribution in a space of the winding total current is represented as mmf of the winding $\dot{F}$ [1–9]. Now, in the equation $\dot{I} = \oint \dot{H}dl$ instead of the current $\dot{I}$, we can use, as noted above, the mmf $\dot{F}$ of the winding, that is

$$\dot{F} = \oint \dot{H}dl \tag{3.22}$$

As it follows from (3.22), the mmf of the winding $\dot{F}$ creates the magnetic field in the electric machine. We consider the case when the magnetic permeabilities of the stator and rotor magnetic circuits take relatively large values. In these conditions, the magnetic field is concentrated in the air gap of the machine. The magnetic field created in the air gap by the mmf $\dot{F}$ can have the radial and the tangential components; therefore, the condition shown in (3.22) should be applied for each magnetic field component separately. In this work, we use the right-handed coordinate system with the coordinate axes x, y, z. In this case, the coordinate axis x indicates the tangential direction, the coordinate axis y is directed in the radial direction, and z coordinate axis coincides with the axial axis of the electric machine. In accordance with the accepted coordinate system, to determine the tangential component of the magnetic field intensity, in the equation $\dot{F} = \oint \dot{H}dl$ instead of dl, we should take the value of the dx. This means that in (3.22) the integration contour, in this case, is oriented in the tangential direction, i.e., $dl \to dx$. When determining the radial component of the magnetic field intensity, in the equation $\dot{F} = \oint \dot{H}dl$ the value for the dl should be replaced by dy. In this case, the integration contour coincides with the radial direction, i.e. $dl \to dy$.

The mmf of the winding $\dot{F}$ using in (3.22) is determined by the following expression, for example, [1–9]

$$\dot{F} = \frac{mwk_w}{\pi p}\sqrt{2}\dot{I} \tag{3.23}$$

The axial component of the electric field intensity $\dot{E}$ used in the equation $\dot{Э} = \oint \dot{E}dl$ is determined by the radial component of the magnetic flux density $\dot{B}$ which is confirmed by the condition (2.12) having the form $\oint \dot{E}dl = j\omega \iint \dot{B}dS$. Here, the radial component of the magnetic flux density $\dot{B}$ arises from the equation $\dot{F} = \oint \dot{H}dl$. In this equation, the integration contour coincides, as mentioned earlier, with the radial direction, that is, $dl \to dy$. Therefore, the vector $\dot{E}$ is represented by the single axial component. In this regard, in the equation $\dot{Э} = \oint \dot{E}dl$, the value for dl should be replaced with dz, i.e. $dl \to dz$.

Now, taking into account the above, as well as the conditions shown in (3.22) and (3.23), the equations obtained in (3.21) can be expressed as

$$\dot{Э} = \int \dot{E}dz \quad \dot{F} = \int \dot{H}dx \tag{3.24}$$

The Eq. (3.24) can be used to transfer the emf and the currents of the machine windings from the phase system of units to the specific system of units and vice versa.

3.4 Currents and emf of Winding: Transfer from One System of Units to Another

The use of the voltage equations represented in the phase and specific system of units forces the solution of the problem associated with the transfer of the quantities applied in them from one system of units to another and vice versa. Let's start by considering the problem of converting the machine winding currents from the phase system of units to the specific one and vice versa. For this purpose, we proceed from the total current law equation shown in (3.24) as $\dot{F} = \int \dot{H}dx$. In this equation, the vector $\dot{H}$ represents the tangential components of the magnetic field intensities defined on the surfaces of the stator or rotor of the electric machine. In this case, we assume that the surface current with the linear density $\dot{\varkappa}$ is distributed on the surfaces of the stator or rotor of the electric machine in the form of the infinitely thin layer, for example, [1–8]. The linear density $\dot{\varkappa}$ of the surface current has a tangential direction. On the surface at the length dx, we will determine the value of the elementary mmf $d\dot{F}$ caused by the tangential component of the magnetic field intensity $\dot{H}$ created by the linear density $\dot{\varkappa}$ of the surface current. The elementary mmf $d\dot{F}$ formed by the linear density of the surface current $\dot{\varkappa}$ is determined as $d\dot{F} = \dot{\varkappa}\, dx$. To determine the elementary mmf $d\dot{F}$ from the tangential component of the magnetic field intensity $\dot{H}$, we can use the total current law, for which we consider the integration contour, the one side of which is located on the surface of the magnetic circuit and has the length equal to dx. The other sides of the integration contour are located in the magnetic circuit with the infinitely high magnetic permeability, and therefore, they can be excluded from further considerations. Then, for the elementary mmf $d\dot{F}$ caused by the tangential component of the magnetic field intensity $\dot{H}$, we can have that $d\dot{F} = \oint \dot{H}dl = \dot{H}dx$. Considering that on the one hand $d\dot{F} = \dot{\varkappa}\, dx$ and on the other hand $d\dot{F} = \dot{H}\, dx$, we can obtain that $\dot{H} = \dot{\varkappa}$, for example, [1–8]. It follows that the tangential component of the magnetic field intensity $\dot{H}$ acting on the surface of the magnetic circuit is equal to the linear density $\dot{\varkappa}$ determined on this surface. The condition $\dot{H} = \dot{\varkappa}$ is valid for the rotor magnetic circuit. Acting in the similar way, for the stator magnetic circuit, we can have the condition of the form $\dot{H} = -\dot{\varkappa}$.

We will determine the value of the fundamental harmonic of the linear density $\dot{\varkappa}$ of the surface current. For this purpose, the fundamental space harmonic of the winding mmf $\dot{F}(x)$ can be considered. The winding mmf $\dot{F}(x)$ is changed along the surface of the magnetic circuit (at the tangential direction) as $\dot{F}(x) = \dot{F}e^{-j(\pi/\tau)x}$, and therefore, it is the function of the coordinate x measured from the winding axis. Consequently, for the fundamental harmonic of the spatial distribution of the winding mmf $\dot{F}(x)$, we have

$$\dot{F}(x) = \dot{F}e^{-j(\pi/\tau)x} \tag{3.25}$$

where $\dot{F} = \dfrac{mwk_w}{\pi p}\sqrt{2}\dot{I}$.

According to (3.25), the winding mmf $\dot{F}$ is expressed through the effective current flowing in the winding.

In the theory of the electric machines, in addition to the winding mmf $\dot{F}(x)$, the concept of the linear density $\dot{\varkappa}(x)$ of the surface current is applied. The linear density $\dot{\varkappa}(x)$ of the surface current is formed as a result of the distribution of the total winding current by the infinitely thin layer on the smooth surface of the magnetic circuit on which the winding with the current is leaned, for example, [1–10]. In this regard, the winding mmf curve $\dot{F}(x) = f(x)$ can be considered as the integral curve of the distribution of the linear density $\dot{\varkappa}(x)$ of the surface current, i.e.

$$\dot{F}(x) = \int \dot{\varkappa}(x)dx \tag{3.26}$$

Therefore, the curve of the spatial distribution of the linear density $\dot{\varkappa}(x)$ of the surface current is the derivative of the winding mmf $\dot{F}(x)$, i.e.

$$\dot{\varkappa}(x) = \frac{d\dot{F}(x)}{dx} \tag{3.27}$$

Based on (3.25) and (3.27), for the fundamental harmonic of the linear density $\dot{\varkappa}$ of the surface current, we have

$$\dot{\varkappa} = \frac{\pi}{\tau}\dot{F} \tag{3.28}$$

Here, the value of the winding mmf $\dot{F}$ is determined by the expressions shown in (3.23) and (3.25).

In the technical literature, the quantity $\dot{\varkappa}$ is often represented through the average value of the linear current load of the winding $\dot{A}$ defined as the sum of the effective values of the winding currents (2 $mw\dot{I}$) referred to the length of the circumference of the magnetic circuit surface (2 $p\tau$), i.e.

$$\dot{A} = \frac{mw\dot{I}}{p\tau} \tag{3.29}$$

With the use of (3.23), (3.28) and (3.29), we can obtain

$$\dot{\varkappa} = \sqrt{2}\dot{A}k_w \tag{3.30}$$

Taking into account that the first space harmonic of the linear density $\dot{\varkappa}$ of the surface current is equal to the tangential component of the magnetic field intensity $\dot{H}$, i.e. $\dot{H} = \dot{\varkappa}$, and also using the expressions given in (3.29), (3.30), it is followed

$$\sqrt{2}\dot{I} = \frac{p\tau}{mwk_w}\dot{H} \tag{3.31}$$

This expression connects the effective value of the current $\dot{I}$ flowing in the winding with the amplitude value of the tangential component of the magnetic field intensity $\dot{H}$ created by the current of this winding.

We turn to the equation shown in (3.24) in the form

$$\dot{Э} = \int \dot{E}dz \tag{3.32}$$

In Eq. (3.32), we consider its right part determined in terms of condition (2.12) as

$$\int \dot{E}dz = \omega wk_w \iint \dot{B}dS \tag{3.33}$$

The electric field intensity $\dot{E}$ using in (3.33) has the single axial component. The magnetic flux density $\dot{B}$ has the single radial component and in terms of (3.25) its spatial change can be represented as $\dot{B}(x) = \dot{B}e^{-j(\pi/\tau)x}$. With the use of this condition, the Eq. (3.33) acquires the form

$$\int \dot{E}dz = 2wk_w l\dot{B}\frac{\omega\tau}{\pi} = 2wk_w l\dot{E} \tag{3.34}$$

where $\dot{E} = \dot{B}(\omega\tau/\pi)$.

From (3.32) and (3.34), it arises

$$\sqrt{2}\dot{Э} = 2wk_w l\dot{E} \tag{3.35}$$

The expression (3.35) allows us to define the value of the electric field intensity $\dot{E}$ using the effective magnitude of the emf $\dot{Э}$ induced in the winding.

Based on the expressions obtained in (3.31) and (3.35), the interconnection between the phase values of the emf and the currents ($\dot{Э}$, $\dot{I}$) and the specific values of the electric and magnetic field intensities ($\dot{E}$, $\dot{H}$) can be represented as

$$\sqrt{2}\dot{Э} = 2wk_w l\dot{E} \quad \sqrt{2}\dot{I} = \frac{p\tau}{mwk_w}\dot{H} \tag{3.36}$$

The ability to represent the electromagnetic quantities in the different systems of units creates favorable conditions for an analysis of the voltage equations obtained by the use of Maxwell's equations which opens up new opportunities for further improving the theory of calculating electric machines [10].

3.5 Ohmic Resistance and Reactance: Physical Meaning

Phase ohmic resistance and reactance. We will consider the voltage equation obtained in (3.7) for the secluded circuit loop. The left and right parts of this equation can be multiplied by the conjugate current complex $\dot{I}^*$. Then, it follows

$$\dot{U}\dot{I}^* = rI^2 + jxI^2 = P + jQ \tag{3.37}$$

where $P = rI^2$; $Q = xI^2$.

Here, the value of rI^2 characterizes the power losses in the circuit loop and the value of xI^2 represents the magnetic energy stored in the circuit loop. Using the Eqs. (3.7) and (3.37) it can be shown that the circuit-loop ohmic resistance r allows us, on the one hand, to find the value of the voltage drop across the resistance r (as the product of the ohmic resistance r and the circuit-loop current $\dot{I}$, i.e. $r\dot{I}$) and, on the other hand, to calculate the power losses in the circuit loop (as the product of the ohmic resistance and the square of the circuit-loop current $\dot{I}^2$, i.e. $r\dot{I}^2$). Similarly, the circuit-loop reactance x allows us, on the one hand, to establish the value of the voltage drop across the reactance x (as the product of the reactance x and the circuit-loop current $\dot{I}$, i.e. $x\dot{I}$) and, on the other hand, to find the magnetic field energy stored in the circuit loop (as the product of the reactance x and the square of the circuit-loop current $\dot{I}^2$, i.e. $x\dot{I}^2$). Thus, the ohmic resistance and reactance of the circuit loop (r and x) must correspond, on the one hand, to the voltage equations (Kirchhoff's equations), and, on the other hand, they must ensure the balance of power losses and magnetic energy stored in the circuit loop [11, 12]. These conditions are uniquely fulfilled for the geometrically linear circuit loops the cross-section dimensions of which are relatively small compared to the length of the circuit loop. In such circuit loops, the direction of the current density vector $\dot{\delta}$ coincides with the direction of the circuit-loop element vector dl. This means, that the spatial orientation of the geometrically linear circuit loop uniquely determines the direction of the current density vector $\dot{\delta}$. Therefore, in this case, the field equations are simplified and the electromagnetic processes in the circuit loop can be described by the phase quantities (the voltage $\dot{U}$, current $\dot{I}$, ohmic resistance r, reactance x, etc.), as it is shown in (3.7).

Specific ohmic resistance and reactance. We will consider the voltage equation obtained in (3.17) for the secluded circuit loop. The left and right parts of this equation can be multiplied by the conjugate complex $\dot{H}^*$. Then, we have

$$0.5\left(\dot{U}^*\dot{H}^*\right) = 0.5\left(r^*H^2\right) + j0.5\left(x^*H^2\right) = P^* + jQ^* \tag{3.38}$$

where $P^* = 0.5(r^*H^2)$; $Q^* = 0.5(x^*H^2)$.

We consider the equation of kind $P^* = 0.5(r^*H^2)$. Here, the $P^* = 0.5(r^*H^2)$ represents the Poynting vector characterizing the power losses per unit surface of the circuit loop. Therefore, the ohmic resistance r^* allows us to determine the value of the power losses per unit surface of the circuit loop. In this regard, the value r^* can be considered as the ohmic resistance of the unit surface of the circuit loop. In accordance with the Eqs. (3.17) and (3.38), the specific ohmic resistance r^* can be used to calculate the value of the voltage drop across the resistance r^* as $\dot{E}_a = r^*\dot{H}$.

In (3.38), we proceed from the equation $Q^* = 0.5(x^*H^2)$. Here, the $Q^* = 0.5(x^*H^2)$ represents the Poynting vector characterizing the magnetic field energy per unit surface of the circuit loop. Therefore, the reactance x^* allows us to determine the value of the magnetic field energy per unit surface of the circuit loop. In this regard, the value of x^* can be considered as the reactance of the unit surface of the circuit loop. According to (3.17) and (3.38), the specific reactance x^* can be used to calculate the voltage drop across the reactance x^* as $\dot{E}_p = x^*\dot{H}$.

Thus, the phase ohmic resistance r and the reactance x reflect the power losses and magnetic energy stored in the volume of the circuit loop. The specific ohmic resistance r^* and the reactance x^* represent the power losses and magnetic energy per unit surface of the circuit loop. The ohmic resistances and reactances used in Kirchhoff's equations allow us, on the one hand, to find the voltage drops and, on other hand, to calculate the power losses and magnetic energy stored in the circuit loop. Therefore, the ohmic resistances and reactances obtained in the phase and specific system of units and used in the voltage equations must meet these two requirements.

3.6 Impedance of Winding: Transfer from One System of Units to Another

For the secluded circuit loop, we consider the voltage equation obtained in (3.7). Based on this equation, the impedance of the secluded circuit loop Z can be represented as

$$Z = \frac{\dot{U}}{\dot{I}} \tag{3.39}$$

where $Z = r + jx$; r and x are the ohmic resistance and reactance of the secluded circuit loop in the phase system of units.

For the definiteness, instead of the secluded circuit loop, we consider the stator winding. Then, the expression (3.7) can be used to determine, for example, the input impedance of the stator winding of the machine operating in the no-load mode. In relation to the voltage applied to the stator winding terminals, the equations shown in (3.36) obtain the form

$$\sqrt{2}\dot{U} = 2wk_w l\dot{E} \quad \sqrt{2}\dot{I} = \frac{p\tau}{mwk_w}\dot{H} \tag{3.40}$$

For the voltage $\dot{U}$ and the current $\dot{I}$ used in (3.39), we can apply the equations given in (3.40). Then, it arises

$$Z = \frac{\dot{U}}{\dot{I}} = \frac{2m(wk_w)^2 l}{p\tau} \frac{\dot{E}}{\dot{H}} = k\frac{\dot{E}}{\dot{H}} \tag{3.41}$$

where $k = \dfrac{2m(wk_w)^2 l}{p\tau}$.

Based on (3.41), we will determine the impedance Z^* which follows as the ratio of the electric field intensity $\dot{E}$ caused by the induced in the winding emf to the magnetic field intensity $\dot{H}$ created by the current flowing in the winding. The impedance obtained in this way is expressed in the specific system of units, and it acquires the form

$$Z^* = \frac{\dot{E}}{\dot{H}} \tag{3.42}$$

Now, by the expressions (3.41) and (3.42), we can have the following conditions

$$Z = kZ^* \quad Z^* = Z/k \tag{3.43}$$

Here, the dimensionless quantity k determined from the expression (3.41) can be considered as the reduction factor of the specific impedance Z^* to the value of the phase impedance Z or the reduction factor of the phase impedance Z to the value of the specific impedance Z^*.

3.7 Summary

1. The electromechanical interaction taking place in the asynchronous machine is characterized by energy conversion. In this case, in the machine, electromagnetic energy can be converted into mechanical energy (motor) or mechanical energy is converted into electromagnetic energy (generator). The electromagnetic and electromechanical processes taking place in the machine are accompanied by various physical phenomena. In this work, as the main phenomena are considered the creation of the magnetic field, the induction of the emf, the force effect of the magnetic field, and the energy transfer by the electromagnetic field. However, the knowledge of these physical phenomena is not sufficient to study the operating properties of the electric machine. To do this, it is necessary to have the equations expressed in terms of the quantities representing these physical phenomena. In the theory of electric machines, the voltage equations (Kirchhoff's equations) are used for this purpose.
2. It is shown that the voltage equations can be obtained based on the use of Ampere's law, Faraday's law, and Ohm's law. In this work, the voltage equations for the secluded circuit loop and for the system of the inductively coupled circuit loops are received. The equations drawn up for the system of the inductively

coupled circuit loops can be considered as the initial ones to obtain the voltage equations suitable to perform the practical calculations of the electric machine.

3. The voltage equations can be obtained for the system of the inductively coupled circuit loops formed from the geometrically linear conductors, i.e. such conductors the cross-section dimensions of which are small in the comparison with their total length. In such circuit loops, the direction of the current density vector is uniquely determined by the spatial arrangement of their conductors, i.e. in other words, in this case, the direction of the current density vector and the spatial orientation of the conductors coincide. This condition is fundamental when drawing up the voltage equations for the circuit loops representing the windings of the electric machine.
4. The voltage equations for the circuit loop contain the ohmic resistances and reactances. In the case of the linear circuit loop, the product of the ohmic resistance and the square of the current and the product of the reactance and the square of the current gives the power losses and magnetic field energy stored in the circuit loop, respectively. The product of the ohmic resistance and the current and the product of the reactance and the current determine the voltage drops in the circuit loop. These provisions are the base conditions for introducing the ohmic resistance and reactance of the circuit loop.
5. For the windings used in the electric machines, the conductors with a limited cross section are often used. For such conductors, the linearity condition is generally not met. In this regard, the determination of the voltage equations for the circuit loops representing the windings made from the conductors with the limited cross section is associated with serious difficulties.
6. The voltage equations obtained in this work are represented in the phase and specific system of units. The quantities used in the voltage equations represented in the phase system of units can be relatively easily determined experimentally which contributes to an increase in the level of reliability of the calculation results. In addition, the results obtained using the voltage equations compiled applying the phase quantities are quite easy to the physical interpretation which is very important for the engineering practice.
7. The voltage equations represented in the specific system of units contain the quantities used in Maxwell's equations. In this regard, the use of the specific system of units can be considered as a factor in stimulating the research aimed at the direct use of Maxwell's equations when drawing up the voltages equations of the machine windings. Thus, the use of the specific quantities (and, consequently, Maxwell's equations) can contribute to the significant expansion of the opportunities for the further improvement of the methods for calculating the electric machines.
8. To increase the efficiency of using the voltage equations represented in the phase and specific system of units, the expressions that make it possible to transfer the values from one system of units to another and back are proposed. The recommended expressions are represented through the corresponding machine design parameters which facilitate their practical use.

9. The ohmic resistances and reactances used in the voltage equations compiled in the phase system of units characterize the power losses and magnetic energy stored in the volume of the circuit-loop conductor. The ohmic resistances and reactances used in the voltage equations represented in the specific system of units reflect the electromagnetic energy released and accumulated on the surface of the circuit-loop conductor. Thus, the phase ohmic resistances and reactances reflect the power losses and magnetic energy stored in the volume of the circuit-loop conductor. The specific ohmic resistances and reactances characterize the power losses and the magnetic energy released on the unit surface of the circuit-loop conductor. Consequently, the use of the voltage equations compiled in the phase and specific system of units allows us to analyze the electromagnetic processes in the electric machine from various point of view which contributes to the increase in visibility and facilitates the physical interpretation of the research results.

References

1. Voldek, A.: Electric Machines, Publishing Energiya, Leningrad (1974)
2. Alger, P.: Induction Machines, 2nd ed., Gordon and Beach, New York (1970)
3. Say, M.: Alternating Current Machines, John Wiley and Sons, New York (1902)
4. Fitzgerald, A., Kingsley, C., Umans, S.: Electric Machinery, McGraw-Hill, 5th ed., New York (1990)
5. Krause, P.: Analysis of Electric Machinery, McGraw-Hill, New York (1986)
6. Kostenko, M., Piotrovskij, L.: Alternating Current Electric Machines, Publishing Energiya, Leningrad (1973)
7. Ivanov-Smolensky, A.: Electric Machines, Publishing Energiya, Moscow (1980)
8. Vazhnov, A.: Electric Machines, Publishing Energiya, Leningrad (1968)
9. Chapman, S.: Electric Machinery Fundamentals, 4th ed., McGraw-Hill, Inc., New York (2005)
10. Asanbayev, V.: Alternating Current Multi-Circuit Electric Machines: A New Approach to Steady-State Parameter Determination, Springer Sham Heidelberg, New York, Dordrecht London (2015)
11. Shimoni, K.: Theoretical Electrical Engineering, (Translation from German), Publishing Mir, Moscow (1964)
12. Seely, S.: Electromechanical Energy Conversion, McGraw-Hill Book Company, Inc., New York (1962)

Chapter 4
Rotating Magnetic Field and Electromagnetic Torque: Practical Expressions for Calculating

4.1 Asynchronous Machine Design: Basic Elements

At present, asynchronous machines have a wide range of applications. The asynchronous machine consists of two parts; the stationary part is called the stator and the moving part represents the rotor. The stator and rotor are assembled from the shits of the electro-technical steel. The rotor shaft rotates in the bearings placed in the end shields attached to the stator frame or the free-standing bearing struts [1–4].

On the inner cylindrical surface of the stator (stator bore surface) and the outer surface of the rotor, the axial slots are milled. In these axial slots, the conductors of the stator and rotor windings are placed [1–10]. In the slots of the stator core, the three-phase winding supplied by the three-phase current is located. The rotor winding can be designed as the three-phase winding the three ends of which are connected, in this case, with the contact rings located on the rotor shaft. As a result, the asynchronous machine with the phase rotor (wound rotor) is obtained. The stator and rotor phase windings are performed with the same number of poles.

The squirrel-cage winding can be installed on the rotor of the electric machine. With this design of the rotor winding, the cooper or aluminum bars are placed in its slots. From both ends of the rotor, the ends of all the bars are short-circuited using the cooper or aluminum rings. The asynchronous machine with this design is called the squirrel-cage asynchronous machine. In asynchronous machines, the air gap is chosen as small as possible based on the condition of the reliable operation.

4.2 Magnetic Flux Density as Rotating Vector

On the stator of the asynchronous machine, the three-phase symmetrical winding is usually applied. To indicate the phases of the stator winding, the letters A, B, C are mainly used. The beginning and end of phase A are designated by the letters A and X.

V. Asanbayev, *Asynchronous Machines*,
https://doi.org/10.1007/978-3-030-92284-9_4

The beginning and end of phase B are represented by the letters B and Y. The beginning and end of phase C are shown by the letters C and Z. The three-phase winding is supplied with the three-phase current system. In this case, each of the phase currents creates the pulsating magnetic field directed along the axis of each winding. We will consider here only the first harmonics of the expansion of the phase magnetic fields of the three-phase winding in the Fourier series. Therefore, the spatial distribution of the phase magnetic flux densities of the three-phase winding can be represented as [3–16].

$$\begin{aligned} B_{A(\alpha,t)} &= G_B i_A \cos\alpha \\ B_{B(\alpha,t)} &= G_B i_B \cos\left(\alpha - \frac{2}{3}\pi\right) \\ B_{C(\alpha,t)} &= G_B i_C \cos\left(\alpha - \frac{4}{3}\pi\right) \end{aligned} \tag{4.1}$$

where $G_B = \frac{\mu_0}{\delta'} \frac{2}{\pi} \frac{w_1 k_{w1}}{p}$; $\alpha = \frac{\pi}{\tau} x$; $\delta' = \delta k_\delta$ is the calculated radial length of the air gap; k_δ is the air gap factor; w_1 is the number of the phase turns; and k_{w1} is winding factor of the stator winding; i_A, i_B, and i_C are the instantaneous values of the phase currents of the stator winding.

The currents flowing in the phases of the three-phase winding represent the symmetrical sinusoidal currents shifted in the phase by one-third of the time period. From here, it follows for the currents of the three-phase winding

$$\begin{aligned} i_A &= \sqrt{2} I_1 \sin\omega_1 t \\ i_B &= \sqrt{2} I_1 \sin\left(\omega_1 t - \frac{2}{3}\pi\right) \\ i_C &= \sqrt{2} I_1 \sin\left(\omega_1 t - \frac{4}{3}\pi\right) \end{aligned} \tag{4.2}$$

The magnetic flux densities used in (4.1) are determined by the currents flowing in the phases of the winding. Therefore, together with the Eqs. (4.1) and (4.2), we can obtain for the phase magnetic flux densities

$$\begin{aligned} B_{A(\alpha,t)} &= B_m \cos\alpha \sin\omega_1 t \\ B_{B(\alpha,t)} &= B_m \cos\left(\alpha - \frac{2}{3}\pi\right) \sin\left(\omega_1 t - \frac{2}{3}\pi\right) \\ B_{C(\alpha,t)} &= B_m \cos\left(\alpha - \frac{4}{3}\pi\right) \sin\left(\omega_1 t - \frac{4}{3}\pi\right) \end{aligned} \tag{4.3}$$

where $B_m = \frac{2\sqrt{2}}{\pi} \frac{\mu_0}{\delta'} \frac{w_1 k_{w1}}{p} I_1$.

In fact, the single common magnetic field of the three-phase winding takes place in the machine. To obtain this field, it is necessary to add the phase magnetic flux densities $B_{A(\alpha,t)}$, $B_{B(\alpha,t)}$, and $B_{C(\alpha,t)}$. Before adding these magnetic flux densities, one usually performs the following system of the operations, for example, [3, 4, 6, 8–

16]. The pulsating wave of the magnetic flux density of each phase is decomposed into the two components rotating in the opposite directions and the amplitudes of which are equal to half of the pulsating wave. The all right-handed waves of the magnetic flux densities are in the phase. As a result, these waves form the single common right-handed wave of the magnetic flux density. The left-handed waves of the magnetic flux densities are shifted in the space by the angle equal to $\alpha = 2\pi/3$. Their total magnetic flux density is equal to zero. Now, for the magnetic flux density of the three-phase winding, we have that $B_{1(\alpha,t)} = B_{A(\alpha,t)} + B_{B(\alpha,t)} + B_{C(\alpha,t)}$. Taking into account this condition and the expression shown in (4.3), the formula for the $B_{1(\alpha,t)}$ acquires the form

$$\begin{aligned} B_{1(\alpha,t)} &= B_m\left[\cos\alpha\sin\omega_1 t + \cos\left(\alpha - \frac{2}{3}\pi\right)\sin\left(\omega_1 t - \frac{2}{3}\pi\right) + \cos\left(\alpha - \frac{4}{3}\pi\right) \times \sin\left(\omega_1 t - \frac{4}{3}\pi\right)\right] \\ &= \frac{B_m}{2}\left[\sin(\omega_1 t - \alpha) + \sin(\omega_1 t - \alpha) + \sin(\omega_1 t - \alpha) + \sin(\omega_1 t + \alpha) + \sin\left(\omega_1 t + \alpha - \frac{4}{3}\pi\right) + \sin\left(\omega_1 t + \alpha - \frac{2}{3}\pi\right)\right] \\ &= \frac{3}{2}B_m\sin(\omega_1 t - \alpha) \end{aligned} \tag{4.4}$$

since $\sin(\omega_1 t + \alpha) + \sin\left(\omega_1 t + \alpha - \frac{4}{3}\pi\right) + \sin\left(\omega_1 t + \alpha - \frac{2}{3}\pi\right) = 0$

In accordance with expression (4.4), it can be affirmed that at any moment of a time t the point with the coordinate α at which the magnetic flux density $B_{1(\alpha,t)}$ has the amplitude value found. This means that for any t such α is found under which the condition is met

$$\omega_1 t - \alpha = \frac{\pi}{2} \tag{4.5}$$

After differentiating this condition in a time, we can receive

$$\frac{d}{dt}(\omega_1 t - \alpha) = 0$$

From here

$$\frac{d\alpha}{dt} = \omega_1 \tag{4.6}$$

This condition can be interpreted so that the angular velocity of the mechanical displacement of the magnetic flux density wave $\frac{d\alpha}{dt}$ expressed in the electric radians per second is equal to the angular frequency of the current chance in the winding ω_1. Therefore, during the one period of the current chance in the winding, the magnetic flux density wave moves the distance equal to the double pole pitch 2τ.

Considering that $\alpha = \pi x/\tau$, we define the linear speed of the rotating magnetic flux density vector υ_1. For this purpose, the condition shown in (4.6) can be used. Then, we have

$$\frac{dx}{dt} = \frac{d\alpha}{dt}\frac{\tau}{\pi} = \frac{\omega_1 \tau}{\pi}$$

From here

$$v_1 = \frac{dx}{dt} = \frac{\omega_1 \tau}{\pi} = 2f_1\tau \tag{4.7}$$

For example, in [8], the demonstration of the rotating vector of the magnetic flux density created by the three-phase stator winding of the asynchronous machine is given.

From (4.3) and (4.4), we can obtain

$$\begin{aligned} B_{1(\alpha,t)} &= \frac{\mu_0}{\delta'}\frac{3w_1k_{w1}}{\pi p}\sqrt{2}I_1 \sin(\omega_1 t - \alpha) \\ &= \lambda_\delta \frac{m_1 w_1 k_{w1}}{\pi p}\sqrt{2}I_1 \sin(\omega_1 t - \alpha) = \lambda_\delta F_{1(\alpha,t)} \end{aligned} \tag{4.8}$$

where $\lambda_\delta = \mu_0/\delta'$; $m_1 = 3$.

Here, the value reflects the number of the stator winding phases of the electric machine.

In (4.8), the value $F_{1(\alpha,t)}$ represents the mmf of the stator winding or else the distribution of the total current of the stator winding. From (4.8), for the mmf $F_{1(\alpha,t)}$, we have

$$F_{1(\alpha,t)} = \frac{m_1 w_1 k_{w1}}{\pi p}\sqrt{2}I_1 \sin(\omega_1 t - \alpha) = F_1 \sin(\omega_1 t - \alpha) \tag{4.9}$$

Here, the mmf F_1 determined from (3.23) and also shown in (4.9) receives the form

$$F_1 = \frac{m_1 w_1 k_{w1}}{\pi p}\sqrt{2}I_1.$$

From (4.8) and (4.9), we can obtain

$$B_1 = \lambda_\delta F_1 \tag{4.10}$$

In accordance with (4.10), we can assert that the magnetic field in the electric machine is created by the mmf of the winding or else by the total current flowing in it. The mmf F_1, as it follows from (4.9), represents the sinusoidally distributed in a space wave of the total current of the stator winding. The wave of the mmf F_1 is rotated with the angular speed equal to ω_1. Therefore, the vector of the magnetic flux density B_1 created by the mmf F_1 of the winding is rotated with the angular speed equal to ω_1.

When used on the rotor of the three-phase winding, the magnetic flux density created by the rotor winding current can be determined by an analogy with (4.10) in the following form

$$B_2 = \lambda_\delta F_2 \tag{4.11a}$$

where $F_2 = \frac{m_2 w_2 k_{w2}}{\pi p} \sqrt{2} I_2$.

The squirrel-cage rotor creates the magnetic flux density that can be determined with the use of the expression (4.11a). In this case, in the expression used in (4.11a) for the mmf F_2, it is necessary to take into account that for the considered rotor winding, we have that $m_2 = Z_2$; $w_2 = 1/2$; $k_{w2} = 1.0$. Then, the expression of the following form is obtained

$$F_2 = \frac{Z_2}{2\pi p} \sqrt{2} I_2 \tag{4.11b}$$

4.3 Asynchronous Machine: Principle of Operating

The rotating magnetic flux created by the stator winding currents crosses the conductors of the rotor winding as a result of which the emf is induced in them. Under the influence of this emf, the currents arise in the rotor winding (if the rotor winding is closed). If the three-phase winding is used on the rotor, then, the three-phase current system flows in it. This current creates the rotating magnetic flux of the rotor winding. The rotor magnetic flux is rotated in the direction of rotation of the stator magnetic flux.

When using the squirrel-cage rotor, a multiphase system of the currents is induced in its slot bars. The phase angle of the shift of the currents flowing in the slot bars is $\gamma = 2p\tau/Z_2$, where Z_2 is the number of the rotor slot bars. The slot bar currents create the rotating magnetic flux of the rotor. The squirrel-cage rotor magnetic flux is rotated in the direction of rotation of the stator flux.

When starting the machine (the rotor is fixed when $n = 0$), the rotation speed of the rotor magnetic flux is $n_2 = n_1$, where n_1 is the rotation speed of the stator magnetic flux. The stator and rotor magnetic fluxes are rotated with the same rotation speed. With the fixed rotor, the frequency of the rotor currents is equal to the frequency of the stator currents, that is, $f_2 = f_1$, where f_1 is the frequency of the stator currents. In these conditions, the single common magnetic flux is formed in the electric machine. This magnetic flux can be considered as the magnetic flux existing mainly in the air gap of the machine. Now, this magnetic flux interacts with the currents flowing in the rotor winding as a result of which the electromagnetic forces effecting on the rotor arise. These forces form the electromagnetic torque of the electric machine.

If the static braking torque on the machine shaft is less than the electromagnetic rotating torque, then, the rotor of the machine will start rotating in the direction of the rotation of the stator magnetic flux. In this case, relative to the stator flux, the rotor will rotate with some slip. The value of the rotor slip is determined from the following condition

$$s = \frac{n_1 - n}{n_1} \tag{4.12}$$

From here, for the rotation speed of the rotor, we receive

$$n = (1 - s)n_1 \tag{4.13}$$

When starting the machine ($n = 0$), for the slip, we have that $s = 1.0$. When the rotor rotates synchronously with the stator magnetic flux ($n = n_1$), for the slip, it follows that $s = 0$. In this case ($n = n_1$), the current has not induced the emf in the rotor winding ($M = 0$) since the stator magnetic flux relative to the rotor is fixed and no crossing the conductors of the rotor winding. When the rotor rotates, the frequency of the current induced in the rotor winding is

$$f_2 = p(n_1 - n)/60 \tag{4.14a}$$

This expression in terms of (4.12) can be represented as

$$f_2 = \frac{pn_1}{60}\frac{(n_1 - n)}{n_1} = f_1 s \tag{4.14b}$$

i.e. the frequency of the current flowing in the rotor winding is proportional to the slip.

From the expression (4.12), for the rotation speed of the rotor magnetic flux relative to the rotor itself, it obtains

$$n_2 = n_1 s \tag{4.15}$$

The rotation speed of the rotor magnetic flux (n + n_2) determined by (4.13) and (4.15) relative to the stator can be found as

$$(n + n_2) = (1 - s)n_1 + n_1 s = n_1 \tag{4.16}$$

The rotation speed of the rotor magnetic flux relative to the stator is equal to the rotation speed of the stator magnetic flux, that is, the stator and the rotor magnetic fluxes are rotated synchronously (at any rotation speed of the rotor), and therefore, they form in the electric machine, as noted above, the common rotating magnetic flux.

The asynchronous machine can operate in three modes: in the motor mode, in the generator mode, and in the electromagnetic brake mode. In the motor mode, the rotor is rotated in the direction of the stator magnetic flux. In this case, the rotation speed of the rotor does not exceed the rotation speed of the stator magnetic flux, i.e., $0 < n < n_1$ and $1 > s > 0$. In the generator mode, the rotation speed of the rotor exceeds (due to the external torque) the rotation speed of the stator magnetic flux, that is ($n > n_1$). In this case, the direction of the electromagnetic torque is changed (compared with the motor mode). The electromagnetic torque becomes braking, and the machine operates in the generator mode giving active power to the supply network. In this mode, $s < 0$. If the rotor is rotated against the direction of the rotation of the stator magnetic flux ($n < 0$), then, the electromagnetic torque will act in the direction of the rotation of the stator magnetic flux, but it will slow down the rotation of the rotor. This mode of the machine is called the electromagnetic brake mode. In this mode, $s > 1.0$.

4.4 Energy Conversion

In this part of the work, we will consider the energy conversion corresponding to the operating mode of the electric machine [6–16]. For this purpose, we will analyze the basis of the processes taking place in the machine connected to the power supply network with the voltage equal to U_1. Assume that the machine is operating in the motor mode, and the current in the stator winding is I_1. Then, an active power consumed from the supply network is determined by the following formula

$$P_1 = m_1 U_1 I_1 \cos\varphi_1 \tag{4.17}$$

where φ_1 is the angle between the voltage and current of the stator winding.

Some part of the power P_1 is spent on covering the stator losses p_1 including the losses in the cooper of the stator winding and the magnetic losses in the stator iron. As a result, to the rotor, the rotating magnetic flux transfers the electromagnetic power $P_{эм}$ for which we have

$$P_{эм} = P_1 - p_1 \tag{4.18}$$

In the rotor, the part of the electromagnetic power $P_{эм}$ goes to cover the losses in the cooper of the rotor winding p_2 (including the losses in the brush contacts if the rotor has the phase winding). The residual part of the electromagnetic power $P_{эм}$ is converted into the mechanical power $P_{мх}$ of the electric machine, and it represents the total mechanical power for which we have

$$P_{мх} = P_{эм} - p_2 \tag{4.19}$$

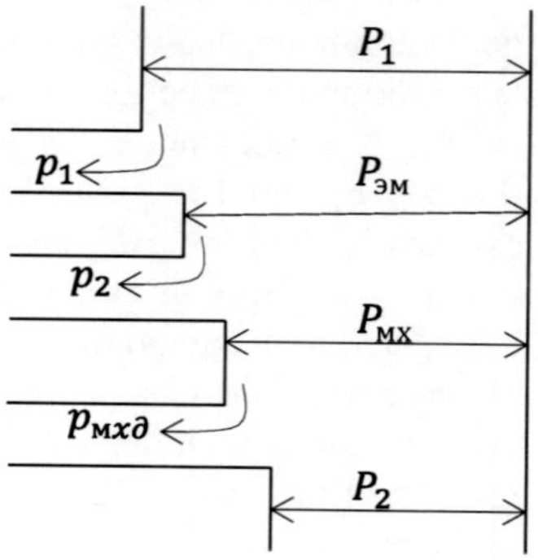

Fig. 4.1 Active power conversion diagram of the asynchronous machine

The part of the mechanical power P_{mx} goes to covering the mechanical losses in the bearings and for the ventilation as well as the additional losses. If the total mechanical losses and the additional losses to denote by $p_{мхд}$, then, for the useful mechanical power on the electric machine shaft, we obtain

$$P_2 = P_{мх} - p_{мхд} \tag{4.20}$$

The simplified energy diagram of the asynchronous machine is shown in Fig. 4.1.

Now, we will consider the diagram of the conversion of the reactive power in the asynchronous machine. The reactive power-consuming from the supply network is

$$Q_1 = m_1 U_1 I_1 \sin \varphi_1 \tag{4.21}$$

The reactive power Q_1 determined by (4.21) is spent on the creation of the leakage fields of the stator and rotor windings and also on the creation of the magnetizing field of the electric machine. Thus, we can have

$$Q_1 = Q_m + q_1 + q_2 \tag{4.22}$$

Here, the reactive power Q_m is spent on the formation of the machine magnetizing field. The reactive powers q_1 and q_2 are spent on the creation of the leakage fields of the stator and rotor windings, respectively. The reactive power diagram of the asynchronous machine is shown in Fig. 4.2. The reactive powers used in Fig. 4.2 are determined by the expressions of the form

$$Q_m = m_1 I_m^2 x_m \quad q_1 = m_1 I_1^2 x_{1\sigma} \quad q_2 = m_2 I_2^2 x_{2\sigma} \tag{4.23}$$

where x_m is the magnetizing reactance of the electric machine; $x_{1\sigma}$ and $x_{2\sigma}$ are the leakage reactances of the stator and rotor windings.

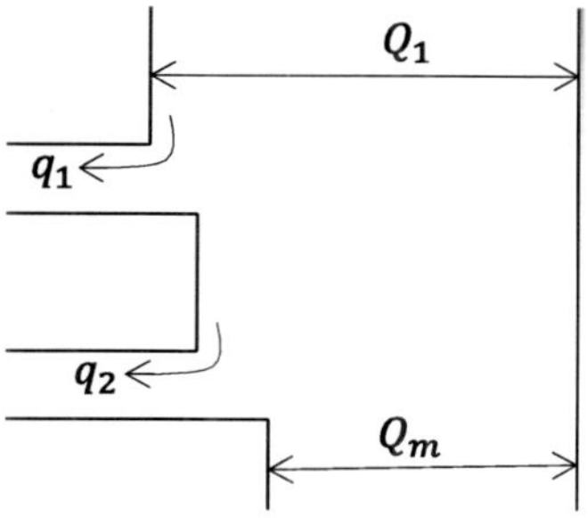

Fig. 4.2 Reactive power conversion diagram of the asynchronous machine

4.5 Electromagnetic Rotating Torque

For the derivation of the formula allowing us to calculate the electromagnetic rotating torque of the electric machine, the electromagnetic and mechanical powers $P_{эм}$ and $P_{мх}$ determined by the expressions shown in (4.18) and (4.19) can be used.

4.5.1 General Analytical Expression

Using the expression of kind $M_{эм} = P_{эм}/\omega_1$ shown in (2.23), we will represent the electromagnetic and mechanical powers ($P_{эм}$ and $P_{мх}$) through the value of the machine rotating electromagnetic torque $M_{эм}$, for example, [6–16]. Then, we can have

$$P_{эм} = \omega_1 M_{эм} \quad P_{мх} = \omega_2 M_{эм} \tag{4.24}$$

Based on the expressions (4.19) and (4.24) and also taking into account the condition of the form $\omega_2 = (1 - s)\omega_1$, the following formula can be obtained

$$\begin{aligned} P_{эм} &= \omega_1 M_{эм} = P_{мх} + p_2 \\ &= \omega_2 M_{эм} + p_2 = (1 - s)\omega_1 M_{эм} + p_2 \end{aligned}$$

From here, it arises that

$$p_2 = s\omega_1 \mathrm{M}_{эм} \tag{4.25}$$

Now, the electromagnetic rotating torque of the electric machine is determined as

$$M_{эм} = \frac{p_2}{s\omega_1} \tag{4.26}$$

4.5.2 *Practical Expressions for Calculating*

The expression (4.26) can be used to analyze the basic electromechanical properties of the asynchronous machine. Below, we will make this expression more convenient for practical use.

4.5.2.1 Phase System of Units

In (4.26), the cooper losses of the rotor winding p_2 are determined as $p_2 = m_1 \left(I_2'\right)^2 r_2'$, for example, [6–14]. In the rotor winding circuit, the additional resistance can take place, if the phase winding is installed on the rotor. Then, in the expression $p_2 = m_1 \left(I_2'\right)^2 r_2'$, the ohmic resistance r_2' is determined in terms of this additional resistance. Now, the expression (4.26) can be represented as

$$M_{эм} = \frac{p_2}{s\omega_1} = \frac{pm_1 \left(I_2'\right)^2}{\omega_1} \frac{r_2'}{s} \tag{4.27}$$

The value of the electromagnetic torque $M_{эм}$ is often expressed through the phase voltage $\dot{U}_1$ applied to the stator winding, the ohmic resistances and the reactances of the machine windings, and also the rotor slip s. For this purpose, for the current I_2' used in (4.27), the expression obtained in (10.18) can be applied. Then, we can receive

$$\dot{I}_2' = \frac{\dot{U}_1}{\sqrt{\left(r_1 + c_1 \frac{r_2'}{s}\right)^2 + \left(x_{\sigma 1} + c_1 x_{\sigma 2}'\right)^2}} \tag{4.28}$$

Now, based on (4.27) and (4.28), it follows

$$M_{эм} = \frac{pm_1 \dot{U}_1^2 \left(r_2'/s\right)}{\omega_1 \left[\left(r_1 + c_1 \frac{r_2'}{s}\right)^2 + \left(x_{\sigma 1} + c_1 x_{\sigma 2}'\right)^2\right]} \tag{4.29}$$

This expression represents the rotating electromagnetic torque of the machine in a phase system of units, and it is often used to calculate the torque curve $M_{эм} = f(s)$ obtained at $\dot{U}_1 = const.$ As follows from (4.29), the machine rotating torque depends on the parameters $r_1, x_{\sigma 1}, r_2'/s$ and $x_{\sigma 2}'$ and also on the factor c_1 arising from (10.17). Below, we will determine these parameters directly from the field equations.

4.5.2.2 Specific System of Units

As noted earlier, the expression obtained in (4.29) for the machine rotating torque is represented in the phase system of units. Here, for the machine rotating torque, we will receive the expression in the specific system of units. For this purpose, the expression obtained in (4.29) can be used. In this expression, we will introduce the specific rotation torque determined as $M^*_{эм} = M_{эм}/S_R$, where S_R is the area of the active zone surface of the rotor. For S_R we have that $S_R = 2p\tau l$. Here, l is the axial length of the rotor active zone. In (4.29), the effective voltage $\dot{U}_1$ can be expressed through the effective electric field intensity $\dot{E}_1$, determined as $\dot{E}_1 = \dot{E}/\sqrt{2}$. Then, based on the condition given in (3.36) for the voltage $\dot{U}_1$, we can have that $\dot{U}_1 = 2w_1 k_{w1} l\dot{E}_1$. In (4.29), the ohmic resistances and reactances expressed in the phase system of units are used. These resistances can be represented in the specific system of units for which we will use condition obtained in (3.43) and shown in the following form $Z = kZ^*$ where k is the factor determined for the stator winding as $k = \left(2m_1 w_1^2 k_{w1}^2 l\right)/p\tau$. Here, Z^* is the impedance represented in the specific system of units. Now, in (4.29), we can use the conditions of the form

$$M^*_{эм} = M_{эм}/S_R \quad \dot{U}_1 = 2w_1 k_{w1} l\dot{E}_1 \quad Z = kZ^* \tag{4.30}$$

where $S_R = 2p\tau l; k = \left(2m_1 w_1^2 k_{w1}^2 l\right)/p\tau$.

Then, the expression (4.29) acquires the form

$$M^*_{эм} = \frac{p\dot{E}_1^2 \left(r_2^*/s\right)}{\omega_1 \left[\left(r_1^* + c_1 \frac{r_2^*}{s}\right)^2 + \left(x_{\sigma 1}^* + c_1 x_{\sigma 2}^*\right)^2\right]} \tag{4.31}$$

The ohmic resistances and reactances used in (4.31) are determined in the twelfth chapter of this work. The coefficient c_1 used in (4.31) can be defined by the expression given in (12.75). These ohmic resistances and reactances are represented in the specific system of units, and therefore, the expression (4.31) allows us to calculate the value of the electric machine rotating torque corresponding to the unit surface of the rotor.

4.6 Summary

1. The electromagnetic and electromechanical processes taking place in the asynchronous machine are caused by the rotating magnetic field created by the currents flowing in the stator and rotor windings. When connecting the terminals of the stator winding to the power supply network, the voltage (rotor winding is open), and the current appears in this winding. This current creates the rotating magnetic field of the stator. The stator rotating magnetic field crosses the

conductors of the rotor winding as a result of which the emf is induced in them. Under the action of this emf, the currents appear in the closed winding of the rotor. These rotor currents create the rotating magnetic field that rotates in the direction of the rotation of the stator magnetic field. The stator and rotor magnetic fields are rotated with the same angle speed. In these conditions, the rotating resulting magnetic field is created in the electric machine. This rotating magnetic field interacts with the current-carrying conductors of the rotor winding as a result of which the electromagnetic forces (as consequently, the electromagnetic rotating torque of the machine) acting on the rotor appears. Under the action of these electromagnetic forces, the rotor comes into rotation in the direction of the rotation of the stator magnetic field. As a feature of the electric machine operation can be considered the fact that the rotor is rotated relative to the stator magnetic field with some slip.

2. In this work, the expressions represented in the phase and specific system of units are obtained to calculate the machine rotating torque. The recommended expressions are written in the form convenient for analysis and practical use. The expressions shown in the specific system of units allow us to calculate rotating torque corresponding to the per unit surface of the active zone of the machine rotor.

References

1. Alekseev, A.: Design of Electric Machines, Publishing Energiya, Leningrad (1960)
2. Sergeev, P., Vinogradov, N., Goryainov, F.: Design of Electric Machines, Publishing Energiya, Moscow (1969)
3. Kopylov, I., Goryainov, F., Klokov, B., Moroskin, V., Tokarev, V.: Design of Electric Machines, Publishing Energiya, Moscow (1980)
4. Dombrovsky, V., Zaichik, V.: Asynchronous Machines: Theory, Calculation, Elements of Design, Publishing Energoatomizdat, Leningrad (1990)
5. Gervais, G.: Windings of Electric Machines, Publishing Energoatomizdat, Leningrad (1989)
6. Kostenko, M., Piotrovskij, L.: Alternating Current Electric Machines, Publishing Energiya, Leningrad (1973)
7. Shuisky, V.: Calculation of Electric Machines (Translated from German), Publishing Energiya, Leningrad (1968)
8. Voldek, A.: Electric Machines, Publishing Energiya, Leningrad (1974)
9. Ivanov-Smolensky, A.: Electric Machines, Publishing Energiya, Moscow (1980)
10. Vazhnov, A.: Electric Machines, Publishing Energiya, Leningrad (1968)
11. Alger, P.: Induction Machines, 2nd ed., Gordon and Beach, New York (1970)
12. Fitzgerald, A., Kingsley, C., Umans, S.: Electric Machinery, McGraw-Hill, 5th ed., New York (1990)
13. Krause, P.: Analysis of Electric Machinery, McGraw-Hill, New York (1986)
14. McPherson, G.: An Introduction to Electrical Machines and Transformers, Wiley, New York (1981)
15. Boldea, I.: Electric Generators Handbook, Vols. 1 and 2, FL: CRC Press. Boca Raton (2006)
16. Dell Toro, V.: Electric Machines and Power Systems, Prentice-Hall, Inc., Englewood Cliffs, New Jersey (1985)

Chapter 5
Voltage Equations Through Space Vectors

5.1 Space Vectors

In the theory of the electric machines, co-called, the space vectors are widely used, for example, [1–15]. Below, we will outline, in general terms, the basics of introducing the space vectors. For this purpose, we proceed from consideration of the symmetrical three-phase asynchronous machine, the stator winding of which is supplied with the symmetrical sinusoidal voltage. The frequency of the winding voltage is constant and equal to the frequency of the power supply network ω_1. The currents flowing in the phases of the winding create the pulsating mmf of the individual phases. The phase mmf, in turn, causes the magnetic fields (magnetic fluxes) in the air gap of the electric machine, the flux densities of which pulsate in the phase with mmf of the individual phases. In what follows, our reasoning will be based on the assumption that the mmf and the magnetic flux densities created in the air gap of the machine by the currents of the individual phases have the sinusoidal distribution in space. In this case, the currents of the individual phases create the sinusoidal distribution in space and the resulting mmf and the resulting magnetic flux densities the amplitudes of which are 3/2 of the amplitudes of the pulsating mmf and magnetic flux densities of the individual phases. The resulting mmf and the resulting magnetic flux density are rotated at the speed corresponding to the frequency of the power supply network ω_1 forming the traveling waves of the mmf and the magnetic flux density. The currents flowing in the rotor phase windings create the analogous rotating in the space the sinusoidal waves of the resulting mmf and the resulting magnetic flux density. These rotating waves of the mmf and flux density can be represented as the rotating space vectors the modules of which are equal to the absolute values of the amplitudes of the phase mmf and the phase flux density. We will note that the direction of the rotation of the space vectors coincides with the direction of the rotation of the space waves of the mmf and the flux density. In practice, it is assumed that mmf and flux density are proportional to the corresponding currents. This gives grounds to assume that there are rotating spatial

V. Asanbayev, *Asynchronous Machines*,
https://doi.org/10.1007/978-3-030-92284-9_5

waves in the machine not only of the mmf and magnetic flux density but also of the voltages, currents, flux linkages, and so on. It is convenient to represent the values sinusoidally distributed in space and a time in the form of the vectors reflecting, in our case, the voltages, currents, and flux linkages of the machine windings. Such vectors have received the conditional name space vectors.

5.1.1 Space Vector of Stator Winding Current

Based on (4.1), (4.4), and (4.10), the mmf of the stator winding can be determined as

$$\begin{aligned}\dot{F}_{1(\alpha,t)} &= \frac{1}{\lambda_\delta}\dot{B}_{1(\alpha,t)} = \frac{1}{\lambda_\delta}\left[\dot{B}_{A(\alpha,t)} + \dot{B}_{B(\alpha,t)} + \dot{B}_{C(\alpha,t)}\right] \\ &= \frac{2}{\pi}\frac{w_1 k_{w1}}{p}\left[i_A \cos\alpha + i_B \cos\left(\alpha - \frac{2}{3}\pi\right) + i_C \cos\left(\alpha - \frac{4}{3}\pi\right)\right]\end{aligned} \tag{5.1}$$

where i_A, i_B, i_c are the phase currents of the stator winding; 0, $(2/3)\pi$, $(4/3)\pi$ are the angles determining the positions of the axes of the phase windings; α represents the initial angle at $t = 0$.

The expression (5.1) can be represented in the complex form. For this, we use Fig. 5.1. In Fig. 5.1, the magnetic axes of the three-phase winding of the stator are given. These axes are spaced apart from one another by the angles $(2/3)\pi$ and $(4/3)\pi$, respectively. In Fig. 5.1, the magnetic axis of the phase A coincides with the real axis (+1) of the complex rectangular coordinate system. The angles α are determined from the magnetic axis of the phase A and, in our case, based on Fig. 5.1, we have that $\alpha = 0$. The imaginary axis (+j) of the rectangular coordinate system is rotated in the counterclockwise direction by angle $\pi/2$. Now, by Fig. 5.1 and the expression (5.1), the mmf of the stator winding can be obtained in the complex form as

$$F_{1(t)} = \frac{2}{\pi}\frac{w_1 k_{w1}}{p}\left(i_A + a i_B + a^2 i_C\right) \tag{5.2}$$

where a and a^2 are the turn operators

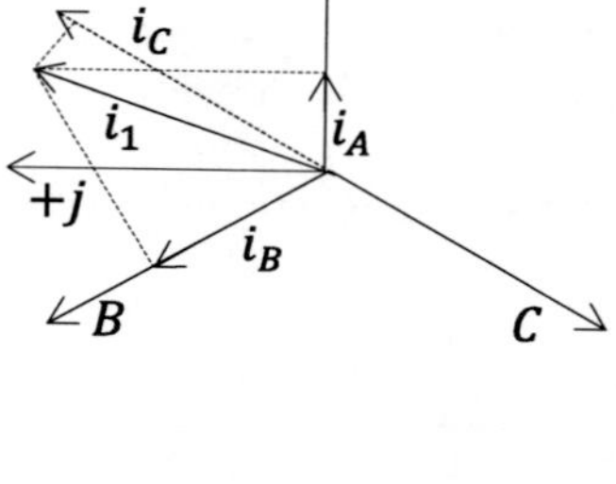

Fig. 5.1 Phase currents of the stator winding as the projections of the space current vector

$$a = e^{j120^\circ} = -\frac{1}{2} + j\frac{\sqrt{3}}{2} \quad a^2 = e^{j240^\circ} = -\frac{1}{2} - j\frac{\sqrt{3}}{2} \tag{5.3}$$

By the expression (4.9) the formula (5.2) can be obtained as

$$F_{1(t)} = \frac{3}{\pi}\frac{w_1 k_{w1}}{p}\frac{2}{3}\left(i_A + a i_B + a^2 i_C\right) = \frac{3}{\pi}\frac{w_1 k_{w1}}{p} i_1 \tag{5.4}$$

where $i_1 = \frac{2}{3}(i_A + a i_B + a^2 i_C)$.

Here, i_1 represents the space vector of the current of the stator winding. In (5.4), the current vector i_1 can be considered as the vector proportional to the mmf of the stator winding $F_{1(t)}$ if the dimensionless quantity $\frac{3}{\pi}\frac{w_1 k_{w1}}{p}$ is taken as the proportionality coefficient. As follows from (5.4), the space vector i_1 is equal to the geometric sum of the instantaneous values of the phase currents of the stator winding. With this in mind, the instantaneous values of the phase currents of the stator winding can be determined by the space vector of the current i_1 calculating by (5.4). For this, we will consider the projections of the current vector i_1 on the axes of the corresponding phases (Fig. 5.1). We will assume that the neutral wire is absent. Therefore the sum of the phase currents is equal to zero at any time, that is, $i_A + i_B + i_C = 0$. Provided that $i_A + i_B + i_C = 0$, the projections of the current vector i_1 on the phase axes represent the instantaneous values of the phase currents of the stator winding, i.e.

$$i_A = Re\left(i_1\right) \quad i_B = Re\left(a^2 i_1\right) \quad i_C = Re\left(a i_1\right) \tag{5.5}$$

Based on conditions (5.5) and considering the expression (5.4) and also keeping in mind that $i_A + i_B + i_C = 0$, we can obtain

$$\begin{aligned} Re\left(i_1\right) &= \frac{2}{3}\, Re\left(i_A + a i_B + a^2 i_C\right) = \frac{2}{3}\left(i_A - \frac{i_B + i_C}{2}\right) = i_A \\ Re\left(a^2 i_1\right) &= \frac{2}{3}\, Re\left(a^2 i_A + a^3 i_B + a^4 i_C\right) = \frac{2}{3}\left(i_B - \frac{i_A + i_C}{2}\right) = i_B \\ Re\left(a i_1\right) &= \frac{2}{3}\, Re\left(a i_A + a^2 i_B + a^3 i_C\right) = \frac{2}{3}\left(i_C - \frac{i_B + i_A}{2}\right) = i_C \end{aligned} \tag{5.6}$$

The operators a^3 and a^4, used in (5.6), are determined with considering the conditions of the form $a^3 = 1.0$ and $a^4 = a$. From the expressions (5.6) it follows that in (5.4) the coefficient 3/2 is used in order of the magnitude of the current space vector coincides with the amplitude of the instantaneous value of the phase current.

5.1.2 Space Vector of Rotor Winding Current

If the rotor has the three-phase symmetrical winding, then, for the space vector of the rotor winding current, we can use the expressions given earlier. In this case, the real axis of the coordinate system is oriented along the phase axis a of the rotor. Now, based on the expression (5.4), the space vector of the rotor winding current i_2 acquires the form

$$i_2 = \frac{2}{3}\left(i_a + ai_b + a^2 i_c\right) \tag{5.7}$$

Acting by analogy with (5.5) and (5.6), that is, projecting the space vector i_2 on the phase axes, the instantaneous values of the phase currents of the rotor winding (i_a, i_b and i_c) can be found.

The concept of the space vector can be applied to the squirrel-cage rotor. In this case, the squirrel-cage as the rotor winding can be replaced with the conventional three-phase winding. Then, all the above-described procedures for introducing the space vectors can be applied to such conventional rotor winding.

5.1.3 Space Vectors of Voltages and Flux Linkages

By analogy with (5.4) and (5.7), the space vectors of the voltages and the flux linkages of the stator and rotor windings are represented as

$$\begin{aligned} \Psi_1 &= \frac{2}{3}\left(\Psi_A + a\Psi_B + a^2\Psi_C\right) \quad & \Psi_2 &= \frac{2}{3}\left(\Psi_a + a\Psi_b + a^2\Psi_c\right) \\ u_1 &= \frac{2}{3}\left(u_A + au_B + a^2 u_C\right) \quad & u_2 &= \frac{2}{3}\left(u_a + au_b + a^2 u_c\right) \end{aligned} \tag{5.8}$$

As it follows from the expressions (5.1) to (5.4), (5.7), and (5.8), the space vectors of the currents, voltages, and flux linkages are directly connected with the existing in the air gap of the machine the rotating spatial waves of the mmf and magnetic flux density. Consequently, the space vectors are due to the space-time vectors of the waves of the mmf and the magnetic flux density that takes place in the machine. In this case, it should be noted again that the projections of the space vectors on the phase axes of the stator and rotor windings represent the instantaneous values of the real phase quantities. Therefore, the space vectors are widely used to analyze the electromagnetic processes taken place in the electric machine.

In the symmetric machines and with the symmetrical sinusoidal power supply of its windings, in the phases of the stator and rotor windings, the same picture of the electromagnetic processes is observed. This means that the analysis of the processes in the machine using the space vectors can be carried out on the single phase since in the other phases the picture of the processes will be the same. In this case, however, it should be noted that the processes in the other phases will occur with the time shift by the phase angle equal to 120° or 240°, respectively.

5.2 Coordinate Transformation

The asynchronous machine is the system with the mutually moving windings located on the stator and the rotating rotor. As a consequence of this, the space vectors related to the stator and rotor will mutually move relative to each other. This situation seriously complicates the analysis of the processes in the electric machine. The analysis of the processes is greatly simplified if the space vectors are represented in the single coordinate system, for example, rigidly connected to the stator or rigidly connected to the rotor [3–15]. In this work, as the unified coordinate system, we choose the coordinate system rigidly connected to the stator. In this unified fixed stator coordinate system, the stator and rotor space vectors will be mutually fixed. In addition, in this case, the stator space vectors will reflect the real values and the rotor space vectors will have the transformed values.

Transition from the rotating rotor coordinate system to the fixed stator one. Let us consider the transition from the rotating rotor coordinate system to the fixed stator coordinate system [3–8, 10–15]. The angle shift between the real coordinate axes represents the current coordinate α. For the accepted coordinate system of the stator and rotor, the value of α is determined as $\alpha = \int_0^t \omega dt + \alpha_0$, where ω is the electric angular speed of the rotor and γ_0 is the initial shift of the coordinate systems at $t = 0$. When the rotor speed is uniform, the value of α is calculated by the formula $\alpha = \omega t + \alpha_0$. The space vector of the rotor current in the fixed coordinate system, we will denote through the i_2 and the same space vector of the rotor current in the rotating coordinate system, we will denote through the i_{2B}. The module of the space vector i_2 can be represented as I_2. In the fixed stator coordinate system, the space vector of the rotor current i_2 can be represented as $i_2 = I_2(\cos\alpha_s + j\sin\alpha_s) = I_2 e^{j\alpha_s}$, where $e^{j\alpha_s} = \cos\alpha_s + j\sin\alpha_s$. In the rotating rotor coordinate system, the space vector of the rotor current i_{2B} can be expressed as $i_{2B} = I_2(\cos\alpha_R + j\sin\alpha_R) = I_2 e^{j\alpha_R}$, where $e^{j\alpha_R} = \cos\alpha_R + j\sin\alpha_R$. Here, α_s and α_R are the angle shifts of the rotor current vector in the fixed stator and the rotating rotor coordinate systems, respectively. Consequently, the angle shift between the fixed stator and rotating rotor coordinate axes α is determined as $\alpha = \alpha_s - \alpha_R$. Now, for the ratio of the space vectors of the currents i_2 and i_{2B} represented, respectively, in the fixed stator and the rotating rotor coordinate systems, we can have that $i_2 / i_{2B} = e^{j(\alpha_s - \alpha_R)} = e^{j\alpha}$, where $\alpha = \alpha_s - \alpha_R$. From the condition of kind i_2/i_{2B}, the formula for the transition from the rotating rotor coordinate system to the fixed stator coordinate system can be obtained in the form

$$i_2 = i_{2B} e^{j\alpha} \tag{5.9}$$

where $\alpha = \omega t + \alpha_0$.

If the space vector of the rotor current in the rotating axes is $i_{2B} = I_2 e^{j(\omega_2 t + \gamma_0)}$, then, in the fixed system of the coordinate for this space vector using the expression (5.9), we can have, that $i_2 = i_{2B} e^{j\alpha} = I_2 e^{j(\omega_2 t + \gamma_0)} e^{j(\omega t + \alpha_0)} =$

$I_2 e^{j(s\omega_1 t+\gamma_0)} e^{j[(1-s)\omega_1 t+\alpha_0]} = I_2 e^{j[\omega_1 t+(\alpha_0+\gamma_0)]}$. Here, it is taken into account that $\omega_2 = s\omega_1$ and $\omega = (1 - s)\omega_1$. Consequently, the space vector of the rotor current in the fixed stator coordinate system has the radian frequency ω_1 of the power supply network current.

From (5.9), the space vectors of the rotor flux linkage and the rotor voltage acquire the form in the fixed stator coordinate system

$$\Psi_2 = \Psi_{2B} e^{j\alpha} \quad u_2 = u_{2B} e^{j\alpha} \tag{5.10}$$

It can be shown that these rotor space vectors are changed with the radian frequency equal to the frequency of the power supply network current ω_1.

5.3 Voltage Equations

We consider the symmetrical three-phase asynchronous machine. Suppose that the machine has on the stator the winding with the number of the turns w_A, w_B, and w_C and on the rotor the winding with the number of the turns w_a, w_b, and w_c. The stator and rotor windings are shifted in the space by the angle 120°. Moreover, we assume that $w_A = w_B = w_C = w_1$ and $w_a = w_b = w_c = w_2$. The stator and rotor windings are powered by the symmetrical system of the sinusoidal voltages. The rotor of the machine is rotated with the angular speed equal to $\omega = d\gamma/dt$. In this case, the stator and rotor windings undergo relative displacement.

5.3.1 *Voltage Equations Through Instantaneous Phase Quantities*

For each of the machine windings, we write down the voltage equations expressed through the instantaneous values of the phase quantities. Then, we have

$$\begin{aligned} u_A &= r_A i_A + d\Psi_A/dt & u_a &= r_a i_a + d\Psi_a/dt \\ u_B &= r_B i_B + d\Psi_B/dt & u_b &= r_b i_b + d\Psi_b/dt \\ u_C &= r_C i_C + d\Psi_C/dt & u_c &= r_c i_c + d\Psi_c/dt \end{aligned} \tag{5.11}$$

where $\Psi_A = L_A i_A + M_{AB} i_B + M_{AC} i_C + M_{Aa} i_a + M_{Ab} i_b + M_{Ac} i_c$.

The flux linkages Ψ_B, Ψ_C, Ψ_a, Ψ_b, and Ψ_c used in (5.11) are written similarly to the flux linkage Ψ_A. In (5.11), u_i are the voltages; i_i are the currents; r_i are the ohmic resistances; Ψ_i are the flux linkages of the stator and rotor phase windings, where $i = A, B, C$ and a, b, c.

The voltages, currents, and flux linkages used in (5.11) have the instantaneous phase values. When substituting the expressions for the flux linkages Ψ_A, Ψ_B, Ψ_C,

Ψ_a, Ψ_b, and Ψ_c into the Eq. (5.11), they are converted into the equations with the periodic coefficients, for example, [3, 5–8]. In the general case, such equations are difficult to solve analytically. A practical approach to solving the equations with the periodic coefficients is reduced to reducing them to the equations with the constant coefficients. As noted earlier, obtaining such machine equations is associated with the use of space vectors.

5.3.2 *Voltage Equations Through Space Vectors*

In (5.11), we will consider the first three equations related to the stator. The first equation is multiplied by (2/3), the second equation is multiplied by (2/3)a, and the third equation is multiplied by (2/3)a^2. Then, the equations obtained in this way can be summed, for example, [4, 9]. As a result, we have

$$\frac{2}{3}\left(u_A + au_B + a^2u_C\right) = \frac{2}{3}r_1\left(i_A + ai_B + a^2i_C\right) + \frac{2}{3}\frac{d}{dt} \times \left(\Psi_A + a\Psi_B + a^2\Psi_C\right) \tag{5.12}$$

where $r_1 = r_A = r_B = r_C$.

The expressions shown in (5.4) and (5.8) can be used to introduce in (5.12) the space vector of the current i_1, the space vector of the voltage u_1, and the space vector of the flux linkage Ψ_1. Then, Eq. (5.12) acquires the form

$$u_1 = r_1 i_1 + \frac{d\Psi_1}{dt} \tag{5.13}$$

In (5.13), the flux linkage Ψ_1 is determined by the joint action of the stator and rotor winding currents. In the fixed stator system of the coordinate, Ψ_1 is represented as

$$\Psi_1 = L_1 i_1 + L_{21} i_{2B} e^{j\alpha} \tag{5.14}$$

where L_1 is the total self-inductance of the stator phase; L_{21} is the mutual inductance; $i_{2B}e^{j\alpha}$ is the space vector of the rotor winding current reduced to the fixed stator.

Acting in a similar way, the voltage equation for the rotor winding can be received in the following form

$$\begin{aligned} u_{2в} &= r_2 i_{2B} + \frac{d\Psi_{2в}}{dt} \\ \Psi_{2в} &= L_2 i_{2B} + L_{12} i_1 e^{-j\alpha} \end{aligned} \tag{5.15}$$

where L_2 is the total self-inductance of the rotor phase; L_{12} is the mutual inductance; $i_1 e^{-j\alpha}$ is the space vector of the stator winding current reduced to the rotating rotor.

As follows from (5.14) and (5.15), the flux linkages Ψ_1 and $\Psi_{2в}$ contain the components varying according to the harmonic law. Therefore, the substitution of the flux linkages Ψ_1 и $\Psi_{2в}$ in the voltage equations of the stator and rotor windings leads to the differential equations with the periodic coefficient which, in general case, can be solved only numerically using the computer.

5.3.3 Voltage Equations in Fixed Stator System of Coordinate

To obtain the machine voltage equations with constant coefficients, it is necessary to represent the voltage equations for the stator and rotor windings in the single coordinate system. As such coordinate system, we choose the fixed stator coordinate system. In this coordinate system, the variables related to the stator do not change. For the rotor winding, the new variables can be introduced for what we will use the conditions given in (5.9) and (5.10). In accordance with these conditions, all terms of the equations given for u_{2B} and Ψ_{2B} in (5.15) can be multiplied by the operator $e^{j\alpha}$. In this case, the rotor flux linkage in the rotating coordinate system Ψ_{2B} can be expressed through its value in the fixed stator coordinate system, that is, $\Psi_{2B} = \Psi_2 e^{-j\alpha}$. Then, Eq. (5.15) acquire the form

$$\begin{aligned} e^{j\alpha} u_{2B} &= r_2 i_{2B} e^{j\alpha} + e^{j\alpha} \frac{d(\Psi_2 e^{-j\alpha})}{dt} \\ e^{j\alpha} \Psi_{2B} &= L_2 i_{2B} e^{j\alpha} + L_{12} i_1 \end{aligned} \tag{5.16}$$

Using in the fixed stator coordinate system the new variables $u_2 = u_{2B} e^{j\alpha}$, $i_2 = i_{2B} e^{j\alpha}$ and $\Psi_2 = \Psi_{2B} e^{j\alpha}$ as well as performing the differentiation operation, we can have

$$\begin{aligned} u_2 &= r_2 i_2 + \frac{d\Psi_2}{dt} - j\omega \Psi_2 \\ \Psi_2 &= L_2 i_2 + L_{12} i_1 \end{aligned} \tag{5.17}$$

The final system of the voltage equations for the stator and rotor windings from (5.13), (5.14), and (5.17) takes the following form

$$\begin{aligned} u_1 &= r_1 i_1 + \frac{d\Psi_{1p}}{dt} & \Psi_{1p} &= L_1 i_1 + L_{21} i_2 \\ u_2 &= r_2 i_2 + \frac{d\Psi_{2p}}{dt} - j\omega \Psi_{2p} & \Psi_{2p} &= L_2 i_2 + L_{12} i_1 \end{aligned} \tag{5.18}$$

Instead of the flux linkages Ψ_1 and Ψ_2 used in the Eqs. (5.13), (5.14), and (5.15), in the system of the Eq. (5.18), the flux linkages with the new designations Ψ_{1p} and

Ψ'_{2p} are applied to emphasize that they are the resulting flux linkages of the stator and rotor windings.

The Eq. (5.18) do not contain the periodic coefficients. They represent the equations with the constant coefficients, and therefore, they are relatively easily solved. In the system of the Eq. (5.18), the rotor quantities are converted, and the stator quantities are real.

5.4 Summary

1. On the stator and rotor of the asynchronous machine, the symmetrical three-phase windings powered by the symmetrical three-phase voltage system can be placed. In this case, the electric machine consists of the six inductively coupled phase windings the mutual inductances of which are changed periodically as the function of the angle position of the rotor. Because of this, the machine equations compiled in relation to the phase quantities contain the periodic coefficients. The solution of the equations with the periodic coefficients and obtaining, in this case, the results suitable for the practice of the engineering calculations represent great difficulties. In the theory of electric machines, the variable change technique is used to simplify the voltage equations. In accordance with this technique, the machine voltage equations are expressed through the fictitious vectors (space vectors) of the currents, voltages, and flux linkages being the functions of the real currents, voltages, and flux linkages of the phase windings of the electric machine. In this case, the number of the voltage equations is shortened. Instead of the six inductively coupled phase windings, we have the two inductively coupled windings. All this greatly simplifies the solution of the machine voltage equations.
2. The machine voltage equations expressed through the space vectors are written in the single coordinate system. In this work, as such coordinate system, the fixed stator coordinate system is taken. In this coordinate system, the variables related to the stator are not changed. For the rotor winding, the new variables are used. Thus, in the fixed stator coordinate system, the rotor quantities are converted, and the stator quantities are real. As a result of this transformation, instead of the equations with the periodic coefficients, the equations with the constant coefficients arise. This opens up new opportunities for analysis.

References

1. Kostenko, M., Piotrovskij, L.: Alternating Current Electric Machines, Publishing Energiya, Leningrad (1973)
2. Voldek, A.: Electric Machines, Publishing Energiya, Leningrad (1974)
3. Vazhnov, A.: Electric Machines. Publishing Energiya, Leningrad (1968)
4. Ivanov-Smolensky, A.: Electric Machines, Publishing Energiya, Moscow (1980)

5. Treshchev, I.: Electromechanical Processes in AC Electric Machines, Energiya, Leningrad (1980)
6. Kopylov, I.: Mathematic Modeling of Electric Machines, Vysshaya Shkola, Moscow (1987)
7. Sipailov, G., Kononenko, E., Kharkov, K.: Electric Machines. Special Course, Vysshaya Shkola, Moscow (1987)
8. Kimbrak, E.: Power System Stability, vol.3. Synchronous Machines, John and Sons, New York - London (1956)
9. Danilevich, Y., Dombrovsky, V., Kazovsky, E.: Parameters of AC Electric Machines, Publishing Nauka, Moscow (1965)
10. McPherson, G.: An Introduction to Electrical Machines and Transformers, Wiley, New York (1981)
11. Adkins, B.: The General Theory of Electrical Machines, Chapman and Hall, London (1959)
12. Fitzgerald, A., Kingsley, C., Umans, S.: Electric Machinery, McGraw-Hill, 5th ed., New York (1990)
13. Dell Toro, V.: Electric Machines and Power Systems, Prentice-Hall, Inc., Englewood Cliffs, New Jersey (1985)
14. Krause, P.: Analysis of Electric Machinery, McGraw-Hill, New York (1986)
15. Chapman, S.: Electric Machinery Fundamentals, 4th ed., McGraw-Hill, Inc., New York (2005)

Chapter 6
Voltage Equations and Equivalent Circuit of Machine: Formation of Fragments

6.1 Voltage Equations: Steady-State Operating Mode

The steady-state voltage equations follow from the system (5.18), if in the first and second equations of this system to accept that $d/dt = j\omega_1$, where ω_1 is the angular frequency of the power supply network voltage. In addition, we assume that the symmetrical three-phase systems of the voltages are applied to the terminals of the stator winding. In this case, for the instantaneous value of the phase voltage, for example, for the voltage of the phase A, we have that $u_A = U_m \cos(\omega_1 t + \varphi_A)$. From here, for the complex time vector of the voltage of the stator phase winding A, it is followed that $u_A = U_{1m}e^{j\varphi_A}e^{j\omega_1 t} = \dot{U}_1 e^{j\omega_1 t}$, where $\dot{U}_1 = U_{1m}e^{j\varphi_A}$. Here, the vector $\dot{U}_1$ represents the complex amplitude of the phase voltage of the stator winding. A similar expression can be obtained for the phase voltage a of the rotor winding in the form $\dot{U}_2 = U_{2m}e^{j\varphi_a}$. The vector $\dot{U}_2$ represents the complex amplitude of the phase voltage of the rotor winding. It can be shown that in the steady-state operating mode, the space vectors of the stator and rotor voltages form in the fixed axes system the vectors rotating with the angular velocity ω_1. These vectors are equal to the complex amplitudes of the phase voltages of the stator A and rotor a windings, respectively. This statement is true for the space vectors of the currents and flux linkages of the stator and rotor windings ($\dot{I}_1 = I_{1m}e^{j\varphi_A}, \dot{\Psi}_1 = \Psi_{1m}e^{j\varphi_A}$ and $\dot{I}_2 = I_{2m}e^{j\varphi_a}, \dot{\Psi}_2 = \Psi_{2m}e^{j\varphi_a}$). In this regard, for the steady-state operating mode, the system of the voltage Eq. (5.18) takes the following form, for example, [1–9, 11].

$$
\begin{aligned}
\dot{U}_1 &= r_1\dot{I}_1 + j\omega_1\dot{\Psi}_{1p} \\
\dot{U}_2 &= r_2\dot{I}_2 + j\omega_1\dot{\Psi}_{2p} - j\omega\dot{\Psi}_{2p} = r_2\dot{I}_2 + j(\omega_1 - \omega)\dot{\Psi}_{2p} \\
\dot{\Psi}_{1p} &= L_1\dot{I}_1 + L_{21}\dot{I}_2; \dot{\Psi}_{2p} = L_{12}\dot{I}_1 + L_{22}\dot{I}_2
\end{aligned}
\tag{6.1}
$$

In (6.1), we take into account that $(\omega_1 - \omega) = (1 - \omega/\omega_1)\omega_1 = s\omega_1$, where $s = (1 - \omega/\omega_1)$. Then, it acquires

V. Asanbayev, *Asynchronous Machines*,
https://doi.org/10.1007/978-3-030-92284-9_6

$$\begin{aligned} \dot{U}_1 &= r_1\dot{I}_1 + j\omega_1\Psi_{1p} \\ \left(-\frac{\dot{U}_2}{s}\right) &= \frac{r_2}{s}\dot{I}_2 + j\omega_1\Psi_{2p} \end{aligned} \tag{6.2}$$

6.2 Resulting emf Equations for Windings of Stator and Rotor

The flux linkages $\dot{\Psi}_{1p}$ and $\dot{\Psi}_{2p}$ used in (6.2) are determined by (6.1). In (6.2), we use the following conditions

$$(-\dot{\text{Э}}_{1p}) = j\omega_1\dot{\Psi}_{1p} \quad (-\dot{\text{Э}}_{2p}) = j\omega_1\dot{\Psi}_{2p} \tag{6.3}$$

Then, from (6.2) and (6.3), for the $\dot{\text{Э}}_{1p}$ and $\dot{\text{Э}}_{2p}$, we can obtain

$$\begin{aligned} (-\dot{\text{Э}}_{1p}) &= -r_1\dot{I}_1 + \dot{U}_1 \\ (-\dot{\text{Э}}_{2p}) &= -\frac{r_2}{s}\dot{I}_2 - \left(+\frac{\dot{U}_2}{s}\right) \end{aligned} \tag{6.4}$$

Based on (6.1) and (6.3), the emf $\dot{\text{Э}}_{1p}$ and $\dot{\text{Э}}_{2p}$ used in (6.4) acquire the form

$$\begin{aligned} (-\dot{\text{Э}}_{1p}) &= j\omega_1 L_1\dot{I}_1 + j\omega_1 L_{21}\dot{I}_2 = jx_1\dot{I}_1 + jx_{21}\dot{I}_2 \\ (-\dot{\text{Э}}_{2p}) &= j\omega_1 L_{12}\dot{I}_1 + j\omega_1 L_2\dot{I}_2 = jx_{12}\dot{I}_1 + jx_2\dot{I}_2 \end{aligned} \tag{6.5}$$

where $x_1 = \omega_1 L_1$; $x_{21} = \omega_1 L_{21}$; $x_{12} = \omega_1 L_{12}$; $x_2 = \omega_1 L_2$.

In the system of the Eq. (6.5), the reactances x_1, x_{21}, x_{12} and x_2 are represented as: x_1 is the total self-induction reactance of the stator winding; x_{21} is the mutual induction reactance of the stator winding; x_{12} is the mutual induction reactance of the rotor winding; and x_2 is the total self-induction reactance of the rotor winding.

In accordance with (6.4) and (6.5), the voltage equations are represented as the two systems of the equations in relation to the emf $\dot{\text{Э}}_{1p}$ and $\dot{\text{Э}}_{2p}$. In this case, the Eq. (6.4) reflect the processes associated with the losses in the stator and rotor windings, and the Eq. (6.5) represent the resulting magnetic field creating in the machine by the currents of the stator and rotor windings. In the technical literature, the resulting magnetic field of the electric machine is determined as a result of the superposition of the magnetic fields caused by the currents of the stator and rotor windings, individually. The emf equations corresponding to this provision will be considered below.

According to (6.3), the emf $(-\dot{\text{Э}}_{1p})$ and $(-\dot{\text{Э}}_{2p})$ used in (6.4) and (6.5) are represented as $(-\dot{\text{Э}}_{1p}) = j\omega_1\dot{\Psi}_{1p}$ and $(-\dot{\text{Э}}_{2p}) = j\omega_1\dot{\Psi}_{2p}$. From here, it is followed that to determine the values of the emf $(-\dot{Э}_{1p})$ and $(-\dot{Э}_{2p})$, it is necessary to find the flux linkages $\dot{\Psi}_{1p}$ and $\dot{\Psi}_{1p}$ of the stator and rotor windings, respectively.

6.3 “Inductively Coupled Circuits” Concept: Features of Application

In (6.4) and (6.5), the values of $\left(-\dot{Э}_{1p}\right)$ and $\left(-\dot{Э}_{2p}\right)$ represent the emf induced by the flux linkages $\dot{\Psi}_{1p}$ and $\dot{\Psi}_{1p}$ of the stator and rotor windings. The flux linkages $\dot{\Psi}_{1p}$ and $\dot{\Psi}_{1p}$ are determined by the resulting magnetic flux of the stator winding $\dot{\Phi}_{1p}$ and by the resulting magnetic flux of the rotor winding $\dot{\Phi}_{2p}$. In the theory of electric machines, the superposition principle is used to determine the magnetic fluxes $\dot{\Phi}_{1p}$ and $\dot{\Phi}_{2p}$. The physical meaning of this principle is known in the technical literature as the “inductively coupled circuits” concept. In this case, the electric machine is represented as consisting of the two inductively coupled windings located on the stator and rotor. In the operating mode, the currents $\dot{I}_1$ and $\dot{I}_2$ flowing in the stator and rotor windings create the resulting magnetic field of the electric machine.

In accordance with the superposition principle, the resulting magnetic field of the electric machine can be represented as a result of the superposition of the magnetic fields created by the currents $\dot{I}_1$ and $\dot{I}_2$, individually [1–12]. So, in the presence of the current $\dot{I}_1$ in the stator winding and in the absence of the current $\dot{I}_2$ in the rotor winding, i.e. $\dot{I}_2 = 0$, according to the total current law, the rotating magnetic flux of the stator $\dot{\Phi}_1$ is created in the machine. In this case, the turns of the stator winding are interlinked with the total self-induction magnetic flux $\dot{\Phi}_1$, and the turns of the rotor winding are embraced by the mutual induction magnetic flux $\dot{\Phi}_{12}$ (Fig. 6.1). Consequently, the flux $\dot{\Phi}_1$ and the flux $\dot{\Phi}_{12}$ are created by the current of the stator winding.

In the presence of the current $\dot{I}_2$ in the rotor winding and in the absence of the current $\dot{I}_1$ in the stator winding, i.e. $\dot{I}_1 = 0$, according to the total current law, the rotating magnetic flux of the rotor $\dot{\Phi}_2$ is created in the machine. In this case, the turns of the rotor winding are interlinked with the total self-induction magnetic flux $\dot{\Phi}_2$, and the turns of the stator winding are embraced by the mutual induction magnetic flux $\dot{\Phi}_{21}$ (Fig. 6.2). So, the flux $\dot{\Phi}_2$ and the flux $\dot{\Phi}_{21}$ are created by the current of the rotor winding.

Now, in the operating mode, when the stator and rotor windings are flowing by the currents $\dot{I}_1$ and $\dot{I}_2$, the resulting magnetic fluxes of the stator and rotor windings can be represented as a result of the superposition of the magnetic fluxes created by the currents flowing in the windings of the electric machine, individually. In this case, the resulting magnetic flux of the stator winding $\dot{\Phi}_{1p}$ is determined as a result of the superposition of the total self-induction magnetic $\dot{\Phi}_1$ created by the stator winding current and the mutual induction magnetic flux $\dot{\Phi}_{21}$ caused by the rotor winding current, i.e. $\dot{\Phi}_{1p} = \dot{\Phi}_1 + \dot{\Phi}_{21}$. The resulting magnetic flux of the rotor winding $\dot{\Phi}_{2p}$ is determined as a result of the superposition of the total self-induction

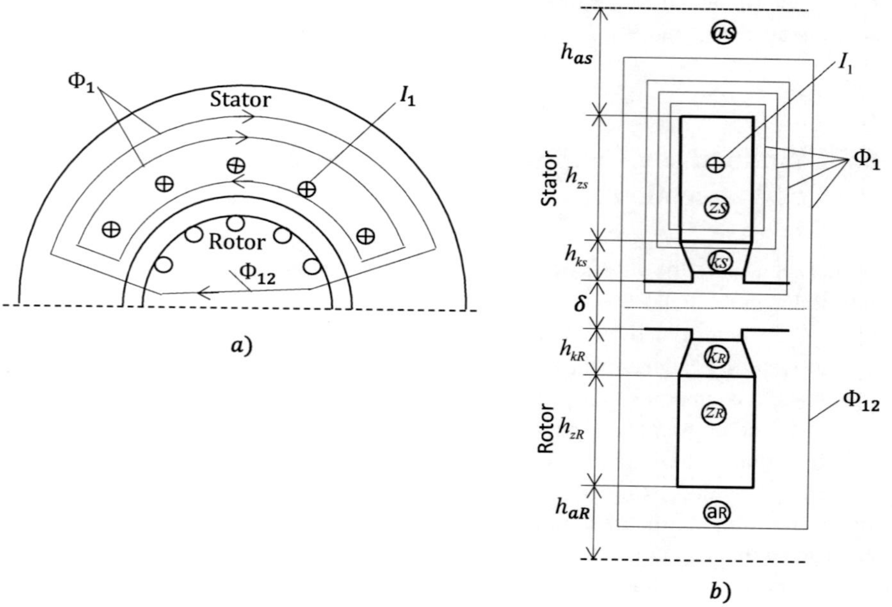

Fig. 6.1 Picture for the conventional lines of the total self-induction magnetic flux of the stator winding: (**a**) conventional lines of the total self-induction magnetic flux created by the stator winding current in the phase zone; (**b**) conventional lines of the total self-induction magnetic flux created by the stator winding current in the single tooth pitch

magnetic flux $\dot{\Phi}_2$ created by the rotor winding current and the mutual induction magnetic flux $\dot{\Phi}_{12}$ caused by the stator winding current, i.e. $\dot{\Phi}_{2p} = \dot{\Phi}_2 + \dot{\Phi}_{12}$. Thus, the resulting magnetic fluxes of the stator and rotor windings $\dot{\Phi}_{1p}$ and $\dot{\Phi}_{2p}$ taken place in the electric machine in the operating mode are supported by the currents $\dot{I}_1$ and $\dot{I}_2$ flowing in the stator and rotor windings, respectively.

6.4 Equations for Resulting Magnetic Fluxes of Windings of Stator and Rotor

Now, for the magnetic fluxes $\dot{\Phi}_{1p}$ and $\dot{\Phi}_{2p}$, we can obtain

$$\begin{aligned} \dot{\Phi}_{1p} &= \dot{\Phi}_1 + \dot{\Phi}_{21} \\ \dot{\Phi}_{2p} &= \dot{\Phi}_2 + \dot{\Phi}_{12} \end{aligned} \tag{6.6}$$

In Eq. (6.6), the magnetic fluxes $\dot{\Phi}_1$, $\dot{\Phi}_{21}$, $\dot{\Phi}_2$, and $\dot{\Phi}_{12}$ are determined as: $\dot{\Phi}_1$ is the total self-induction magnetic flux created in the stator winding by the stator winding current; $\dot{\Phi}_{21}$ is the mutual induction magnetic flux caused in the stator winding by

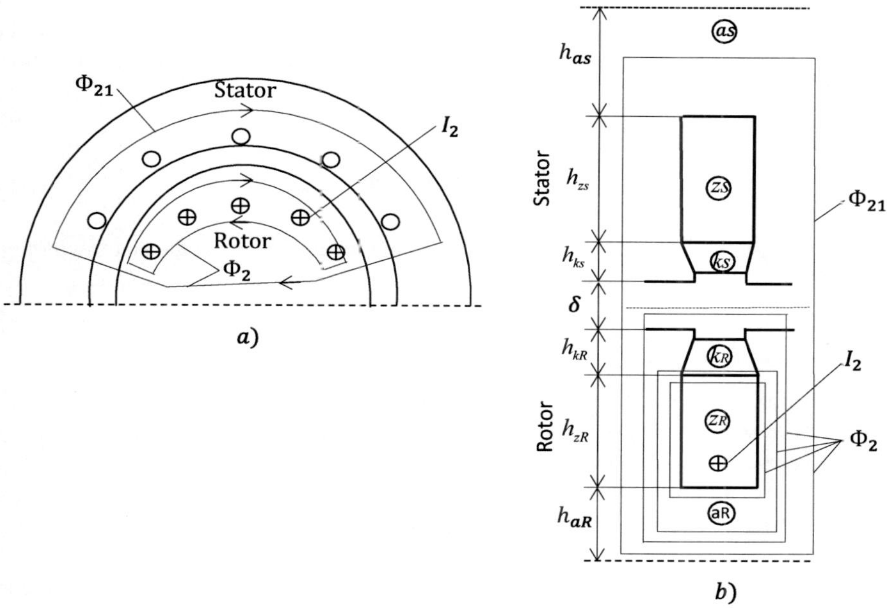

Fig. 6.2 Picture for the conventional lines of the total self-induction magnetic flux of the rotor winding: (**a**) conventional lines of the total self-induction magnetic flux created by the rotor winding current in the phase zone; (**b**) conventional lines of the total self-induction magnetic flux created by the rotor winding current in the single tooth pitch

the rotor winding current; $\dot{\Phi}_2$ is the total self-induction magnetic flux created in the rotor winding by the rotor winding current; and $\dot{\Phi}_{12}$ is the mutual induction magnetic flux caused in the rotor winding by the stator winding current.

Based on the Eq. (6.6), the following conditions can be introduced

$$\begin{aligned}\dot{\Phi}_{1\sigma} &= \dot{\Phi}_1 - \dot{\Phi}_{12} = \dot{\Phi}_1 - \dot{\Phi}_{11}\\ \dot{\Phi}_{2\sigma} &= \dot{\Phi}_2 - \dot{\Phi}_{21} = \dot{\Phi}_2 - \dot{\Phi}_{22}\end{aligned} \tag{6.7}$$

Here, for the self- and mutual induction magnetic fluxes $\dot{\Phi}_{11}$ and $\dot{\Phi}_{12}$, the conditions of the form $\dot{\Phi}_{12} = \dot{\Phi}_{11}$ are met, since these fluxes are created by the stator winding current, and they are closed along the main magnetic circuit (magnetizing circuit) of the electric machine. The double indexing used for the magnetic fluxes $\dot{\Phi}_{11}$ and $\dot{\Phi}_{12}$ means that the flux $\dot{\Phi}_{11}$ is created by the stator winding current, and it interlinks with the stator winding, and the flux $\dot{\Phi}_{12}$ is created by the stator winding current, and it embraces the rotor winding. For the self- and mutual induction magnetic fluxes $\dot{\Phi}_{22}$ and $\dot{\Phi}_{21}$, the conditions of the form $\dot{\Phi}_{21} = \dot{\Phi}_{22}$ are met, since these fluxes are created by rotor winding current, and they are closed along the main magnetic circuit (magnetizing circuit) of the electric machine. The double indexing used for the magnetic fluxes $\dot{\Phi}_{22}$ and $\dot{\Phi}_{21}$ means that the flux $\dot{\Phi}_{22}$ is created by the rotor winding current, and it interlinks with the rotor winding, and the flux $\dot{\Phi}_{21}$ is created by the rotor winding current, and it embraces the stator winding, for example, [5, 6, 10].

The magnetic flux $\dot{\Phi}_{1\sigma}$ is determined from (6.7) as

$$\dot{\Phi}_{1\sigma} = \dot{\Phi}_1 - \dot{\Phi}_{11} \tag{6.8}$$

The magnetic flux $\dot{\Phi}_{1\sigma}$ represents the leakage magnetic flux of the stator winding. The magnetic flux $\dot{\Phi}_{2\sigma}$ is given in the form in accordance with (6.7)

$$\dot{\Phi}_{2\sigma} = \dot{\Phi}_2 - \dot{\Phi}_{22} \tag{6.9}$$

The flux $\dot{\Phi}_{2\sigma}$ reflects the leakage magnetic flux of the rotor winding. In accordance with (6.8) and (6.9), the leakages fluxes $\dot{\Phi}_{1\sigma}$ and $\dot{\Phi}_{2\sigma}$ are determined as a result of the separation from the total self-induction magnetic fluxes caused by the currents of the stator and rotor windings, individually. In our case, the leakage fluxes $\dot{\Phi}_{1\sigma}$ and $\dot{\Phi}_{2\sigma}$ represent, as it was noted earlier, the slot leakage fluxes of the stator and rotor windings.

By the expressions (6.6), (6.8), and (6.9), the equations for the resulting magnetic fluxes of the stator and rotor windings can be shown as

$$\begin{aligned} \dot{\Phi}_{1p} &= \dot{\Phi}_{1\sigma} + \dot{\Phi}_{11} + \dot{\Phi}_{21} \\ \dot{\Phi}_{2p} &= \dot{\Phi}_{2\sigma} + \dot{\Phi}_{12} + \dot{\Phi}_{22} \end{aligned} \tag{6.10}$$

Based on (6.10), the picture of the distribution of the resulting magnetic fluxes of the stator and rotor windings has been given in Fig. 6.3.

6.5 Equations for Resulting Flux Linkages of Windings of Stator and Rotor

We will proceed from the equations for the magnetic fluxes received in (6.10). Here, the slot leakage flux $\dot{\Phi}_{1\sigma}$ is interlinked with the stator winding, and the slot leakage flux $\dot{\Phi}_{2\sigma}$ embraces the rotor winding. In this regard, the slot leakage fluxes $\dot{\Phi}_{1\sigma}$ and $\dot{\Phi}_{2\sigma}$ form with the stator and rotor windings the slot leakage flux linkage of the stator winding $\dot{\Psi}_{1\sigma}$ and the slot leakage flux linkage of the rotor winding $\dot{\Psi}_{2\sigma}$ determined as

$$\Psi_{1\sigma} = w_1 k_{w1} \Phi_{1\sigma} \quad \Psi_{2\sigma} = w_2 k_{w2} \Phi_{2\sigma} \tag{6.11}$$

The magnetic fluxes $\dot{\Phi}_{11}$ and $\dot{\Phi}_{21}$ used in the Eq. (6.10) create the flux linkages $\dot{\Psi}_{11}$ and $\dot{\Psi}_{21}$ with the stator winding. The magnetic fluxes $\dot{\Phi}_{22}$ and $\dot{\Phi}_{12}$ form the flux linkages $\dot{\Psi}_{22}$ and $\dot{\Psi}_{12}$ with the rotor winding. Therefore, for the flux linkages $\dot{\Psi}_{11}, \dot{\Psi}_{21}, \dot{\Psi}_{22}$ and $\dot{\Psi}_{12}$, we have

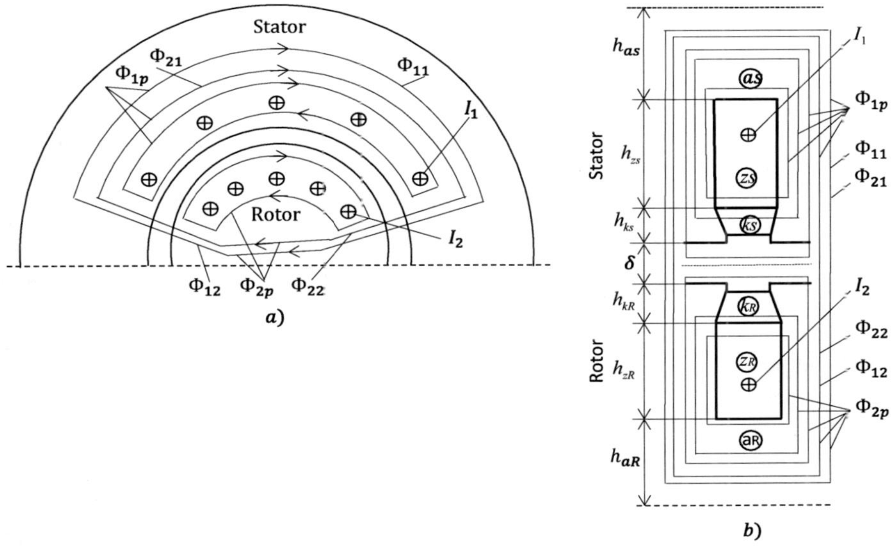

Fig. 6.3 Picture for the conventional lines of the resulting magnetic flux of the stator and rotor windings: (**a**) conventional lines of the resulting magnetic flux in the stator and rotor phase zones; (**b**) conventional lines of the resulting magnetic flux in the stator and rotor single tooth pitch

$$\begin{aligned} \Psi_{11} &= w_1 k_{w1} \Phi_{11} \quad & \Psi_{12} &= w_2 k_{w2} \Phi_{12} \\ \Psi_{21} &= w_1 k_{w1} \Phi_{21} \quad & \Psi_{22} &= w_2 k_{w2} \Phi_{22} \end{aligned} \tag{6.12}$$

In the expressions (6.12), the flux linkages $\dot{\Psi}_{11}, \dot{\Psi}_{21}, \dot{\Psi}_{22}$ and $\dot{\Psi}_{12}$ are determined as: $\dot{\Psi}_{11}$ is the flux linkage formed by the self-induction magnetic flux of the stator winding; $\dot{\Psi}_{21}$ is the flux linkage caused by the mutual induction magnetic flux of the stator winding; $\dot{\Psi}_{22}$ is the flux linkage associated with the self-induction magnetic flux of the rotor winding; $\dot{\Psi}_{12}$ is the flux linkage created by the mutual induction magnetic flux of the rotor winding. In terms of the expressions (6.11) and (6.12), the equations for the resulting magnetic fluxes given in (6.10) can be represented through the corresponding flux linkages of the stator and rotor windings in the form

$$\begin{aligned} \dot{\Psi}_{1p} &= \dot{\Psi}_{1\sigma} + \dot{\Psi}_{11} + \dot{\Psi}_{21} \\ \dot{\Psi}_{2p} &= \dot{\Psi}_{2\sigma} + \dot{\Psi}_{12} + \dot{\Psi}_{22} \end{aligned} \tag{6.13}$$

6.6 Resulting emf Equations Through Slot Leakage and Self- and Mutual Induction Reactances

We will use the equations for the resulting flux linkages Ψ_{1p} and Ψ_{2p} obtained in (6.13). In the stator and rotor windings, the slot leakage flux linkages $\Psi_{1\sigma}$ and $\Psi_{2\sigma}$ applied in (6.13) cause the slot leakage emf $\dot{Э}_{1\sigma}$ and $\dot{Э}_{2\sigma}$ for which it is followed

$$\begin{aligned}\left(-\dot{Э}_{1\sigma}\right) &= j\omega_1\dot{\Psi}_{1\sigma} = jx_{1\sigma}\dot{I}_1 \\ \left(-\dot{Э}_{2\sigma}\right) &= j\omega_1\dot{\Psi}_{2\sigma} = jx_{2\sigma}\dot{I}_2\end{aligned} \tag{6.14}$$

In the system of the Eq. (6.14), the slot leakage reactance of the stator winding $x_{1\sigma}$ and the slot leakage reactance of the rotor winding $x_{2\sigma}$ are used. These reactances represent the slot leakage magnetic fluxes associated with the stator winding and with the rotor winding. They ($x_{1\sigma}$ and $x_{2\sigma}$) are a result of the separation of the slot leakages from the total self-induction magnetic fluxes created by the currents of the stator and rotor windings, individually. In this definition, there are no specific indications on how these leakage fluxes interlink only with the stator winding or embrace only the rotor winding. This definition does not contain specific information on the nature of the leakage flux distribution in the electric machine. Therefore, the slot leakage reactances $x_{1\sigma}$ and $x_{2\sigma}$ used in Eq. (6.14) are not have the strong enough physical grounding which causes the difficulties associated with ensuring the necessary level of reliability in determining these leakage reactances. The issues of determining the slot leakage reactances $x_{1\sigma}$ and $x_{2\sigma}$ will be discussed in more detail below.

In stator and rotor windings, the self-and mutual induction flux linkages $\dot{\Psi}_{11}, \dot{\Psi}_{21}, \dot{\Psi}_{22}$, and $\dot{\Psi}_{12}$ used in (6.13) create the emf $\dot{Э}_{11}, \dot{Э}_{21}, \dot{Э}_{22}$, nd $\dot{Э}_{12}$ which can be expressed in terms of the reactances x_{11}, x_{21}, x_{22}, and x_{12} in the following form

$$\begin{aligned}\left(-\dot{Э}_{11}\right) &= j\omega_1\Psi_{11} = jx_{11}\dot{I}_1 \quad \left(-\dot{Э}_{12}\right) = j\omega_1\Psi_{12} = jx_{12}\dot{I}_1 \\ \left(-\dot{Э}_{21}\right) &= j\omega_1\Psi_{21} = jx_{21}\dot{I}_2 \quad \left(-\dot{Э}_{22}\right) = j\omega_1\Psi_{22} = jx_{22}\dot{I}_2\end{aligned} \tag{6.15}$$

In the system of the Eq. (6.15), the reactances x_{11} and x_{22} are determined as: x_{11} is the self-induction reactance of the stator winding; x_{22} is the self-induction reactance of the rotor winding. Now, by the expressions (6.13) to (6.15), for the resulting emf of the stator and rotor windings, we have

$$\begin{aligned}\left(-\dot{Э}_{1p}\right) &= j(x_{1\sigma} + x_{11})\dot{I}_1 + jx_{21}\dot{I}_2 \\ \left(-\dot{Э}_{2p}\right) &= jx_{12}\dot{I}_1 + j(x_{2\sigma} + x_{22})\dot{I}_2\end{aligned} \tag{6.16}$$

Based on the system of the Eqs. (6.5) and (6.16), the total self-induction reactances of the stator and rotor windings acquire

$$\begin{aligned} x_1 &= x_{1\sigma} + x_{11} \\ x_2 &= x_{2\sigma} + x_{22} \end{aligned} \tag{6.17}$$

The Eq. (6.16) can be represented as

$$\begin{aligned} \left(-\dot{Э}_{1p}\right) &= jx_{1\sigma}\dot{I}_1 + jx_{11}\dot{I}_1 + jx_{21}\dot{I}_2 = jx_{1\sigma}\dot{I}_1 + \left(-\dot{Э}_1\right) \\ \left(-\dot{Э}_{2p}\right) &= jx_{2\sigma}\dot{I}_2 + jx_{12}\dot{I}_1 + jx_{22}\dot{I}_2 = jx_{2\sigma}\dot{I}_2 + \left(-\dot{Э}_2\right) \end{aligned} \tag{6.18}$$

where $\left(-\dot{Э}_1\right) = jx_{11}\dot{I}_1 + jx_{21}\dot{I}_2$; $\left(-\dot{Э}_2\right) = jx_{12}\dot{I}_1 + jx_{22}\dot{I}_2$.

In (6.18), the values of the emf $\dot{Э}_1$ and $\dot{Э}_2$ represent the emf induced in the stator and rotor windings by the self- and mutual induction magnetic fluxes created by the currents of the stator and rotor windings individually.

6.7 Resulting emf Equations for Windings of Stator and Rotor: Formation of Fragments

The Eq. (6.18) can be represented in relation to emf $\dot{Э}_1$ and $\dot{Э}_2$ induced in the stator and rotor windings by the self- and mutual induction magnetic fluxes. In this case, from the Eq. (6.18), the two systems of the equations can be obtained in relation to the emf $\dot{Э}_1$ and $\dot{Э}_2$. From (6.18), for the first system of the equations, we have

$$\begin{aligned} \left(-\dot{Э}_1\right) &= jx_{11}\dot{I}_1 + jx_{21}\dot{I}_2 \\ \left(-\dot{Э}_2\right) &= jx_{12}\dot{I}_1 + jx_{22}\dot{I}_2 \end{aligned} \tag{6.19}$$

As it was noted earlier, the emf $\dot{Э}_1$ and $\dot{Э}_2$ used in (6.19) are determined by the self- and mutual induction magnetic fluxes closed on the main magnetic circuit (magnetizing circuit) of the electric machine. When drawing up the voltage equations, the reluctances of the sections (regions) of the stator and rotor magnetic circuits are not taken into account assuming that $\mu_c = \infty$. In these conditions, the machine's magnetic field is mainly concentrated in its air gap. Therefore, in the future, the Eq. (6.19) we will consider as the air gap emf equations of the electric machine.

From (6.18), for the second system of the equations representing in relation to the emf $\dot{Э}_1$ and $\dot{Э}_2$, we can obtain

$$\begin{aligned} \left(-\dot{Э}_1\right) &= -jx_{1\sigma}\dot{I}_1 + \left(-\dot{Э}_{1p}\right) \\ \left(-\dot{Э}_2\right) &= -jx_{2\sigma}\dot{I}_2 + \left(-\dot{Э}_{2p}\right) \end{aligned} \tag{6.20}$$

Here, we use the equations received in (6.4). As a result, the Eq. (6.20) acquire the form

$$\begin{aligned}(-\dot{Э}_1) &= -(r_1 + jx_{1\sigma})\dot{I}_1 + \dot{U}_1 \\ (-\dot{Э}_2) &= -\left(\frac{r_2}{s} + jx_{2\sigma}\right)\dot{I}_2 + \left(-\frac{\dot{U}_2}{s}\right)\end{aligned} \tag{6.21}$$

In this work, the ohmic resistances of the stator and rotor windings are not specifically addressed. Therefore, in the future, the Eq. (6.21) we will consider as the slot leakage emf equations of the stator and rotor windings.

According to (6.19) and (6.21) and in relation to the emf $\dot{Э}_1$ and $\dot{Э}_2$, the voltage equations obtained in (6.1) and (6.2) are represented as the three emf equations: the air gap emf equations, the stator slot leakage emf equation, and the rotor slot leakage emf equation. Below, we will consider these emf equations.

6.8 Air Gap emf Equations

We will consider the system of the equations shown in (6.19). These equations can be obtained from the voltage Eqs. (6.4) and (6.5), if in them the ohmic resistances r_1 and r_2/s and also the slot leakage reactances $x_{1\sigma}$ and $x_{2\sigma}$ to accept equal to zero. In these conditions, for the emf $\dot{Э}_1$ and $\dot{Э}_2$ used in the Eq. (6.19), the conditions of the form $\dot{U}_1 = (-\dot{Э}_1)$ and $\dot{U}_2/s = (-\dot{Э}_2)$ are implemented. In the Eq. (6.19), the values of the emf $\dot{Э}_1$ and $\dot{Э}_2$ are determined by the self- and mutual induction magnetic fluxes closed on the main magnetic circuit (magnetizing circuit) of the electric machine. Consequently, the Eq. (6.19) reflect the processes associated with magnetizing the main magnetic circuit of the machine. As mentioned earlier, when compiling the voltage equations, the reluctances of the sections (regions) of the stator and rotor magnetic circuits are taken to be zero, which means that $\mu_c = \infty$. In these conditions, the magnetic field is only distributed in the machine air gap, and the emf $\dot{Э}_1$ and $\dot{Э}_2$ caused by the magnetic fluxes of the self- and mutual induction are determined by the parameters of the magnetic circuit corresponding to the machine air gap. In this regard, the Eq. (6.19), as it was noted above, we will consider in the future as the air gap emf equations. The air gap emf equations describe the phenomena associated with the presence of inductive coupling between the stator and the rotor windings. Therefore, these equations reflect the basic processes associated with converting electromagnetic energy in the electric machine.

6.9 Slot Leakage emf Equations of Stator and Rotor

In the system of Eq. (6.21), the first equation reflects the processes associated with the slot leakage and the losses in the stator winding. The second equation represents the processes associated with the slot leakage and the losses in the rotor winding. In this work, the issues associated with the losses in the machine windings are not considered. In this regard, we will proceed from the fact that the first equation of the system (6.21) represents the processes taken place in the stator slot leakage magnetic circuits and the second equation of this system describes the processes in the rotor slot leakage magnetic circuits. Therefore, the first equation of the system (6.21) is considered as the stator slot leakage emf equation. The second equation of the system (6.21) is represented as the rotor slot leakage emf equation.

The windings of the machine are located in the axial slots of the stator and rotor. Providing that $\mu_c = \infty$, the leakage magnetic fields exist in the axial slots of the stator and rotor, that is, the leakage magnetic fluxes creating the emf $\dot{Э}_1$ and $\dot{Э}_2$ are determined by the parameters of the slot leakage magnetic circuits of the stator and rotor. In this regard, the emf Eq. (6.21) represent the slot areas of the stator and rotor. Thus, the emf Eqs. (6.19) and (6.21) following as a result of the fragmentation of the system of the Eqs. (6.4) and (6.5) describe the processes that have taken place in the air gap of the machine and also in the stator and rotor slots.

6.10 Fragments of Equivalent Circuit: "Inductively Coupled Circuits" Concept

We will consider the emf Eqs. (6.19) and (6.21). These emf equations can be represented as the corresponding equivalent circuits. The emf Eq. (6.19) reflect the processes associated with magnetizing the air gap of the electric machine. Therefore, the equivalent circuit corresponding to the emf Eq. (6.19) is determined as the air gap equivalent circuit. The first equation of the system (6.21) represents the processes associated with the slot leakage of the stator winding. In this regard, the equivalent circuit following from the first equation of the system (6.21) can be represented as the stator slot leakage equivalent circuit. The second equation of the system (6.21) reflects the processes associated with the slot leakage of the rotor winding. Consequently, the equivalent circuit arising from the second Eq. (6.21) can be considered as the rotor slot leakage equivalent circuit.

6.10.1 Equivalent Circuit of Air Gap

The air gap emf equations are shown in (6.19). Using the Eq. (6.9), the air gap equivalent circuit receives the form in Fig. 6.4a. As follows from Fig. 6.4a, the air gap equivalent circuit contains the inductively coupled elements.

Fig. 6.4 Fragments of machine equivalent circuit: (**a**) air gap equivalent circuit with the inductively coupled elements: (**b**) stator slot leakage equivalent circuit; (**c**) rotor slot leakage equivalent circuit

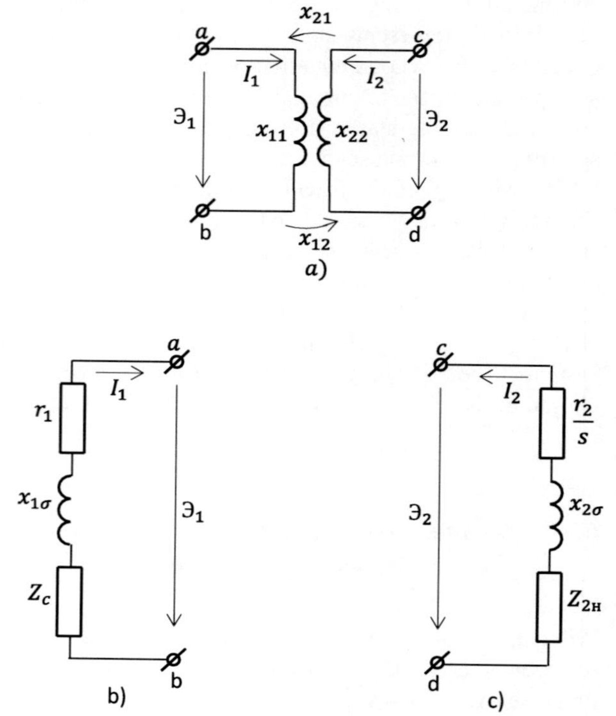

6.10.2 Slot Leakage Equivalent Circuit of Stator

From the emf Eq. (6.21), we can have that the first equation describing the processes in the stator slot leakage magnetic circuits is represented as

$$(-\dot{Э}_1) = -(r_1 + jx_{1\sigma})\dot{I}_1 + \dot{U}_1 \tag{6.22}$$

In Eq. (6.22), the voltage value on the terminals of the stator winding can be replaced by the voltage drop across the power supply network impedance Z_c, i.e. $(-\dot{U}_1) = Z_c\dot{I}_1$. Considering the condition $\dot{U}_1 = -Z_c\dot{I}_1$, the impedance used in the circuit loop of the stator winding can be expressed as $(r_1 + jx_{1\sigma} + Z_c)$. As a result, the equation shown in (6.22) can be obtained in the form

$$(-\dot{Э}_1) = (r_1 + jx_{1\sigma} + Z_c)(-\dot{I}_1) \tag{6.23}$$

The stator slot leakage equivalent circuit corresponding to the Eq. (6.24) is shown in Fig. 6.4b.

6.10.3 *Slot Leakage Equivalent Circuit of Rotor*

The second equation of the system (6.21) describes the processes in the rotor slot leakage magnetic circuits. This equation has the form

$$(-\dot{Э}_2) = -\left(\frac{r_2}{s} + jx_{2\sigma}\right)\dot{I}_2 + \left(-\frac{\dot{U}_2}{s}\right) \tag{6.24}$$

In Eq. (6.24), the voltage value on the terminals of the rotor winding can be replaced by the voltage drop across the load impedance $Z_{2н}$, i.e. $\dot{U}_2/s = Z_{2н}\dot{I}_2$. Keeping in mind that $\dot{U}_2/s = Z_{2н}\dot{I}_2$, the impedance used in the circuit loop of the rotor winding can be expressed as $(r_2/s + jx_{2\sigma} + Z_{2н})$. Now, the equation shown in Eq. (6.24) can be represented in the form

$$(-\dot{Э}_2) = (r_2/s + jx_{2\sigma} + Z_{2н})(-\dot{I}_2) \tag{6.25}$$

The rotor slot leakage equivalent circuit corresponding to the Eq. (6.25) is given in Fig. 6.4c.

Thus, in relation to the emf $\dot{Э}_1$ and $\dot{Э}_2$ induced in the stator and rotor windings, the machine voltage equations obtained in Eqs. (6.4) and (6.5) can be fragmentized into the three types of the emf equations: the stator slot leakage emf equation, the air gap emf equations, and the rotor slot leakage emf equation. Based on these equations, the stator slot leakage equivalent circuit (Fig. 6.4b), the air gap equivalent circuit (Fig. 6.4a), and the rotor slot leakage equivalent circuit (Fig. 6.4c) have been obtained.

The equivalent circuits shown in Fig. 6.4 have been received based on the emf equations represented in (6.19) and (6.21) and reflect the separate fragments of the voltage equations for the machine windings. In this regard, we can proceed from the fact that the equivalent circuits in Fig. 6.4 represent the separate fragments of the machine equivalent circuit. Therefore, the equivalent circuits in Fig. 6.4 can be used to build the machine equivalent circuit.

Separately, the equivalent circuits in Fig. 6.4 represent the less complex processes that have taken place in the electric machine. As will be shown below, such a feature of the equivalent circuits in Fig. 6.4 will greatly simplify an analysis.

6.11 Equivalent Circuit with Inductively Coupled Elements: Application of Cascade Way of Connecting

The equivalent circuit of the electric machine can be built by the use of the fragment equivalent circuits shown in Fig. 6.4. As it follows from Fig. 6.4, the equivalent circuit of the machine can be obtained as a result of connecting the input terminals $c - d$ of the rotor slot leakage equivalent circuit (Fig. 6.4c) with the output terminals

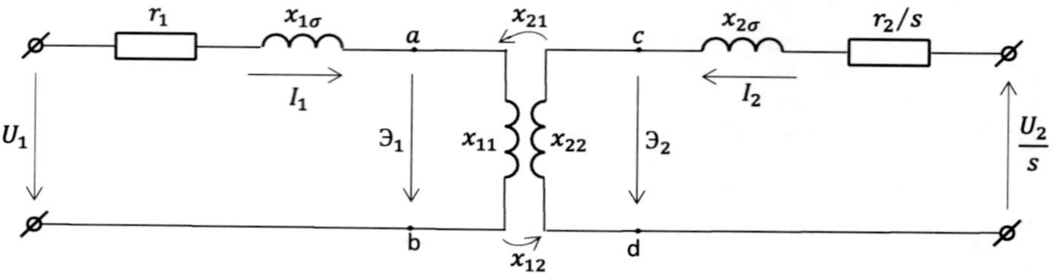

Fig. 6.5 Machine equivalent circuit with the inductively coupled elements

$c - d$ of the air gap equivalent circuit (Fig. 6.4a), and also as a result of connecting the input terminals $a - b$ of the air gap equivalent circuit (Fig. 6.4a) with the input terminals $a - b$ of the stator slot leakage equivalent circuit (Fig. 6.4b). Thus, the machine equivalent circuit can be built as a result of the cascade way of connecting the air gap equivalent circuit (Fig. 6.4a), the stator slot leakage equivalent circuit (Fig. 6.4b), and the rotor slot leakage equivalent circuit (Fig. 6.4c). The machine equivalent circuit obtained in this way has the inductively coupled elements, and it is given in Fig. 6.5.

As will be shown below, the use of the "fragmentation" concept and the cascade way of connecting opens up new opportunities for the further development of the equivalent circuits and the voltage equations of the electric machine. In this regard, when constructing the machine equivalent circuit, instead of Kirchhoff's equations representing the sum of the voltage drops across the ohmic resistances and the reactances of the stator and rotor windings, we will use the cascade way of connecting the equivalent circuits reflected in relation to emf $\dot{Э}_1$ and $\dot{Э}_2$ induced in the stator and rotor windings the corresponding design regions of the electric machine (in our case, the air gap region, and the stator and rotor slot regions). Below in this work, this provision will be considered in more detail.

The equivalent circuit obtained in Fig. 6.5 represents the stator and rotor windings as the inductively coupled two circuit loops, and therefore, in this form, it is not suitable for practical use. Besides, the features of the practical application of the equivalent circuit in Fig. 6.5 are associated with the difficulties of ensuring the required level of reliability of the determination of the parameters used in it [12]. To eliminate these difficulties, the equivalent circuit in Fig. 6.5 should be transformed. The physical meaning of the transformation used in the theory of the electric machines is reduced to the elimination of the inductively coupled elements taking place in the machine equivalent circuit in Fig. 6.5, that is, is reduced to the construction of the machine equivalent circuit with the simplified structure and also with the parameters that are amenable to the sufficiently accurate calculation and the experimental determination.

The equivalent circuit in Fig. 6.5 reflects the diverse and rather complex processes occurring in the electric machine (for example, the processes associated with the slot leakages, the losses in the windings, magnetizing phenomena, etc.). Meanwhile, the transformation of the equivalent circuit in Fig. 6.5 is associated with the elimination in it of the inductively coupled elements. In this regard, the direct use for this purpose of the equivalent circuit in Fig. 6.5 representing the electromagnetic processes in the electric machine as a whole should be recognized as irrational. In addition, with this approach, certain difficulties in providing the necessary level of consistency, clarity, and methodicality in the presentation of the processes associated with the transformation of the machine equivalent circuit arise.

In this work, the machine equivalent circuit (Fig. 6.5) follows based on the use of the equivalent circuits representing the machine design fragments shown in Fig. 6.4. As follows from Fig. 6.4, the inductively coupled elements are used in the air gap equivalent circuit (Fig. 6.4a). In this regard, for solving the problem of the elimination in the machine equivalent circuit of the inductively coupled elements, the air gap equivalent circuit shown in Fig. 6.4a can be used in our case. Therefore, we can restrict ourselves to transforming this equivalent circuit (the air gap equivalent circuit), as a result of which we can get the equivalent circuit for the air gap without the inductively coupled elements. Now, constructing the machine equivalent circuit can be implemented by the cascade way of connecting the converted air gap equivalent circuit and the stator and rotor slot leakage equivalent circuits adapted to this transformation. Obtained in this way the machine equivalent circuit will not contain the inductively coupled elements, and therefore, it will have a simplified structure. Below, the technique for constructing the machine equivalent circuit by the cascade way of connecting, we will consider in more detail, and we will start this construction process by transforming the air gap equivalent circuit.

6.12 "Fragmentation" Concept: Features of Application

As shown above, the machine equivalent circuit can be obtained as a result of the cascade way of connecting the equivalent circuits for the fragments reflecting the corresponding elements of the machine design. In our considered case, the fragments of the machine equivalent circuit are determined in relation to the emf induced in the stator and rotor windings, and they represent the separate regions of the machine magnetic circuits corresponding, in our case, to the air gap region, the stator slot-teeth region, and the rotor slot-teeth region. In the engineering practice, the machine magnetizing circuit and the stator and rotor slot leakage magnetic circuits are considered, and they are analyzed as a whole. In this case, the machine magnetizing circuit and the stator and rotor slot leakage magnetic circuits are broken into the regions, and this process can be considered as corresponding to the fragmentation of the machine magnetizing circuit and the stator and rotor slot leakage magnetic circuits. By the use of the machine magnetizing circuit and the stator and rotor slot leakage magnetic circuits fragmentized in this way into the corresponding regions, the machine magnetizing flux and the stator and rotor slot leakage magnetic fluxes can be determined. For this, the total current law equation can be used. Now, by the electromagnetic induction law, the machine magnetizing emf and the stator and rotor slot leakage emf are established. In this case, the machine magnetizing reactance arises from the magnetizing emf equation, and the stator and rotor slot leakage reactances follow from the stator and rotor slot leakage emf equations. The machine magnetizing reactance and the stator and rotor slot leakage reactances obtained in this way are used in the machine equivalent circuit.

However, to determine these reactances (the machine magnetizing reactance and the stator and rotor slot leakage reactances), a different approach can be used. The techniques described above are based on the determination of the "partial" magnetic fields (the machine magnetizing field and the stator and rotor slot leakage magnetic fields). The machine magnetizing reactance and the stator and rotor slot leakage reactance can be determined based on the consideration of the machine resulting magnetic field [10]. For this, using the machine magnetizing circuit and the stator and rotor slot leakage magnetic circuits fragmentized into the corresponding regions, the machine magnetizing equivalent circuit and the stator and rotor slot leakage equivalent circuits can be constructed. Now, applying the superposition principle for these equivalent circuits, the "unfolded" machine equivalent circuit can be developed. The "unfolded" machine equivalent circuit obtained in this way corresponds to the resulting magnetic field taking place in the electric machine. Then, to receive the machine magnetizing reactance and the stator and rotor slot leakage reactances, the "unfolded" machine equivalent circuit is reduced to the "folded" machine equivalent circuit for what the multiple transfer technique can be used [10].

In this case, the "unfolded" machine equivalent circuit can be considered as a result of the cascade way of connecting the equivalent circuits for the fragments reflecting the corresponding regions into which the machine magnetic circuits are broken. With this approach, the "fragmentation" concept can be considered as the way of the subdivision into the regions of the machine magnetic circuits. Below, in this work, such interpretation of the "fragmentation" concept has received further development.

In this chapter, the machine equivalent circuit in Fig. 6.5 is obtained with the use of the superposition principle, the "inductively coupled circuits" concept, the "fragmentation" concept, as well as with the application of the cascade way of connecting. However, as follows from Fig. 6.5, the machine equivalent circuit arising based on the use of these provisions contains the inductively coupled elements, and therefore, it is not suitable for practical application. In [10], it is shown that the machine equivalent circuit suitable for practical use follows based on the application of the superposition principle, the "fragmentation" concept, and the "no-load and under-load" concept as well as the use of the cascade way of connecting and the multiple transfer technique. Below, we will consider the features of the application of these provisions used to construct the machine equivalent circuit suitable for practical calculations.

6.13 Summary

1. In the theory of electric machines, the determination of the voltage equations for the machine windings is based on the use of the "inductively coupled circuits" concept. These equations are written in relation to the voltage values obtained on the terminals of the stator and rotor windings. The voltage equations drawing up by the use of the "inductively coupled circuits" concept contain the inductively coupled elements. These voltage equations are considered as the initial equations to obtain the voltage equations suitable for engineering calculations.
2. In this work, instead of the voltage equations written in relation to the value of the voltages determined on the terminals of the stator and rotor windings, the equations drew up in relation to the emf induced in the stator and rotor windings are used. These emf equations can be obtained as a result of breaking the voltage equations into separate fragments ("fragmentation" concept). An important feature of the emf equations obtained in this way is that they reflect separately the more homogenous and less complex processes taking place in the electric machine, which helps to simplify analysis.
3. The emf equations representing the separate fragments of the machine voltage equations can be visualized by the corresponding equivalent circuits. The equivalent circuits obtained in this way are built in relation to the emf induced in the stator and rotor windings, and therefore, they can be considered as separate fragments of the machine equivalent circuit. In this regard, the fragment equivalent circuits can be used to construct the machine equivalent circuit for what the cascade way of connecting them can be applied. Consequently, the "fragmentation" concept allows us to construct the machine equivalent circuit by applying the cascade way of connecting which opens up, as will be shown below, the new opportunities for the further improvement of the theory of the machine calculations.
4. In this work, the stator slot leakage equivalent circuit, the air gap equivalent circuit, and the rotor slot leakage equivalent circuit are represented as the fragments of the machine equivalent circuit. In this regard, the machine equivalent circuit arises as a result of the cascade way of connecting the stator slot leakage equivalent circuit, the air gap equivalent circuit, and the rotor slot leakage equivalent circuit. Consequently, the process of constructing the machine equivalent circuit is greatly simplified when the fragment equivalent circuits are connected in the cascade way.

References

1. Kostenko, M., Piotrovskij, L.: Alternating Current Electric Machines, Publishing Energiya, Leningrad (1973)
2. Treshchev, I.: Electromechanical Processes in AC Electric Machines, Energiya, Leningrad (1980)

3. Vazhnov, A.: Electric Machines, Publishing Energiya, Leningrad (1968)
4. Ivanov-Smolensky, A.: Electric Machines, Publishing Energiya, Moscow (1980)
5. Seely, S.: Electromechanical Energy Conversion, McGraw-Hill Book Company, Inc., New York (1962)
6. Say, M.: Alternating Current Machines, John and Wiley, New York (1902)
7. Krause, P.: Analysis of Electric Machinery, McGraw-Hill Book Company, New York (1986)
8. Sarema, M.: Electric Machines. Steady-State Theory and Dynamic Performance, West Publishing Company, St. Paul, New York (1986)
9. Chapman, S.: Electric Machinery Fundamentals, 4th ed., McGraw-Hill, Inc., New York (2005)
10. Asanbayev, V.: Alternating Current Multi-Circuit Electric Machines: A New Approach to Steady-State Parameter Determination, Springer Cham Heidelberg, New York, Dordrecht London (2015).
11. Kazovsky, E.: Transient Processes in AC Electric Machines, Publishing of Academy of Science USSR, Moscow-Leningrad (1962)
12. Voldek, A.: Electric Machines, Publishing Energiya, Leningrad (1974)

Chapter 7
EMF Equations of Air Gap: Self- and Mutual Induction Reactances

7.1 Self- and Mutual Induction Reactances: Features of Determination

We will proceed from the air gap emf equations given in (6.19) in the form

$$
\begin{aligned}
(-\dot{Э}_1) &= jx_{11}\dot{I}_1 + jx_{21}\dot{I}_2 \\
(-\dot{Э}_2) &= jx_{12}\dot{I}_1 + jx_{22}\dot{I}_2
\end{aligned}
$$

This system of the equations has been written for the fragment located in the machine equivalent circuit (Fig. 6.5) between the terminals $a - b$ and the terminals $c - d$ and containing the inductively coupled elements (x_{11}, x_{12}, x_{21}, and x_{22}). The Eq. (6.19) are expressed in relation to the emf of the stator and rotor windings ($\dot{Э}_1$ and $\dot{Э}_2$) caused by the self- and mutual induction magnetic fluxes.

The Eq. (6.19) can be also obtained, as it was noted above, from the system of the Eq. (6.4) and (6.5) if in them to accept that the ohmic resistances and the leakage reactances of the stator and rotor windings are equal to zero, i.e. $r_1 = 0$, $r_2/s = 0$ and $x_{1\sigma} = 0$, $x_{2\sigma} = 0$. Consequently, in accordance with the Eq. (6.19) written in relation to the emf $\dot{Э}_1$ and $\dot{Э}_2$, we can remind again that the currents of the stator and rotor windings create the self- and mutual induction magnetic fluxes that are closed along the main magnetic circuit (magnetizing circuit) of the electric machine. Moreover, when deriving the Eq. (6.19), it was assumed that the magnetic permeabilities of the stator and rotor steel are relatively large, i.e. $\mu_c = \infty$. This indicates that there is no magnetic flux in the stator and rotor steel. In these conditions, the magnetic flux exists only in the air gap of the machine, and it can be determined as a result of the superposition of the self- and mutual induction magnetic fluxes created by the currents flowing in the stator and rotor windings. In this regard, we can assert that the determination of the reactances x_{11}, x_{12}, x_{21}, and x_{22} used in the Eq. (6.19) is reduced to the calculation of the self- and mutual induction magnetic fluxes existing

V. Asanbayev, *Asynchronous Machines*,
https://doi.org/10.1007/978-3-030-92284-9_7

in the air gap of the electric machine. Below, we will use these provisions when determining the reactances applied in the air gap emf equations.

7.2 Approaches to Determination of Reactances

We proceed from the fact that the air gap equivalent circuit arises from the linear system of the emf equations obtained in (6.19). The circuit visualization of the Eq. (6.19) is shown in Fig. 6.4a. In accordance with Fig. 6.4a, the stator and rotor windings are expressed as the inductively coupled two circuit loops. In our case, the reactances x_{11}, x_{12}, x_{21}, and x_{22} used in the Eq. (6.19) and in the equivalent circuit in Fig. 6.4a can be obtained by the use of the superposition principle, for example, [1–8].

In the technical literature, when determining the reactance x instead of the emf $(-\dot{Э})$ induced in the winding by the magnetic flux created by the current $\dot{I}$, the voltage drop across the reactance x, i.e. $\dot{Э} = x\dot{I}$, is used. The value of this voltage is equal to the module of the emf $(-\dot{Э})$, and in the phase, this voltage is directed opposite to the emf $(-\dot{Э})$. Then, for the reactance x, we have that $x = \dot{Э}/\dot{I}$. Below, the reactances x_{11}, x_{12}, x_{21}, and x_{22} of the system of the Eq. (6.19), we will determine taking into account this provision.

In accordance with (6.15), the reactance x_{11} is calculated as the self-induction reactance determined in relation to the emf $\dot{Э}_{11}$ induced in the stator winding by the self-induction magnetic flux created by current of the stator winding $\dot{I}_1$, when the rotor winding is open $\left(\dot{I}_2 = 0\right)$. Consequently, we have

$$x_{11} = \left(\frac{\dot{Э}_{11}}{\dot{I}_1}\right)_{\dot{I}_2=0} \tag{7.1}$$

According to (6.15), the reactance x_{21} is calculated as the mutual induction reactance determined in relation to the emf $\dot{Э}_{21}$ induced in the stator winding by the mutual induction magnetic flux created by current of the rotor winding $\dot{I}_2$, when the stator winding is open $\left(\dot{I}_1 = 0\right)$. Then, the following expression arises

$$x_{21} = \left(\frac{\dot{Э}_{21}}{\dot{I}_2}\right)_{\dot{I}_1=0} \tag{7.2}$$

In accordance with the expression given in (6.15), the reactance x_{12} is calculated as the mutual induction reactance determined in relation to the emf $\dot{Э}_{12}$ induced in the rotor winding by the mutual induction magnetic flux created by the current of the stator winding $\dot{I}_1$, when the rotor winding is open $\left(\dot{I}_2 = 0\right)$. Now, the expression of the following form can be obtained

$$x_{12} = \left(\frac{\dot{Э}_{12}}{\dot{I}_1}\right)_{\dot{I}_2=0} \tag{7.3}$$

With considering (6.15), the reactance x_{22} is found as the self-induction reactance determined in relation to the emf $\dot{Э}_{22}$ induced in the rotor winding by the self-induction magnetic flux created by current of the rotor winding $\dot{I}_2$, when the stator winding is open $\left(\dot{I}_1 = 0\right)$. As a result, we can obtain

$$x_{22} = \left(\frac{\dot{Э}_{22}}{\dot{I}_2}\right)_{\dot{I}_1=0} \tag{7.4}$$

In accordance with (7.1)–(7.4), the reactances x_{11}, x_{12}, x_{21}, and x_{22} are determined by the values of the emf $\dot{Э}_{11}, \dot{Э}_{21}, \dot{Э}_{12}$, and $\dot{Э}_{22}$ inducing by the self-and mutual induction magnetic fluxes created by the currents of the stator and rotor windings, individually. Based on the expressions given in (6.15) for the emf $\dot{Э}_{11}, \dot{Э}_{21}, \dot{Э}_{12}$, and $\dot{Э}_{22}$, we can receive

$$\begin{aligned} \dot{Э}_{11} &= \omega_1 w_1 k_{w1} \Phi_{11} \quad & \dot{Э}_{21} &= \omega_1 w_1 k_{w1} \Phi_{21} \\ \dot{Э}_{12} &= \omega_1 w_2 k_{w2} \Phi_{12} \quad & \dot{Э}_{22} &= \omega_1 w_2 k_{w2} \Phi_{22} \end{aligned} \tag{7.5}$$

As follows from (7.5), to determine the emf $\dot{Э}_{11}, \dot{Э}_{21}, \dot{Э}_{12}$, and $\dot{Э}_{22}$ it is necessary to have the values of the magnetic fluxes Φ_{11}, Φ_{21}, Φ_{12}, and Φ_{22} creating by the currents of the stator and rotor windings, individually. In Fig. 7.1, the conditional lines of the self-induction magnetic flux Φ_{11} and the mutual induction magnetic flux Φ_{12} created by the current of the stator winding $\dot{I}_1$ are represented. Here, for the fluxes Φ_{11} and Φ_{12}, we have that $\Phi_{11} = \Phi_{12}$. In Fig. 7.2, the conditional lines of the self-induction magnetic flux Φ_{22} and mutual induction magnetic flux Φ_{21} created by the current of the rotor winding $\dot{I}_2$ are shown. For the fluxes Φ_{22} and Φ_{21}, the condition of kind $\Phi_{22} = \Phi_{21}$ is true. In accordance with the assumptions made earlier, the magnetic fluxes Φ_{11}, Φ_{21}, Φ_{12}, and Φ_{22} can be determined as a result of the calculation of the magnetic fluxes taken place in the machine air gap.

7.3 Flat Model of Air Gap

The magnetic field in the machine air gap is created by the currents flowing in the stator and rotor windings. The windings used in the electric machines are located in the axial slots milled on the stator and rotor surfaces. Taking into account that the magnetic permeabilities of the stator and rotor steel are assumed to be conditionally equal to the infinity, i.e. $\mu_c = \infty$, the magnetic fluxes created in the machine air gap by the currents flowing in machine windings weakly depend on the size of the winding conductors and the location of these conductors in the slots. According to

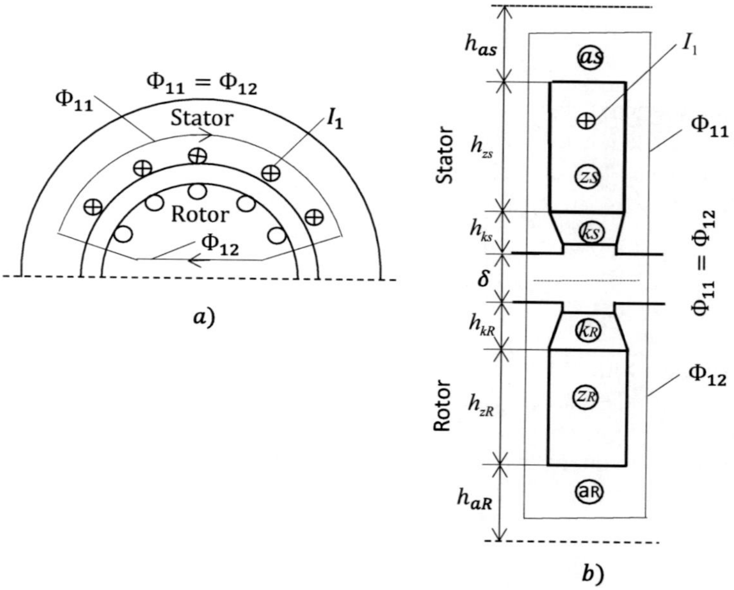

Fig. 7.1 Conditional line of the mutual induction magnetic flux created by the stator winding current: (**a**) conditional line of the mutual induction magnetic flux created in the phase zone by the stator winding current; (**b**) conditional line of the mutual induction magnetic flux created in the single tooth pitch zone by the stator winding current

[5], this statement is the basis for presenting the slot conductors as tightened into the one thread of the infinity small diameter. This thread of the infinity small diameter is located in the center of the slot slits and has the current equal to the total current of all conductors lying in the slot. In this case, the calculation length of the machine air gap determined as $\delta' = k_\delta \delta$ is introduced. Here, the value of k_δ represents the air gap factor. In this regard, the toothed surfaces of the stator and rotor are replaced by the conventional smooth surfaces. It is assumed that the length of the air gap is small compared to the pole pitch. Therefore, in this work, the flat air gap is considered, and it is assumed that the magnetic lines cross the machine air gap radially, i.e. we are dealing with the one-dimensional magnetic field in the air gap of the electric machine. The calculated length of the pole pitch τ_0 determined on the surface passing through the middle of the radial length of the machine air gap is introduced. In relation to this surface, the machine air gap is represented as the flat layer. The axial length of this surface is l_0, and it presents the calculated axial length of the electric machine [1–12]. In these conditions, the concept of the magnetomotive force (mmf) of the winding can be introduced. The use of the winding mmf contributes to the significant simplification of the process of analyzing the magnetic field in the machine air gap. The mmf represents the distribution on the stator and rotor smooth surfaces of the total slot currents of the windings. In accordance with (3.23), the mmf provides the passage of the magnetic flux through the single air gap. Now, to

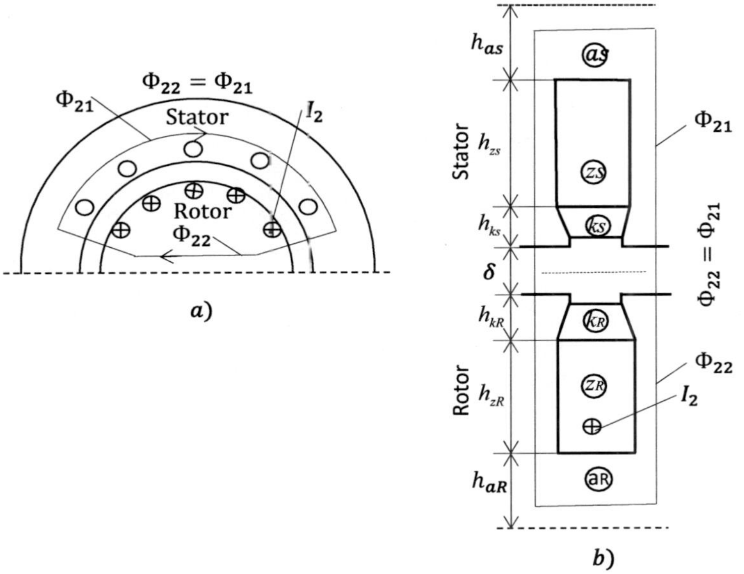

Fig. 7.2 Conditional line of the mutual induction magnetic flux created by the rotor winding current (**a**) conditional line of the mutual induction magnetic flux created in the phase zone by the rotor winding current; (**b**) conditional line of the mutual induction magnetic flux created in the single tooth pitch zone by the rotor winding current

determine the magnetic flux density in the air gap (on the surface passing through the middle of the radial length of the air gap) the total current law can be used, for example, [1–11]. In this case, the air gap model can be considered as the flat layer having the smooth surfaces on which the stator and rotor winding mmf act.

7.4 Magnetic Fluxes in Air Gap

We will consider the case when the stator winding is flowed by the current. In the presence of the current in the stator winding ($\dot{I}_1 \neq 0$ and $\dot{I}_2 = 0$), the magnetic flux density in the air gap is determined by the following expression obtained in (4.10) in the form

$$B_1 = \frac{\mu_0}{\delta'} F_1 \tag{7.6}$$

where $F_1 = \dfrac{m_1\sqrt{2}}{\pi} \dfrac{w_1 k_{w1}}{p} I_1$.

Here, the value F_1 represents the stator winding mmf acting on the stator bore surface. With considering (7.6), the self-induction magnetic flux $\dot{\Phi}_{11}$ created in the air gap by the stator winding current is calculated as

$$\Phi_{11} = \frac{2}{\pi} B_1 \tau_0 l_0 \tag{7.7}$$

The magnetic flux Φ_{11} can be represented through the stator winding mmf F_1. Then, from (7.6) and (7.7), we can have

$$\Phi_{11} = \frac{2}{\pi} B_1 \tau_0 l_0 = \frac{2}{\pi} \frac{\mu_0}{\delta'} \tau_0 l_0 F_1 \tag{7.8}$$

We will consider the case when the rotor winding is flowed by the current, i.e. $\dot{I}_2 \neq 0$. The current in the stator winding is absent, i.e. $\dot{I}_1 = 0$. In this case, the total current law can be used. As a result, by an analogy with (4.11a), we can obtain the expression for the magnetic flux density created in the air gap by the rotor winding current

$$B_2 = \frac{\mu_0}{\delta'} F_2 \tag{7.9}$$

where $F_2 = \frac{m_2 \sqrt{2}}{\pi} \frac{w_2 k_{w2}}{p} I_2$.

Here, the value F_2 represents the rotor winding mmf acting on the rotor surface. The flux density B_2 creates in the air gap the self-induction magnetic flux Φ_{22} determined in accordance with (7.7) as

$$\Phi_{22} = \frac{2}{\pi} B_2 \tau_0 l_0 \tag{7.10}$$

The magnetic flux Φ_{22} can be represented through the rotor winding mmf F_2. Then, from (7.7) to (7.10), it follows

$$\Phi_{22} = \frac{2}{\pi} B_2 \tau_0 l_0 = \frac{2}{\pi} \frac{\mu_0}{\delta'} \tau_0 l_0 F_2 \tag{7.11}$$

The expressions (7.8) and (7.11) for the self-induction magnetic fluxes $\dot{\Phi}_{11}$ and $\dot{\Phi}_{22}$ can be used to determine the mutual induction magnetic fluxes $\dot{\Phi}_{12}$ and $\dot{\Phi}_{21}$. These magnetic fluxes are calculated by the use of the conditions of the form $\dot{\Phi}_{12} = \dot{\Phi}_{11}$ and $\dot{\Phi}_{21} = \dot{\Phi}_{22}$. Therefore, from (7.9) to (7.11), we receive

$$\dot{\Phi}_{12} = \frac{2}{\pi} \frac{\mu_0}{\delta'} \tau_0 l_0 \dot{F}_1 \quad \dot{\Phi}_{21} = \frac{2}{\pi} \frac{\mu_0}{\delta'} \tau_0 l_0 \dot{F}_2 \tag{7.12}$$

The magnetic fluxes Φ_{11}, Φ_{21}, Φ_{12}, and Φ_{22} arising from the expressions given in (7.8), (7.11), and (7.12) can be used to determine the emf $\dot{Э}_{11}, \dot{Э}_{21}, \dot{Э}_{12}$ and $\dot{Э}_{22}$. The expressions for the emf $\dot{Э}_{11}, \dot{Э}_{21}, \dot{Э}_{12}$, and $\dot{Э}_{22}$ obtained in this way allow us to calculate the self- and mutual induction reactances x_{11}, x_{21}, x_{12}, and x_{22} used in the air gap emf equations.

7.5 Self- and Mutual Induction Reactances: Expressions for Calculating

As it was noted above, the air gap emf equations contain the self-and mutual induction reactances x_{11}, x_{21},x_{12}, and x_{22}. To determine these reactances it is necessary to define the values of the emf $\dot{Э}_{11}, \dot{Э}_{21,}\dot{Э}_{12}$, nd $\dot{Э}_{22}$ induced in the stator and rotor windings by the magnetic fluxes Φ_{11}, $\Phi_{21,}\Phi_{12}$ and Φ_{22}. For this purpose, the expressions obtained in (7.5) can be used. The magnetic fluxes Φ_{11}, $\Phi_{21,}\Phi_{12}$, and Φ_{22} used in the expressions (7.5) are determined by the formulas received in (7.8), (7.11), and (7.12).

Self-induction reactance of stator winding. Based on the expressions (7.1), (7.5), and (7.8), the self-induction emf $\dot{Э}_{11}$ of the stator winding can be determined as

$$\dot{Э}_{11} = \frac{\omega_1 \mu_0 \tau_0^2}{\pi^2 \delta'} k_1 \dot{I}_1 = x_{11} \dot{I}_1 \tag{7.13}$$

where $k_1 = \dfrac{2m_1 (w_1 k_{w1})^2 l_0}{p \tau_0}$.

From (7.1) and (7.13), the self-induction reactance x_{11} of the stator winding acquires the form

$$x_{11} = \frac{\omega_1 \mu_0 \tau_0^2}{\pi^2 \delta'} k_1 \tag{7.14}$$

Mutual induction reactance of the rotor winding. The magnetic flux $\dot{\Phi}_{11}$ being the self-induction magnetic flux of the stator winding is also the mutual induction magnetic flux by its nature. Therefore, this flux interlinks with the stator winding, and at the same time, it crosses the rotor winding creating with the rotor winding the mutual induction flux linkage determined in accordance with (6.10) as $\dot{\Psi}_{12} = w_2 k_{w2} \dot{\Phi}_{11} = w_2 k_{w2} \dot{\Phi}_{12}$. Here, we have the condition of the form $\dot{\Phi}_{11} = \dot{\Phi}_{12}$. The double indexing of the flux $\dot{\Phi}_{12}$ emphasizes that it is created by the stator winding current, and it crosses the rotor winding inducing in it the emf $\dot{Э}_{12}$. From (7.4), (7.5), and (7.12), the emf $\dot{Э}_{12}$ is defined as

$$\dot{Э}_{12} = \frac{\omega_1 \mu_0 \tau_0^2}{\pi^2 \delta'} k_{12} \dot{I}_1 = x_{12} \dot{I}_1 \tag{7.15}$$

where $k_{12} = \dfrac{2m_1 (w_1 k_{w1} w_2 k_{w2}) l_0}{p \tau_0}$.

From (7.4) and (7.15), the mutual induction reactance x_{12} of the rotor winding is received as

$$x_{12} = \frac{\omega_1 \mu_0 \tau_0^2}{\pi^2 \delta'} k_{12} \tag{7.16}$$

Self-induction reactance of rotor winding. The self-induction magnetic flux $\dot{\Phi}_{22}$ induces in the rotor winding the self-induction emf $\dot{Э}_{22}$ the value of which based on the expressions (7.2), (7.5), and (7.11) acquires

$$\dot{Э}_{22} = \frac{\omega_1 \mu_0 \tau_0^2}{\pi^2 \delta'} k_2 \dot{I}_2 = x_{22} \dot{I}_2 \tag{7.17}$$

where $k_2 = \frac{2m_2 (w_2 k_{w2})^2 l_0}{p\tau_0}$.

By the expressions (7.2) and (7.17), we have for the self-induction reactance x_{22} of the rotor winding

$$x_{22} = \frac{\omega_1 \mu_0 \tau_0^2}{\pi^2 \delta'} k_2 \tag{7.18}$$

Mutual induction reactance of the stator winding. The magnetic flux $\dot{\Phi}_{22}$ being the self-induction flux of the rotor winding is the mutual induction magnetic flux by its nature. Therefore, this flux interlinks with the rotor winding, and at the same time, it crosses the stator winding creating with the stator winding the mutual induction flux linkage determined in accordance with (6.10) as $\dot{\Psi}_{21} = w_1 k_{w1} \dot{\Phi}_{22} = w_1 k_{w1} \dot{\Phi}_{21}$. Here, we have the condition of the form $\dot{\Phi}_{22} = \dot{\Phi}_{21}$. The double indexing of the flux $\dot{\Phi}_{21}$ emphasizes that $\dot{\Phi}_{21}$ is created by the rotor winding current and it crosses the stator winding inducing in it the emf $\dot{Э}_{21}$. From (7.3), (7.5) and (7.12), the emf $\dot{Э}_{21}$ is obtained as

$$\dot{Э}_{21} = \frac{\omega_1 \mu_0 \tau_0^2}{\pi^2 \delta'} k_{21} \dot{I}_2 = x_{21} \dot{I}_2 \tag{7.19}$$

where $k_{21} = \frac{2m_2 (w_1 k_{w1} w_2 k_{w2}) l_0}{p\tau_0}$.

From (7.3) and (7.19), the mutual induction reactance x_{21} of the stator winding is determined in the form

$$x_{21} = \frac{\omega_1 \mu_0 \tau_0^2}{\pi^2 \delta'} k_{21} \tag{7.20}$$

Based on the expressions given in (7.14), (7.16), (7.18), and (7.20), the self- and mutual induction reactances used in the air gap emf equations can be represented as

$$x_{11} = \frac{\omega_1 \mu_0 \tau_0^2}{\pi^2 \delta'} k_1 \quad x_{21} = \frac{\omega_1 \mu_0 \tau_0^2}{\pi^2 \delta'} k_{21}$$
$$x_{12} = \frac{\omega_1 \mu_0 \tau_0^2}{\pi^2 \delta'} k_{12} \quad x_{22} = \frac{\omega_1 \mu_0 \tau_0^2}{\pi^2 \delta'} k_2 \qquad (7.21)$$

Here, the coefficients k_1, k_{21}, k_{12}, and k_2 are determined by the expressions given in (7.13), (7.15), (7.17), and (7.19). In accordance with the expressions obtained in (7.14), (7.16), (7.18), and (7.20), the self- and mutual induction reactances x_{11}, x_{21}, x_{12}, and x_{22} are defined by the self- and mutual induction magnetic fluxes Φ_{11}, Φ_{12}, Φ_{21}, and Φ_{22} creating in the machine air gap by the currents of the stator and rotor windings, individually. In this regard, the expressions received in (7.21) do not take into account the saturation of the main magnetic circuit (magnetizing circuit) of the electric machine. We will consider the issues of taking into account the influence of the saturation below.

The air gap emf Eq. (6.19) represent the stator and rotor windings as the inductively coupled circuit loops are not used in practice.

As follows from (7.21), the mutual induction reactances x_{21} and x_{12} depend on the number of the phases m_1 and m_2 of the stator and rotor windings since the reactance x_{21} is determined with considering magnetic flux created by the rotor winding and the reactance x_{12} represents a result of the action of the stator winding magnetic flux. Therefore, when $m_1 \neq m_2$, the reactances x_{21} and x_{12} get the different values, i.e. $x_{21} \neq x_{12}$.

7.6 EMF Equations of Air Gap Through Referred Quantities

We will consider the air gap emf equations shown in (6.19). In this system of the equations, the referred values of the emf and current of the rotor winding can be used, i.e. $\dot{Э}_2' = \dot{Э}_2 k_Э$ and $\dot{I}_2' = \dot{I}_2 / k_I$, where $k_Э = \frac{w_1 k_{w1}}{w_2 k_{w2}}$ and $k_I = \frac{m_1}{m_2} \frac{w_1 k_{w1}}{w_2 k_{w2}}$. Then, in terms of (7.21), the air gap emf Eq. (6.19) acquire the form

$$\dot{Э}_1 = x_{11} \dot{I}_1 + x_{21} k_I \dot{I}_2' = \frac{\omega_1 \mu_0 \tau_0^2}{\pi^2 \delta'} k_1 (\dot{I}_1 + \dot{I}_2') = x_0 \dot{I}_0$$
$$\dot{Э}_2' = x_{12} k_Э \dot{I}_1 + x_{22} k_Э k_I \dot{I}_2' = \frac{\omega_1 \mu_0 \tau_0^2}{\pi^2 \delta'} k_1 (\dot{I}_1 + \dot{I}_2') = x_0 \dot{I}_0 \qquad (7.22)$$

where $x_0 = \frac{\omega_1 \mu_0 \tau_0^2}{\pi^2 \delta'} k_{1A}; k_1 = \frac{2 m_1 (w_1 k_{w1})^2 l_0}{p \tau_0}; \dot{I}_0 = (\dot{I}_1 + \dot{I}_2')$.

The system of the Eq. (7.22) can be visualized by the air gap equivalent circuit in Fig. 7.3. As follows from Figs. 6.4a and 7.3, the air gap equivalent circuit in Fig. 7.3 does not contain the inductively coupled elements and, therefore, has a very simple structure. Now, we can build the machine equivalent circuit using the air gap equivalent circuit shown in Fig. 7.3 and also the stator and rotor slot leakage

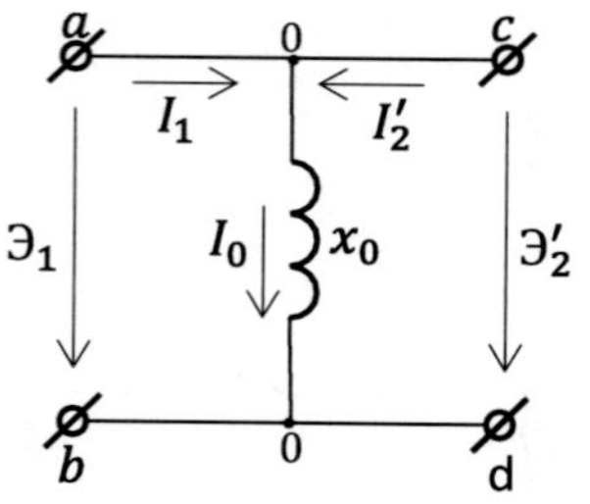

Fig. 7.3 Single-element air gap equivalent circuit

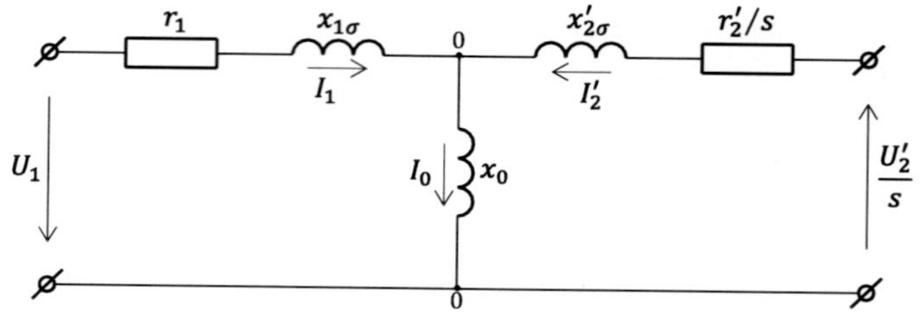

Fig. 7.4 Machine equivalent circuit (“inductively coupled circuits” concept)

equivalent circuits shown in Fig. 6.4b and c, respectively. The output terminals of the equivalent circuit in Fig. 7.3 have the referred values of the rotor winding emf and current. In this regard, the rotor slot leakage equivalent circuit (Fig. 6.4c) is expressed through the referred rotor quantities. Then, for the air gap equivalent circuit (Fig. 7.3) and for the stator and rotor slot leakage equivalent circuits (Fig. 6.4b and c) can be used the cascade way of connecting, as a result of which the machine equivalent circuit acquires the form in Fig. 7.4. The machine equivalent circuit obtained in this way does not contain the inductively coupled elements and has the simplified structure. Consequently, the use of the referred rotor quantities allows us to solve the problem of eliminating the inductively coupled elements in the machine equivalent circuit. The machine equivalent circuit in Fig. 7.4 has the traditional form. However, this equivalent circuit cannot be used effectively in the practice. The machine equivalent circuit in Fig. 7.4 has been obtained based on the use of the “inductively coupled circuits” concept, and in this regard, the slot leakage reactances $x_{1\sigma}$ and $x'_{2\sigma}$ in it is determined by the equations shown in (6.14). In these equations, the slot leakage flux linkages of the stator and rotor windings are created by the slot leakage magnetic fluxes determined using the expressions received in (6.8) and (6.9). The expressions (6.8) and (6.9) do not provide the sufficiently strict determination of the slot leakage magnetic fluxes of the stator and rotor windings. Therefore, the slot leakage reactances $x_{1\sigma}$ and $x'_{2\sigma}$ used in the machine equivalent circuit in Fig. 7.4 are calculated with the large error. Thus, within the “inductively coupled circuits” concept, although it is possible to obtain the machine equivalent circuit of the traditional form, however, its practical use seems difficult due to the presence of the problem associated with the calculation of the slot leakage reactances $x_{1\sigma}$ and $x'_{2\sigma}$.

As will be shown below, the solution to this problem is associated with the use of the "no-load and under-load" concept. Below, we will consider the features of the implementation of the "no-load and under-load" concept.

7.7 Summary

1. In this work, the air gap emf equations obtained in relation to the emf induced in the stator and rotor windings have been considered. These equations represent the stator and rotor windings as the inductively coupled circuit loops. In this regard, the air gap emf equations contain the self- and mutual induction reactances x_{11}, x_{21},x_{12}, and x_{22} the determination of which is associated with the calculation of the magnetic field taking place in the machine air gap.
2. It is considered the electric machine with the symmetrical design of the windings in which the symmetrical system of the sinusoidal currents is flowed out. The currents flowing in the stator and rotor windings create in the air gap sinusoidally distributed in space and in a time the waves of the flux density and mmf. In these conditions, the currents flowing in the stator and rotor windings create in the air gap the self- and mutual induction magnetic fluxes passing through the main machine magnetic circuit (magnetizing circuit). In this case, the lines of the self- and mutual induction magnetic flux densities are interlinked with the conditional turns of the stator and rotor windings. The determination of the magnetic fluxes created in the air gap by the currents of the stator and rotor windings is implemented by the use of the superposition principle.
3. In this work, the machine air gap is considered as the flat layer obtained in relation to the pole pitch τ_0 determined on the surface passing through the middle of the radial length of the air gap. The axial length of this surface is equal to l_0. It is assumed that the stator and rotor surfaces are smooth, and on these surfaces, the currents of the stator and rotor windings are placed in the form of infinitely thin layers. The space distribution of the total slot currents of the stator and rotor windings represents the mmf of these windings. It is assumed that the lines of the flux densities created by the mmf of the stator and rotor windings cross the air gap of the machine radially. In this case, the one-dimensional magnetic field takes place in the air gap.
4. In this work, the analytical expressions for the reactances x_{11}, x_{21},x_{12}, and x_{22} used in the air gap emf equations are recommended. The recommended expressions for the reactances x_{11}, x_{21},x_{12}, and x_{22} have the traditional forms. These reactances are obtained as a result of the determination of the magnetic field created in the air gap by the currents of the stator and rotor windings, individually.
5. The air gap emf equations obtained by the use of the "inductively coupled circuits" concept cannot be used in engineering practice. Therefore, in these equations, the inductively coupled elements are eliminated to obtain the new equations of the form convenient to the use. It is shown that for this purpose, in the air gap emf equations with the inductively coupled elements, the referred rotor

quantities can be applied. When applying the referred rotor quantities, the machine equivalent circuit obtained based on the “inductively coupled circuits” concept can be reduced to the new machine equivalent circuit having the traditional form. However, the slot leakage reactances used in this new machine equivalent circuit are difficult to calculate accurately. In the technical literature, the machine equivalent circuits suitable for practical use are built using the “no-load and under-load” concept.

References

1. Kostenko, M., Piotrovskij, L.: Alternating Current Electric Machines, Publishing Energiya, Leningrad (1973)
2. Piotrovskij, L.: Electric Machines, Publishing Energiya, Leningrad (1975)
3. Voldek, A.: Electric Machines, Publishing Energiya, Leningrad (1974)
4. Asanbayev, V.: Alternating Current Multi-Circuit Electric Machines: A New Approach to Steady-State Parameter Determination, Springer Cham Heidelberg, New York, Dordrecht London (2015).
5. Richter, R.: Electric Machines, vol. 1-5, United Scientific and Technical Publishing (Gosenergoizdat), Moscow (1934-19397).
6. Say, M.: Alternating Current Machines, John and Wiley, New York (1902)
7. White, D., Woodson, H.: Electromechanical Energy Conversion, John Wiley and Sons, Inc., New York (1064)
8. Seely, S.: Electromechanical Energy Conversion, McGraw-Hill Book Company, Inc., New York (1962)
9. Krause, P.: Analysis of Electric Machinery, McGraw-Hill Book Company, New York (1986)
10. Sarema, M.: Electric Machines. Steady-State Theory and Dynamic Performance, West Publishing Company, St. Paul, New York (1986)
11. Chapman, S.: Electric Machinery Fundamentals, 4th ed., McGraw-Hill, Inc., New York (2005)
12. Ivanov-Smolensky, A.: Electric Machines, Publishing Energiya, Moscow (1980)

Chapter 8
Equivalent Circuits for Design Fragments of Machine: "No-Load and Under-Load" Concept

8.1 "No-Load and Under-Load" Concept: Features of Application

In the operating mode, the resulting magnetic flux created by the currents flowing in the stator and rotor windings acts in the electric machine. In the sixth chapter, the "inductively coupled circuits" concept was used to describe the process of forming the resulting magnetic flux in the machine. In accordance with this concept, the machine resulting magnetic flux represents as a result of the superposition of the magnetic fluxes created by the currents of the stator and rotor winding, individually. In this chapter, for this purpose, we will apply another technique based on the use of the "no-load and under-load" concept.

8.1.1 "No-Load and Under-Load" Concept: Physical Meaning

The operating mode of the electric machine can be considered as a result of the superposition of the no-load mode and the under-load mode, for example, [1–10]. In this regard, the resulting magnetic fluxes of the stator and rotor windings can be represented as a result of the superposition of the magnetic fluxes taken place in the machine in the no-load mode and the magnetic fluxes created by the currents of the stator and rotor windings in the under-load mode ("no-load and under-load" concept). In this work, as it was noted earlier, the "no-load and under-load" concept is implemented in relation to the emf induced in the stator winding. Therefore, the no-load mode of the electric machine is reached when the rotor winding is open, and there is no current in it ($I_2 = 0$). In the under-load mode of the machine, the rotor winding is closed to the load impedance [1–13, 16–20]. In the technical literature, such scheme for implementing the "no-load and under-load" concept is most

V. Asanbayev, *Asynchronous Machines*,
https://doi.org/10.1007/978-3-030-92284-9_8

widespread. Below, we will give a brief description of the processes taking place in the electric machine in the no-load mode and under-load mode.

In the no-load mode (when there is no current in the rotor winding, i.e., $I_2 = 0$), the no-load current $\dot{I}_0$ is flowed in the stator winding. In accordance with the total current law, this current $\dot{I}_0$ causes the rotating magnetic flux $\dot{\Phi}_0$ in the machine. This rotating magnetic flux $\dot{\Phi}_0$ is interlinked with the stator and rotor windings, and in accordance with the electromagnetic induction law, it induces the emf $\dot{\text{Э}}_1$ in the stator winding and the emf $\dot{\text{Э}}_2$ in the rotor winding. When the rotor winding is connected to the load impedance, under the influence of the emf $\dot{\text{Э}}_2$ induced in the rotor winding in the no-load mode, the load current I_2 arises in it. This load current I_2 forms the mmf of the rotor winding F_2 which creates the rotor magnetic flux. The rotor magnetic flux induces in the stator winding emf under the influence of which the load component of the stator winding current $I_{1\text{н}}$ is created. On the condition that $\dot{\text{Э}}_1 = const$, the component of the stator winding current $I_{1\text{н}}$ will have such value and orientation that the mmf generated by this stator winding current completely balances the value of the rotor winding mmf F_2, i.e. $F_{1\text{н}} = -F_2$. This means that in the under-load mode, the currents of the stator and rotor windings do not create the total magnetic flux of the mutual induction. Thus, in the under-load mode, the currents of the stator and rotor windings form the leakage flux $\Phi_{1\sigma}$ embracing the stator winding and the leakage flux $\Phi_{2\sigma}$ interlinking with the rotor winding. Consequently, in this case, the resulting magnetic fluxes of the stator and rotor windings Φ_{1p} and Φ_{2p} can be represented as a result of the superposition of the magnetic flux Φ_0 taking place in the machine in the no-load mode and the leakage magnetic fluxes of the stator and rotor windings ($\Phi_{1\sigma}$ and $\Phi_{2\sigma}$) existing in the machine in the under-load mode. So, we have

$$
\begin{aligned}
\Phi_{1p} &= \Phi_{1\sigma} + \Phi_0 \\
\Phi_{2p} &= \Phi_{2\sigma} + \Phi_0
\end{aligned}
\tag{8.1}
$$

The Eq. (8.1) for the magnetic fluxes Φ_{1p} and Φ_{2p} are obtained based on the use of the "no-load and under-load" concept. The Eq. (6.10) for the magnetic fluxes Φ_{1p} and Φ_{2p} are determined by applying the "inductively coupled circuits" concept. As follows from the Eq. (8.1), the leakage magnetic fluxes of the stator and rotor windings $\Phi_{1\sigma}$ and $\Phi_{2\sigma}$ are used in them. In Eq. (6.10), the magnetic fluxes $\Phi_{1\sigma}$ and $\Phi_{2\sigma}$ reflect the leakage fluxes separated from the total self-induction magnetic fluxes of the stator and rotor windings. The magnetic fluxes $\Phi_{1\sigma}$ and $\Phi_{2\sigma}$ used in the Eq. (8.1) represent the leakage fluxes from the volume of the currents flowing in the stator and rotor slots. Thus, the leakage fluxes $\Phi_{1\sigma}$ and $\Phi_{2\sigma}$ applying in the Eqs. (6.10) and (8.1) have the different physical nature of the forming.

8.2 Current Equation for Winding of Stator

Above, the "no-load and under-load" concept was used to determine the equations for the resulting magnetic fluxes of the stator and rotor windings Φ_{1p} and Φ_{2p}. Here, we will consider the equation for the stator winding current. In accordance with the "no-load and under-load" concept the stator winding current I_1 can be represented as consisting of the two components: the no-load current (magnetizing current) I_0 and the under-load current $I_{1н}$, i.e. $I_1 = I_0 + I_{1н}$. Below, we will determine the components I_0 and $I_{1н}$ of the stator winding current I_1 and compile the equation for the stator winding currents.

8.2.1 Magnetizing Current

We will proceed from the fact that in the absence of the current in the rotor winding, i.e. $\dot{I}_2 = 0$, under the influence of the emf $\dot{Э}_1$ the magnetizing current $\dot{I}_0$ flows in the stator winding. The current $\dot{I}_0$ creates the rotating magnetic flux Φ_0. In Fig. 8.1, the picture of the conditional line of the magnetic flux density is given in the no-load

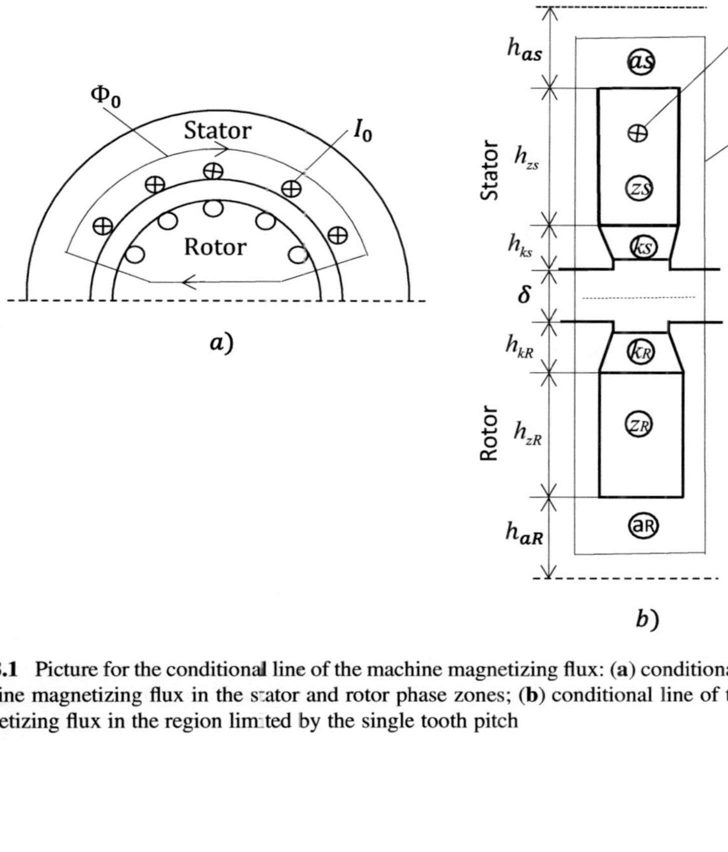

Fig. 8.1 Picture for the conditional line of the machine magnetizing flux: (**a**) conditional line of the machine magnetizing flux in the stator and rotor phase zones; (**b**) conditional line of the machine magnetizing flux in the region limited by the single tooth pitch

mode. The magnetic flux Φ_0 passes through the main magnetic circuit (magnetizing circuit) of the electric machine. We consider the case, when the magnetic permeability of the stator and rotor iron is relatively large, i.e. $\mu_c = \infty$. In these conditions, the magnetic flux Φ_0 is basically concentrated in the machine air gap. Therefore, the flux Φ_0 can be represented as the magnetizing flux of the machine air gap. In this regard, the emf $\dot{Э}_0$ induced by the flux Φ_0 can be expressed as the magnetizing emf of the air gap. The magnetic flux Φ_0 is determined on the surface passing through the middle of the radial length of the machine air gap. Therefore, the emf $\dot{Э}_0$ induced by the flux Φ_0 corresponds to the machine air gap surface. In our case, we have that $\dot{Э}_0 = \dot{Э}_1$, where $\dot{Э}_1$ is the emf induced in the stator winding in the no-load mode. Under the influence of the emf $\dot{Э}_0$, the magnetizing current of the air gap $\dot{I}_0$ corresponding to the no-load mode flows in the stator winding. Now, in accordance with (7.13) and (7.14), the emf $\dot{Э}_0$ can be expressed through the self-induction reactance ($x_{11} = x_0$) of the stator winding, i.e. $\dot{Э}_0 = x_0\dot{I}_0$. From here, for the magnetizing current $\dot{I}_0$ flowing in the stator winding, we have

$$\dot{I}_0 = \frac{\dot{Э}_0}{x_0} \tag{8.2}$$

8.2.2 Under-Load Current

Under the influence of the emf $\dot{Э}_2$ induced in the rotor winding in the no-load mode, the under-load current $\dot{I}_2$ arises in the rotor winding closed to the load impedance. The current $\dot{I}_2$ causes the mmf F_2 of the rotor winding. МДС F_2 creates the rotor magnetic flux Φ_{22}. The rotor flux Φ_{22} is interlinked with the stator winding and induces in it the emf under influence of which the load component of the stator winding current $\dot{I}_{1н}$ appears in the stator winding. The current $\dot{I}_{1н}$ creates the mmf $F_{1н}$ of the stator winding. Thus, in the under-load mode, the currents of the stator and rotor windings form the mmf $F_{1н}$ and F_2, respectively. МДС $F_{1н}$ and F_2 created by the currents of the stator and rotor windings cause the mutual induction magnetic fluxes in the air gap. However, in the under-load mode, the currents flowing in the stator and rotor windings form, in addition to the mutual induction magnetic fluxes, also the leakage fluxes of the stator and rotor windings. In this part, the phenomena associated with the leakage fluxes are not considered. They will be considered below. Here, we will analyze the mutual induction magnetic fluxes created by the currents of the stator and rotor windings in the under-load mode.

As it was noted earlier, the "no-load and under-load" concept is implemented in relation to the emf $\dot{Э}_1$ of the stator winding. Consequently, we can proceed from the fact that in the operating mode the mmf F_1 of the stator winding balances the mmf of the stator winding F_0 created by the no-load current $\dot{I}_0$ and the mmf $F_{1н}$ of the stator winding formed by the load component of the stator winding current $\dot{I}_{1н}$. Thus, we can obtain the condition of the form

$$F_1 = F_0 + F_{1н} \tag{8.3}$$

In accordance with (8.3), the components of the stator winding mmf F_0 and $F_{1н}$ are determined based on the various operating conditions of the electric machine. The mmf F_0 is created by the no-load current $\dot{I}_0$ and the МДС $F_{1н}$ is formed by the current of the stator winding $\dot{I}_{1н}$ taking place in the under-load mode. Although, the mmf F_0 and $F_{1н}$ represent the different operating conditions of the machine, according to (8.3), they determine in the sum the value of the stator winding mmf necessary to maintain the machine operating mode.

In the technical literature, the emf equations and the machine equivalent circuit are represented through the currents flowing in the machine windings. For this, the mmf equations for stator winding in the form of (8.3) can be used. The mmf equation obtained in (8.3) can be expressed through the stator winding current components $\dot{I}_0$ and $\dot{I}_{1н}$, forming the mmf F_0 and $F_{1н}$, respectively. According to (8.3), in the operating mode, the stator winding mmf F_1 balances the value of the stator winding mmf F_0 created by the current $\dot{I}_0$ flowing in the stator winding in the no-load mode and the value of the stator winding mmf $F_{1н}$, forming by the current $\dot{I}_{1н}$ flowing in the stator winding in the under-load mode, i.e. $F_1 = F_0 + F_{1н}$. Taking into account that $F_1 = k_{F1}\dot{I}_1$, $F_{1н} = k_{F1}\dot{I}_{1н}$ and $F_0 = k_{F1}\dot{I}_0$, where $k_{F1} = \frac{\sqrt{2}m_1 w_1 k_{w1}}{\pi p}$, the mmf equation of the form $F_1 = F_0 + F_{1н}$ can be represented through corresponding components of the stator winding current in the following form

$$\dot{I}_1 = \dot{I}_0 + \dot{I}_{1н} \tag{8.4}$$

In (8.4), the current $\dot{I}_0$ is calculated by the expression obtained in (8.2). The current $\dot{I}_{1н}$ presents the under-load component of the stator winding current, and it is needed to be determined.

The currents $\dot{I}_0$ and $\dot{I}_{1н}$ used in (8.4) flow in the stator winding, and therefore, these currents can be considered caused by the stator winding emf $\dot{Э}_1$. However, the current $\dot{I}_{1н}$ arises physically under the influence of the emf $\dot{Э}_2$ induced in the rotor winding in the no-load mode, and the current $\dot{I}_0$ is created under the influence of the stator winding emf $\dot{Э}_0$. In practice, the condition of kind $\dot{Э}_0 \neq \dot{Э}_2$ is usually satisfied for the emf induced in the stator and rotor windings. In these conditions, Eq. (8.4) obtained for the stator winding currents has no physical meaning. Equation (8.4) acquires the physical meaning when the condition of the form $\dot{Э}_0 = \dot{Э}_2$ is satisfied for the emf induced in the stator and rotor windings. To implement this condition, the stator and rotor windings must have the same number of turns. However, in practice, the stator and rotor windings are made with different number of turns. In these conditions, it is assumed that the calculated winding is placed on the rotor. The calculated potor winding and the stator winding acquire the same number of the turns. In practice, this technique is known as reducing the rotor winding to the number of turns of the stator winding. For the referred rotor winding, the referred value of the rotor winding emf $\dot{Э}_2'$ is used. In this case, the condition of the form $\dot{Э}_0 = \dot{Э}_2'$ is satisfied. In this regard, the value of the current $\dot{I}_{1н}$ applied in (8.4) is determined for the referred value of the rotor winding emf $\dot{Э}_2'$. Below, we will show it.

8.2.3 Under-Load Current: Determination

The load component of the stator winding current $\dot{I}_{1н}$ arises as the reaction to the appearance of the load current $\dot{I}_2$ in the closed rotor winding. The load current $\dot{I}_2$ is caused under the influence of the emf $\dot{Э}_2$ induced in the rotor winding in the no-load mode. The load current $\dot{I}_2$ creates the mutual induction magnetic flux Φ_{22} of the rotor winding. The magnetic flux Φ_{22} crosses the stator winding and induces in it the emf under influence of which the load component of the stator winding current $\dot{I}_{1н}$ arises in the stator winding. The load component of the stator winding current $\dot{I}_{1н}$ causes the mutual induction magnetic flux Φ_{11} of the stator winding. In accordance with Lenz's law, the stator flux Φ_{11} acts towards the rotor flux Φ_{22}, and it is strived to destroy it. The picture of the flux density lines of the magnetic fluxes Φ_{11} and Φ_{22} is shown in Fig. 8.2. In the conditions of the constancy of the stator winding emf, i.e. $\dot{Э}_1 = \dot{Э}_0 = const$, the stator flux Φ_{11} completely compensates the action of the rotor winding flux Φ_{22}, i.e. $\Phi_{11} = -\Phi_{22}$ as a result of which the stator winding mmf $F_{1н}$ and the rotor winding mmf F_2 do not create the mutual induction common magnetic field in the machine. Due to this, for the mmf $F_{1н}$ and F_2, the condition of the following form is satisfied

$$F_{1н} = -F_2 \tag{8.5}$$

It follows from (8.5) that the stator and rotor winding mmf $F_{1н}$ and F_2 are equal in the modulus and opposite in the phase. Thus, due to the fact that in the under-load mode the currents of the stator and rotor windings do not create the mutual induction common magnetic flux, this machine under-load mode does not lead to the change in the values of the emf $\dot{Э}_1$ and $\dot{Э}_2$ induced in the stator and rotor windings in the no-load mode.

The condition $F_{1н} = -F_2$ can be used to determine the load component of the stator winding current $\dot{I}_{1н}$. The value of the mmf $F_{1н}$ directly proportional to the load component of the stator winding current $\dot{I}_{1н}$, i.e. $F_{1н} = k_{F1}\dot{I}_{1н}$, where $k_{F1} = \frac{\sqrt{2}m_1 w_1 k_{w1}}{\pi p}$. The mmf F_2 directly proportional to the load current of the rotor winding $\dot{I}_2$, i.e. $F_2 = k_{F2}\dot{I}_2$, where $k_{F2} = \frac{\sqrt{2}m_2 w_2 k_{w2}}{\pi p}$. Here, the values k_{F1} and k_{F2} as follows from the given expressions, represent the dimensionless coefficients that depend on the design parameters of the machine windings. With considering these conditions,

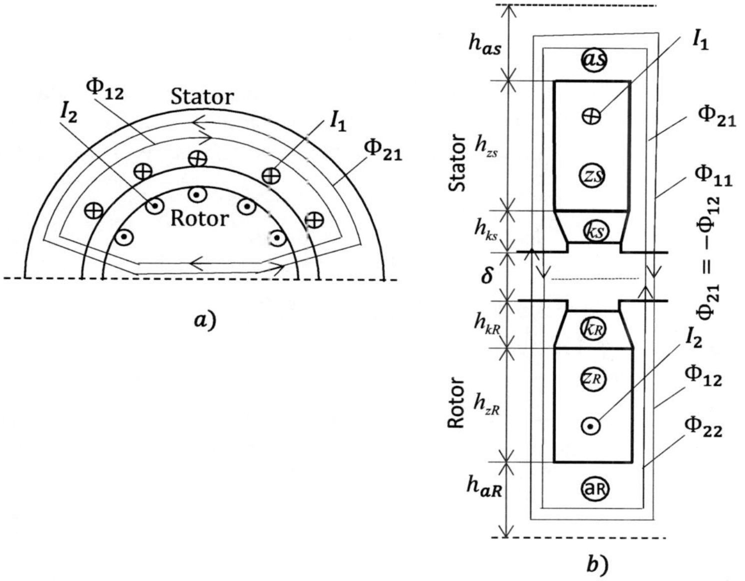

Fig. 8.2 Picture for the conditional lines of the mutual induction magnetic fluxes in the machine under-load mode: (**a**) conditional lines of the mutual induction magnetic fluxes in the stator and rotor phase zones; (**b**) conditional lines of the mutual induction magnetic fluxes in the region limited by the single tooth pitch

for the stator and rotor winding mmf $F_{1н}$ and F_2, expression (8.5) represented as $F_{1н} = -F_2$ can be given in the form

$$k_{F1}\dot{I}_{1н} = -k_{F2}\dot{I}_2$$

From here, we obtain

$$\dot{I}_{1н} = -\frac{k_{F2}}{k_{F1}}\dot{I}_2 = -\frac{\dot{I}_2}{k_I} = (-\dot{I}'_2) \tag{8.6}$$

where $k_I = \dfrac{k_{F1}}{k_{F2}}; \dot{I}'_2 = \dfrac{\dot{I}_2}{k_I}$.

In (8.6), $\dot{I}'_2$ referred to the stator winding value of the rotor winding current. The dimensionless factor k_I is the reduction factor of rotor winding current determined as

$$k_I = \frac{k_{F1}}{k_{F2}} = \frac{m_1 w_1 k_{w1}}{m_2 w_2 k_{w2}} \tag{8.7}$$

Consequently, from (8.6), we have

$$\dot{I}_{1н} = (-\dot{I}_2') \tag{8.8}$$

In accordance with the condition (8.8), the load component of the stator winding current is represented through the referred value of the rotor winding current and it can be considered as determined under the influence of the referred value of the rotor winding emf $\dot{Э}_2'$.

8.2.4 *Current Equation for Winding of Stator: Operating Mode*

Using the conditions (8.4) and (8.8), the equation for the stator winding current flowing in the stator winding in the operating mode can be expressed as

$$\dot{I}_1 = \dot{I}_0 + (-\dot{I}_2') \tag{8.9}$$

According to (8.9), the "no-load and under-load" concept is represented by the current equation for the stator winding. In this case, the value of the stator winding current $\dot{I}_1$ is considered as a result of the sum of the currents $\dot{I}_0$ and $(-\dot{I}_2')$ reflecting the components of the stator winding current taking place in the machine in the no-load mode and under-load mode, respectively. Consequently, based on Eq. (8.9), we can note that the implementation of the "no-load and under-load" concept is reduced to establishing the components of the stator winding current $\dot{I}_0$ and $(-\dot{I}_2')$ corresponding to the no-load mode and under-load mode of the machine. As will be shown below, Eq. (8.9) is very important for the calculation theory of the asynchronous machine.

8.3 Equivalent Circuits of Air Gap

The air gap equivalent circuit shown in Fig. 6.4a represents the fragment of the machine equivalent circuit obtained based on the "inductively coupled circuits" concept. This equivalent circuit is received in relation to the stator winding emf and current $(\dot{Э}_1, \dot{I}_1)$ and the rotor winding emf and current $(\dot{Э}_2, \dot{I}_2)$. We will transform this equivalent circuit shown in Fig. 6.4a for that the "no-load and under-load" concept can be used. In this work, the "no-load and under-load" concept is implemented by the use of the Eq. (8.9) obtained for the stator winding current. In accordance with this equation, the transformation of the air gap equivalent circuit in Fig. 6.4a is reduced to the determination by this equivalent circuit the stator winding current components $\dot{I}_0$ and $\dot{I}_2'$ arising from the consideration of the no-load mode and the under-load mode of the electric machine. As a result, the new air gap equivalent circuit represented through the stator winding current components $\dot{I}_0$ and $\dot{I}_2'$ follows. Below, the transformation of the air gap equivalent circuit given in Fig. 6.4a, we will implement proceed from the physical considerations.

8.3.1 Magnetizing Equivalent Circuit of Air Gap

In the no-load mode of the electric machine, the magnetic flux Φ_0 induces in the stator and rotor windings the emf $\dot{Э}_1$ and $\dot{Э}_2$ determining in accordance with (7.5) as

$$\dot{Э}_1 = \omega_1 w_1 k_{w1} \Phi_0 \quad \dot{Э}_2 = \omega_1 w_2 k_{w2} \Phi_0 \tag{8.10}$$

We will consider the ratio of the stator winding emf $\dot{Э}_1$ to the rotor winding emf $\dot{Э}_2$, i.e. $\dot{Э}_1/\dot{Э}_2 = k_Э$. From here, for the referred value of the rotor winding emf $\dot{Э}_2'$, it arises

$$\dot{Э}_1 = \dot{Э}_2 k_Э = \dot{Э}_2' \tag{8.11}$$

In (8.11), the dimensionless quantity $k_Э$ represents the reduction factor of the rotor winding emf determined from (8.10) and (8.11) as

$$k_Э = \frac{\dot{Э}_1}{\dot{Э}_2} = \frac{w_1 k_{w1}}{w_2 k_{w2}} \tag{8.12}$$

With considering (8.11) and (8.12), for the stator and rotor winding emf, we can have from (8.10)

$$\dot{Э}_1 = \omega_1 w_1 k_{w1} \Phi_0 \quad \dot{Э}_2' = \dot{Э}_2 k_Э = \omega_1 w_1 k_{w1} \Phi_0 \tag{8.13}$$

From here, it is established the condition $\dot{Э}_1 = \dot{Э}_2'$ under which the Eq. (8.2) is satisfied.

We consider the emf equations given in (8.13). The magnetic flux Φ_0 used in (8.13) takes place in the machine in the no-load mode. In accordance with (7.10), the flux Φ_0 is calculated by the expression of the following form

$$\Phi_0 = \frac{2\mu_0}{\delta'} \frac{m_1 w_1 k_{w1} \tau_0 l_0}{\pi^2 p} \dot{I}_0 \tag{8.14}$$

Here, $\dot{I}_0$ is considered as the magnetizing current of the electric machine. By the use of (8.14), the first equation of the system (8.13) obtains the form

$$\dot{Э}_1 = \frac{\omega_1\mu_0\tau_0^2}{\pi^2\delta'}\frac{2m_1w_1^2k_{w1}^2l_0}{p\tau_0}\dot{I}_0 = \frac{\omega_1\mu_0\tau_0^2}{\pi^2\delta'}k_1\dot{I}_0 \tag{8.15}$$

where $k_1 = \frac{2m_1w_1^2k_{w1}^2l_0}{p\tau_0}$; $\delta' = \delta k_\delta$.

Based on (8.8) and (8.15), the equations for the emf induced in the stator and rotor windings in the no-load mode of the machine take the form

$$\begin{aligned}\dot{Э}_1 &= \frac{\omega_1\mu_0\tau_0^2}{\pi^2\delta'}k_1\dot{I}_0 = x_0\dot{I}_0\\ \dot{Э}_2' &= \frac{\omega_1\mu_0\tau_0^2}{\pi^2\delta'}k_1\dot{I}_0 = x_0\dot{I}_0\end{aligned} \tag{8.16}$$

where $x_0 = \frac{\omega_1\mu_0\tau_0^2}{\pi^2\delta'}k_1$.

In (8.16), the value x_0 represents the magnetizing reactance of the machine air gap. From the system of the Eq. (8.16), for the magnetizing current of the air gap $\dot{I}_0$, we receive

$$\dot{I}_0 = \frac{\dot{Э}_1}{x_0} = \frac{\dot{Э}_2'}{x_0} \tag{8.17}$$

Based on the system of the Eq. (8.16), the air gap magnetizing equivalent circuit corresponding to the no-load mode of the machine acquires the form in Fig. 8.3a. In the equivalent circuit in Fig. 8.3a, the referred value of the rotor winding emf $\dot{Э}_2'$ is applied. This circumstance greatly simplifies the process of determining the actual value of the emf $\dot{Э}_2$ induced in the rotor winding. For this, in accordance with (8.11) and (8.12), the value of the emf $\dot{Э}_1$ induced in the stator winding, it is necessary to divide by the reduction factor $k_Э$ since $\dot{Э}_1 = \dot{Э}_2'$. For the electric machines having multiple windings on the rotor, this provision is very important.

8.3.2 *Equivalent Circuit of Air Gap: Operating Mode*

The air gap magnetizing equivalent circuit corresponding to the no-load mode of the machine is given in Fig. 8.3a. The magnetizing current $\dot{I}_0$ flowing in this equivalent circuit is determined by the expression obtained in (8.7). In the operating mode of the machine, the magnetizing current $\dot{I}_0$ arises from the equation shown in (8.9) and it is defined as

$$\dot{I}_0 = \dot{I}_1 + \dot{I}_2' \tag{8.18}$$

In accordance with the "no-load and under-load" concept, the operating mode of the machine can be represented as a result of the superposition of the no-load mode

Fig. 8.3 Fragments of the machine equivalent circuit: (**a**) air gap magnetizing equivalent circuit (no-load mode); (**b**) air gap equivalent circuit (operating mode); (**c**) stator slot leakage equivalent circuit; (**d**) rotor slot leakage equivalent circuit

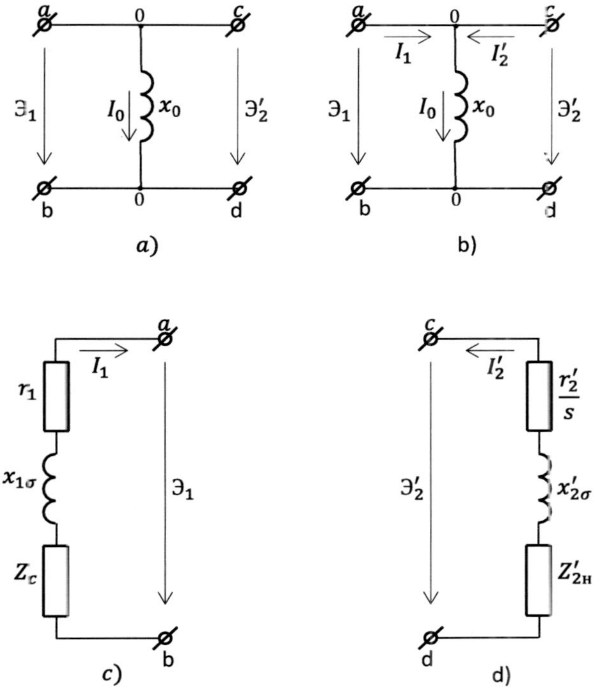

and under-load mode of the electric machine. In the under-load mode of the machine, as it was noted above, the currents flowing in the stator and rotor windings do not create the mutual induction common magnetic flux since the mmf from these currents acting towards each other compensate each other, i.e. $F_{1н} = \; - F_2$. Due to this, magnetizing flux existing in the machine in the no-load mode is retained in the operating mode of the electric machine. From here, it follows that in the operating mode of the machine the magnetizing current $\dot{I}_0$ arising from (8.18) creates the same magnetizing flux as the magnetizing current $\dot{I}_0$ determined in the no-load mode by the Eq. (8.17). Thus, the value of the magnetizing current $\dot{I}_0$ used in the air gap magnetizing emf Eq. (8.16) can be determined by the equation given in (8.18). Now, using the Eq. (8.18) in the system of the Eq. (8.16), we can obtain the new system of the equations suitable for the description of the operating mode of the electric machine. In this regard, the air gap emf equations suitable for an analysis of the operating mode of the machine based on the equations shown in (8.16) and (8.18) acquire the form

$$\begin{aligned} \dot{\Im}_1 &= x_0 \dot{I}_0 \\ \dot{\Im}_2' &= j x_0 \dot{I}_0 \\ \dot{I}_0 &= \dot{I}_1 + \dot{I}_2' \end{aligned} \tag{8.19}$$

As it follows from the system of Eqs. (6.19) and (8.19), the air gap emf equations obtained based on the "no-load and under-load" concept have a fairly simple structure.

The air gap emf equations shown in (8.19) are obtained based on the physical considerations. These equations can be derived from the air gap emf Eq. (6.19) drawing up with the use of the "inductively coupled circuits" concept. For this in (6.19), it is necessary to apply the referred values of the rotor winding emf and the current received by (8.6) and (8.11). This opportunity will be discussed in more detail below.

Based on Eq. (8.19), the air gap equivalent circuit acquires the form in Fig. 8.3b. This equivalent circuit has the common branch $0-0$ which is flowed around by the sum of the currents of the stator and rotor windings, i.e. $(I_1 + I_2')$. This means that in the equivalent circuit in Fig. 8.3b, the stator and rotor windings are represented as the conditionally electrically connected circuits. The air gap equivalent circuit (Fig. 8.3b) obtained based on the "no-load and under-load" concept is more simple and suitable for the practical use form than the air gap equivalent circuit built with the use of the "inductively coupled circuits" concept and given in Fig. 6.4a. In this regard, for the air gap, we will use the equivalent circuit shown in Fig. 8.3b.

The reactance x_0 used in the air gap equivalent circuit (Fig. 8.3b) is determined by the expression given in (8.16). This expression corresponds to the condition when the only single radial component of the magnetizing flux calculated by (8.14) takes place in the air gap.

8.4 Equivalent Circuits for Fragments of Stator and Rotor: Matching Quantities

In this work, the stator and rotor fragments are considered their slot zones in which the slot leakage magnetic fluxes exist. The equivalent circuits shown in Fig. 6.4b and c for these fragments are represented as the stator slot leakage equivalent circuit and the rotor slot leakage equivalent circuit. In connection with the introduction of the converted air gap equivalent circuit obtained in Fig. 8.3b based on the use of the "no-load and under-load" concept, the quantities applied in the rotor slot leakage equivalent circuit (Fig. 6.4c) and in the air gap equivalent circuit (Fig. 8.3b) are needed to be matched. Below, we will demonstrate it.

8.4.1 Slot Leakage Equivalent Circuit of Stator

Based on the stator slot leakage equivalent circuit given in Fig. 6.4b and the air gap equivalent circuit shown in Fig. 8.3b, it follows that on the terminals $a - b$ of these equivalent circuits act the same values of the emf $\dot{Э}_1$ and the current $\dot{I}_1$ of the stator winding. This means that under the application of the "no-load and under-load" concept, the quantities used in the stator slot leakage equivalent circuit (Fig. 6.4b) and in the stator slot leakage emf equation obtained in (6.24) are not changed. In this regard, from (6.24) for the stator slot leakage emf equation, we can receive

$$(-\dot{Э}_1) = -(r_1 + jx_{1\sigma} + Z_c)\dot{I}_1 \tag{8.20}$$

This equation can be considered as corresponding to the "no-load and under-load" concept. The stator slot leakage equivalent circuit following from the Eq. (8.20) takes the form in Fig. 8.3c. This equivalent circuit is similar to the stator slot leakage equivalent circuit obtained on the basis of the "inductively coupled circuits" concept in Fig. 6.4b. Consequently, in our case of applying the "no-load and under-load" concept, the stator slot leakage equivalent circuit is not changed, and it does not need to be transformed.

8.4.2 Slot Leakage Equivalent Circuit of Rotor

Now, we consider the terminals $c - d$ of the rotor slot leakage equivalent circuit shown in Fig. 6.4c and the terminals $c - d$ of the converted air gap equivalent circuit given in Fig. 8.3b. On the terminals $c - d$ of the equivalent circuit in Fig. 6.4c, the values of the emf $\dot{Э}_2$ and the current $\dot{I}_2$ of the rotor winding act. Corresponding to these conditions, the rotor slot leakage emf equations are given in (6.25). On the terminals $c - d$ of the converted air gap equivalent circuit in Fig. 8.3b, the referred values of the emf $\dot{Э}'_2$ and the current $\dot{I}'_2$ act. In this regard, the rotor slot leakage emf equations obtained in (6.25) should be expressed through the referred values of the emf $\dot{Э}'_2$ and the current $\dot{I}'_2$ of the rotor winding. Now, based on (6.25) for the rotor slot leakage emf equations, we can receive through the referred rotor quantities

$$(-\dot{Э}'_2) = -(r'_2/s + jx'_{2\sigma} + Z'_{2H})\dot{I}'_2 \tag{8.21}$$

where $r'_2 = r_2k_2; x'_{2\sigma} = x_{2\sigma}k_2; Z'_{2\text{H}} = Z_{2\text{H}}k_2; k_2 = k_Э k_I$.

The rotor slot leakage equivalent circuit in Fig. 8.3d corresponds to the equation (8.21). This equivalent circuit is similar in the structure to the rotor slot leakage equivalent circuit obtained based on the "inductively coupled circuits" concept in Fig. 6.4c. However, in the equivalent circuit in Fig. 8.3d, the referred values of the emf $\dot{Э}'_2$ and the current $\dot{I}'_2$ of the rotor winding are used. This indicates that the rotor slot leakage equivalent circuit in Fig. 8.3d fits the "no-load and under-load" concept. Thus, the transformation of the rotor slot leakage equivalent circuit is reduced to the representation through the referred rotor quantities.

8.5 Leakage Reactances of Stator and Rotor: Features of Determination

The stator and rotor slot leakage emf equations obtained by the use of the "no-load and under-load" concepts are given in (8.20) and (8.21). The stator and rotor slot leakage equivalent circuits corresponding to these equations are shown in Fig. 8.3c and Fig. 8.3d. These emf equations and equivalent circuits contain the slot leakage reactance of the stator winding $x_{1\sigma}$, and they also include the slot leakage reactance of the rotor winding $x'_{2\sigma}$. Consequently, in the case of applying the "no-load and under-load" concept, the rotor slot leakage emf equation and rotor slot leakage equivalent circuit contain the stator winding value of the rotor slot leakage reactance $x'_{2\sigma}$. When using the "inductively coupled circuits" concept, the slot leakage reactances of the stator winding $x_{1\sigma}$ and the rotor winding $x_{2\sigma}$ are applied in the stator and rotor slot leakage emf equations and equivalent circuits. The slot leakage reactances $x_{1\sigma}$ and $x_{2\sigma}$ can be determined by the expressions (6.14) obtained based on the "inductively coupled circuits" concept in the form

$$x_{1\sigma} = \frac{\dot{Э}_{1\sigma}}{\dot{I}_1} \quad x_{2\sigma} = \frac{\dot{Э}_{2\sigma}}{\dot{I}_2} \tag{8.22}$$

In accordance with (8.22), to determine the slot leakage reactances $x_{1\sigma}$ and $x_{2\sigma}$, it is necessary to establish the stator slot leakage emf $\dot{Э}_{1\sigma}$ and the rotor slot leakage emf $\dot{Э}_{2\sigma}$. The slot leakage emf $\dot{Э}_{1\sigma}$ and $\dot{Э}_{2\sigma}$ are caused by the stator slot leakage magnetic flux $\Phi_{1\sigma}$ and by the rotor slot leakage magnetic flux $\Phi_{2\sigma}$ calculated in accordance with the "inductively coupled circuits" concept by the expressions given in (6.8) and (6.9) in the form $\Phi_{1\sigma} = \Phi_1 - \Phi_{11}$ and $\Phi_{2\sigma} = \Phi_2 - \Phi_{22}$. Here, Φ_1, Φ_2 and Φ_{11}, Φ_{22} represent the total self-and mutual induction magnetic fluxes of the stator and rotor windings, respectively. The separation of the slot leakage magnetic fluxes from the total self-induction magnetic fluxes represents to some extent an uncertain process, as a result of which it becomes difficult to ensure the required level of reliability in determining the slot leakage reactances of the electric machine, for example, [3].

In the technical literature, another approach to determining the slot leakage reactances is widespread. This approach is based on the use of the "no-load and under-load" concept, and it assumes the definition of the slot leakage reactances based on the direct calculation of the slot leakage magnetic fluxes caused by the volume of the stator slot or rotor slot currents of the electric machine [1–20].

In accordance with the "no-load and under-load" concept, the resulting magnetic flux taking place in the electric machine in the operating mode can be represented as a result of the superposition of the magnetizing flux Φ_0 caused by the magnetizing current $\dot{I}_0$ flowing in the stator winding in the no-load mode and the slot leakage magnetic fluxes forming by the currents $\dot{I}_{1н}$ and $\dot{I}_2$ flowing in stator and rotor windings in the under-load mode. In the under-load mode, the rotor winding current $\dot{I}_2$ causes the magnetic flux. The one part of this flux Φ_{22} is closed along the main magnetic circuit of the electric machine, and the other part of the flux $\Phi_{2\sigma}$ embracing the rotor winding represents the rotor winding slot leakage magnetic flux. In the under-load mode, the stator winding current $\dot{I}_{1н}$ causes the magnetic flux Φ_{11} closing along the main magnetic circuit of the electric machine, and the other part of the flux $\Phi_{1\sigma}$ embracing the stator winding represents the stator winding slot leakage magnetic flux. As it was noted earlier, in the under-load mode, the stator magnetic flux Φ_{11} compensates the rotor magnetic flux Φ_{22}, i. e. $\Phi_{11} = -\Phi_{22}$ and therefore, they do not form the common mutual induction magnetic flux in the machine, for example, [1–3, 5, 10–14, 17, 18, 20]. In Appendix B (Fig. B.4), it is shown that for the symmetrical electric machine, the condition $\Phi_{11} = -\Phi_{22}$ is implemented on the surface $0-0$ passing through the middle of the radial length of the air gap, since on this air gap surface the radial component of the flux density is equal to zero. Consequently, in the case of considering the machine active zone, we can proceed from the fact that in the under-load mode the stator current $\dot{I}_{1н}$ creates the stator winding slot leakage flux $\Phi_{1\sigma}$ which embraces the stator slot-tooth region and the air gap sublayer limited by the air gap surface $0-0$ and the rotor current $\dot{I}_2$ causes the rotor winding slot leakage flux $\Phi_{2\sigma}$ which embraces the rotor slot-tooth region and the air gap sublayer limited by the air gap surface $0-0$. Now, knowing the regions of the distribution of the stator and rotor slot leakage magnetic fluxes, we can determine the reactances $x_{1\sigma}$ and $x_{2\sigma}$ of the stator and rotor windings. For this, we can calculate the slot leakage magnetic fluxes caused by the volume of the stator slot and the rotor slot currents of the electric machine and also determine the slot leakage emf $\dot{Э}_{1\sigma}$ and $\dot{Э}_{2\sigma}$ induced in the machine windings by these slot leakage fluxes. Then, the slot leakage reactances $x_{1\sigma}$ and $x_{2\sigma}$ are defined by the expressions obtained in (8.22) in the following form $x_{1\sigma} = \dot{Э}_{1\sigma}/\dot{I}_{1н}$ and $x_{2\sigma} = \dot{Э}_{2\sigma}/\dot{I}_2$. Here, the second equation is represented in relation to the referred rotor quantities. Therefore, for the reactance $x'_{2\sigma}$, we have that $x'_{2\sigma} = \dot{Э}'_{2\sigma}/\dot{I}'_2$.

Thus, the introduction in the machine air gap of the surface $0-0$ allows us to distinguish the slot leakage magnetic flux created by the stator winding current and the slot leakage magnetic flux created by the rotor winding current. In this regard, the determination of the slot leakage reactances $x_{1\sigma}$ and $x_{2\sigma}$ is realized separately for each stator and rotor windings that simplify the calculation process.

The leakage reactances $x_{1\sigma}$ and $x_{2\sigma}$ can be defined by the equations obtained in (8.20) and (8.21) in the form

$$\begin{aligned} \dot{Э}_1 &= (r_1 + jx_{1\sigma} + Z_c)\dot{I}_1 \\ \dot{Э}'_2 &= (r'_2/s + jx'_{2\sigma} + Z'_{2н})\dot{I}'_2 \end{aligned} \tag{8.23}$$

For this purpose, in Eq. (8.23), the condition $\dot{Э}_1 = \dot{Э}'_2 = 0$ (the condition of the absence of the total mutual induction magnetic flux in the electric machine in the under-load mode) can be used. Taking into account that $(-\dot{U}_1) = Z_c\dot{I}_1$ and $(-\dot{U}'_2/s) = Z'_{2н}\dot{I}'_2$, the Eq. (8.23), acquire the form

$$\begin{aligned} \dot{U}_1 &= (r_1 + jx_{1\sigma})\dot{I}_1 \\ \dot{U}'_2/s &= (r'_2/s + jx'_{2\sigma})\dot{I}'_2 \end{aligned} \tag{8.24}$$

To determine the ohmic resistances of the stator and rotor windings r_1 and r'_2/s used in (8.24), the losses dissipated in the stator and rotor windings can be calculated or the corresponding experiments are performed. These studies go beyond the volume of this book, and therefore, they are omitted here. Now, in the Eq. (8.24), the impedances Z_1 and Z'_2 determined as $Z_1 = r_1 + jx_{1\sigma}$ and $Z'_2 = r'_2/s + jx'_{2\sigma}$ can be used. Then, we have that

$$Z_1 = \frac{\dot{U}_1}{\dot{I}_1} \quad Z'_2 = \frac{\dot{U}'_2/s}{\dot{I}'_2} \tag{8.25}$$

From here, it arises that

$$x_{1\sigma} = \sqrt{Z_1^2 - r_1^2} \quad x'_{2\sigma} = \sqrt{(Z'_2)^2 - (r'_2/s)^2} \tag{8.26}$$

The conditions taking place in the electric machine in the under-load mode can be realized, in principle, in the practice. For this, the stator and rotor windings should be supplied with the currents in such a way that to provide in the machine the equality to zero of the total mutual induction magnetic flux. In this case, the stator and rotor winding mmf obtain the same values, and they take the opposite sign. Therefore, in these conditions, the stator and rotor winding currents create the leakage magnetic fluxes in the electric machine. Then, the expressions shown in (8.25) can be used to determine the leakage reactances of the stator and rotor windings. This method for obtaining the leakage reactances is known as the opposition balancing method proposed by the German electrical engineer Rogowski [14]. Due to the inequality of the number of the turns of the stator and rotor windings ($w_1 \neq w_2$), the practical implementation of this method meets serious difficulties. Despite this, the idea of the opposition balancing method lies at the heart of obtaining the leakage magnetic fluxes taken place in the electric machine.

In accordance with (8.22), to determine the slot leakage reactances $x_{1\sigma}$ and $x_{2\sigma}$, it is necessary to find the values of the emf induced in the stator and rotor windings by the slot leakage magnetic fluxes. In the stator and rotor slot areas, the slot leakage magnetic fluxes pass across the slots. If the phase of the winding is laid in the q slots, then all these q slots can be represented as covered by the one common transverse leakage magnetic flux. This common picture of the transverse leakage magnetic flux can be simplified if to proceed from the following considerations. The mmf of the q slots increases in q times in comparison with the mmf of the single slot. But the reluctance across of the q slots increases, in this case, in the same time. Therefore, the value of the common transverse leakage magnetic flux remains unchanged and equal to the leakage flux of the single slot. In this regard, in practice, the leakage flux calculations are performed per slot, then, the flux linkages and the leakage emf are increased in q times, for example, [1–13, 15–18, 20]. In Appendix E, this technique is used for defining the slot leakage reactances of the stator and rotor windings (Fig. E.3).

8.6 "No-Load and Under-Load" Concept: Way of Transformation

As it was shown earlier, in the air gap emf equations obtained based on the "inductively coupled circuits" concept, the stator and rotor windings are represented as the inductively coupled two circuit loops, for example, [1–13, 16–18, 20]. Therefore, the air gap equivalent circuit contains, in this case, the inductively coupled elements. When using the "no-load and under-load" concept, in the air gap emf equations, the stator and rotor windings are represented as the conditionally electrically connected two circuit loops, for example, [1–13, 16–18, 20]. Therefore, in this case, we have the single-element equivalent circuit for the air gap of the electric machine. Thus, when using the "no-load and under-load" concept, the air gap emf equations and the air gap equivalent circuit are simplified, and they obtain the form suitable for practical application. In this regard, the "no-load and under-load" concept can be considered as a way of the transformation of the emf equations and the equivalent circuits.

The "no-load and under-load" concept is usually realized in relation to the emf of the separated single winding of the electric machine. In this work, as such winding, we choose, as it was noted above, the stator winding. In this regard, the "no-load and under-load" concept is represented in accordance with (8.9) in the form of the equation obtained for the stator winding current as $\dot{I}_1 = \dot{I}_0 + (-\dot{I}_2')$. Here, the current $\dot{I}_0$ represents the magnetizing current flowing in the stator winding under influence of the emf $\dot{Э}_1$ induced in the no-load mode. The current $(-\dot{I}_2')$ flows in the stator winding in the under-load mode, and it is created under the influence of the referred value of the emf $\dot{Э}_2'$ induced in the rotor winding in the no-load mode.

The transformation of the emf equations and equivalent circuits of the electric machine can be implemented by the use of the equation $\dot{I}_1 = \dot{I}_0 + (-\dot{I}_2')$ obtained for the stator winding current. In accordance with this equation, the transformation of the emf equations and equivalent circuits of the electric machine is reduced to the determination of the components $\dot{I}_0$ and $(-\dot{I}_2')$ of the stator winding current for what the machine initial emf equations and the initial equivalent circuits are needed to be transformed is used. In this case, to determine the current $\dot{I}_0$ in the initial emf equations or in the initial equivalent circuits, the no-load mode ($\dot{I}_2 = 0$) should be implemented, and as a result in relation to the current $\dot{I}_0$, the new machine emf equations or the equivalent circuits should be obtained.

Based on Eq. (8.9), for the current $(-\dot{I}'_2)$, we can receive that $(-\dot{I}'_2) = \dot{I}_1 - \dot{I}_0$. From here follows that to determine of the current $(-\dot{I}'_2)$ it is necessary to establish the currents $\dot{I}_0$ and $\dot{I}_1$. The current $\dot{I}_0$ is determined by the emf equations or the equivalent circuits drawing up for the machine no-load mode. The expression for the current $\dot{I}_1$ can be established by the initial machine emf equations or the initial equivalent circuits. Now, when the currents $\dot{I}_1$ and $\dot{I}_0$ are found, the expression for the current $(-\dot{I}'_2)$ follows from the condition of the form $(-\dot{I}'_2) = \dot{I}_1 - \dot{I}_0$. In relation to the current $(-\dot{I}'_2)$, the new emf equations or the equivalent circuits corresponding to the machine under-load mode should be obtained. Further, by the use of the emf equations or the equivalent circuits corresponding to the no-load mode and the emf equations or the equivalent circuits reflecting the under-load mode, the transformed emf equations or the equivalent circuits are composed for what in accordance with the condition $\dot{I}_1 = \dot{I}_0 + (-\dot{I}'_2)$ the superposition principle can be implemented. In the technical literature, this way of determining the components $\dot{I}_0$ and $(-\dot{I}'_2)$ of the stator winding current and obtaining the transformed emf equations or the equivalent circuits is known as the reduction or transfer techniques for example, [1–13, 16–18, 20]. Below, we will consider the formal transformation techniques based on the use of the stator winding current equation of kind $\dot{I}_1 = \dot{I}_0 + (-\dot{I}'_2)$.

8.7 Summary

1. In this book, the air gap layer, the stator slot-tooth region, and the rotor slot-tooth region are considered as machine design fragments. For these fragments, the emf equations and the equivalent circuits are obtained. For this purpose, the "inductively coupled circuits" concept and the "no-load and under-load" concept are used. It is shown that the emf equations and the equivalent circuits obtained for the machine design fragments by the use of the "no-load and under-load" concept are suitable for practical application. Therefore, in this work, this concept has received more detailed consideration.
2. The machine operating mode can be considered as a result of the superposition of the machine no-load mode and the under-load mode. This no-load mode and under-load mode of the electric machine can be implemented in relation to the

emf $\dot{Э}_1$ induced in the stator winding. In this regard, the machine no-load mode is implemented when the rotor winding is open and there is no current in it. In view of this, the stator winding current $\dot{I}_0$ arises under the influence of the emf $\dot{Э}_1$ induced in the stator winding in the no-load mode. The current $\dot{I}_0$ flowing in the stator winding represents the magnetizing current of the electric machine.

3. The machine under-load mode is implemented when the rotor winding is closed to the load impedance, and the current in the rotor winding arises under the influence of the emf $\dot{Э}_2$ induced in the rotor winding in the no-load mode. The current flowing in the rotor winding in the under-load mode creates the mutual induction magnetic flux which induces in the stator winding the emf. Under the influence of this emf, the under-load current appears in the stator winding. Thus, in the operating mode, the value of the stator winding current can be determined as sum of the magnetizing current taken place in the stator winding in the no-load mode and the load current, flowing in the stator winding in the under-load mode. This formula for the stator winding current reflects the "no-load and under-load" concept.
4. In the "no-load and under-load" concept, the components of the stator winding current represent the no-load current arising in the stator winding under the influence of the emf $\dot{Э}_1$ taken place in the no-load mode and the load current caused in the stator winding under the influence of the emf $\dot{Э}_2$ induced in the rotor winding in the no-load mode. The emf $\dot{Э}_1$ and $\dot{Э}_2$ have, in general case, different values, i.e. $\dot{Э}_1 \neq \dot{Э}_2$. To have the ability to add the stator no-load current and the stator under-load current and, as a result of this, to get the stator winding current corresponding to the machine operating mode, it is necessary to comply with the condition of the form $\dot{Э}_1 = \dot{Э}_2$. In practice, this condition is achieved by reduction of the rotor winding emf $\dot{Э}_2$ to the value of the stator winding emf $\dot{Э}_1$ as a result of which the condition $\dot{Э}_1 = \dot{Э}_2'$ arises. Here, $\dot{Э}_2'$ is referred to the number of the turns of the stator winding value of the rotor winding emf. In this regard, the implementation of the "no-load and under-load" concept, and consequently, the determination of the equation for the stator winding current is associated with the use of the referred value of the rotor winding emf.
5. In the machine under-load mode, the load current $\dot{I}_2$ creates the mmf of the rotor winding $\dot{F}_2$. This rotor mmf $\dot{F}_2$ causes the mutual induction magnetic flux which induces in the stator winding emf. Under the influence of this emf, the load current $\dot{I}_{1н}$ flows in the stator winding. The load current $\dot{I}_{1н}$ forms the mmf of the stator winding $\dot{F}_{1н}$ which causes in the machine the mutual induction magnetic flux. According to Lenz's law, the stator winding magnetic flux created by the mmf $\dot{F}_{1н}$ aspires to destroy the rotor winding magnetic flux caused by the mmf $\dot{F}_2$. Provided that $\dot{Э}_1 = const$ (for the considered operating mode of the machine), the mmf $\dot{F}_{1н}$ created by the load current of the stator winding acquires the value sufficient to fully compensate the mmf of the rotor winding $\dot{F}_2$,

i.e. $\dot{F}_{1н} = -\dot{F}_2$. In this work, this condition is used to determine the load current of the stator winding $\dot{I}_{1н}$ for which we have that $\dot{I}_{1н} = (-\dot{I}'_2)$, where $\dot{I}'_2$ is the referred to the stator winding value of the rotor winding load current. Now, for the stator winding current corresponding to the machine operating mode, the condition of kind $\dot{I}_1 = \dot{I}_0 + (-\dot{I}'_2)$ is obtained. The equation for the stator winding current in the form of $\dot{I}_1 = \dot{I}_0 + (-\dot{I}'_2)$ takes the central position in the theory of calculating the asynchronous machine.

6. The implementation of the "no-load and under-load" concept is associated with the use of the machine no-load mode and the machine under-load mode. In the no-load mode, the magnetizing magnetic flux takes place in the machine, and in the under-load mode, the currents flowing in the stator and rotor windings do not create the total mutual induction magnetic flux in the machine, since in accordance with the condition of kind $\dot{F}_{1н} = -\dot{F}_2$, the mutual induction magnetic fluxes of the stator and rotor windings mutually compensate each other. In this regard, in the machine under-load mode, the stator and rotor winding currents create only the leakage magnetic fluxes. Consequently, when using the "no-load and under-load" concept, the machine resulting magnetic flux can be represented as a result of the superposition of the magnetizing flux and the leakage magnetic fluxes of the stator and the rotor windings that greatly simplifies an analysis.
7. In this work, the "no-load and under-load" concept is considered as a way of the transformation of the machine emf equations and the equivalent circuits. For this, it is necessary to use the equation for the stator winding current having the form $\dot{I}_1 = \dot{I}_0 + (-\dot{I}'_2)$. In accordance with this equation, the use of the "no-load and under-load" concept as a way to transform the machine emf equations and the equivalent circuits is reduced to the determination of the components $\dot{I}_0$ and $(-\dot{I}'_2)$ of the stator winding current. The currents $\dot{I}_0$ and $(-\dot{I}'_2)$ can be defined by the initial emf equations or the initial equivalent circuits to be subjected to the appropriate transformation. Then, the superposition principle is used for the obtained in relation to the currents $\dot{I}_0$ and $(-\dot{I}'_2)$ emf equations or the equivalent circuits as a result of which the transformed emf equations or the equivalent circuits suitable for the practical application acquire.
8. In this chapter, the fragments of the machine equivalent circuit are represented as the stator slot leakage equivalent circuit, the air gap equivalent circuits and the rotor slot leakage equivalent circuit. These equivalent circuits are obtained in relation to the emf induced in the stator and rotor windings. In this case, the stator slot leakage equivalent circuit is represented in relation to the stator winding emf $\dot{Э}_1$. The air gap magnetizing equivalent circuits are built in relation to the stator and rotor winding emf $\dot{Э}_1$ and $\dot{Э}_2$. The rotor slot leakage equivalent circuit is obtained in relation to the rotor winding emf $\dot{Э}_2$. The transformation of

these equivalent circuits is implemented by the use of the "no-load and under-load" concept. The "no-load and under-load" concept is implemented in relation to the stator winding emf $\dot{Э}_1$. Consequently, the transformation made in this work is reduced to the representation of the stator slot leakage equivalent circuit, the air gap equivalent circuit, and the rotor slot leakage equivalent circuit in relation to the stator winding emf $\dot{Э}_1$.

9. In the case of using the "inductively coupled circuits" concept, the air gap equivalent circuit as the fragment of the machine equivalent circuit contains the inductively coupled elements. In this regard, the transformation of the air gap equivalent circuit is reduced to solving the problem associated with the elimination of the inductively coupled elements. For this, the air gap equivalent circuit containing the inductively coupled elements is represented through the currents $\dot{I}_0$ and $(-\dot{I}'_2)$ reflecting the components of the stator winding current. As a result, the air gap equivalent circuit obtained in this way is sweated out from the inductively coupled elements and acquires the simplified structure represented through the referred rotor quantities. In this chapter, the implementation of the "no-load and under-load" concept, and consequently, the determination of the currents $\dot{I}_0$ and $(-\dot{I}'_2)$ are made proceed from the physical considerations.
10. The air gap equivalent circuit arising on the basis of the use of the "no-load and under-load" concept has the common parallel branch $0 - 0$ that it is flowed around by the sum of the currents $(\dot{I}_1 + \dot{I}'_2)$ of the stator and rotor windings. It means that in this equivalent circuit, the stator and rotor windings are represented as the conditionally electrically connected circuit loops as a result of which the structure of the air gap equivalent circuit is simplified and acquires the form suitable for practical use. Therefore, the "no-load and under-load" concept is considered, in this work, as the main way of constructing the equivalent circuit of the electric machine.
11. The stator slot leakage equivalent circuit is obtained in relation to the stator winding emf $\dot{Э}_1$, and therefore, it is not subject to the transformation. The rotor slot leakage equivalent circuit is represented in relation to the rotor winding emf $\dot{Э}_2$. In this regard, within the "no-load and under-load" concept, this equivalent circuit is represented through the referred rotor quantities $(\dot{Э}'_2, \dot{I}'_2)$. This procedure does not lead to the change of the structure of the rotor slot leakage equivalent circuit.
12. It is shown that for the symmetrical electric machine, the condition $\Phi_{12} = -\Phi_{21}$ is implemented on the surface $0 - 0$ passing through the middle of the radial length of the machine air gap, since on this air gap surface the radial component of the magnetic flux density is equal to zero. In this work, the stator and rotor slot leakage magnetic fluxes $\Phi_{1\sigma}$ and $\Phi_{2\sigma}$ are determined in relation to air gap surface $0 - 0$. The stator slot leakage magnetic flux $\Phi_{1\sigma}$ embraces the stator slot-tooth region and the air gap sublayer adjoined to the stator bore surface is

limited by the surface 0 − 0. The rotor slot leakage magnetic flux $\Phi_{2\sigma}$ embraces the rotor slot-tooth region and the air gap sublayer adjoined to the rotor surface is limited by the surface 0 − 0. Consequently, the introduction in the machine air gap of the surface 0 − 0 allows us to establish the slot leakage magnetic flux created by the stator winding current and the slot leakage magnetic flux created by the rotor winding current. Now, the determination of these slot leakage magnetic fluxes and also the slot leakage reactances can be implemented for each stator and rotor windings individually that simplified the computational process.

13. The equivalent circuits of the machine design fragments (the stator slot leakage equivalent circuit, the air gap equivalent circuits, and the rotor slot leakage equivalent circuit) reflect the more homogeneous and the less complex processes taking place in the electric machine. This provision contributes to the significant simplification of analysis.

References

1. Kostenko, M., Piotrovsky, L.: Electric Machines, Energiya, Leningrad (1973)
2. Piotrovskij, L.: Electric Machines, Publishing Energiya, Leningrad (1975)
3. Voldek. A.: Electric Machines, Energiya, Leningrad (1974)
4. McPherson, G., Laramore, R.: Electric Machines and Transformers, 2nd ed., John Wiley and Sons, Inc., New York (1990)
5. Say, M.: Alternating Current Machines, John Wiley and Sons, New York (1902)
6. Fitzgerald, A., Kingsley, C., Umans, S.: Electric Machinery, McGraw-Hill, Inc., 5th ed., New York (1990)
7. Alger, P.: Induction Machines, 2nd ed., Gordon and Beach (1970)
8. Del Toro, V.: Electric Machines and Power Systems, Prentice-Hall, Inc., Englewood Cliffs, New Jersey (1985)
9. Krause, P.: Analysis of Electric Machinery, McGraw-Hill, New York (1986)
10. Chapman, S.: Electric Machinery Fundamentals, 4th ed., McGraw-Hill, Inc., New York (2005)
11. Vazhnov, A.: Electric Machines, Energiya, Leningrad (1968)
12. Asanbayev, V.: Alternating Current Multi-Circuit Electric Machines. A New Approach to Steady-State Parameter Determination, Springer Cham Heidelberg, New York, Dordrecht London (2015)
13. Seely, S.: Electromechanical Energy Conversion, McGraw Book Company Inc., New York (1962)
14. Rogowski, B.: About the Leakage Field and the Leakage Induction Coefficient of a Transformer with the Disc Winding and Split End Coils, Mittetlungen uber Forschungsarbeitten, H.71, 1-36 (1909)
15. Dombrovsky, V., Zaichik, V.: Asynchronous Machines: Theory, Calculation, Elements of Design, Publishing Energoatomizdat, Leningrad (1990)
16. Bragstad, O.: Theory of AC Machines with an Introduction to the Theory of Steady-State Alternating Currents (Translated from German), Izdatelstvo Lensoveta i Oblispolkoma, Leningrad (1933)

17. Antonov, L.: Alternating Current Machines, Naval Publishing, Leningrad (1968)
18. Kostenko, M.: Electric Machines. Special Part, Gosenergoizdat, Moscow (1949)
19. Voldek, A.: Induction Magneto-hydrodynamic Machines with Liquid-Metal Working Substance, Publishing Energiya, Leningrad (1970)
20. Ivanov-Smolensky, A.: Electric Machines, Publishing Energiya, Moscow (1980)

Chapter 9
"No-Load and Under-Load" Concept: Techniques for Implementation

9.1 Stator Winding Current Decomposition Technique

In accordance with the "no-load and under-load" concept, the stator winding current $\dot{I}_1$ can be represented from (8.9) in the following form

$$\dot{I}_1 = \dot{I}_0 + (-\dot{I}_2') \tag{9.1}$$

As it follows from (9.1), the components of the stator winding current are the current $\dot{I}_0$ flowing in the stator winding in the machine no-load mode and the current $(-\dot{I}_2')$ taking place in the stator winding in the machine under-load mode [1–19]. For the implementation of the "no-load and under-load" concept, it is necessary to find the currents $\dot{I}_0$ and $(-\dot{I}_2')$. For this purpose, we use the air gap emf Eq. (6.19) composed in the form of Z and given below as

$$\begin{aligned}(-\dot{Э}_1) &= jx_{11}\dot{I}_1 + jx_{21}\dot{I}_2\\ (-\dot{Э}_2) &= jx_{12}\dot{I}_1 + jx_{22}\dot{I}_2\end{aligned} \tag{9.2}$$

Analytical expressions for the reactances x_{11}, x_{12}, x_{21}, and x_{22} used in (9.2) are shown in (7.21).

No-load mode. In the system of the emf Eq. (9.2), we implement the no-load mode for what the condition of kind $\dot{I}_2 = 0$ can be taken in it. Then, with considering that $\dot{I}_1 = \dot{I}_0$, the following conditions arise from (9.2)

$$\begin{aligned}(-\dot{Э}_1) &= jx_{11}\dot{I}_0\\ (-\dot{Э}_2) &= jx_{12}\dot{I}_0\end{aligned} \tag{9.3}$$

Here, we will exclude the current $\dot{I}_0$. As a result, for the emf $\dot{Э}_1$ and $\dot{Э}_2$, the condition follows

V. Asanbayev, *Asynchronous Machines*,
https://doi.org/10.1007/978-3-030-92284-9_9

$$\dot{Э}_1 = \frac{x_{11}}{x_{12}}\dot{Э}_2 = k_Э\dot{Э}_2 = \dot{Э}_2' \quad (9.4)$$

where $\dot{Э}_2' = k_Э\dot{Э}_2; k_Э = x_{11}/x_{12}$.

In (9.4), the emf $\dot{Э}_2'$ represents the referred to the stator winding value of the rotor winding emf for which we can have

$$\dot{Э}_2' = k_Э\dot{Э}_2 \quad (9.5)$$

The dimensionless quantity $k_Э$ is the emf reduction factor determined in accordance with (9.4) by the expressions given in (7.21), as

$$k_Э = \frac{x_{11}}{x_{12}} = \frac{w_1 k_{w1}}{w_2 k_{w2}} \quad (9.6)$$

In Eq. (9.3), the emf $\dot{Э}_2'$ can be applied. For this purpose, the expressions given in (9.5) can be used. Then, we obtain

$$\begin{aligned}(-\dot{Э}_1) &= jx_{11}\dot{I}_0 = jx_0\dot{I}_0\\(-\dot{Э}_2') &= jx_{12}k_Э\dot{I}_0 = jx_{11}\dot{I}_0 = jx_0\dot{I}_0\end{aligned} \quad (9.7)$$

where $x_0 = x_{11}$ and $x_{11} = k_Э x_{12}$.

From here, for the current $\dot{I}_0$, the following expression arises

$$\dot{I}_0 = \frac{(-\dot{Э}_1)}{jx_0} = \frac{(-\dot{Э}_2')}{jx_0} \quad (9.8)$$

Under-load mode. Now, we will consider the under-load current $(-\dot{I}_2')$ flowing in the stator winding. From (9.1), the under-load current $(-\dot{I}_2')$ is determined as

$$(-\dot{I}_2') = \dot{I}_1 - \dot{I}_0 \quad (9.9)$$

Here, the current $\dot{I}_0$ can be found by the expression (9.8), and the current $\dot{I}_1$ is needed to be established for what the system of the Eq. (9.2) is used. Believing in the system of the Eq. (9.2) that is $\dot{Э}_2 = Z_{2н}\dot{I}_2$, where $Z_{2н}$ is the load impedance, the current $\dot{I}_1$ acquires.

$$\dot{I}_1 = (-\dot{Э}_1)\frac{Z_{2н} + jx_{22}}{jx_{11}(Z_{2н} + jx_{22}) + x_{12}x_{21}.}$$

We substitute this expression in (9.9) and also in (9.9), we take into account the expression (9.8). Then, from (9.9), the under-load current $(-\dot{I}_2')$ can be represented as

$$
\begin{aligned}
(-\dot{I}'_2) &= (-\dot{Э}_1)\left[\frac{Z_{2\text{H}} + jx_{22}}{jx_{11}(Z_{2\text{H}} + jx_{22}) + x_{12}x_{21}} - \frac{1}{jx_{11}}\right] \\
&= (-\dot{Э}_1)\frac{-x_{12}x_{21}}{jx_{11}\left[Z_{2\text{H}}(\,jx_{11}) - x_{11}x_{22}\left(1 - \frac{x_{12}x_{21}}{x_{11}x_{22}}\right)\right]}
\end{aligned}
\tag{9.10}
$$

Based on Eq. (9.2), the current $\left(-\dot{I}_2\right)$ acquires the form

$$
\left(-\dot{I}_2\right) = (-\dot{Э}_1)\frac{jx_{12}}{Z_{2H}(\,jx_{11}) - x_{11}x_{22}\left(1 - \frac{x_{12}x_{21}}{x_{11}x_{22}}\right)}
\tag{9.11}
$$

By expressions (9.10) and (9.11), the ratio for the currents $\dot{I}'_2/\dot{I}_2$ obtains.

$$
\frac{\dot{I}'_2}{\dot{I}_2} = \frac{x_{21}}{x_{11}}.
$$

Now, for the current $\dot{I}'_2$, we have

$$
\dot{I}'_2 = \frac{x_{21}}{x_{11}}\dot{I}_2 = \frac{\dot{I}_2}{k_I}
\tag{9.12}
$$

where $k_I = x_{11}/x_{21}$.

In (9.12), $\dot{I}'_2$ represents the referred to the stator winding value of the rotor winding current. The dimensionless quantity k_I is the current reduction factor of the rotor winding. From (9.12) and also from the expressions obtained in (7.21), we have for the factor k_I

$$
k_I = \frac{x_{11}}{x_{21}} = \frac{m_1}{m_2}\frac{w_1 k_{w1}}{w_2 k_{w2}}
\tag{9.13}
$$

We consider the expression (9.10). In (9.10), as it can be shown by the expressions given in (7.21), the value of the multiplier $(1 - x_{12}x_{21}/x_{11}x_{22})$ can be taken equal to zero. In this regard, the expression (9.10) obtains the form

$$
(-\dot{I}'_2) = (-\dot{Э}_1)\frac{x_{12}x_{21}}{Z_{2H}x_{11}^2}
\tag{9.14}
$$

Taking into account that $k_Э = x_{11}/x_{12}$ and $k_I = x_{11}/x_{21}$, the Eq. (9.14) receives

$$
(-\dot{I}'_2) = \frac{(-\dot{Э}_1)}{Z_{2H}k_Э k_I} = \frac{(-\dot{Э}_1)}{Z_{2H}k_2} = \frac{(-\dot{Э}_1)}{Z'_{2H}}
\tag{9.15}
$$

where $Z'_{2\text{H}} = Z_{2\text{H}}k_2; k_2 = k_Э k_I$.

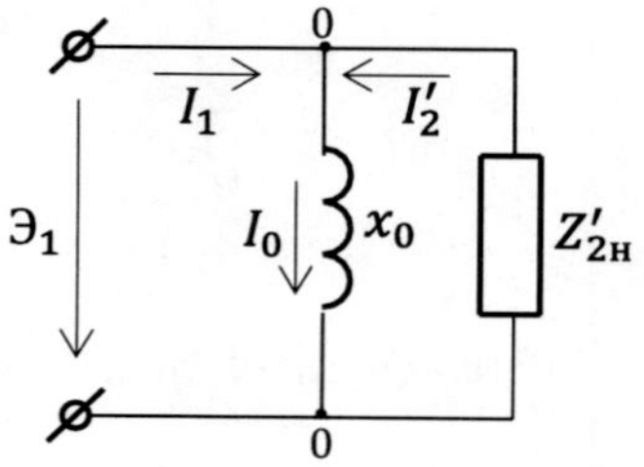

Fig. 9.1 Single-element air gap equivalent circuit closed to the external impedance

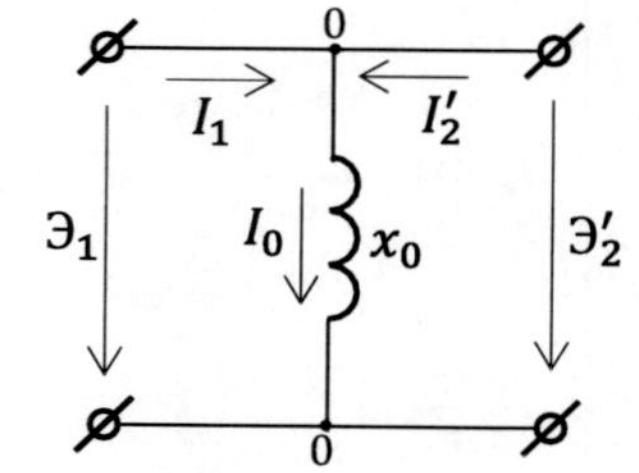

Fig. 9.2 Single-element air gap equivalent circuit

Operating mode. The currents $\dot{I}_0$ and $(-\dot{I}'_2)$ are determined by the expression obtained in (9.8) and (9.15). Based on these expressions and also taking into account the condition (9.1), the stator winding current obtains the form of the following equation

$$\dot{I}_1 = \dot{I}_0 + (-\dot{I}'_2) = (-\dot{Э}_1)\left(\frac{1}{jx_0} + \frac{1}{Z'_{2H}}\right) \tag{9.16}$$

By the Eq. (9.16), the air gap equivalent circuit acquires the form shown in Fig. 9.1. In this equivalent circuit, we use the condition $\dot{Э}'_2 = Z'_{2H}\dot{I}'_2$, where $\dot{Э}'_2 = \dot{Э}_1$. Now, the air gap equivalent circuit in Fig. 9.1 can be reduced to the final form in Fig. 9.2. According to the equivalent circuit in Fig. 9.2, the air gap emf equations can be represented as

$$\begin{aligned} (-\dot{Э}_1) &= jx_0\dot{I}_0 \\ (-\dot{Э}'_2) &= jx_0\dot{I}_0 \\ \dot{I}_0 &= \dot{I}_1 + \dot{I}'_2 \end{aligned} \tag{9.17}$$

From (9.1) and (9.17), it can be shown that the structure of the air gap emf equations is substantially simplified as a result of the transformation implemented by the use of the "no-load and under-load" concept. Therefore, from Figs. 6.4a and 9.2, we can see that the structure of the air gap equivalent circuit is also simplified.

9.2 Transfer Technique

We proceed from Eq. (9.2). These equations contain the four reactances x_{11}, x_{21}, x_{22} and x_{12}. We will replace Eq. (9.2) by conditional electric equivalent circuit, for example, [20]. For this purpose, it is necessary to introduce some common transverse branch flowing around by the some current determined as the sum of the currents flowing in the stator and rotor windings, i.e., $(\dot{I}_1 + \dot{I}_2)$. For this, to the first and the second equations of the system (9.2), the term $jx_{21}\dot{I}_1$ should be added and subtracted from them. In addition to the second equation of the system (9.2), the term $jx_{21}\dot{I}_2$ should be added and subtracted from it. Then, the Eq. (9.2) can be expressed as

$$\begin{aligned}(-\dot{Э}_1) &= j(x_{11} - x_{21})\dot{I}_1 + jx_{21}(\dot{I}_1 + \dot{I}_2)\\ (-\dot{Э}_2) &= j(x_{22} - x_{21})\dot{I}_2 + j(x_{12} - x_{21})\dot{I}_1 + jx_{21}(\dot{I}_1 + \dot{I}_2)\end{aligned} \tag{9.18}$$

To the Eq. (9.18), the equivalent circuit shown in Fig. 9.3 corresponds. Now, we will determine the currents $\dot{I}_0$ and $(-\dot{I}_2')$ by the equivalent circuit in Fig. 9.3.

No-load mode. The current $\dot{I}_0$ can be determined when the condition $\dot{I}_2 = 0$ is implemented in the equivalent circuit in Fig. 9.3. When $\dot{I}_2 = 0$, the equivalent circuit in Fig. 9.3 acquires the form in Fig. 9.4. Based on Fig. 9.4, for the emf $\dot{Э}_1$, we have

$$(-\dot{Э}_1) = jx_{11}\dot{I}_0 \tag{9.19}$$

When $\dot{I}_2 = 0$, for the rotor winding emf $\dot{Э}_2$, we can obtain from the equivalent circuit in Fig. 9.4

$$(-\dot{Э}_2) = jx_{12}\dot{I}_0 \tag{9.20}$$

From (9.19) and (9.20), the emf $(-\dot{Э}_1)$ and $(-\dot{Э}_2)$ follow

$$\begin{aligned}(-\dot{Э}_1) &= jx_{11}\dot{I}_0\\ (-\dot{Э}_2) &= jx_{12}\dot{I}_0\end{aligned} \tag{9.21}$$

Now, based on (9.21), it arises

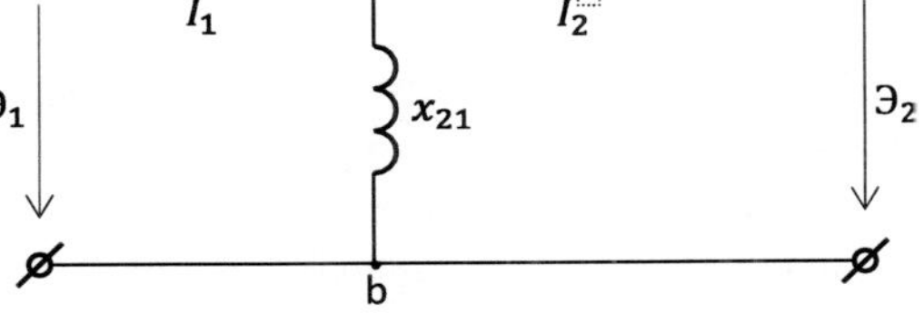

Fig. 9.3 Multi-element air gap equivalent circuit

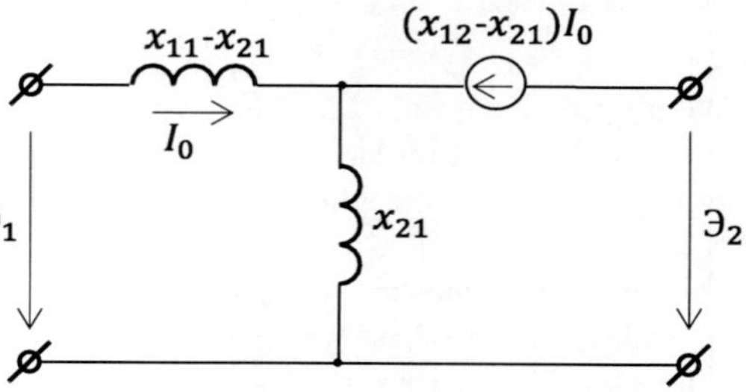

Fig. 9.4 Multi-element air gap equivalent circuit for the no-load mode $(\dot{I}_2 = 0)$

$$\frac{\dot{Э}_1}{\dot{Э}_2} = \frac{x_{11}}{x_{12}}$$

Or

$$\dot{Э}_1 = \frac{x_{11}}{x_{12}} \dot{Э}_2 = k_Э \dot{Э}_2 = \dot{Э}_2' \tag{9.22}$$

where $k_Э = x_{11}/x_{12}$.

From here, for the referred value of the rotor winding emf, we receive

$$\dot{Э}_2' = k_Э \dot{Э}_2 \tag{9.23}$$

In (9.23), $\dot{Э}_2'$ represents the referred to the stator winding value of the rotor winding emf. The dimensionless quantity of $k_Э$ represents the reduction factor of the rotor winding emf determined from (9.2).

In the system of Eq. (9.21), we use the referred to the stator winding value of the rotor winding emf $\dot{Э}_2'$. Then, this system of the equation acquires

$$\begin{aligned} (-\dot{Э}_1) &= jx_{11}\dot{I}_0 = jx_0\dot{I}_0 \\ (-\dot{Э}_2') &= jx_{12}k_Э\dot{I}_0 = jx_0\dot{I}_0 \end{aligned} \tag{9.24}$$

since $x_0 = x_{12}k_Э$.

From (9.24), the current $\dot{I}_0$ can be represented as

$$\dot{I}_0 = \frac{(-\dot{Э}_1)}{jx_0} = \frac{(-\dot{Э}_2')}{jx_0} \tag{9.25}$$

Under-load mode. We will establish the load current $(-\dot{I}_2')$, for which from (9.9) the condition of kind $(-\dot{I}_2') = \dot{I}_1 - \dot{I}_0$ arises. From here, it follows that to determine the load current $(-\dot{I}_2')$, it is necessary to find the values of the currents $\dot{I}_1$ and $\dot{I}_0$. The current $\dot{I}_0$ follows from the expression (9.25). To obtain the current $\dot{I}_1$, we use the equivalent circuit shown in Fig. 9.3. In this case, in the equivalent circuit in Fig. 9.3, the value of the emf $\dot{Э}_2$ can be replaced by the voltage drop across the load impedance $Z_{2н}$, i.e. $\dot{Э}_2 = Z_{2н}\dot{I}_2$. As a result, the current $\dot{I}_1$ acquires

$$\dot{I}_1 = (-\dot{Э}_1)\frac{Z_{2н} + jx_{22}}{jx_{11}(Z_{2н} + jx_{22}) + x_{12}x_{21}}$$

This expression we substitute in (9.9), and also in (9.9), the expression (9.25) can be taken into account for the current $\dot{I}_0$. Then, from (9.9) for the current $(-\dot{I}'_2)$, the following formula arises

$$(-\dot{I}'_2) = (-\dot{Э}_1)\frac{-x_{12}x_{21}}{jx_{11}\left[Z_{2Н}(\ jx_{11}) - x_{11}x_{22}\left(1 - \frac{x_{12}x_{21}}{x_{11}x_{22}}\right)\right]} \tag{9.26}$$

The current $\left(-\dot{I}_2\right)$ can be obtained based on the equivalent circuit in Fig. 9.3. Then, we have

$$\left(-\dot{I}_2\right) = (-\dot{Э}_1)\frac{jx_{12}}{Z_{2н}(\ jx_{11}) - x_{11}x_{22}\left(1 - \frac{x_{12}x_{21}}{x_{11}x_{22}}\right)} \tag{9.27}$$

By the expressions (9.26) and (9.27), the ratio for the currents $\dot{I}'_2/\dot{I}_2$ can be represented. Then, for the current $\dot{I}'_2$, we can obtain

$$\frac{\dot{I}'_2}{\dot{I}_2} = \frac{x_{21}}{x_{11}}\ \dot{I}'_2 = \frac{x_{21}}{x_{11}}\dot{I}_2 = \frac{\dot{I}_2}{k_I} \tag{9.28}$$

where $k_I = x_{11}/x_{21}$.

It can be shown that the reduction factor k_I is determined by the expression (9.13).

We consider the expression (9.26). In (9.26), the value of the multiplier $(1 - x_{12}x_{21}/x_{11}x_{22})$ can be taken equal to zero. In this regard, the expression (9.26) can be reduced to the form

$$(-\dot{I}'_2) = (-\dot{Э}_1)\frac{x_{12}x_{21}}{Z_{2н}x_{11}^2} \tag{9.29}$$

Taking into account that $k_Э = x_{11}/x_{12}$ and $k_I = x_{11}/x_{21}$, the Eq. (9.29) acquires the form

$$(-\dot{I}'_2) = \frac{(-\dot{Э}_1)}{Z_{2н}k_Э k_I} = \frac{(-\dot{Э}_1)}{Z_{2н}k_2} = \frac{(-\dot{Э}_1)}{Z'_{2н}} \tag{9.30}$$

where $Z'_{2Н} = Z_{2Н}k_2; k_2 = k_Э k_I$.

Operating mode. The currents $\dot{I}_0$ and $(-\dot{I}_2')$ are determined by the expressions obtained in (9.25) and (9.30), respectively. Based on these expressions and also the condition (9.1), for the stator winding current $\dot{I}_1$, we receive

$$\dot{I}_1 = \dot{I}_0 + (-\dot{I}_2') = (-\dot{Э}_1)\left(\frac{1}{jx_0} + \frac{1}{Z_{2\text{н}}'}\right) \tag{9.31}$$

By Eq. (9.31), the air gap equivalent circuit takes the form in Fig. 9.1. In this equivalent circuit, we use the condition of kind $\dot{Э}_2' = Z_{2н}'\dot{I}_2'$, where $\dot{Э}_2' = \dot{Э}_1$. Now, the equivalent circuit in Fig. 9.1 can be reduced to the final form in Fig. 9.2. According to the equivalent circuit in Fig. 9.2, the air gap emf equations acquire the form shown in (9.17).

Based on the equivalent circuits shown in Fig. 9.2 and Fig. 9.3, it can be concluded that the transformation of the equivalent circuit in Fig. 9.3 is associated with the transferring its parallel branch $a - b$ to the terminals of the stator winding emf $\dot{Э}_1$. In this case, this transformation technique is called the transfer technique, for example, [1, 4, 20]. As follows from the comparison of the equivalent circuits shown in Fig. 9.2 and Fig. 9.3, the transfer technique allows us to simplify the structure of the equivalent circuit. In this regard, in this work, the transfer technique is used to transform the machine equivalent circuit [20].

9.3 Reduction Technique

In accordance with Eq. (9.1), the current $\dot{I}_1$ flowing in the stator winding in the machine operating mode is represented as a result of the superposition of the no-load current $\dot{I}_0$ and the load current $(-\dot{I}_2')$, i.e. $\dot{I}_1 = \dot{I}_0 + (-\dot{I}_2')$. The no-load current $\dot{I}_0$ is determined when the current in the rotor winding is absent, i.e. $\dot{I}_2 = 0$. In the machine under-load mode, the currents flowing in the stator and rotor windings do not create the common mutual induction magnetic flux (magnetizing flux). Therefore, in the machine under-load mode, the mutual induction emf is not induced in the stator winding, i.e. $\dot{Э}_1 = 0$. Thus, to transform the air gap emf equations obtained in (9.2), we can use the conditions $\dot{I}_2 = 0$ and $\dot{Э}_1 = 0$. For this purpose, we will implement the conditions $\dot{Э}_1 = 0$ and $\dot{I}_2 = 0$ one by one in the Eq. (9.2), for example, [20].

No-load mode. In the system of the Eq. (9.2), the condition $\dot{I}_2 = 0$ can be implemented. Then, it follows

$$\begin{aligned} (-\dot{Э}_1) &= jx_{11}\dot{I}_0 \\ (-\dot{Э}_2) &= jx_{12}\dot{I}_0 \end{aligned} \tag{9.32}$$

From here, we have

$$\dot{\ni}_1 = \frac{x_{11}}{x_{12}}\dot{\ni}_2 = k_{\ni}\dot{\ni}_2 = \dot{\ni}_2'$$

where $k_{\ni} = x_{11}/x_{12}$.

The referred value of the rotor winding emf acquires the form

$$\dot{\ni}_2' = k_{\ni}\dot{\ni}_2 \tag{9.33}$$

The expressions describing the referred value of the rotor winding emf $\dot{\ni}_2'$ and the emf reduction factor $k_{\ni}$ are similar to the formulas obtained earlier in other ways.

Under-load mode. In the first equation of the system (9.2), the condition $\dot{\ni}_1 = 0$ can be taken into account. As a result, we can have

$$0 = x_{11}\dot{I}_1 + x_{21}\dot{I}_2$$

From here, for the current $\dot{I}_1$, we have that $\dot{I}_1 = \frac{x_{21}}{x_{11}}(-\dot{I}_2) = \frac{(-\dot{I}_2)}{k_I} = (-\dot{I}_2')$, i.e. $\dot{I}_1 = (-\dot{I}_2')$. This condition is implemented when the following expressions are true

$$\dot{I}_2' = \frac{\dot{I}_2}{k_I} \quad k_I = \frac{x_{11}}{x_{21}} \tag{9.34}$$

The formulas (9.34) received for the referred value of the rotor winding current $\dot{I}_2'$ and for the current reduction factor k_I are fully agree with the expressions obtained above in the other ways.

Operating mode. The emf equations describing the machine operating mode can be obtained based on the system of the Eq. (9.2). For this, as it was shown in the sixth chapter, in the Eq. (9.2), the emf $\dot{\ni}_2$ and the current $\dot{I}_2$ should be replaced by the referred values of the emf $\dot{\ni}_2'$ and the current $\dot{I}_2'$. Then, we have

$$\begin{aligned}(-\dot{\ni}_1) &= jx_{11}\dot{I}_1 + jx_{21}k_I\dot{I}_2' \\ (-\dot{\ni}_2') &= jx_{12}k_{\ni}\dot{I}_1 + jx_{22}k_{\ni}k_I\dot{I}_2'\end{aligned} \tag{9.35}$$

Based on the expressions obtained in (7.21), it can be shown that for the reactances $x_{21}k_I$, $x_{12}k_{\ni}$ and $x_{22}k_{\ni}k_I$ of the system (9.35), the conditions of the form $x_{21}k_I = x_{12}k_{\ni} = x_{22}k_{\ni}k_I = x_{11}$ and $x_{11} = x_0$ are true. Then, Eq. (9.35) can be represented as

$$\begin{aligned}(-\dot{Э}_1) &= jx_0(\dot{I}_1+\dot{I}_2') = jx_0\dot{I}_0\\ (-\dot{Э}_2') &= jx_0(\dot{I}_1+\dot{I}_2') = jx_0\dot{I}_0\\ \dot{I}_0 &= \dot{I}_1+\dot{I}_2'\end{aligned} \tag{9.36}$$

The air gap equivalent circuit shown in Fig. 9.2 corresponds to these equations. Above, the reduction technique was implemented based on the use of the air gap emf Eq. (9.2). Here, for implementing the reduction technique, the air gap equivalent circuit obtained in Fig. 9.3 is used. In accordance with the reduction technique, the conditions $\dot{I}_2 = 0$ and $\dot{Э}_1 = 0$ should be implemented in the equivalent circuit in Fig. 9.3.

No-load mode. The equivalent circuit corresponding to the machine no-load mode $(\dot{I}_2 = 0)$ is shown in Fig. 9.4. This equivalent circuit can be used to determine the referred stator winding value of the rotor winding emf. For emf $\dot{Э}_1$ and $\dot{Э}_2$, we can have by the equivalent circuit in Fig. 9.4

$$\begin{aligned}(-\dot{Э}_1) &= jx_{11}\dot{I}_0\\ (-\dot{Э}_2) &= jx_{12}\dot{I}_0\end{aligned} \tag{9.37}$$

From (9.37), for the ratio of the emf $\dot{Э}_1$ and $\dot{Э}_2$, it follows

$$\frac{\dot{Э}_1}{\dot{Э}_2} = \frac{x_{11}}{x_{12}} = k_Э$$

From here, the referred to the stator winding value of the rotor winding emf can be expressed as

$$Э_2' = k_Э\dot{Э}_2 \tag{9.38}$$

Under-load mode. The equivalent circuit corresponding to the machine under-load mode $(\dot{Э}_1 = 0)$ is shown in Fig. 9.5. This equivalent circuit arises from the equivalent circuit in Fig. 9.3, and it can be used to determine the load component of the stator winding current. Really, based on the first loop of the equivalent circuit in Fig. 9.5, we can have the condition of the form $(x_{11} - x_{21})(-\dot{I}_1) = x_{21}(\dot{I}_2 - \dot{I}_1)$. In another form $x_{11}\dot{I}_1 = x_{21}(-\dot{I}_2)$. From here, it follows

$$\dot{I}_1 = \frac{x_{21}}{x_{11}}(-\dot{I}_2) = \frac{(-\dot{I}_2)}{k_I} = (-\dot{I}_2') \tag{9.39}$$

where $\dot{I}_2' = \dot{I}_2/k_I; k_I = x_{11}/x_{21}$.

Now, by the equivalent circuit in Fig. 9.3, the equations for the emf $\dot{Э}_1$ and $\dot{Э}_2$ can be compiled. Obtained in this way equations for the emf $\dot{Э}_1$ and $\dot{Э}_2$ can be expressed through the referred rotor quantities $\dot{Э}_2'$ and $(-\dot{I}_2')$. As a result, the air gap emf

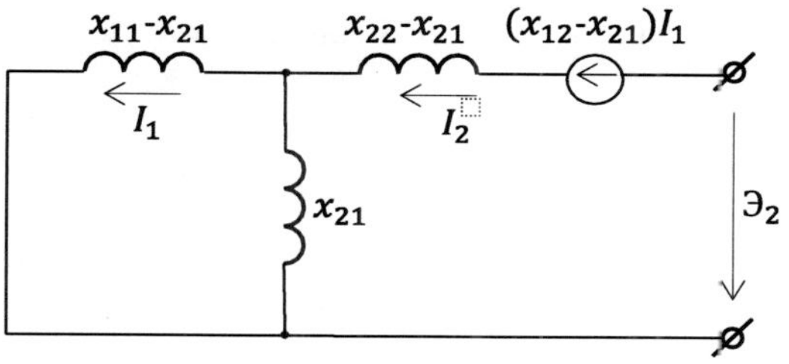

Fig. 9.5 Multi-element air gap equivalent circuit for the under-load mode ($\dot{Э}_1 = 0$)

equations shown in (9.36) follow. The equivalent circuit corresponding to these equations acquires the form in Fig. 9.2.

Thus, the considered techniques (stator winding current decomposition technique, transfer technique, and reduction technique) lead to the same result in relation to the air gap emf equations and to the air gap equivalent circuit. This is due to the fact that the techniques used in this work are based on the representation of the stator winding current as a result of the superposition of the no-load current and the load current.

The techniques given earlier are developed based on the use of the emf equations for the air gap of the machine. In this case, these equations are presented in the form of Z. When analyzing energy problems, it is more convenient to use the emf equations presented in the form of A. In this regard, in Appendix C, the features of the application of the techniques developed above when converting the air gap emf equations are presented in the form of A. In this case, the emf equations for the air gap of the machine are considered.

9.4 Transformation Techniques: Brief Analysis

The techniques for implementing the "no-load and under-load" concept described above are developed based on the use of the air gap emf equations given in (9.2). These emf equations are obtained by the use of the "inductively coupled circuits" concept. The equivalent circuit visualization of the air gap emf Eq. (9.2) is shown in Fig. 6.4a. As follows from the emf Eq. (9.2) and the equivalent circuit in Fig.6.4a, their practical use is problematic. In this regard, the emf Eq. (9.2) and the equivalent circuit in Fig.6.4a can be transformed in such a way that obtained the new air gap emf equations and the air gap equivalent circuit has the form convenient for practical use. As it was noted above, the proposed transformation techniques are based on the idea that the machine operating mode is considered as a result of the superposition of the no-load mode and the under-load mode ("no-load and under-load" concept). In this case, the "partial" machine modes (no-load mode and under-load mode) are implemented in relation to the value of the stator winding emf. Therefore, the stator winding current can be represented, in this case, as a result of the superposition of the no-load current and the load current. We will explain this provision in more detail.

The fact is that the load component of the stator winding current arises under the influence of the emf $\dot{Э}_2$ induced in the rotor winding in the machine no-load mode. The stator and rotor windings are usually implemented with the different number of the turns; therefore, the stator and rotor winding emf $\dot{Э}_1$ and $\dot{Э}_2$ have the different values, i.e. $\dot{Э}_1 \neq \dot{Э}_2$. In these conditions, the stator winding current cannot be represented as a result of the superposition of the no-load current and the load current. The transformation techniques proposed above use the referred to the stator winding values of the rotor winding emf $\dot{Э}'_2$ and the current $\dot{I}'_2$. In this case, the condition $\dot{Э}_1 = \dot{Э}'_2$ is realized. Therefore, the process of the superposition of the "partial" machine modes (machine no-load mode and machine under-load mode) is simplified, and it is reduced to the superposition of the no-load current $\dot{I}_0$ and the load current $(-\dot{I}'_2)$. As a result, the equation for the stator winding current is represented in the form given in (9.1), i.e. $\dot{I}_1 = \dot{I}_0 + (-\dot{I}'_2)$. According to this equation for the stator winding current, the transformation of the initial air gap emf Eq. (9.2) and the initial air gap equivalent circuit (Fig.6.4a) is reduced to finding the values of the currents $\dot{I}_0$ and $(-\dot{I}'_2)$ and to represent these initial air gap equivalent circuit and the air gap emf equations through the current $\dot{I}_0$ and the current $(-\dot{I}'_2)$. This provision is the basis for the transformation techniques described above.

Stator winding current decomposition technique. This transformation technique involves the determination of the currents $\dot{I}_0$ and $(-\dot{I}'_2)$, representing the components of the stator winding current. The currents $\dot{I}_0$ and $(-\dot{I}'_2)$ reflect the machine no-load mode and the machine under-load mode for the determination of which the air gap emf equations given in (9.2) are used in our case. To determine the no-load current $\dot{I}_0$, it is assumed that the rotor winding current is equal to zero in the system of the Eq. (9.2), i.e. $\dot{I}_2 = 0$. The load component of the stator winding current $(-\dot{I}'_2)$ is calculated by the equation $(-\dot{I}'_2) = \dot{I}_1 - \dot{I}_0$. Here, the stator winding current $\dot{I}_1$ can be defined by the use of the system of the Eq. (9.2). As a result of these manipulations, the referred to the stator winding values of the rotor winding emf $\dot{Э}'_2$ and the current $\dot{I}'_2$ are defined, and the reduction factors of the rotor winding emf and current are established. Now, as a result of using the referred to the stator winding values of the rotor winding emf $\dot{Э}'_2$ and the current $\dot{I}'_2$, the system of the Eq. (9.2) acquires the form shown in (9.17). In this case, the air gap equivalent circuit in Fig.6.4a is converted into the equivalent circuit given in Fig. 9.2. In the equivalent circuit in Fig. 9.2, the stator and rotor windings are represented as the conditionally electrically connected circuit loops thanks to which the air gap equivalent circuit gets the form convenient for practical use.

Transfer technique. For implementing the transfer technique, the equation of the form $\dot{I}_1 = \dot{I}_0 + (-\dot{I}'_2)$ is used for the stator winding current. In this case, the currents $\dot{I}_0$ and $(-\dot{I}'_2)$ is determined by the equivalent circuit in Fig. 9.3 obtained by the system of the emf Eq. (9.2). This technique differs from the decomposition technique of the stator winding current in that the currents $\dot{I}_0$ and $(-\dot{I}'_2)$ are defined by the equivalent circuit. The realization of this transformation technique is associated with transferring the transverse branch $a - b$ of the equivalent circuit in Fig. 9.3 to the terminals of the stator winding emf $\dot{Э}_1$ (transfer technique). As a result of applying the transfer technique, the equivalent circuit in Fig. 9.3 acquires the simple structure shown in Fig. 9.2. This feature of the transfer technique can be used to simplify the structure of the multi-loop equivalent circuits with the mixed connection of the elements. In this case, the transfer technique can be represented as the multiple transfer technique [20].

Reduction technique. The reduction technique, like the techniques discussed above, is based on the use of the equation for the stator winding current in the form $\dot{I}_1 = \dot{I}_0 + (-\dot{I}_2')$. In this case, the determination of the current $\dot{I}_0$ and the current $(-\dot{I}_2')$ is implemented by the use of the air gap emf equations obtained in (9.2). To determine the current $\dot{I}_0$, it is assumed that the rotor winding current is equal to zero in the emf Eq. (9.2), i.e. $\dot{I}_2 = 0$. The load component of the stator winding current $(-\dot{I}_2')$ corresponds to the machine under-load mode. As it was shown above, in the machine under-load mode, the mutual induction common magnetic flux (magnetizing flux) is absent in the electric machine. Consequently, in the machine under-load mode, the mutual induction emf is not induced in the stator winding, i.e. $\dot{Э}_1 = 0$. Now, to determine the current $(-\dot{I}_2')$, the condition of the form $\dot{Э}_1 = 0$ can be used in the emf Eq. (9.2). Thus, the conditions $\dot{I}_2 = 0$ and $\dot{Э}_1 = 0$ can be applied to transform the air gap emf equations obtained in (9.2). In general, the conditions of the form $\dot{I}_2 = 0$ and $\dot{Э}_1 = 0$ are the basis for the reduction technique. Using these conditions in the emf Eq. (9.2) allows us to establish the reduction factors and to define the referred stator winding values of the rotor winding emf $\dot{Э}_2'$ and the current $\dot{I}_2'$ and as a result to reduce the emf Eq. (9.2) to the form shown in (9.17).

The reduction technique can be implemented by the use of the air gap equivalent circuit. Such equivalent circuit is shown in Fig. 9.3. In accordance with the reduction technique, the conditions of the form $\dot{I}_2 = 0$ and $\dot{Э}_1 = 0$ should be implemented in the equivalent circuit shown in Fig. 9.3. Using in the equivalent circuit in Fig. 9.3 these conditions ($\dot{I}_2 = 0$ and $\dot{Э}_1 = 0$) allows us to establish the reduction factors, to determine the referred values of the rotor winding emf $\dot{Э}_2'$ and the current $\dot{I}_2'$. As a result, the equivalent circuit in Fig. 9.3 can be reduced to the simple form in Fig. 9.2.

Thus, the transformation techniques considered above lead to the same results, since they all perform the reduction function. However, it should be noted that the transfer technique and reduction technique are more preferable for practical use. In this case, the transfer technique is convenient for converting the equivalent circuits and the reduction technique is preferable for converting the voltage equations.

In the theory of the electric machine, the transfer technique and reduction technique are mainly considered as independent transformation techniques. This state of affairs does not contribute to the further successful development of the methods suitable for the engineering practice.

9.5 Summary

1. The transformation techniques used in the theory of the electric machines to convert the voltage equations and the equivalent circuits are based on applying the "no-load and under-load" concept. In accordance with this concept, the stator winding current is represented as a result of the superposition of the no-load current and the load current, i.e. $\dot{I}_1 = \dot{I}_0 + (-\dot{I}_2')$. This equation for the stator winding current forms the basis for the transformation techniques recommended in this work (the stator winding current decomposition technique, transfer technique, and reduction technique).
2. The "no-load and under-load" concept is implemented in relation to the single winding. In the technical literature, the stator winding is usually considered as such winding. In this regard, the components of the stator winding current $\dot{I}_0$ and $(-\dot{I}_2')$ are determined in relation to the stator winding emf. The transformation techniques proposed in this work are based on the implementation of this provision.
3. The transformation techniques are developed based on the consideration of the air gap equivalent circuit and the air gap emf equations following from the "inductively coupled circuits" concept. The air gap equivalent circuit and the air gap emf equations obtained in this way contain the inductively coupled elements. This means that the stator and rotor windings are represented in them as the inductively coupled circuit loops. Because of this, the air gap equivalent circuit and the air gap emf equations having the inductively coupled elements are transformed. The transformation techniques considered in this chapter are directed to eliminate the inductively coupled elements and to give to the air gap equivalent circuit and the air gap emf equations the form suitable for practical use.
4. In this work, the description of the three transformation techniques is given (the stator winding current decomposition technique, transfer technique, and reduction technique). In practice, the transfer technique and the reduction technique are widespread. The reduction technique is most effectively used when transforming the voltage equations. The transfer technique is usually applied when transforming the equivalent circuit. In this case, the simplification of the voltage equations and the equivalent circuit is achieved. It can be noted that the transformation techniques considered above have different forms of implementation. However, these techniques all implement the single reduction function. Therefore, they provide the same result.
5. The general scheme for the implementation of the "no-load and under-load" concept can be represented as consisting of two stages. At the first stage, based on the initial emf equations and the initial equivalent circuits, the reduction factors and the referred to the stator winding values of the rotor winding emf $\dot{Э}_2'$ and the current $\dot{I}_2'$ are established. In our case, as the initial emf equations and the initial equivalent circuits are considered the air gap equivalent circuit and the air gap emf equations. At the second stage, the initial emf equations and the initial

equivalent circuits are expressed through the referred stator winding values of the rotor winding emf $\dot{Э}'_2$ and the current $\dot{I}'_2$. As a result of this action, the transformed emf equations and the equivalent circuits are followed. The transformed emf equations and the equivalent circuits obtained in this way have the forms suitable for the engineering practice.

References

1. Kostenko, M.: Electric Machines. Special Part, Gosenergoizdat, Moscow (1949)
2. Bragstad, O.: Theory of AC Machines with an Introduction to the Theory of Steady-State Alternating Currents (Translated from German), Izdatelstvo Lensoveta i Oblispolkoma, Leningrad (1933)
3. Richter, R.: Electric Machines, vol. 1-5. United Scientific and Technical Publishing, Moscow (1934-1939)
4. Voldek. A.: Electric Machines, Energiya, Leningrad (1974)
5. Ivanov-Smolensky, A.: Electric Machines, Energiya, Moscow (1980)
6. Shenfer, K.: Asynchronous Machines, State United Scientific and Technical Publishing, Moscow/Leningrad (1938)
7. Vazhnov, A.: Electric Machines, Energiya, Leningrad (1968)
8. Antonov, L.: Alternating Current Machines, Naval Publishing, Leningrad (1968)
9. Piotrovskij, L.: Electric Machines, Publishing Energiya, Leningrad (1975)
10. Alger, P.: Induction Machines, 2nd ed., Gordon and Beach, New York (1970)
11. Say, M.: Alternating Current Machines, D. Van No strand Company, Inc., New York (1961)
12. McPherson, G., Laramore, R.: Electrical Machines and Transformers, 2nd ed., John Wiley and Sons, Inc., New York (1990)
13. Riff, P., Platnick, D., Karnas, J.: Electrical Machines and Transformers. Principles and Applications, Prentice-Hall, Inc., New Jersey (1987)
14. Sen, P.: Principles of Electric Machines and Power Electronics, John Wiley and Sons, New York (1989)
15. Fitzgerald, A., Kingsley, C., Umans, S.: Electric Machinery, McGraw-Hill, Inc., New York (1990)
16. Del Toro, V.: Electric Machines and Power Systems, Prentice-Hall, Inc., Englewood Cliffs, New Jersey (1985)
17. Krause, P.: Analysis of Electric Machinery, McGraw-Hill, New York (1986)
18. Sarema, M.: Electric Machines. Steady-State Theory and Dynamic Performance, West Publishing Company, St. Paul, New York (1986)
19. Chapman, S.: Electric Machinery Fundamentals, 4th ed., McGraw-Hill, Inc., New York (2005)
20. Asanbayev, V.: Alternating Current Multi-Circuit Electric Machines: A New Approach to Steady-State Parameter Determination, Springer Cham Heidelberg, New-York, Dordrecht London (2015).

Chapter 10
Equivalent Circuit of Machine in Phase System of Units: Techniques for Construction

10.1 Equivalent Circuit of Machine

We will proceed from the fragment equivalent circuits shown in Fig. 8.3. In Fig. 8.3, the air gap equivalent circuit is represented as the two equivalent circuits shown in Fig. 8.3a and b. The equivalent circuit in Fig. 8.3a corresponds to the machine no-load mode $\left(\dot{I}_2 = 0\right)$. The equivalent circuit in Fig. 8.3b is obtained in relation to the stator winding current $\dot{I}_1$ and referred to the stator winding value of the rotor winding current $\dot{I}_2'$, and it reflects the machine operating mode. The equivalent circuits in Fig. 8.3a and b represent the various machine modes; however, as it was shown above, they represent the magnetic field (magnetizing field) taking place in the machine air gap in the no-load mode. Therefore, these equivalent circuits include the single reactance x_0 known, in this case, as the air gap magnetizing reactance. In this regard, the equivalent circuits in Fig. 8.3a and b can be used to build the machine equivalent circuit. With the above in mind, the two ways of constructing the machine equivalent circuit can be used. The first way is based on the application of the stator slot leakage equivalent circuit (Fig. 8.3c), the rotor slot leakage equivalent circuit (Fig. 8.3d), and also the air gap equivalent circuit (Fig. 8.3a) corresponding to the machine no-load mode. In this case, the construction of the machine equivalent circuit assumes, as will show below, the mixed (parallel-cascade) connection of the fragment equivalent circuits given in Fig. 8.3a, c, and d. The second way is based on the application of the stator slot leakage equivalent circuit (Fig. 8.3c), the rotor slot leakage equivalent circuit (Fig. 8.3d), and also the air gap equivalent circuit (Fig. 8.3b) corresponding to the machine operating mode. In this case, the construction of the machine equivalent circuit assumes the cascade connection of the fragment equivalent circuits given in Fig. 8.3b, c and d. Below, we will consider the features of implementing these two ways of building the machine equivalent circuit.

V. Asanbayev, *Asynchronous Machines*,
https://doi.org/10.1007/978-3-030-92284-9_10

10.1.1 Mixed Way of Connecting: Application

We proceed from the fragment equivalent circuits shown in Fig. 8.3. In Fig. 8.3, the equivalent circuits represented in Fig. 8.3a, c, d can be considered. These equivalent circuits are shown in Fig. 10.1. Below, the equivalent circuits in Fig. 10.1 are used to construct the machine equivalent circuit. We will start building the machine equivalent circuit by considering the air gap equivalent circuit shown in Fig. 10.1a and the rotor slot leakage equivalent circuit in Fig. 10.1c. The output terminals $c - d$ of the air gap equivalent circuit in Fig. 10.1a can be connected to the input terminals $c - d$ of the rotor slot leakage equivalent circuit in Fig. 10.1c which means the parallel connection of the impedance $(r_2'/s + jx_{2\sigma}' + Z_{2H}')$ used in the equivalent circuit in Fig. 10.1c with the air gap magnetizing reactance x_0 (Fig. 10.1a). As a result, the equivalent circuit shown in Fig. 10.2a arises. In this equivalent circuit, the condition characteristic for the stator winding current in the form $\dot{I}_1 = \dot{I}_0 + (-\dot{I}_2')$ is performed on its input terminals $a - b$. Thus, the values of the stator winding emf $\dot{Э}_1$ and the current $\dot{I}_1$ exist on the input terminals $a - b$ of the equivalent circuit in Fig. 10.2a. As follows from Fig. 10.1b, at its input terminals $a - b$, the same values of the stator winding emf $\dot{Э}_1$ and the current $\dot{I}_1$ take place. Consequently, to obtain the machine equivalent circuit, the input terminals $a - b$ of the equivalent circuit in Fig. 10.2a can be connected to the input terminals $a - b$ of the stator slot leakage equivalent circuit

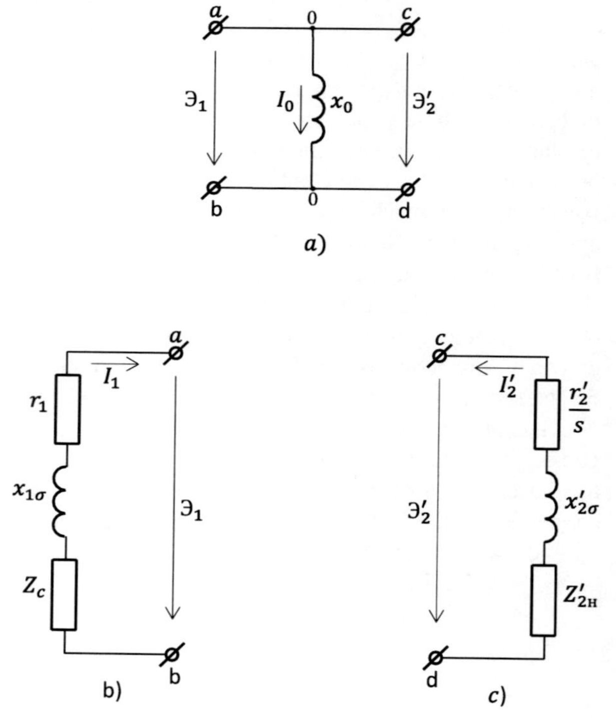

Fig. 10.1 Fragments of machine equivalent circuit: (**a**) air gap equivalent circuit (no-load mode); (**b**) stator slot leakage equivalent circuit; (**c**) rotor slot leakage equivalent circuit

Fig. 10.2 Construction of machine equivalent circuit: (**a**) machine equivalent circuit without the stator circuit; (**b**) machine equivalent circuit taking into account the stator circuit

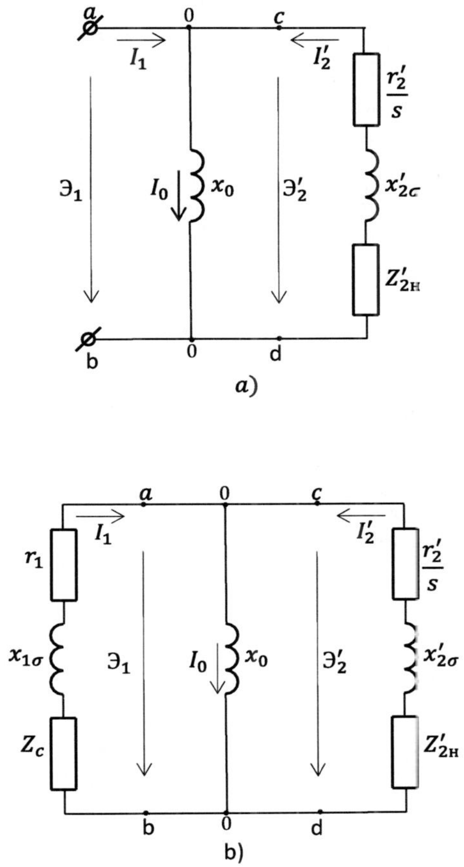

in Fig. 10.1b (the cascade way of connecting). As a result, the machine equivalent circuit can be reduced to the form in Fig. 10.2b. Now, if for the impedance Z_c used in the equivalent circuit in Fig. 10.2b the condition of the form $\left(-\dot{U}_1\right) = Z_c\dot{I}_1$ to apply and also for the impedance Z'_{2H} used in the equivalent circuit in Fig. 10.2b the condition of the form $\dot{U}'_2/s = Z'_{2H}\dot{I}'_2$ to use, then, the machine equivalent circuit can be reduced to the form in Fig. 10.3. Thus, the construction of the machine equivalent circuit shown in Fig. 10.3 is implemented, in our case, by the air gap equivalent circuit corresponding to the no-load mode (Fig. 10.2a), the stator slot leakage equivalent circuit (Fig. 10.1b), and the rotor slot leakage equivalent circuit (Fig. 10.1c). In this case, the mixed (parallel-cascade) way of connecting these equivalent circuits is applied.

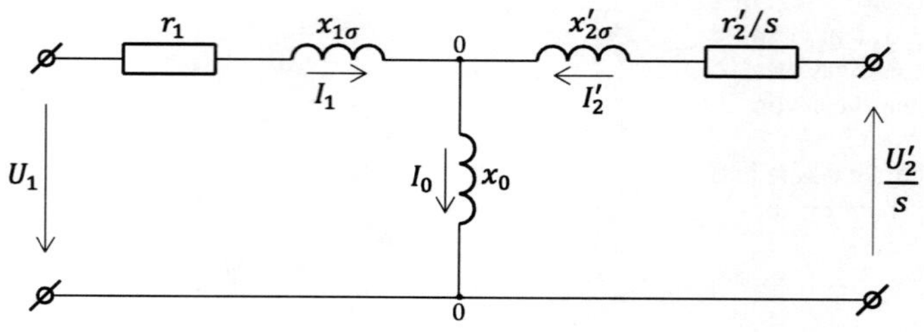

Fig. 10.3 Machine equivalent circuit

Fig. 10.4 Fragments of machine equivalent circuit (operating mode). (**a**) air gap equivalent circuit (**b**) stator slot leakage equivalent circuit (**c**) rotor slot leakage equivalent circuit

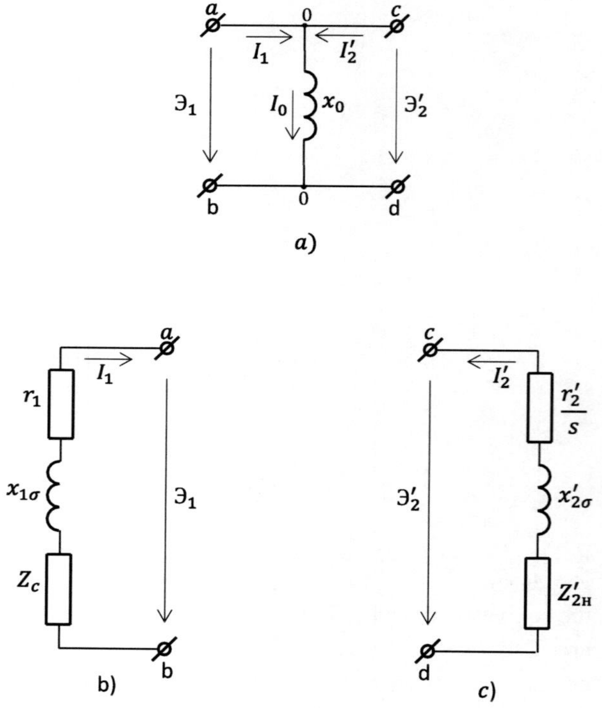

10.1.2 Cascade Way of Connecting: Application

We proceed from the fragment equivalent circuits given in Fig. 8.3. In Fig. 8.3, the equivalent circuits shown in Fig. 8.3b, c, and d can be applied. These equivalent circuits are shown in Fig. 10.4. The equivalent circuit obtained in Fig. 10.4a represents the transformed air gap equivalent circuit, and it corresponds to the machine operating mode. The stator equivalent circuit is given in Fig. 10.4b. The rotor equivalent circuit in Fig. 10.4c is expressed through the referred to the stator winding values of the rotor winding emf and the current.

The fragment equivalent circuits obtained in Fig. 10.4 can be applied to construct the machine equivalent circuit. In this case, the machine equivalent circuit construction is reduced to the connection of the input terminals $c - d$ of the rotor slot leakage equivalent circuit (Fig. 10.4c) with the output terminals $c - d$ of the transformed air gap equivalent circuit (Fig. 10.4a) and also to the connection of the input terminals $a - b$ of the transformed air gap equivalent circuit (Fig. 10.4a) with the input terminals $a - b$ of the stator slot leakage equivalent circuit (Fig. 10.4b). As it was noted above, such connection is known as the cascade way of connecting. From the above, it follows that the machine equivalent circuit arises as a result of the cascade way of connecting the input terminals $c - d$ of the rotor slot leakage equivalent circuit (Fig. 10.4c) with the output terminals $c - d$ of the transformed air gap equivalent circuit (Fig. 10.4a) and also the input terminals $a - b$ of the transformed air gap equivalent circuit (Fig. 10.4a) with the input terminals $a - b$ of the stator slot leakage equivalent circuit (Fig. 10.4b). As a result, the machine equivalent circuit acquires the form shown in Fig. 10.2b. As it was shown above, the equivalent circuit in Fig. 10.2b can be reduced to the form in Fig. 10.3.

10.1.3 Application of Equivalent Circuits of Stator and Rotor

Above, the two ways of the construction of the machine equivalent circuit have been demonstrated. Recall that the first way of the machine equivalent circuit construction is based on the use of the fragment equivalent circuits shown in Fig. 10.1. In Fig. 10.1a, the air gap equivalent circuit corresponds to the machine no-load mode. In this case, to construct the machine equivalent circuit, the mixed way of connecting the fragment equivalent circuits given in Fig. 10.1 is used. The second way of the machine equivalent circuit construction is implemented using the fragment equivalent circuits shown in Fig. 10.4. In Fig. 10.4a, the transformed air gap equivalent circuit corresponds to the machine operating mode. Here, to construct the machine equivalent circuit, the cascade way of connecting the fragment equivalent circuits shown in Fig. 10.4 is applied. Below, we will consider another implementation version for the second way of construction of the machine equivalent circuit.

The stator and rotor slot leakage equivalent circuits are given in Fig. 10.4b and Fig. 10.4c, respectively. The stator slot leakage equivalent circuit (Fig. 10.4b) is represented in relation to the stator winding emf $\dot{Э}_1$. The rotor slot leakage equivalent circuit (Fig. 10.4c) is obtained in relation to the referred stator winding value of the rotor winding emf $\dot{Э}_2'$. Now, we consider the air gap equivalent circuit in Fig. 10.4a. The input terminals of this equivalent circuit are under the influence of the stator winding emf $\dot{Э}_1$, and the output terminals have the referred rotor winding emf $\dot{Э}_2'$. The air gap equivalent circuit in Fig. 10.4a can be decomposed into the two sub-circuits. For this purpose, the reactance x_0 used in the equivalent circuit in

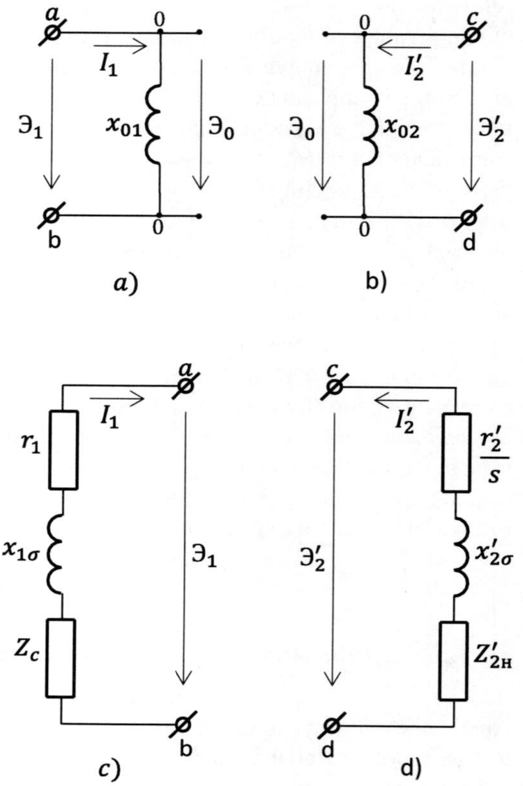

Fig. 10.5 Fragments of machine equivalent circuit when the air gap is divided into two sublayers: (**a**) equivalent circuit of the air gap sublayer adjoined to the stator bore surface; (**b**) equivalent circuit of the air gap sublayer adjoined to the rotor surface; (**c**) stator slot leakage equivalent circuit; (**d**) rotor slot leakage equivalent circuit

Fig. 10.4a and being under the influence of the emf $\dot{Э}_0$ can be expressed through the parallel connected reactances x_{01} and x_{02} [1]. This means that the machine air gap can be considered as consisting of the two sublayers having the calculated radial lengths equal to δ_1 and δ_2 defined as $\delta_1 = \delta k_{1\delta}/2$ and $\delta_2 = \delta k_{2\delta}/2$, where $k_{1\delta}$ and $k_{2\delta}$ are the air gap sublayers factors [1]. The sublayer with the radial length equal to δ_1 is adjoined to the stator bore surface, and the sublayer with the radial length δ_2 is adjoined to the rotor surface. From (8.16), for the reactances x_{01} and x_{02}, we can obtain that $x_{01} = \frac{\omega_1 \mu_0 \tau_0^2}{\pi^2 \delta_1} k_1$ and $x_{02} = \frac{\omega_1 \mu_0 \tau_0^2}{\pi^2 \delta_2} k_1$. Then, in relation to the emf $\dot{Э}_0$, the air gap equivalent circuit in Fig. 10.4a can be represented as the two sub-circuits shown in Fig. 10.5a and b. Here, the emf $\dot{Э}_0$ refers to the surface passing through the middle of the radial length of the air gap (Appendix B). Now, the equivalent circuits given in Fig. 10.4b, c can be represented in Fig. 10.5. The sub-circuit in Fig. 10.5a reflects the air gap sublayer adjoined to the stator bore surface. The sub-circuit in Fig. 10.5b represents the air gap sublayer adjoined to the rotor surface. Thus, the sub-circuits in Fig. 10.5a, b can be considered as the equivalent circuits of the air gap fragments (the air gap sublayers). The equivalent circuits obtained in Fig. 10.5 can be used to construct the stator and rotor equivalent circuits. We will show it below.

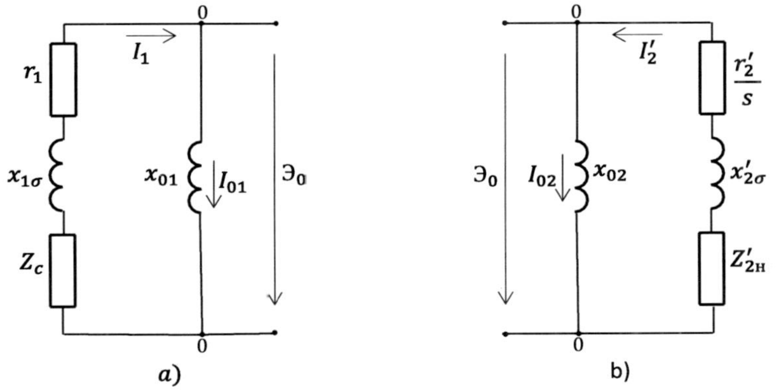

Fig. 10.6 Stator and rotor equivalent circuits: (**a**) stator equivalent circuit; (**b**) rotor equivalent circuit

Stator equivalent circuit. The stator equivalent circuit can be constructed by the fragment equivalent circuits given in Fig. 10.5a, c. The terminals $a - b$ of these equivalent circuits are under the influence of the stator winding emf $\dot{Э}_1$ and the current $\dot{I}_1$. Therefore, the terminals $a - b$ of the equivalent circuits in Fig. 10.5a, c can be connected in the cascade way. As a result, the stator equivalent circuit shown in Fig. 10.6a arises.

Rotor equivalent circuit. The rotor equivalent circuit can be constructed by the fragment equivalent circuits obtained in Fig. 10.5b, d. These equivalent circuits are expressed through the referred values of the rotor winding emf $\dot{Э}_2'$ and the current $\dot{I}_2'$. In this regard, the terminals $c - d$ of these equivalent circuits are under the influence of the rotor winding emf $\dot{Э}_2'$ and the current $\dot{I}_2'$. Therefore, the terminals $c - d$ of the equivalent circuits in Fig. 10.5b, d can be connected in the cascade way. Then, the rotor equivalent circuit acquires the form in Fig. 10.6b.

Machine equivalent circuit. To construct the machine equivalent circuit, the stator and rotor equivalent circuits received in Fig. 10.6a, b can be used. For this, in the stator and rotor equivalent circuits (Fig. 10.6a, b), their parallel branches having the reactances x_{01} and x_{02}, respectively, can be combined into one common branch with the common reactance x_0 determined as $x_0 = x_{01}x_{02}/(x_{01} + x_{02})$. As a result, the machine equivalent circuit acquires the form in Fig. 10.3. In this work, this technique for constructing the machine equivalent circuit has got further development.

Based on the machine equivalent circuit in Fig. 10.3, the machine voltage equations obtain the form.

$$\begin{gathered}\dot{U}_1 = (r_1 + jx_{1\sigma})\dot{I}_1 + jx_0\dot{I}_0 \\ \left(-\frac{\dot{U}_2'}{s}\right) = \left(\frac{r_2'}{s} + jx_{2\sigma}'\right)\dot{I}_2' + jx_0\dot{I}_0 \\ \dot{I}_0 = \dot{I}_1 + \dot{I}_2'\end{gathered} \tag{10.1}$$

Thus, when using the cascade way of connecting, the machine voltage equations are written out based on the machine equivalent circuit. Note that in the theory of the electric machines, the machine voltage equations are usually first compiled, and then, the machine equivalent circuit is built using these machine voltage equations.

In our case, the construction of the machine equivalent circuit became possible due to the use of the transformed equivalent circuit (Fig. 10.4a) obtained for the machine air gap by the "no-load and under-load" concept, to the application of the rotor slot leakage equivalent circuit (Fig. 10.4c) expressed through the referred rotor quantities, to the use of the stator slot leakage equivalent circuit (Fig. 10.4b) and also to the use of the cascade way of connecting them. As shown in the Chap. 8, the "no-load and under-load" concept can be considered as the way to transform the air gap equivalent circuit having the inductively coupled elements (Fig. 6.5a). In the transformed air gap equivalent circuit, the stator and rotor windings are represented as the conditionally electrically coupled two circuit loops (Fig. 8.3b and 10.4). As a result, the air gap equivalent circuit is simplified, and it takes the form suitable for practical use.

In the theory of electric machines, the machine equivalent circuit is obtained by the use of the "no-load and under-load" concept (Appendix D). In this case, the "no-load and under-load" concept is used for the equivalent circuit built in relation to the magnitude of the voltage given at the stator winding terminals. It should be noted that at the same time the simplicity, clarity, and practicality of the construction of the machine equivalent circuit is achieved. However, with this method of the construction of the machine equivalent circuit, the use of the field equations is difficult. Besides, additional difficulties arise in ensuring the consistency and methodically in the presentation of the provisions associated with the construction of the machine equivalent circuit, for example, [2–20].

The machine equivalent circuit shown in Fig. 10.3 follows as a result of the cascade way of connecting the air gap equivalent circuit (Fig. 10.4a), the stator slot leakage equivalent circuit (Fig. 10.4b), the rotor slot leakage equivalent circuit (Fig. 10.4c) considered as the separate fragments of the machine equivalent circuit. As shown in [1], the idea of the construction of the machine equivalent circuit based on the cascade way of connecting the fragment equivalent circuits has the prospect of the practical implementation in the case of the description of the machine resulting magnetic field by Maxwell's equations. Therefore, this idea of the construction of the machine equivalent circuit forms the basis of the content of this work, and below, it will be considered in more detail.

10.2 Features of Applying Various Techniques

We will consider the machine equivalent circuit shown in Fig. 7.4 and 10.3. The machine equivalent circuit shown in Fig. 7.4 is obtained based on the use of the "inductively coupled circuits" concept. The machine equivalent circuit shown in Fig. 10.3 follows based on applying the "no-load and under-load" concept. The machine equivalent circuit shown in Fig. 7.4 is expressed through the referred rotor quantities. Therefore, the equivalent circuits obtained in Fig. 7.4 and Fig. 10.3 have the same structure, and they do not differ from each other externally. The difference between these equivalent circuits lies in the ways of determining used in them the leakage reactances of the stator and rotor windings $x_{1\sigma}$ and $x'_{2\sigma}$. The leakage reactances $x_{1\sigma}$ and $x'_{2\sigma}$ applying in the equivalent circuit shown in Fig. 7.4 are calculated by the expressions given in (8.22) in the following form $x_{1\sigma} = \dot{Э}_{1\sigma}/\dot{I}_1$ and $x'_{2\sigma} = \dot{Э}'_{2\sigma}/\dot{I}'_2$ or $x_{2\sigma} = \dot{Э}_{2\sigma}/\dot{I}_2$. In these expressions, the leakage emf $\dot{Э}_{1\sigma}$ and $\dot{Э}_{2\sigma}$ are determined by the leakage fluxes of the stator and rotor windings $\dot{\Phi}_{1\sigma}$ and $\dot{\Phi}_{2\sigma}$, for which we have that $\dot{\Phi}_{1\sigma} = \dot{\Phi}_1 - \dot{\Phi}_{11}$ and $\dot{\Phi}_{2\sigma} = \dot{\Phi}_2 - \dot{\Phi}_{22}$. Here, $\dot{\Phi}_1$ and $\dot{\Phi}_2$ reflect the total self-induction magnetic fluxes of the stator and rotor windings and $\dot{\Phi}_{11}$ and $\dot{\Phi}_{22}$ represent the mutual induction magnetic fluxes of the stator and rotor windings. As it was noted above, the conditions $\dot{\Phi}_{1\sigma} = \dot{\Phi}_1 - \dot{\Phi}_{11}$ and $\dot{\Phi}_{2\sigma} = \dot{\Phi}_2 - \dot{\Phi}_{22}$ do not provide the required level of reliability in determining the leakage reactances of the stator and rotor windings $x_{1\sigma}$ and $x'_{2\sigma}$. For this reason, the equivalent circuit shown in Fig. 7.3 cannot be recommended for engineering practice.

The leakage reactances of the stator and rotor windings $x_{1\sigma}$ and $x'_{2\sigma}$ applying in the equivalent circuit in Fig. 10.3 follow based on the "no-load and under-load" concept. In this case, the leakages fluxes $\dot{\Phi}_{1\sigma}$ and $\dot{\Phi}_{2\sigma}$ are considered created by the volume of the currents flowing in the stator and rotor slots. Therefore, the leakages fluxes $\dot{\Phi}_{1\sigma}$ and $\dot{\Phi}_{2\sigma}$ are determined as a result of the direct calculation of the leakages fluxes caused by the total currents in the stator and rotor slots. It should be noted that, in this case, the leakage fluxes are often determined by the use of the total current equation. As the practice shows, the quite acceptable for the engineering calculations level of the reliability in determining the leakage reactances of the stator and rotor windings $x_{1\sigma}$ and $x'_{2\sigma}$ is achieved. In this regard, the equivalent circuit in Fig. 10.3 has become widespread.

Now, we can note that the machine equivalent circuit arising from the "inductively coupled circuits" concept and corresponding to the requirements of the engineering calculation practice must be expressed through the referred rotor quantities, and the leakage reactances of the stator and rotor windings $x_{1\sigma}$ and $x'_{2\sigma}$ used in it must correspond to the conditions of the "no-load and under-load" concept.

10.3 Magnetizing Reactance of Machine

The machine equivalent circuit in Fig. 10.3 contains the transverse branch 0 – 0 having the reactance x_0. The reactance x_0 is determined by the expression given in (8.14), and it represents the air gap magnetizing reactance. The expression (8.14) reflects the air gap reluctance, and it corresponds to the case when the reluctances of the magnetic circuit regions of the stator and rotor are neglected. To take into account the reluctances of the stator and rotor magnetic circuit regions, it is necessary to implement the calculation of the machine magnetic circuit.

10.3.1 Saturation Factor

The calculation of the machine magnetic circuit is implemented by the use of the total current law, described as $0.5 \oint \dot{H}dl = \dot{F}_m$, where $\dot{F}_m$ is the mmf necessary to create the machine magnetizing flux Φ_m. The flux Φ_m is usually determined for the given value of the stator winding emf $\dot{Э}_1 = \dot{Э}_m$, where $\dot{Э}_m$ is the emf corresponding to the surface passing through the middle of the radial length of the air gap. The coefficient 0.5 in the formula for the total current law means that the mmf $\dot{F}_m$ is determined on one pole, and therefore, only the half of the total length of the machine magnetic circuit is taken into account. In Fig. 10.7, the magnetic circuit located in the machine cross section is shown. In Fig. 10.7a, the machine magnetic

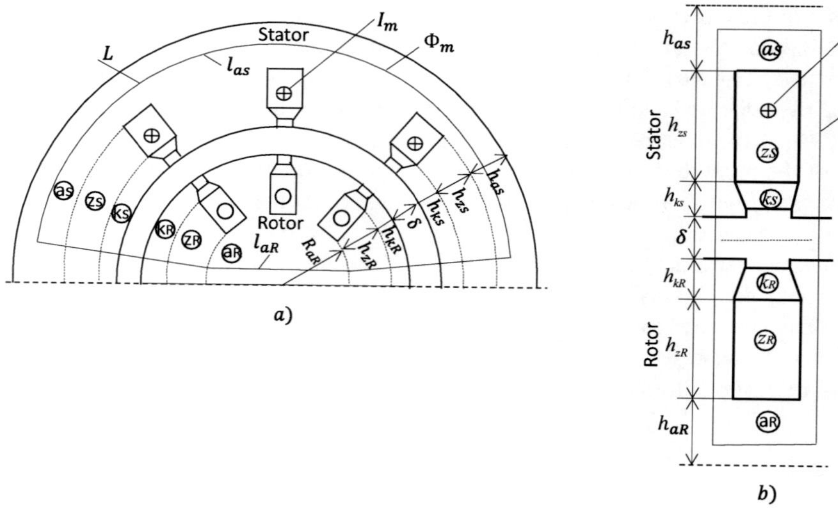

Fig. 10.7 To the calculation of the machine magnetizing circuit: (**a**) conditional line of the magnetizing flux within the pole pitch; (**b**) conditional line of the magnetizing flux within the single tooth pitch

circuit is located within the single-pole pitch. In Fig. 10.7b, the machine magnetic circuit is represented within the single tooth pitch. In these figures, the middle line of the magnetizing flux Φ_m is shown. In order to simplify the calculations, the magnetic circuit is divided into sections in such a way that within the selected sections the value of the magnetic field intensity remains constant, i.e. $H = const$. In practice, the sections of the machine magnetic circuit represent the air gap area, the slot-teeth areas including the slot-wedge areas, the stator and rotor winding location areas, as well as the stator and rotor yoke areas. Thus, the machine magnetic circuit can be divided into seven sections the one of which represents the air gap layer and the rest belong to the stator and rotor sections.

When calculating the machine magnetic circuit, the stator and rotor surfaces are assumed smooth, for example, [1–20]. The presence of the teeth on the stator and rotor surfaces is taken into account using the air gap factor k_δ. In this case, the calculation air gap length is determined as $\delta' = \delta k_\delta$. The linear integral $0.5 \oint \dot{H}dl = \dot{F}_m$ can be used to the magnetic circuit shown in Fig. 10.7a. Replacing the linear integral $0.5 \oint \dot{H}dl$ with the sum of the magnetic tensions in the accepted sections of the magnetic circuit, we can obtain.

$$0.5 \oint \dot{H}dl = \dot{F}_{as} + \dot{F}_{zs} + \dot{F}_{ks} + \dot{F}_0 + \dot{F}_{kR} + \dot{F}_{zR} + \dot{F}_{aR} = \dot{F}_m \tag{10.2}$$

where $\dot{F}_{as}$ and $\dot{F}_{aR}$ are the magnetic tensions of the sections of the stator and rotor yoke areas, $\dot{F}_{ks}, \dot{F}_{zs}$ and $\dot{F}_{kR}, \dot{F}_{zR}$ are the magnetic tensions of the sections of the stator and rotor slot-wedge areas and the winding location areas, respectively, and $\dot{F}_0$ is the magnetic tension of the machine air gap.

The saturation level of the machine magnetic circuit is characterized by the coefficient $k_н$ determined as

$$\dot{F}_m = \dot{F}_0 k_н \tag{10.3}$$

where $k_н = \frac{\dot{F}_m}{\dot{F}_0} = 1 + \frac{\dot{F}_{as} + \dot{F}_{zs} + \dot{F}_{ks} + \dot{F}_{kR} + \dot{F}_{zR} + \dot{F}_{aR}}{\dot{F}_0}$.

Here $k_н$ reflects the saturation factor of the machine magnetic circuit. The saturation of the sections of the stator and rotor magnetic circuits can be taken into account by introducing the calculated length of the air gap defined as $\delta' = \delta k_н k_\delta$. Then, from (8.16) the machine magnetizing reactance x_m follows

$$x_m = \frac{\omega_1 \mu_0 \tau_0^2}{\pi^2 \delta'} k_1 \tag{10.4}$$

where $k_1 = \frac{2m_1 (w_1 k_{w1})^2 l_0}{p\tau_0}$; $\delta' = \delta k_\delta k_н$.

10.3.2 Approximate Saturation Accounting

Now, for the reactance used in the parallel branch $0-0$ of the equivalent circuit in Fig. 10.3, the expression given in (10.4) can be applied. This expression takes into account the reluctances of the sections of the machine magnetic circuit (the air gap and the sections of the stator and rotor magnetic circuits). Now, if in the equivalent circuit in Fig. 10.3 to use the expression (10.4), then, this equivalent circuit can be reduced to the form shown in Fig. 10.8. Based on the equivalent circuit in Fig. 10.8, the voltage equations acquire the form

$$\begin{gathered}\dot{U}_1=(r_1+jx_{1\sigma})\dot{I}_1+jx_m\dot{I}_m\\ \left(-\frac{\dot{U}_2'}{s}\right)=\left(\frac{r_2'}{s}+jx_{2\sigma}'\right)\dot{I}_2'+jx_m\dot{I}_m\\ \dot{I}_m=\dot{I}_1+\dot{I}_2'\end{gathered}\tag{10.5}$$

The calculations of the machine magnetic circuit can be implemented for the various values of the emf $\dot{Э}_m$ as a result of which the machine magnetizing curve described in the form $\dot{Э}_m=f(\dot{I}_m)$ can be obtained. The magnetizing curve allows us to find the machine magnetizing current $\dot{I}_m$, and also, it can be used to establish the machine magnetizing reactance determined as $x_m=\dot{Э}_m/\dot{I}_m$. Magnetizing reactance x_m defined in this way takes into account the different level of the magnetic circuit saturation. In practice, the magnetizing curve $\dot{Э}_m=f(\dot{I}_m)$ is calculated or determined experimentally usually for the stator winding. This curve can be also used to find the value of the emf $\dot{Э}_2$ induced in the rotor winding in the no-load mode. For this, the value of the emf $\dot{Э}_m$ defined by the magnetizing curve $\dot{Э}_m=f(\dot{I}_m)$ should be divided by the reduction factor $k_Э$. Consequently, to determine the emf induced in the stator and rotor windings, we can restrict ourselves to the one magnetizing curve that simplifies the calculations. This provision is very important for the multi-winding electric machines.

In the equivalent circuit in Fig. 10.8, the stator slot leakage reactance $x_{1\sigma}$ and the referred to the stator winding value of the rotor slot leakage reactance $x_{2\sigma}'$ are used. The features of determining the stator and rotor slot leakage reactances were considered in the Chap. 8.

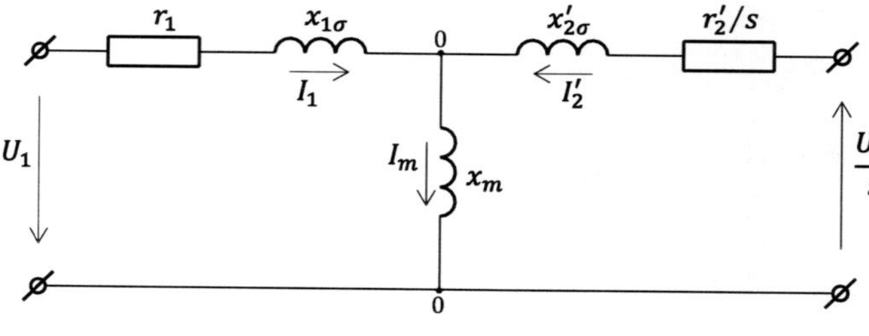

Fig. 10.8 Machine equivalent circuit taking into account the saturation

10.3.3 Equivalent Circuits of Stator and Rotor

Above, the technique for constructing the machine equivalent circuit using the stator and rotor equivalent circuits was described. In this case, the stator and rotor equivalent circuits were obtained based on the machine equivalent circuit shown in Fig. 10.3. Here, the stator and rotor equivalent circuits are obtained based on the machine equivalent circuit in Fig. 10.8. For this purpose, the equivalent circuit in Fig. 10.8 can be decomposed in relation to the emf $\dot{Э}_1 = \dot{Э}_2'$ into the separate fragments. Besides, the magnetizing reactance x_m used in the equivalent circuit in Fig. 10.8 can be expressed through the parallel connected two reactances x_{m1} and x_{m2}. This procedure corresponds to subdividing the machine air gap into the two sublayers. Then, the magnetic tension of the air gap $\dot{F}_0$ used in (10.2) can be, in this case, represented as $\dot{F}_0 = \dot{F}_{01} - \dot{F}_{02}$, where $\dot{F}_{01}$ is the magnetic tension of the air gap sublayer adjoined to the stator bore surface and $\dot{F}_{02}$ is the magnetic tension of the air gap sublayer adjoined to the rotor surface. Now, with considering the condition of kind $\dot{F}_0 = \dot{F}_{01} + \dot{F}_{02}$, the expression (10.2) can be obtained in the following form $\dot{F}_m = \dot{F}_{01}\left(\frac{\dot{F}_{ks}+\dot{F}_{zs}+\dot{F}_{as}}{\dot{F}_{01}}\right) + \dot{F}_{02}\left(\frac{\dot{F}_{kR}+\dot{F}_{zR}+\dot{F}_{aR}}{\dot{F}_{02}}\right) = \dot{F}_{01}k_{H1} + \dot{F}_{02}k_{H2}$. The values of k_{1_H} and k_{2_H} are determined as $k_{1_H} = \frac{\dot{F}_{ks}+\dot{F}_{zs}+\dot{F}_{as}}{\dot{F}_{01}}$ and $k_{2_H} = \frac{\dot{F}_{kR}+\dot{F}_{zR}+\dot{F}_{aR}}{\dot{F}_{02}}$, where k_{1_H} and k_{2_H} are the saturation factors of the stator and rotor. Using these conditions, the machine air gap can be considered as consisting of the two sublayers having the calculated radial lengths equal to δ_1 and δ_2 defined, in this case, as $\delta_1 = \delta k_{1\delta} k_{1_H}/2$ and $\delta_2 = \delta k_{2\delta} k_{2_H}/2$. Now, from (10.4) for the reactances x_{m1} and x_{m2}, we can obtain that $x_{m1} = \frac{\omega_1 \mu_0 \tau_0^2}{\pi^2 \delta_1} k_1$ and $x_{m2} = \frac{\omega_1 \mu_0 \tau_0^2}{\pi^2 \delta_2} k_1$. Then, the fragment equivalent circuits arising from the equivalent circuit in Fig. 10.8 acquire the forms shown in Fig. 10.9. In Fig. 10.9a, the stator magnetizing equivalent circuit is shown, and in Fig. 10.9b, the rotor magnetizing equivalent circuit is obtained. These equivalent circuits reflect the stator and rotor magnetic circuits. In Fig. 10.9c and d, the stator and rotor slot leakage equivalent circuits obtained earlier in Fig. 10.5 are given. The equivalent circuits in Fig. 10.9 can be used to construct the stator and rotor equivalent circuits.

Stator equivalent circuit. The stator equivalent circuit can be constructed by the fragment equivalent circuits given in Fig. 10.9a, c. For this, the terminals $a - b$ of the equivalent circuits in Fig. 10.9a, c can be connected in the cascade way. As a result, the stator equivalent circuit shown in Fig. 10.10a is obtained.

Rotor equivalent circuit. The rotor equivalent circuit can be built by the fragment equivalent circuits given in Fig. 10.9b, d. For this, the terminals $c - d$ of the equivalent circuits in Fig. 10.9b, d can be connected in the cascade way. Now, the rotor equivalent circuit given in Fig. 10.10b is received.

Machine equivalent circuit. .ts shown in Fig. 10.10. For this purpose, in the stator and rotor equivalent circuits in Fig. 10.10, the parallel branches $0 - 0$ having the reactances x_{m1} and x_{m2} can be combined into one common branch with the magnetizing reactance x_m determined as $x_m = x_{m1}x_{m2}/(x_{m1} + x_{m2})$. As a result, the machine equivalent circuit obtains the form in Fig. 10.8.

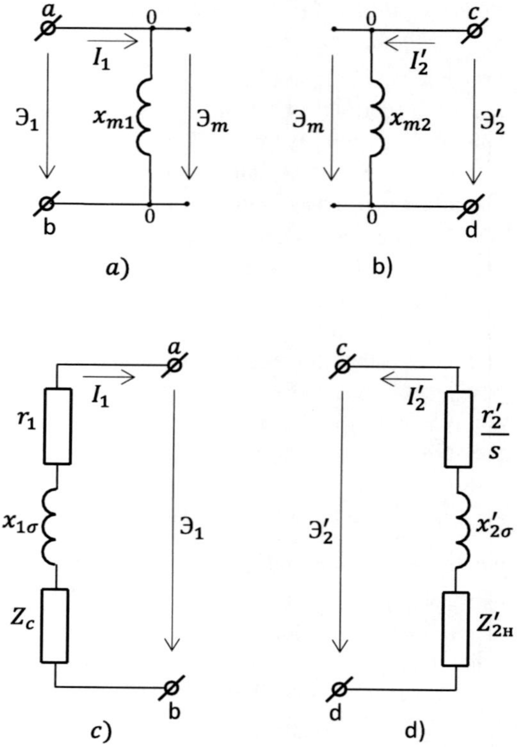

Fig. 10.9 Fragments of machine equivalent circuit: (**a**) stator magnetizing equivalent circuit taking into account the saturation; (**b**) rotor magnetizing equivalent circuit taking into account the saturation; (**c**) stator slot leakage equivalent circuit; (**d**) rotor slot leakage equivalent circuit

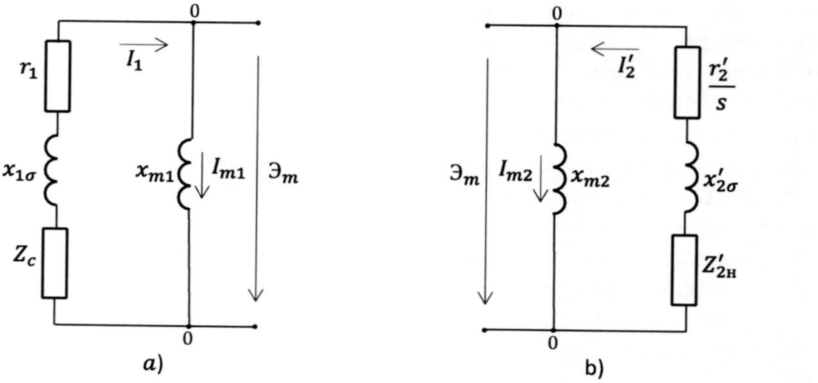

Fig. 10.10 Stator and rotor equivalent circuits: (**a**) stator equivalent circuit taking into account the saturation; (**b**) rotor equivalent circuit taking into account the saturation

In this chapter, the three ways of constructing the machine equivalent circuit are considered. Besides, in Appendix D, the machine equivalent circuit used in the technical literature is given. The ways of constructing the machine equivalent circuit shown above are based on the use of the "partial" magnetic fields to which the machine's resulting magnetic field is decomposed in accordance with the "no-load and under-load" concept. As it was noted above, the third way assuming the construction of the stator and rotor equivalent circuits is considered more preferable when building the machine equivalent circuit. In this regard, in the future, we will use the thirst way of constructing the machine equivalent circuit.

10.4 L-Shaped Equivalent Circuit of Machine

The equivalent circuit given in Fig. 10.8 is known as the T-shaped equivalent circuit of the asynchronous machine. In practice, L-shaped machine equivalent circuit is often used. This equivalent circuit greatly simplifies the determination of the machine load current. The L-shaped machine equivalent circuit can be obtained by the equivalent circuit shown in Fig. 10.8. For this, in the equivalent circuit in Fig. 10.8, its parallel branch $0-0$ (magnetizing branch) can be reduced to the stator winding terminals with the voltage $\dot{U}_1$. Below, we provide the description of this way of the reduction [20]. For the purpose of simplification, in the equivalent circuit in Fig. 10.8, we use the following conditions $Z_1 = r_1 + jx_{1\sigma}; Z_2' = r_2'/s + jx_{2\sigma}'$ and $\dot{U}_2'/s = 0$. Then, the equivalent circuit in Fig. 10.8 acquires the form in Fig. 10.11. In this equivalent circuit, its parallel magnetizing branch $0-0$ can be transferred to the terminals of the voltage $\dot{U}_1$. For this purpose, we can use the transfer technique described in the Chap. 9. In accordance with this technique, the stator current $\dot{I}_1$ can be represented as $\dot{I}_1 = \dot{I}_m' + (-\dot{I}_2'')$. Here, the current $\dot{I}_m'$ is determined by the equivalent circuit in Fig. 10.11 when in it the condition of the form $\dot{I}_2' = 0$ is implemented (no-load mode). For the current $(-\dot{I}_2'')$, we have that $(-\dot{I}_2'') = \dot{I}_1 - \dot{I}_m'$. In this equation, the stator current $\dot{I}_1$ is determined by the equivalent circuit in Fig. 10.11, and it obtains the following form

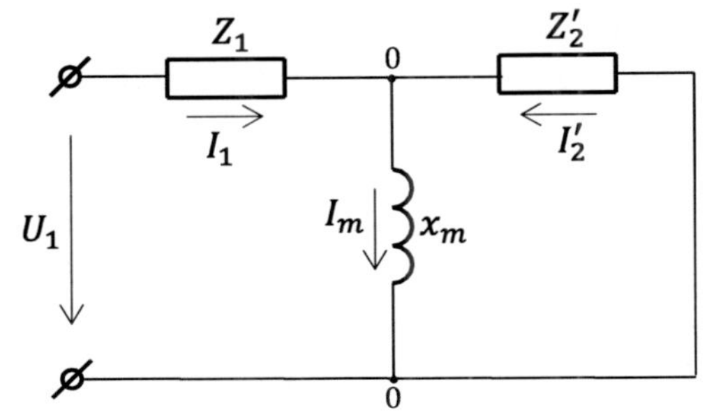

Fig. 10.11 T-shaped machine equivalent circuit

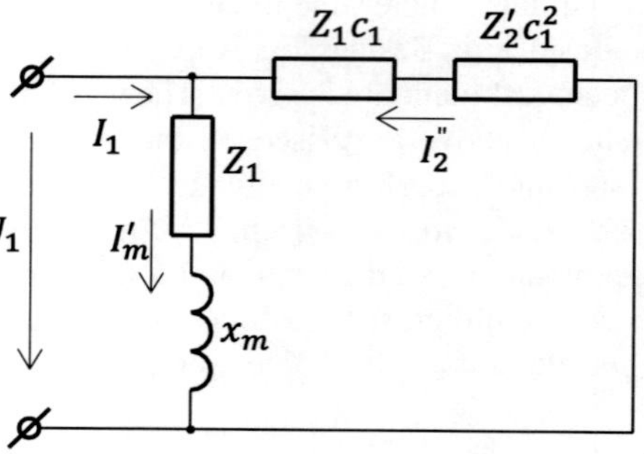

Fig. 10.12 L-shaped machine equivalent circuit

$$\dot{I}_1 = \dot{U}_1 \frac{Z_2' + jx_m}{Z_1(\,jx_m) + (\,jx_m)Z_2' + Z_1 Z_2'} = \dot{U}_1 \frac{Z_2' + jx_m}{jx_m(Z_1 + Z_2'\dot{c}_1)} \tag{10.6}$$

where $\dot{c}_1 = 1 + Z_1/jx_m$.

We will find the currents $\dot{I}_m'$ and $(-\dot{I}_2'')$. When $\dot{I}_2' = 0$, from the equivalent circuit in Fig. 10.11, the current $\dot{I}_m'$ acquires

$$\dot{I}_m' = \frac{\dot{U}_1}{Z_1 + jx_m} = \frac{\dot{U}_1}{jx_m\dot{c}_1} \tag{10.7}$$

The current $(-\dot{I}_2'')$ follows from the equation $(-\dot{I}_2'') = \dot{I}_1 - \dot{I}_m'$. Therefore, from (10.6) and (10.7), we have

$$(-\dot{I}_2'') = \dot{I}_1 - \dot{I}_m' = \frac{\dot{U}_1}{jx_m\dot{c}_1}\left[\frac{Z_2'\dot{c}_1 + jx_m\dot{c}_1}{(Z_1 + Z_2'\dot{c}_1)} - 1\right] = \frac{\dot{U}_1}{Z_1\dot{c}_1 + Z_2'\dot{c}_1^2} \tag{10.8}$$

Based on the Eqs. (10.7) and (10.8), the stator winding current $\dot{I}_1$ arises

$$\dot{I}_1 = \dot{I}_m' + (-\dot{I}_2'') = \frac{\dot{U}_1}{Z_1 + jx_m} + \frac{\dot{U}_1}{Z_1\dot{c}_1 + Z_2'\dot{c}_1^2} \tag{10.9}$$

From Eq. (10.9), the L-shaped machine equivalent circuit shown in Fig. 10.12 follows. The L-shaped machine equivalent circuit is often used to determine the load current of the rotor winding $\dot{I}_2'$. The current $\dot{I}_2'$ flows in the T-shaped machine equivalent circuit in Fig. 10.10. However, in the L-shaped machine equivalent circuit (Fig. 10.12), the current $\dot{I}_2''$ determined by the expression (10.8) takes place. In order to have the possibility the results of the calculations performed for the current $\dot{I}_2''$ by the L-shaped equivalent circuit to use for determining the rotor load current $\dot{I}_2'$, it is necessary to establish the mutual connection between the currents $\dot{I}_2'$ and $\dot{I}_2''$. The current $\dot{I}_2''$ is determined by the expression obtained in (10.8). To find the current $\dot{I}_2'$, we can use the T-shaped equivalent circuit in Fig. 10.10 and the expression (10.6) for the current $\dot{I}_1$. As a result, it follows

$$(-\dot{I}_2') = \frac{\dot{U}_1 - Z_1\dot{I}_1}{Z_2'} = \frac{\dot{U}_1}{Z_2'}\left[1 - \frac{Z_1Z_2' + (jx_m)Z_1}{(jx_m)Z_1 + (jx_m)Z_2'\dot{c}_1)}\right] = \frac{\dot{U}_1}{Z_1 + Z_2'\dot{c}_1} \quad (10.10)$$

Now, based on the expressions (10.8) and (10.10), we have

$$\dot{I}_2' = \dot{c}_1\dot{I}_2'' \quad (10.11)$$

The complex transfer factor $\dot{c}_1$ in another form

$$\dot{c}_1 = c_1e^{-j\theta} \quad (10.12)$$

The imaginary part of the transfer factor $\dot{c}_1$ is usually negative, therefore, its argument θ is also negative. For the machine of the normal design, it can be assumed that $\theta \approx 0$. Then, the complex transfer factor $\dot{c}_1$ can be represented as the real number c_1 determined from the expression (10.6) as

$$c_1 \approx 1 + \frac{x_{1\sigma}}{x_m} \quad (10.13)$$

On the basis of (10.8), (10.11), and (10.13), and also keeping in mind that $Z_1 = r_1 + jx_{1\sigma}$ and $Z_2' = r_2'/s + jx_{2\sigma}'$, the current $\dot{I}_2'$ acquires

$$\dot{I}_2' = c_1\dot{I}_2'' = \frac{\dot{U}_1}{\sqrt{\left(r_1 + c_1\frac{r_2'}{s}\right)^2 + \left(x_{1\sigma} + c_1x_{2\sigma}'\right)^2}} \quad (10.14)$$

As follows from (10.14), when using the L-shaped machine equivalent circuit, the determination of the load current $\dot{I}_2'$ is greatly simplified.

10.5 Voltage Equations Through Referred Quantities: "Inductively Coupled Circuits" Concept

As it was noted above, the voltage equations obtained by the use of the "inductively coupled circuits" concept and given in (6.15) cannot be effectively applied in the engineering practice. The difficulties in using the Eq. (6.15) are due to the fact that the reactances applied in these equations cannot be determined with the required accuracy. In addition, in a view of the inequality of the numbers of the turns of the stator and rotor windings ($w_1 \neq w_2$), the quantities used in the voltage Eq. (6.15) take very different values making an analysis difficult.

In practice, this problem of the Eq. (6.15) is solved by the use in them the referred to the stator winding rotor quantities arising from applying the "no-load and under-load" concept. For this, the voltage Eq. (10.5) compiling on the basis of the "no-load and under-load" concept can be reduced to the form of the voltage Eq. (6.15). Then, the voltage Eq. (10.5) can be represented in the following form

$$\begin{aligned} \dot{U}_1 &= r_1\dot{I}_1 + j(x_{1\sigma} + x_m)\dot{I}_1 + jx_m\dot{I}'_2 = r_1\dot{I}_1 + jx_1\dot{I}_1 + jx_{21}\dot{I}'_2 \\ \left(-\frac{\dot{U}'_2}{s}\right) &= \frac{r'_2}{s}\dot{I}'_2 + jx_m\dot{I}_1 + j(x'_{2\sigma} + x_m)\dot{I}'_2 = \frac{r'_2}{s}\dot{I}'_2 + jx_{12}\dot{I}_1 + jx_2\dot{I}'_2 \end{aligned} \quad (10.15)$$

where $x_1 = x_{1\sigma} + x_m$; $x_2 = x'_{2\sigma} + x_m$; $x_{12} = x_{21} = x_m$.

As follows from Eq. (6.15) and (10.15), the parameters used in the voltage equations arising as a result of applying the "no-load and under-load" concept and expressing through the referred rotor quantities consist of the magnetizing reactance and the slot leakage reactances of the stator and rotor windings. In comparison with the voltage Eq. (6.15), the voltage Eq. (10.5) do not have the fundamental difficulties in determining the reactances used in them. Therefore, the voltage equations of the form (10.5) are widely used in practice.

10.6 Determination of Parameters: "Open Circuit and Short Circuit" Technique

In practice, the "open circuit and short circuit" technique is often used to determine the machine parameters, for example, [2–20]. Below, we consider some features of applying this technique.

"Open circuit" mode. In these conditions, the stator winding is connected to the power supply network with the voltage $\dot{U}_1$ and the rotor winding is open ($\dot{I}_2 = 0$). Under the influence of the voltage $\dot{U}_1$ supplied to the stator winding, the no-load current $\dot{I}_m$ flows in it. The no-load current $\dot{I}_m$ creates the magnetic flux the one part of which (main flux) is coupled with the stator and rotor windings and another part (leakage flux) is embraced the stator winding. The main flux induces the emf $\dot{Э}_1$ and $\dot{Э}_2$ in the stator and rotor windings and the leakage flux induces the leakage emf $\dot{Э}_{1\sigma}$ in the stator winding. In addition, when the current flows in the stator winding, the voltage drop across the ohmic resistance takes place. In this regard, in the no-load mode, the stator winding voltage equation can be expressed as $\dot{U}_1 = Z_1\dot{I}_m + (-\dot{Э}_1)$, where $Z_1 = r_1 + jx_{1\sigma}$. Here, $\dot{I}_m$ represents the no-load current of the machine. In practice, it is assumed that the stator winding leakage emf $(-\dot{Э}_{1\sigma}) = jx_{1\sigma}\dot{I}_m$ and the voltage drop across the ohmic resistance of the stator winding $(-\dot{Э}_{1r}) = r_1\dot{I}_m$ is characterized by relatively small values. Therefore, we can assume with some approximation that the stator winding voltage $\dot{U}_1$ is almost completely compensated by the emf $\dot{Э}_1$ i.e. $\dot{U}_1 + \dot{Э}_1 \approx 0$. The ЭДС $\dot{Э}_1$ and $\dot{Э}_2$ induced in the stator and rotor

windings in the no-load mode can be determined by the expressions given in (6.11) and (6.13). By these expressions, for the emf reduction factor $k_Э$, we can have

$$k_Э = \frac{\dot{U}_1}{\dot{U}_2} \approx \frac{\dot{Э}_1}{\dot{Э}_2} = \frac{w_1 k_{w1}}{w_2 k_{w2}}$$

From here

$$\frac{\dot{U}_1}{\dot{U}_2} \approx \frac{w_1 k_{w1}}{w_2 k_{w2}} = k_Э \quad \dot{U}_1 \approx k_Э \dot{U}_2 = \dot{U}_2' \tag{10.16}$$

where $k_Э = w_1 k_{w1}/w_2 k_{w2}$; $\dot{U}_2' = k_Э \dot{U}_2$ is the referred value of the rotor winding voltage.

Now, for the stator winding reactance in the no-load mode (machine magnetizing reactance), we receive

$$x_m = \frac{\dot{U}_1}{\dot{I}_m} \tag{10.17}$$

In a view of the fact that $\dot{U}_1 + \dot{Э}_1 \approx 0$, the reactance x_m determined by (10.17) is very close to the machine magnetizing reactance obtained in relation to the stator winding emf $\dot{Э}_1$.

Short circuit mode. The machine short circuit conditions are carried out when the rotor winding is short-circuited, and the rotor winding voltage is equal to zero, i.e. $\dot{U}_2 = 0$. The current $\dot{I}_2$ flowing in the rotor winding in the short circuit mode creates the mutual induction magnetic flux, and the leakage flux embracing the rotor winding and inducing in it the leakage emf $\dot{E}_{2\sigma}$. The current $\dot{I}_{1k}$ flowing in the stator winding in the short circuit mode creates the mutual induction magnetic flux, and the leakage flux interlinking with the stator winding and inducing in it the leakage emf $\dot{E}_{1\sigma}$. The currents $\dot{I}_{1k}$ and $\dot{I}_2$ create the stator winding mmf $\dot{F}_{1k}$ and rotor winding mmf $\dot{F}_2$, which form the resulting mmf $\dot{F}_{0k}$. In this regard, the mmf equation reflecting the short circuit mode acquires form

$$\dot{F}_{1k} + \dot{F}_2 = \dot{F}_{0k} \tag{10.18}$$

With consideration of (4.9) and (4.11a), this equation can be reduced to the following form

$$\frac{m_1\sqrt{2}}{\pi}\frac{w_1 k_{w1}}{p}\dot{I}_{1k} + \frac{m_2\sqrt{2}}{\pi}\frac{w_2 k_{w2}}{p}\dot{I}_2 = \frac{m_1\sqrt{2}}{\pi}\frac{w_1 k_{w1}}{p}\dot{I}_{0k}$$

From here, the equation for the currents arises

$$\dot{I}_{1k} = \dot{I}_{0k} + (-\dot{I}_2') \tag{10.19}$$

where $\dot{I}_2' = \dfrac{\dot{I}_2}{k_I}; k_I = \dfrac{m_1}{m_2}\dfrac{w_1 k_{w1}}{w_2 k_{w2}}$.

From (10.18), it follows that the stator winding mmf $\dot{F}_{1k}$ and rotor winding mmf $\dot{F}_2$ act towards each other in the short circuit mode. Therefore, the resulting mmf creating by the stator and rotor currents is relatively small, i.e. $\dot{F}_{0k} \approx 0$. The main magnetic flux creating by the mmf $\dot{F}_{0k}$ in the short circuit mode is small, i.e. $\Phi_0 \approx 0$. As a result of this, from (10.18) and (10.19), we have

$$\dot{F}_{1k} \approx (-\dot{F}_2) \quad \dot{I}_{1k} \approx (-\dot{I}_2') \tag{10.20}$$

On the basis of (10.5), the asynchronous machine voltage equations acquire the form in the short circuit mode ($\dot{U}_2 = 0$)

$$\begin{aligned} \dot{U}_1 &= (r_1 + jx_{1\sigma})\dot{I}_{1k} + (-\dot{Э}_{1k}) \\ (-\dot{Э}_{2k}') &= (r_2'/s + jx_{2\sigma}')\dot{I}_2' \end{aligned} \tag{10.21}$$

Taking into account that $\dot{I}_{1k} \approx (-\dot{I}_2')$, as well as $\dot{Э}_{1k} = \dot{Э}_{2k}'$, for the current $\dot{I}_{1k}$ from (10.21), we can obtain

$$\dot{I}_{1k} \approx \frac{\dot{U}_1}{(r_1 + r_2'/s) + j(x_{1\sigma} + x_{2\sigma}')} = \frac{\dot{U}_1}{r_k + jx_k} \tag{10.22}$$

where $r_k = r_1 + r_2'/s; x_k = x_{1\sigma} + x_{2\sigma}'$; r_k and x_k represent the short circuit parameters of the electric machine.

As follows from (10.22), the short circuit parameters are represented as the sum of the ohmic resistances and the leakage reactances of the stator and rotor windings. In practice, we use the ohmic resistances and the leakage reactances reflecting the stator and rotor windings separately. In this regard, the need to separate the short circuit parameters into the ohmic resistances and the leakage reactances of the stator and rotor windings arises. In this case, the recommendations of the form $r_1 \approx r_2'$ and $x_1 \approx x_2'$ can be used.

10.7 Summary

1. In this chapter, the description of the various techniques for constructing the machine equivalent circuit is given. The recommended techniques are based on the use of the superposition principle, the "fragmentation" concept, the "no-load and under-load" concept, and the application of mixed and cascade ways of connecting. In accordance with the accepted in this work "fragmentation"

concept, the air gap equivalent circuit, the stator slot leakage equivalent circuit, and the rotor slot leakage equivalent circuit are considered. These equivalent circuits represent the separate fragments of the machine equivalent circuit, and they are obtained in accordance with the "no-load and under-load" concept in relation to the stator winding emf $\dot{Э}_1$. It means that the air gap equivalent circuit and the rotor slot leakage equivalent circuit are expressed through the referred rotor quantities.

2. In this work, the three groups of the equivalent circuits reflecting the fragments of the machine equivalent circuit are considered. The first group of the fragment equivalent circuits includes the air gap equivalent circuit corresponding to the machine no-load mode, the stator slot leakage equivalent circuit, and the rotor slot leakage equivalent circuit. In this case, the machine equivalent circuit arises from the mixed (parallel-cascade) ways of connecting the fragment equivalent circuits. The second group of the fragment equivalent circuits consists of the air gap equivalent circuit corresponding to the machine operating mode, the stator slot leakage equivalent circuit, and the rotor slot leakage equivalent circuit. In this case, the cascade way of connecting the fragment equivalent circuits is used to construct the machine equivalent circuit. The third group of the fragment equivalent circuits contains the air gap upper sublayer equivalent circuit, the stator slot leakage equivalent circuit, and also the air gap lower sublayer equivalent circuit, the rotor slot leakage equivalent circuit. In this case, the air gap upper sublayer equivalent circuit is connected in cascade with the stator slot leakage equivalent circuit. As a result, the stator equivalent circuit arises. The air gap lower sublayer equivalent circuit is connected in cascade with the rotor slot leakage equivalent circuit and then the rotor equivalent circuit follows. Now, the machine equivalent circuit follows as a result of combining the stator equivalent circuit and the rotor equivalent circuit. In this work, the third technique for constructing the machine equivalent circuit has received further development.
3. The idea of constructing the machine equivalent circuit as a result of the cascade way of connecting the fragment equivalent circuits representing the corresponding machine design regions can be realized in the case of determining the machine resulting magnetic field by Maxwell's equations. In this case, the implementation of this opportunity of constructing the machine equivalent circuit forms the basis of this work and will be considered in more detail in the future.
4. In the theory of electric machines, the problem of constructing the machine equivalent circuit is often solved using the voltage equations written in relation to the terminals of the stator and rotor windings. These voltage equations contain the ohmic resistances and leakage reactances that are not directly related to the phenomena characteristic of the inductively coupled stator and rotor windings. In this regard, the application of these voltage equations causes certain difficulties in ensuring the consistency and the methodicality in the presentation of the main provisions of the techniques for constructing the machine equivalent circuit.
5. In practice, the determination of the machine parameters is often implemented by the use of the "open circuit and short circuit" technique. For the "open circuit and

short circuit" technique, we have that $\dot{U}_1 \approx (-\dot{Э}_1)$ and $\dot{F}_{1k} \approx (-\dot{F}_2)$ and for the "no-load and under-load" concept arises that $\dot{U}_1 = (-\dot{Э}_1)$ and $\dot{F}_{1н} = (-\dot{F}_2)$. From here, it can be noted that the results obtained from the use of "open circuit and short circuit" technique are quite acceptable for the practice of engineering calculations.

References

1. Asanbayev, V.: Alternating Current Multi-Circuit Electric Machines: A New Approach to Steady-State Parameter Determination, Springer Cham Heidelberg, New York, Dordrecht London (2015)
2. Richter, R.: Electric Machines, vol. 1–5. United Scientific and Technical Publishing, Moscow (1934–1939)
3. Kostenko, M, Piotrovskij, L.: Electric Machines, Publishing Energiya, Leningrad (1973)
4. Voldek. A.: Electric Machines, Energiya, Leningrad (1974)
5. Ivanov-Smolensky, A.: Electric Machines, Energiya, Moscow (1980)
6. Bragstad, O.: Theory of AC Machines with an Introduction to the Theory of Steady-State Alternating Currents (Translated from German), Izdatelstvo Lensoveta i Oblispolkoma, Leningrad (1933)
7. Livshits, A.: Electric Machines, vol. 1–3.United Scientific and Technical Publishing, Moscow (1936)
8. Alger, P.: Induction Machines. Their Behavior and Uses, Gordon and Beach, New York (1970)
9. Antonov, L.: Alternating Current Machines, Naval Publishing, Leningrad (1968)
10. Piotrovskij, L.: Electric Machines, Publishing Energiya, Leningrad (1975)
11. Shenfer, K.: Asynchronous Machines, State United Scientific and Technical Publishing, Moscow/Leningrad (1938)
12. Vazhnov, A.: Electric Machines, Energiya, Leningrad (1968)
13. McPherson, G., Laramore, R.: Electrical Machines and Transformers, second ed., John Wiley and Sons, New York (1990)
14. Say, M.: Alternating Current Machines, John Wiley and Sons, New York (1902)
15. Ryff, P., Platnick, D., Karnas, J.: Electrical Machines and Transformers. Principles and Applications, Prentice-Hall, Inc., New Jersey (1987)
16. Sen, P.: Electric Machines and Power Electronics, John Wiley and Sons, New York (1989)
17. Fitzgerald, A., Kingsley, C., Umans, S.: Electric Machinery, McGraw-Hill, Inc., fifth ed., New York (1990)
18. Del Toro, V.: Electric Machine and Power Systems, Prentice-Hall, Inc., Englewood Cliffs, New Jersey (1985)
19. Krause, P.: Analysis of Electric Machinery, McGraw-Hill, New York (1986)
20. Kostenko, M.: Electric Machines. Special Part, Gosenergoizdat, Moscow (1949)

Chapter 11
Basic Calculation Elements Using Resulting Magnetic Field: Phase System of Units

11.1 "Unfolded" Magnetizing Equivalent Circuits of Stator and Rotor

For calculation purposes, the machine magnetizing field is represented in the form of the one-dimensional magnetic field spreading in the machine cross section in the radial direction. For this case, the picture for the conditional line of the magnetizing flux density is shown in Fig. 10.7. In Appendix E, the machine magnetizing equivalent circuit is built (Fig. E.1). In accordance with [1], the machine magnetizing equivalent circuit can be decomposed into the stator magnetizing equivalent circuit and the rotor magnetizing equivalent circuit.

As follows from [1], to obtain the stator and rotor magnetizing equivalent circuits, the machine air gap is represented decomposed into the two sublayers. These air gap sublayers are determined in relation to the surface $0 - 0$ passing through the middle of the radial length of the machine air gap. The air gap upper sublayer is adjoined to the stator bore surface, and the air gap lower sublayer is adjoined to the rotor surface. Below, for the air gap sublayers, we use the calculated values of the radial lengths δ_1 and δ_2 determined as $\delta_1 = \delta k_{1\delta}/2$ and $\delta_2 = \delta k_{2\delta}/2$, where $k_{1\delta}$ and $k_{2\delta}$ are the air gap sublayer factors. In these conditions, the stator magnetizing circuit includes the air gap sublayer adjoining to the stator bore surface, the stator slot-wedge area, the stator winding location area, and the stator yoke area. The rotor magnetizing circuit contains the air gap sublayer adjoining to the rotor surface, the slot-wedge area, the winding location area, and the rotor yoke area. The equivalent circuits obtained in Appendix E and shown in Fig. E.2a, b correspond to such a way of decomposing into the sections of the stator and rotor magnetizing circuits.

The equivalent circuits received in Fig. E.2 for the stator and rotor magnetizing circuits are represented in Figs. 11.1a and 11.2a, respectively. In the equivalent circuits in Figs. 11.1a and 11.2a, the reactances x_{01} and x_{02} are followed as a result of decomposing the magnetizing reactance x_0 into the two parallel connected reactances. The reactance x_{01} represents the magnetizing reactance of the air gap

V. Asanbayev, *Asynchronous Machines*,
https://doi.org/10.1007/978-3-030-92284-9_11

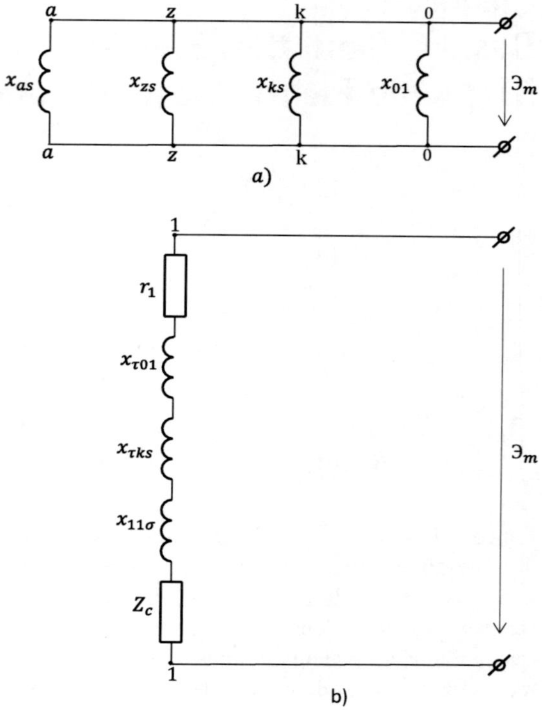

Fig. 11.1 "Unfolded" stator magnetizing and "unfolded" stator slot leakage equivalent circuits: (**a**) "unfolded" stator magnetizing equivalent circuit; (**b**) "unfolded" stator slot leakage equivalent circuit

sublayer adjoining to the stator bore surface, and the reactance x_{02} reflects the magnetizing reactance of the air gap sublayer adjoining to the rotor surface. The equivalent circuits in Figs. 11.1a and 11.2a contain the reactances x_{01}, x_{ks}, x_{zs} and x_{02}, x_{kR}, x_{zR} representing the longitudinal magnetizing circuit loops reflecting the air gap sublayers and also the corresponding sections of the stator and rotor magnetic circuits. The reactances x_{as} and x_{aR} used in the equivalent circuits in Figs. 11.1a and 11.2a reflect the transverse magnetizing circuit loops representing the stator and rotor yoke areas.

The expressions for the reactances used in the equivalent circuits in Figs. 11.1a and 11.2a are obtained in Appendix E. In accordance with (E.7), the reactance of the longitudinal magnetizing circuit loop representing the air gap sublayer adjoining to the stator bore surface and the reactances of the longitudinal magnetizing circuit loops reflecting the stator slot-wedge area, and the stator winding location area (Fig. 11.1a) are determined as

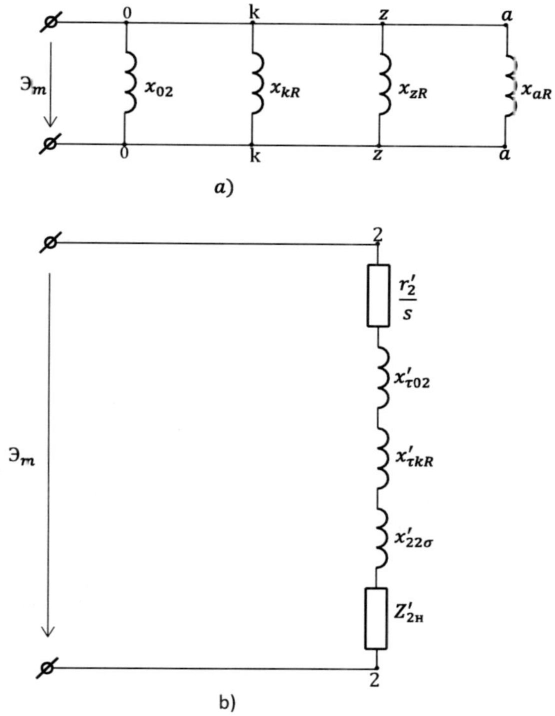

Fig. 11.2 “Unfolded” rotor magnetizing and “unfolded” rotor slot leakage equivalent circuits: (**a**) “unfolded” rotor magnetizing equivalent circuit; (**b**) “unfolded” rotor slot leakage equivalent circuit

$$x_{01} = \frac{\omega_1 \mu_0 \tau_0^2}{\pi^2 \delta_1} k_1$$
$$x_{ks} = \frac{\omega_1 \mu_{ks} \tau_{ks}^2}{\pi^2 h_{ks}} \frac{b_{ks}}{t_{ks}} \frac{\xi_{Lks}}{\xi_{\tau ks}} k_1 \tag{11.1}$$
$$x_{zs} = \frac{\omega_1 \mu_{zs} \tau_{zs}^2}{\pi^2 h_{zs}} \frac{b_{zs}}{t_{zs}} \frac{\xi_{Lzs}}{\xi_{\tau zs}} k_1$$

where $k_1 = \frac{2m_1(w_1 k_{w1})^2 l_0}{p\tau_0}$; δ_1 is the calculated radial length of the air gap sublayer adjoined to the stator bore surface.

In the equivalent circuit in Fig. 11.1a, the reactance of the transverse magnetizing circuit loop reflecting the stator yoke area can be obtained from (E.9) as

$$x_{as} = \omega_1 \mu_{as} t_0 \frac{h_{as}}{t_{as}} \xi_{Las} k_1 \tag{11.2}$$

where $t_0 = 2p\tau_0/Z_1$; $t_{as} = 2p\tau_{as}/Z_1$.

Here, t_{as} is the conditional tooth pitch determined on the stator yoke surface in relation to the number of the stator teeth.

The reactance of the longitudinal magnetizing circuit loop reflecting the air gap sublayer adjoining to the rotor surface, the reactances of the longitudinal magnetizing circuit loops representing the rotor slot-wedge, and the rotor winding location areas (Fig. 11.2a) are defined from (E.8) as

$$\begin{aligned} x_{02} &= \frac{\omega_1 \mu_0 \tau_0^2}{\pi^2 \delta_2} k_1 \\ x_{kR} &= \frac{\omega_1 \mu_{kR} \tau_{kR}^2}{\pi^2 h_{kR}} \frac{b_{kR}}{t_{kR}} \frac{\xi_{LkR}}{\xi_{\tau kR}} k_1 \\ x_{zR} &= \frac{\omega_1 \mu_{zR} \tau_{zR}^2}{\pi^2 h_{zR}} \frac{b_{zR}}{t_{zR}} \frac{\xi_{LzR}}{\xi_{\tau zR}} k_1 \end{aligned} \tag{11.3}$$

where δ_2 is the calculated radial length of the air gap sublayer adjoined to the rotor surface.

In the equivalent circuit in Fig. 11.2a, the reactance of the transverse magnetizing circuit loop representing the rotor yoke area can be obtained from (E.10) in the following form

$$x_{aR} = \omega_1 \mu_{aR} t_0 \frac{h_{aR}}{t_{aR}} k_R \xi_{LaR} k_1 \tag{11.4}$$

where $t_{aR} = 2p\tau_{aR}/Z_2$; $k_R = Z_1/Z_2$.

Here, t_{aR} is the conditional tooth pitch determined on the rotor yoke surface in relation to the number of the rotor teeth.

Below, we will use the equivalent circuits received for the stator and rotor magnetizing circuits in Figs. 11.1a and 11.2a to obtain the "unfolded" stator and rotor equivalent circuits.

11.2 "Unfolded" Slot Leakage Equivalent Circuits of Stator and Rotor

We consider the slot leakage reactance of the stator winding $x_{1\sigma}$ and the slot leakage reactance of the rotor winding $x'_{2\sigma}$ used in the equivalent circuit shown in Fig. 10.6. In Appendix E, the description of the ways of determining the reactances $x_{1\sigma}$ and $x'_{2\sigma}$ is given. For the referred to the stator winding value of the slot leakage reactance of the rotor winding $x'_{2\sigma}$, we have from (E.27)

$$x'_{2\sigma} = x'_{\tau 02} + x'_{\tau kR} + x'_{22\sigma} \tag{11.5}$$

The slot leakage reactance of the rotor winding $x'_{2\sigma}$ represents the sum of the three reactances: the leakage reactance $x'_{\tau 02}$ of the air gap sublayer adjoining to the rotor surface, the leakage reactance $x'_{\tau kR}$ of the rotor slot-wedge section, the leakage reactance $x'_{22\sigma}$ of the rotor winding location section. With considering the expressions (E.28) obtained in Appendix E, the components of the leakage reactance $x'_{2\sigma}$ can be received

$$\begin{aligned} x'_{\tau02} &= \omega_1\mu_0 t_0 \frac{\delta}{2t_{02}} k_{2\sigma}k_R k_1 \\ x'_{\tau kR} &= \omega_1\mu_0 t_0 \frac{h_{kR}}{b_{kR}} k_R \xi_{LkR} k_1 \\ x'_{22\sigma} &= \omega_1\mu_0 t_0 \frac{h_{zR}}{3b_{nR}} k_R \xi_{LzR} k_1 \end{aligned} \tag{11.6}$$

Similar expressions can be obtained for the components of slot leakage reactance of the stator winding $x_{1\sigma}$. According to (E.29), the slot leakage reactance of the stator winding $x_{1\sigma}$ represents the sum of the three reactances: the leakage reactance $x_{\tau C1}$ of the air gap sublayer adjoining to the stator bore surface, the leakage reactance $x_{\tau ks}$ of the stator slot-wedge section, the leakage reactance $x_{11\sigma}$ of the stator winding location section. In this regard, we have

$$x_{1\sigma} = x_{\tau01} + x_{\tau ks} + x_{11\sigma} \tag{11.7}$$

Based on the expressions (E.30), the leakage reactances $x_{\tau01}$, $x_{\tau ks}$, and $x_{11\sigma}$ shown in Appendix E can be represented as

$$\begin{aligned} x_{\tau01} &= \omega_1\mu_0 t_0 \frac{\delta}{2t_0} k_{1\sigma} k_1 \\ x_{\tau ks} &= \omega_1\mu_0 t_0 \frac{h_{ks}}{b_{ks}} \xi_{Lks} k_1 \\ x_{11\sigma} &= \omega_1\mu_0 t_0 \frac{h_{zs}}{3b_{ns}} \xi_{Lzs} k_1 \end{aligned} \tag{11.8}$$

As follows from Appendix E, the components of the leakage reactances $x_{1\sigma}$ and $x'_{2\sigma}$ arise based on the representation of the stator and rotor slot leakage magnetic circuits subdivided into the sections (Fig. E.3). As the sections of the slot leakage magnetic circuits are usually considered the air gap sublayers forming in relation to the surface $0-0$ passing through the middle of the radial length of the air gap, the stator and rotor slot-wedge areas, and the stator and rotor winding location areas.

From (8.21) and (11.5), the equation for the emf induced in the rotor winding takes the form

$$\begin{aligned} \dot{Э}_m &= \left(r'_2/s + jx'_{2\sigma} + Z'_{2H}\right)\dot{I}'_2 \\ &= \left[j\left(x'_{\tau02} + x'_{\tau kR} + x'_{22\sigma}\right) + r'_2/s + Z'_{2H}\right]\dot{I}'_2 \end{aligned} \tag{11.9}$$

where $x'_{2\sigma} = x'_{\tau02} + x'_{\tau kR} + x'_{22\sigma}$.

Here, the condition of kind $\dot{Э}'_2 = \dot{Э}_m$ is used. The rotor slot leakage circuit loop is described by the Eq. (11.9). In Appendix E, the stator and rotor slot leakage equivalent circuits are given (Fig. E.4). With considering Fig. E.4b, the rotor slot leakage equivalent circuit is shown in Fig. 11.2b.

From (8.20) and (11.7), the equation for the emf induced in the stator winding acquires

$$\begin{aligned}\dot{Э}_m &= (r_1 + jx_{1\sigma} + Z_c)\dot{I}_1 \\ &= [j(x_{\tau 01} + x_{\tau ks} + x_{11\sigma}) + r_1 + Z_c]\dot{I}_1\end{aligned} \tag{11.10}$$

where $x_{1\sigma} = x_{\tau 01} + x_{\tau ks} + x_{11\sigma}$.

In (11.10), the condition of kind $\dot{Э}_1 = \dot{Э}_m$ is used. Based on the Eq. (11.10), the stator slot leakage circuit loop can be visualized. According to Fig. E.4a, the stator slot leakage equivalent circuit is given in Fig. 11.1b. As follows from Figs. 11.1b and 11.2b, the stator and rotor slot leakage equivalent circuits have the same structure.

The equivalent circuits given in Figs. 11.1 and 11.2 can be used to construct the equivalent circuits corresponding to the stator and rotor resulting in magnetic fields. For this purpose, the stator and rotor magnetizing equivalent circuits (Figs. 11.1a and 11.2a) and the stator and rotor slot leakage equivalent circuits (Figs. 11.1b and 11.2b) can be considered. The equivalent circuits in Figs. 11.1a and 11.2a correspond to the machine no-load mode, and they represent the stator and rotor magnetizing fields. The equivalent circuits in Figs. 11.1b and 11.2b reflect the machine under-load mode, and they represent the stator and rotor slot leakage magnetic fields. Consequently, in accordance with the "no-load and under-load" concept, these equivalent circuits can be used to build the equivalent circuits reflecting the resulting magnetic fields taking place in the stator and rotor of the electric machine. In this case, the equivalent circuit corresponding to the stator resulting magnetic field can be obtained as a result of the superposition of the stator magnetizing equivalent circuit (Fig. 11.1a) and the stator slot leakage equivalent circuit (Fig. 11.2a). The equivalent circuit corresponding to the rotor resulting magnetic field can be obtained as a result of the superposition of the rotor magnetizing equivalent circuit (Fig. 11.1b) and the rotor slot leakage equivalent circuit (Fig. 11.2b). The stator and rotor equivalent circuits obtained in this way represent the "unfolded" (multi-circuit) equivalent circuits with the parallel connection of the elements. Below, we will show it.

11.3 "Unfolded" Equivalent Circuit of Stator

We consider the stator magnetizing equivalent circuit (Fig. 11.1a) and the stator slot leakage equivalent circuit (Fig. 11.2a). As it was noted earlier, the stator magnetizing equivalent circuit (Fig. 11.1a) reflects the machine no-load mode. The stator slot leakage equivalent circuit (Fig. 11.2a) represents the machine under-load mode. In the stator magnetizing equivalent circuit (Fig. 11.1a), the parallel branch $z - z$

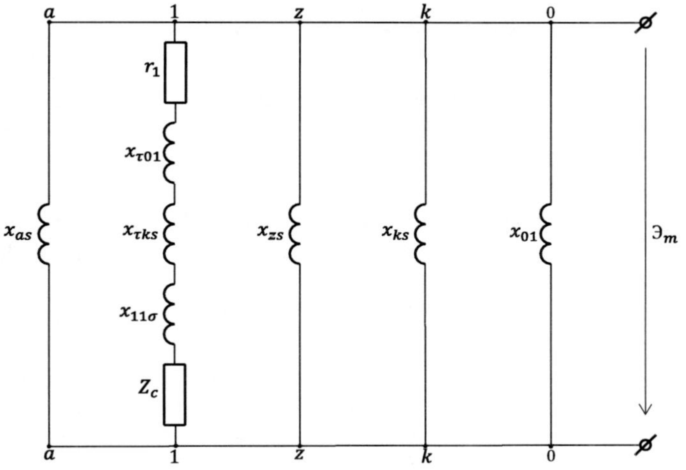

Fig. 11.3 "Unfolded" stator equivalent circuit with the parallel connection of elements

reflects the stator winding location area. In these conditions, to obtain the stator equivalent circuit corresponding to the resulting magnetic field in it, it is enough to the terminals of the parallel branch $z - z$ of the equivalent circuit in Fig. 11.1a to connect the terminals of the branch $1 - 1$ of the equivalent circuit in Fig. 11.2a. As a result, the stator equivalent circuit can be reduced to the form in Fig. 11.3. This equivalent circuit corresponds to the stator resulting magnetic field.

In the stator equivalent circuit in Fig. 11.3, its transverse branches $0 - 0$, $k - k$, $z - z$, and $a - a$ represent the own circuit loops of the longitudinal magnetizing sections reflecting the air gap sublayer adjoining to the stator bore surface, the stator slot-wedge area, the stator winding location area, and the own circuit loop of the transverse magnetizing stator yoke section. The transverse branch $1 - 1$ containing the impedance $[r_1 + j(x_{\tau 01} + x_{\tau ks} + x_{11\sigma}) + Z_c]$ represents the slot leakage circuit loop of the stator winding.

The equivalent circuit in Fig. 11.3 represents the "unfolded" stator equivalent circuit with the parallel connection of the elements. Such structure of the stator equivalent circuit is due to the fact that here the one-dimensional machine magnetizing field and the one-dimensional stator slot leakage magnetic field are considered. Thus, the stator equivalent circuit corresponding to the resulting magnetic field has a rather complex structure, and therefore, it is not convenient for practical use. In this regard, the equivalent circuit in Fig. 11.3 needs to be transformed which will be discussed below.

11.4 "Unfolded" Equivalent Circuit of Rotor

We proceed from the consideration of the rotor magnetizing equivalent circuit (Fig. 11.1b) and the rotor slot leakage equivalent circuit (Fig. 11.2b). The rotor magnetizing equivalent circuit (Fig. 11.1b) corresponds to the no-load mode. The rotor slot leakage equivalent circuit (Fig. 11.2b) reflects the under-load mode. Note that in the rotor magnetizing equivalent circuit (Fig. 11.1b), the parallel branch $z - z$ represents the rotor winding location area. In these conditions, to obtain the rotor equivalent circuit corresponding to the resulting magnetic field in it, it is enough to the terminals of the parallel branch $z - z$ of the equivalent circuit in Fig. 11.1b to connect the terminals of the branch $2 - 2$ of the equivalent circuit in Fig. 11.2b. As a result, the rotor equivalent circuit can be reduced to the form in Fig. 11.4. This equivalent circuit corresponds to the rotor resulting magnetic field.

In the rotor equivalent circuit in Fig. 11.4, its transverse branches $0 - 0, k - k, z - z$ and $a - a$ represent the own circuit loops of the longitudinal magnetizing sections reflecting the air gap sublayer adjoining to the rotor surface, the rotor slot-wedge area, the rotor winding location area, and the own circuit loop of the transverse magnetizing rotor yoke area. The transverse branch $2 - 2$ containing the impedance $\left[r'_2/s + j\left(x'_{\tau 02} + x'_{\tau kR} + x'_{22\sigma}\right) + Z'_{2н}\right]$ represents the slot leakage circuit loop of the rotor winding.

The equivalent circuit in Fig. 11.4 represents the "unfolded" rotor equivalent circuit with the parallel connection of the elements. Such structure of the rotor equivalent circuit is due to the fact that here the one-dimensional machine magnetizing field and the one-dimensional rotor slot leakage magnetic field are considered. Thus, the rotor equivalent circuit corresponding to the resulting magnetic field has a rather complex structure, and therefore, it is not convenient for practical use. To

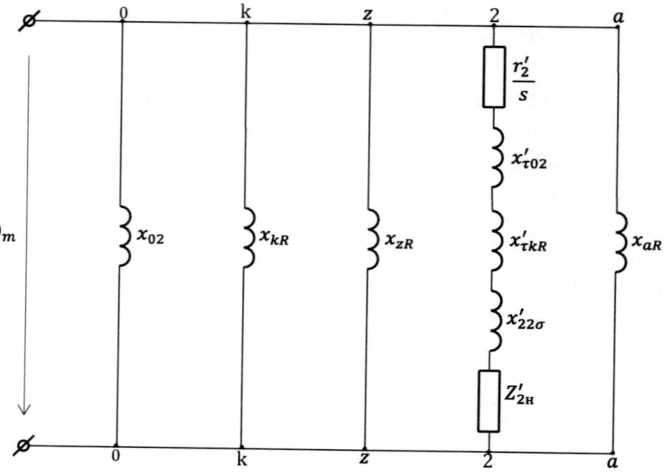

Fig. 11.4 "Unfolded" rotor equivalent circuit with the parallel connection of elements

make it look convenient for practical use, the equivalent circuit in Fig. 11.4 needs to be transformed which will be considered below.

11.5 Magnetizing and Slot Leakage Circuit Loops: “Folded” Equivalent Circuit of Stator

We consider the equivalent circuit shown in Fig. 11.3. In this equivalent circuit, the parallel branches $0 - 0$, $k - k$, $z - z$, and $a - a$ represent the magnetizing circuit-loops reflecting the corresponding stator magnetic circuit sections. Therefore, they can be combined into the one common circuit loop that will reflect stator the magnetizing circuit loop. The reactance used in this common circuit loop (stator magnetizing circuit loop) represents the stator magnetizing reactance.

Based on the equivalent circuit in Fig. 11.3, for the stator magnetizing reactance we have

$$x_{m1} = \frac{1}{\frac{1}{x_{01}} + \frac{1}{x_{ks}} + \frac{1}{x_{zs}} + \frac{1}{x_{as}}} = \frac{x_{01}}{1 + x_{01}\left(\frac{1}{x_{ks}} + \frac{1}{x_{zs}} + \frac{1}{x_{as}}\right)} = \frac{x_{01}}{k_{\text{H}1}} \tag{11.11}$$

where $k_{\text{H}1} = 1 + x_{01}\left(\frac{1}{x_{ks}} + \frac{1}{x_{zs}} + \frac{1}{x_{as}}\right)$.

Here, x_{m1} represents the stator magnetizing reactance, $k_{н1}$ reflects the stator magnetic circuit saturation coefficient.

In the equivalent circuit in Fig. 11.3, the parallel branch $1 - 1$ represents the stator slot leakage circuit loop. In accordance with (11.7), the sum of the reactances $x_{\tau 01}$, $x_{\tau ks}$, and $x_{11\sigma}$ used in the circuit loop $1 - 1$ reflects the stator slot leakage reactance $x_{1\sigma}$ for which we have that $x_{1\sigma} = x_{\tau 01} + x_{\tau ks} + x_{11\sigma}$. Applying in the parallel branch $1 - 1$ of the equivalent circuit in Fig. 11.3 the condition (11.7) for reactance $x_{1\sigma}$ and taking into account the equation for the reactance x_{m1} obtained in (11.11) and also using for the impedance Z_c the equation of kind $\left(-\dot{U}_1\right) = Z_c\dot{I}_1$, the equivalent circuit in Fig. 11.3 can be reduced to the form given in Fig. 11.5. In this case, the equivalent circuit in Fig. 11.5 represents the “folded” stator equivalent circuit which is followed from the “unfolded” stator equivalent circuit with parallel connection of

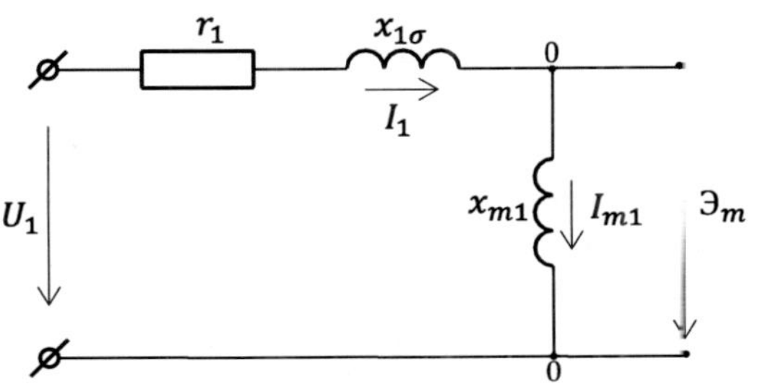

Fig. 11.5 “Folded” stator equivalent circuit

the elements (Fig. 11.3). As follows from the equivalent circuit in Fig. 11.5, in relation to the emf $\dot{Э}_m$, the stator magnetic circuits are represented as the two parallel circuit loops: the stator magnetizing circuit loop containing the reactance x_{m1} and the stator slot leakage circuit loop having the reactance $x_{1\sigma}$. The stator equivalent circuit obtained in this way (Fig. 11.5) repeats the structure of the stator equivalent circuit received in Fig. 10.9a. The "folded" stator equivalent circuit in Fig. 11.5 has a simple structure, and therefore, it is convenient for practical use.

In the "folded" stator equivalent circuit in Fig. 11.5, the reactance $x_{1\sigma}$ which represents the stator slot leakage reactance is used. This reactance consists of three components: the leakage reactance $x_{\tau01}$ of the air gap sublayer adjoining to the stator bore surface, the leakage reactance $x_{\tau ks}$ of the section reflecting the stator slot-wedge area, and the leakage reactance $x_{11\sigma}$ of the section representing the stator winding location area. As follows from (11.8), the reactances $x_{\tau01}$, $x_{\tau ks}$, and $x_{11\sigma}$ are determined as

$$
\begin{aligned}
x_{\tau01} &= \omega_1\mu_0 t_0 \frac{\delta}{2t_0} k_{1\sigma} k_1 = \omega_1\mu_0 t_0 \lambda_{\tau01} k_1 \\
x_{\tau ks} &= \omega_1\mu_0 t_0 \frac{h_{ks}}{b_{ks}} \xi_{Lks} k_1 = \omega_1\mu_0 t_0 \lambda_{\tau ks} k_1 \\
x_{11\sigma} &= \omega_1\mu_0 t_0 \frac{h_{zs}}{3b_{ns}} \xi_{Lzs} k_1 = \omega_1\mu_0 t_0 \lambda_{11\sigma} k_1
\end{aligned}
\tag{11.12}
$$

where $\lambda_{\tau01} = \frac{\delta}{2t_0} k_{1\sigma}$; $\lambda_{\tau ks} = \frac{h_{ks}}{b_{ks}} \xi_{Lks}$; $\lambda_{11\sigma} = \frac{h_{zs}}{3b_{ns}} \xi_{Lzs}$.

In (11.12), the dimensionless quantities $\lambda_{\tau01}$, $\lambda_{\tau ks}$, and $\lambda_{11\sigma}$ represent the relative magnetic conductivities for the tangential components of the magnetic fluxes in the sections reflecting the air gap sublayer adjoining to the stator bore surface, the stator slot-wedge area, and the stator winding location area, respectively.

Now, from (11.7) and (11.12), the start leakage reactance $x_{1\sigma}$ acquires.

$$
\begin{aligned}
x_{1\sigma} &= x_{\tau01} + x_{\tau ks} + x_{11\sigma} \\
&= \omega_1\mu_0 t_0(\lambda_{\tau01} + \lambda_{\tau ks} + \lambda_{11\sigma})k_1 = \omega_1\mu_0 t_0 \lambda_{1\sigma} k_1
\end{aligned}
\tag{11.13}
$$

where $\lambda_{1\sigma} = \lambda_{\tau01} + \lambda_{\tau ks} + \lambda_{11\sigma}$.

Here, the value of $\lambda_{1\sigma}$ reflects the relative magnetic conductivity for the slot leakage magnetic flux of the stator winding.

11.6 Magnetizing and Slot Leakage Circuit Loops: "Folded" Equivalent Circuit of Rotor

In the equivalent circuit in Fig. 11.4, the parallel branches $0-0$, $k-k$, $z-z$, and $a-a$ represent the magnetizing circuit loops reflecting the corresponding sections of the rotor magnetic circuit. Therefore, these parallel branches $0-0$, $k-k$, $z-z$, and

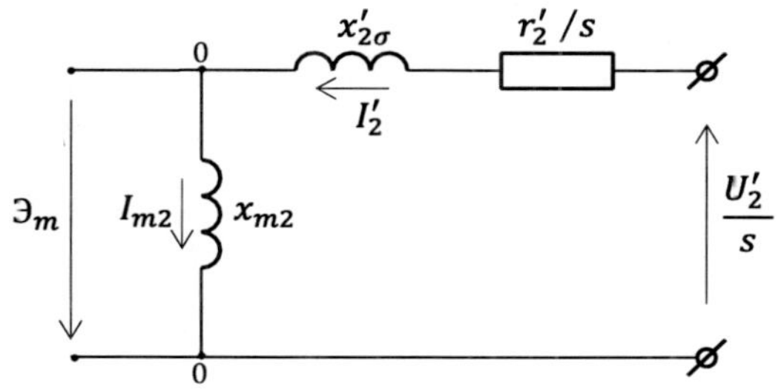

Fig. 11.6 "Folded" rotor equivalent circuit

$a - a$ can be combined into the one common circuit loop that will reflect the rotor magnetizing circuit loop. The reactance used in this common circuit loop (rotor magnetizing circuit loop) represents the rotor magnetizing reactance.

Based on the equivalent circuit in Fig. 11.4, for the rotor magnetizing reactance, we have

$$x_{m2} = \frac{1}{\frac{1}{x_{02}} + \frac{1}{x_{kR}} + \frac{1}{x_{zR}} + \frac{1}{x_{aR}}} = \frac{x_{02}}{1 + x_{02}\left(\frac{1}{x_{kR}} + \frac{1}{x_{zR}} + \frac{1}{x_{aR}}\right)} = \frac{x_{02}}{k_{\text{н}2}} \tag{11.14}$$

where $k_{\text{н}2} = 1 + x_{02}\left(\frac{1}{x_{kR}} + \frac{1}{x_{zR}} + \frac{1}{x_{aR}}\right)$.

Here, x_{m2} represents the rotor magnetizing reactance, $k_{н2}$ reflects the rotor magnetic circuit saturation coefficient.

In the equivalent circuit in Fig. 11.4, the parallel branch $2 - 2$ represents the rotor slot leakage circuit loop. In accordance with (11.5), the sum of the reactances $x'_{\tau 02}, x'_{\tau kP}$, and $x'_{22\sigma}$ used in the circuit loop $2 - 2$ reflects the rotor slot leakage reactance $x'_{2\sigma}$ for which we have that $x'_{2\sigma} = x'_{\tau 02} + x'_{\tau kR} + x'_{22\sigma}$. Applying in the parallel branch $2 - 2$ of the equivalent circuit in Fig. 11.4 the condition (11.9) for reactance $x'_{2\sigma}$ and taking into account the equation for the reactance x_{m2} obtained in (11.14) and also using for the impedance $Z'_{2н}$ the equation of kind $U'_2/s = Z'_{2н}\dot{I}'_2$, the equivalent circuit in Fig. 11.4 can be reduced to the form given in Fig. 11.6. In this case, the equivalent circuit in Fig. 11.6 represents the "folded" rotor equivalent circuit which is followed from the "unfolded" rotor equivalent circuit with the parallel connection of the elements (Fig. 11.4). As follows from the equivalent circuit in Fig. 11.6, in relation to the emf $\dot{Э}_m$, the rotor magnetic circuits are represented as the two parallel circuit loops: the rotor magnetizing circuit loop containing the reactance x_{m2} and the rotor slot leakage circuit loop having the reactance $x'_{2\sigma}$. The rotor equivalent circuit obtained in this way (Fig. 11.6) repeats the structure of the rotor equivalent circuit received in Fig. 10.9d. The "folded" rotor equivalent circuit in Fig. 11.6 has a simple structure, and therefore, it is convenient for practical use.

In the "folded" rotor equivalent circuit in Fig. 11.6, the reactance $x'_{2\sigma}$ which represents the rotor slot leakage reactance is used. This reactance consists of three components: the leakage reactance $x'_{\tau 02}$ of the air gap sublayer adjoining to the rotor surface, the leakage reactance $x'_{\tau kR}$ of the section reflecting the rotor slot-wedge area, the leakage reactance $x'_{22\sigma}$ of the section representing the rotor winding location area. As follows from (11.6), the reactances $x^*_{\tau 02}, x^*_{\tau kR}$ and $x^*_{22\sigma}$ are determined as

$$\begin{aligned} x'_{\tau 02} &= \omega_1 \mu_0 t_0 \frac{\delta}{2t_{02}} k_{2\sigma} k_R k_1 = \omega_1 \mu_0 t_0 \lambda_{\tau 02} k_1 \\ x'_{\tau kR} &= \omega_1 \mu_0 t_0 \frac{h_{kR}}{b_{kR}} k_R \xi_{LkR} k_1 = \omega_1 \mu_0 t_0 \lambda_{\tau kR} k_1 \\ x'_{22\sigma} &= \omega_1 \mu_0 t_0 \frac{h_{zR}}{3b_{nR}} k_R \xi_{LzR} k_1 = \omega_1 \mu_0 t_0 \lambda_{22\sigma} k_1 \end{aligned} \tag{11.15}$$

where $\lambda_{\tau 02} = \frac{\delta}{2t_{02}} k_{2\sigma} k_R; \lambda_{\tau kR} = \frac{h_{kR}}{b_{kR}} k_R \xi_{LkR}; \lambda_{22\sigma} = \frac{h_{zR}}{3b_{nR}} k_R \xi_{LzR}.$

In (11.15), the dimensionless quantities $\lambda_{\tau 02}$, $\lambda_{\tau kR}$, and $\lambda_{22\sigma}$ represent the relative magnetic conductivities for the tangential components of the slot leakage magnetic fluxes in the layers reflecting the air gap sublayer adjoining to the rotor surface, the rotor slot-wedge area, and the rotor winding location area, respectively.

Now, from (11.5) and (11.15), the rotor leakage reactance $x'_{2\sigma}$ acquires

$$\begin{aligned} x'_{2\sigma} &= x'_{\tau 02} + x'_{\tau kR} + x'_{22\sigma} \\ &= \omega_1 \mu_0 t_0 (\lambda_{\tau 02} + \lambda_{\tau kR} + \lambda_{22\sigma}) k_1 = \omega_1 \mu_0 t_0 \lambda_{2\sigma} k_1 \end{aligned} \tag{11.16}$$

where $\lambda_{2\sigma} = \lambda_{\tau 02} + \lambda_{\tau kR} + \lambda_{22\sigma}$.

Here, the value of $\lambda_{2\sigma}$ reflects the relative magnetic conductivity for the slot leakage magnetic flux of the rotor winding.

11.7 Equivalent Circuit of Machine: Voltage Equations

We consider the "folded" stator and rotor equivalent circuits given in Figs. 11.5 and 11.6. The terminals 0 – 0 of the equivalent circuit in Fig. 11.6 can be connected to the terminals 0 – 0 of the equivalent circuit in Fig. 11.5. Then, we can introduce the machine magnetizing reactance x_m determined as $x_m = x_{m1} x_{m2} / (x_{m1} + x_{m2})$. Now, in relation to the terminals 0– 0, the equivalent circuits given in Figs. 11.5 and 11.6 can be combined into the one equivalent circuit as a result of which the machine equivalent circuit of the traditional form shown in Fig. 11.7 arises. In relation to the emf $\dot{Э}_m$, the machine equivalent circuit in Fig. 11.7 consists of the three circuit loops: the machine magnetizing circuit loop, the stator slot leakage circuit loop, and the rotor slot leakage circuit loop. It means that the machine equivalent circuit (Fig. 11.7) obtained based on considering the machine resulting magnetic field and the machine equivalent circuit (Fig. 10.8) constructed based on the decomposition of the machine resulting magnetic field into the "partial" magnetic fields (machine

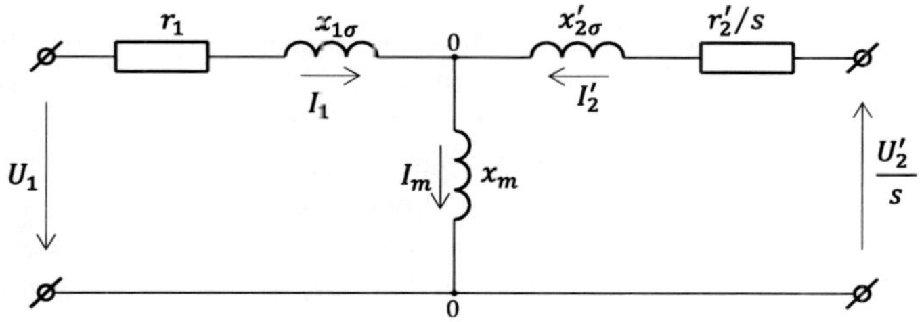

Fig. 11.7 Machine equivalent circuit

magnetizing field, stator slot leakage magnetic field, rotor slot leakage magnetic field) take the same form.

Using the machine equivalent circuit in Fig. 11.7, the voltage equations acquire

$$\begin{aligned} \dot{U}_1 &= (r_1 + jx_{1\sigma})\dot{I}_1 + (-\dot{Э}_m) \\ -\frac{\dot{U}'_2}{s} &= \left(\frac{r'_2}{s} + jx'_{2\sigma}\right)\dot{I}'_2 + (-\dot{Э}_m) \end{aligned} \tag{11.17}$$

Here

$$(-\dot{Э}_m) = jx_m\dot{I}_m \quad \dot{I}_m = \dot{I}_1 + \dot{I}'_2 \tag{11.18}$$

In the system of the Eq. (11.17), the conditions given in (11.18) can be used. Then, the voltage equations corresponding to the "inductively coupled circuits" concept and expressing through the referred rotor quantities can be obtained. As a result, we have

$$\begin{aligned} \dot{U}_1 &= r_1\dot{I}_1 + j(x_{1\sigma} + x_m)\dot{I}_1 + jx_m\dot{I}'_2 = r_1\dot{I}_1 + jx_1\dot{I}_1 + jx_{21}\dot{I}'_2 \\ -\frac{\dot{U}'_2}{s} &= \frac{r'_2}{s}\dot{I}'_2 + jx_m\dot{I}_1 + j(x'_{2\sigma} + x_m)\dot{I}'_2 = \frac{r'_2}{s}\dot{I}'_2 + jx_{12}\dot{I}_1 + jx_2\dot{I}'_2 \end{aligned} \tag{11.19}$$

where $x_1 = x_{1\sigma} + x_m; x_2 = x'_{2\sigma} + x_m; x_{21} = x_m; x_{12} = x_m$.

In accordance with (10.17)–(10.19), the voltage equations used in practice are expressed through the referred quantities.

11.8 Summary

1. In this chapter, the description of the technique for determining the basic calculation elements of the machine is given. The recommended technique is based on the use of the "unfolded" equivalent circuits for the stator and rotor of the electric

machine. These "unfolded" stator and rotor equivalent circuits reflect the machine resulting magnetic field. The implementation of the recommended technique is reduced to the transformation of the "unfolded" stator and rotor equivalent circuits into the "folded" ones. The "folded" stator and rotor equivalent circuits can be used to construct the machine equivalent circuit.

2. The construction of the "folded" stator and rotor equivalent circuits is carried out in the two stages. In the first stage, the "unfolded" stator and rotor equivalent circuits are obtained. Such stator and rotor equivalent circuits arise as a result of the superposition of the "unfolded" stator and rotor magnetizing equivalent circuits and the "unfolded" stator and rotor slot leakage equivalent circuits. In the second stage, the stator and rotor magnetizing circuit loops and the stator and rotor slot leakage circuit loops are established. As a result of the introduction of these circuit loops, the "folded" stator and rotor equivalent circuits are determined.
3. The "folded" stator and rotor equivalent circuits can be used to establish the machine magnetizing circuit loop and to determine the magnetizing reactance of the machine. As a result of this, the machine equivalent circuit can be obtained. In relation to the air gap emf, the machine equivalent circuit consists of the three circuit loops: the machine magnetizing circuit loop, the stator slot leakage circuit loop, and the rotor slot leakage circuit loop.
4. In this chapter, the one-dimensional machine magnetizing field and the one-dimensional stator and rotor slot leakage magnetic fields are considered, and also the phase system of units is used. In these conditions, the machine magnetizing circuit and the stator and rotor slot leakage magnetic circuits are considered subdivided into sections. This procedure greatly simplifies the constriction of the "unfolded" stator and rotor equivalent circuits.
5. The stator and rotor magnetizing circuits can be represented as the "unfolded" stator and rotor magnetizing equivalent circuits. The stator and rotor slot leakage magnetic circuits can be represented as the "unfolded" stator and rotor slot leakage equivalent circuits. The "unfolded" stator and rotor magnetizing equivalent circuits and the "unfolded" stator and rotor slot leakage equivalent circuits are represented in relation to the air gap emf which makes it easier to build the stator and rotor equivalent circuits corresponding to the machine resulting magnetic field.
6. The "unfolded" stator and rotor magnetizing equivalent circuits reflect the machine no-load mode. The "unfolded" stator and rotor slot leakage equivalent circuits correspond to the machine under-load mode. In accordance with the "no-load and under-load" concept, the superposition principle can be applied to these equivalent circuits as a result of which the stator and rotor equivalent circuits reflecting the machine resulting magnetic field are obtained. When using the machine resulting magnetic field, the machine equivalent circuit should be first built, and then, the voltage equations can be compiled with the use of this equivalent circuit.

7. When considering the machine resulting magnetic field and using the phase system of units, the determination of the basic machine calculation elements can be reduced to applying the following provisions:

   - decomposition of the stator and rotor magnetizing circuits and also the stator and rotor slot leakage magnetic circuits into the sections;
   - building the “unfolded” stator and rotor magnetizing equivalent circuits and the “unfolded” stator and rotor slot leakage equivalent circuits;
   - building the “unfolded” stator and rotor equivalent circuits corresponding to the machine resulting magnetic field;
   - introduction of the stator and rotor magnetizing circuit loops and the stator and rotor slot leakage circuit loops, and determination of the magnetizing and the slot leakage reactances used in them;
   - building the “folded” stator and rotor equivalent circuits;
   - introduction of the machine magnetizing circuit loop and determination of the machine magnetizing reactance;
   - building the machine equivalent circuit by the use of the “folded” stator and rotor equivalent circuits;
   - compiling the machine voltage equations by the use of the machine equivalent circuit.

These provisions correspond to the case when the machine resulting magnetic field is determined as a result of the superposition of the one-dimensional machine magnetizing field and the one-dimensional stator and rotor slot leakage magnetic fields and the phase system of units is used.

When using the machine resulting magnetic field, the “fragmentation” concept, the superposition principle, the “no-load and under-load” concept, the cascade way of connecting, and the system of the reduction of the “unfolded” equivalent circuits into the “folded” ones can be applied.

Reference

1. Asanbayev, V.: Alternating Current Multi-Circuit Electric Machines: A New Approach to Steady-State Parameter Determination, Springer Cham Heidelberg, New York, Dordrecht London (2015)

Chapter 12
Basic Calculation Elements in Specific System of Units: Techniques for Determination

12.1 Currents and emfs of Windings of Stator and Rotor: Specific System of Units

The emf and the currents of the stator and rotor windings can be represented in the phase $(\dot{Э}, \dot{I})$ and in the specific $(\dot{E}, \dot{H})$ systems of units. In this regard, we will consider the expressions connecting the values of the emf and the currents represented in the phase and specific system of units.

12.1.1 Linear Density of Surface Current and mmf

In the theory of the electric machines, the currents of the stator and rotor windings are set by two ways. In the first way, to set the machine winding currents, the concept of the winding mmf is used. The mmf of the winding is designated as $\dot{F}(x)$, and it represents the distribution in a space of the total currents of the winding, for example, [1–16]. The introduction of the mmf greatly simplifies the procedure for determining the machine magnetic field. In accordance with (3.25), the fundamental harmonic of the winding mmf is determined as $\dot{F}(x) = \dot{F}_i e^{-j(\pi/\tau_i)x}$, where τ_i is pole pitch on the surface on which the considered winding is leaned. Here, for the amplitude of the mmf $\dot{F}_i$, we have

$$\dot{F}_i = \frac{m_i}{\pi} \frac{w_i k_{wi}}{p} \sqrt{2} \dot{I}_i \tag{12.1}$$

where $i = 1, 2$.

In another way, the currents of the machine windings are represented as the linear densities of the surface currents [1–17]. This way of representation of the current lies in the fact that the total current flowing in each slot can be distributed on the surface

V. Asanbayev, *Asynchronous Machines*,
https://doi.org/10.1007/978-3-030-92284-9_12

determined by the slot width in the form of the thin current sheet. For calculation purposes, it is necessary to have the fundamental harmonic of the linear density $\dot{\varkappa}(x)$ of the surface current for which the spatial distribution of the surface current is decomposed in the Fourier series. This way of determining the fundamental harmonic of the linear density of the surface current is not entirely rational in the practice due to its sufficient laboriousness, for example, [12].

The definition of the linear density of the surface current $\dot{\varkappa}(x)$ is greatly simplified, if for this purpose, we use the distribution function for the winding mmf. As shown in (3.25), the mmf of the winding $\dot{F}(x)$ is the function of the angular coordinate directed along the circumference of the stator or rotor surfaces, and it represents the integral distribution curve of the linear density of the surface current $\dot{\varkappa}(x)$. Therefore, the linear density of the surface current $\dot{\varkappa}(x)$ is the derivative of the mmf of the winding $\dot{F}(x)$, i.e. $\dot{\varkappa}(x) = d\dot{F}(x)/dx$. Differentiating the condition $\dot{F}(x) = \dot{F}_i e^{-j(\pi/\tau_i)x}$ with respect to x, for the amplitude of the fundamental harmonic of the linear density of the surface current, the expression of kind $\dot{\varkappa}_i = (\pi/\tau_i)\dot{F}_i$ can be obtained. Based on this condition and the expression (12.1), for the $\varkappa_i$, we have

$$\dot{\varkappa}_i = \frac{m_i}{\tau_i} \frac{w_i k_{wi}}{p} \sqrt{2} \dot{I}_i \tag{12.2}$$

12.1.2 Magnetic Field Intensities and Currents of Windings of Stator and Rotor

We consider the tangential component of the magnetic field intensity on the stator surface on which the surface current with the linear density $\dot{\varkappa}_i$ is distributed as the infinitely thin layer. In the Chap. 3, it is shown that the condition of kind $\dot{H}_{xi} = \dot{\varkappa}_i$ is implemented on the stator surface. Thus, the tangential component of the magnetic field intensity $\dot{H}_{xi}$ acting on the stator surface is equal to the linear density of the surface current $\dot{\varkappa}_i$, for example, [12, 13, 16]. With considering (12.2), for $\dot{H}_{xi}$, we have

$$\dot{H}_{xi} = \frac{m_i}{\tau_i} \frac{w_i k_{wi}}{p} \sqrt{2} \dot{I}_i \tag{12.3}$$

A more detailed derivation of this condition is given in the Chap. 3.

When constructing the machine equivalent circuit, the value of the magnetic field intensity $\dot{H}_{xi}$ should be reduced to the air gap surface $0 - 0$ having the calculated pole pitch τ_0 and the calculated axial length equal to l_0, for example, [17]. Then, for the referred to the surface $0 - 0$ value of $\dot{H}_i$, we obtain

$$\dot{H}_i = \dot{H}_{xi}\frac{\tau_i}{\tau_o} = \dot{H}_{xi}\xi_{\tau i} \tag{12.4}$$

where $\xi_{\tau i} = \tau_i/\tau_o$.

Based on the expressions (12.3) and (12.4), for the amplitude of the winding current, we can have

$$\sqrt{2}\dot{I}_i = \frac{\tau_0 p}{m_i w_i k_{wi}}\dot{H}_i \tag{12.5}$$

The expression of kind (12.5) is given also in (3.36).

This condition can be used for the stator and rotor windings. As a result, it follows

$$\sqrt{2}\dot{I}_1 = \frac{\tau_0 p}{m_1 w_1 k_{w1}}\dot{H}_1 \quad \sqrt{2}\dot{I}_2 = \frac{\tau_0 p}{m_2 w_2 k_{w2}}\dot{H}_2 \tag{12.6}$$

where $\dot{H}_1 = \dot{H}_{x1}\frac{\tau_1}{\tau_o} = \dot{H}_{x1}\xi_{\tau 1}; \dot{H}_2 = \dot{H}_{x2}\frac{\tau_2}{\tau_o} = \dot{H}_{x2}\xi_{\tau 2}; \xi_{\tau 1} = \frac{\tau_1}{\tau_o}; \xi_{\tau 2} = \frac{\tau_2}{\tau_o}$.

The amplitude of the squirrel-cage rotor current ($w_2 = 1/2; m_2 = Z_2; k_{w2} = 1.0$) is determined as

$$\sqrt{2}\dot{I}_2 = \frac{2\tau_0 p}{Z_2}\dot{H}_2 \quad \sqrt{2}\dot{I}_2 = \dot{H}_2 t_{02} \tag{12.7}$$

since $t_{02} = 2p\tau_0/Z_2$. Here, the value $\sqrt{2}\dot{I}_2$ represents the amplitude of the current flowing in the rotor slot, i.e. in the bar of the squirrel-cage winding.

12.1.3 Electric Field Intensities and emfs of Windings of Stator and Rotor

The emf induced in the machine winding can be expressed through the value of the axial component of the electric field intensity which is demonstrated in the Chap. 3. On the basis of the results obtained in the Chap. 3, we have

$$\sqrt{2}\left(-\dot{Э}_i\right) = j\omega_1\Psi_i \quad \dot{Э}_i = 2w_i k_{wi} l_i \dot{E}_{zi} \tag{12.8}$$

Further, in this chapter, we will use the value of the emf $\left(-\dot{Э}_i\right)$ without the negative sign, i.e. the module value of the emf $\dot{Э}_i$. In (12.8), the value $\dot{E}_{zi}$ can be reduced to the calculated axial length of the air gap surface $0 - 0$. Then, we can obtain

$$\dot{E}_i = \dot{E}_{zi}\frac{l_i}{l_0} = \dot{E}_{zi}\xi_{Li} \tag{12.9}$$

where $\xi_{Li} = l_i/l_0$.

Now, with considering (12.8) and (12.9), the amplitude value of emf acquires

$$\sqrt{2}\dot{Э}_i = 2w_i k_{wi} l_0 \dot{E}_i \tag{12.10}$$

This condition is received in the Chap. 3 and given in (3.36).

The expression (12.10) can be written for the stator and the rotor windings. Now, it follows

$$\sqrt{2}\dot{Э}_1 = 2w_1 k_{w1} l_0 \dot{E}_1 \quad \sqrt{2}\dot{Э}_2 = 2w_2 k_{w2} l_0 \dot{E}_2 \tag{12.11}$$

where $\dot{E}_1 = \dot{E}_{z1}\frac{l_1}{l_0} = \dot{E}_{z1}\xi_{L1}; \dot{E}_2 = \dot{E}_{z2}\frac{l_2}{l_0} = \dot{E}_{z2}\xi_{L2}; \xi_{L1} = \frac{l_1}{l_0}; \xi_{L2} = \frac{l_2}{l_0}$.

The amplitude of the squirrel-cage rotor emf ($w_2 = 1/2; m_2 = Z_2; k_{w2} = 1.0$) can be determined as

$$\sqrt{2}\dot{Э}_2 = l_0 \dot{E}_2 \tag{12.12}$$

When determining the electric field intensity, it is necessary to distinguish between the electric field caused by the emf induced in the winding and the electric field associated with the voltage drop from the current flowing in the winding. As is known in both cases, the emf at the terminals of the winding can be determined by the same expression.

By the expressions (12.6) and (12.11), the emf and the currents of the stator and rotor phase windings can be represented as

$$\begin{aligned} \sqrt{2}\dot{Э}_1 &= k_{1Э}\dot{E}_1 \quad \sqrt{2}\dot{Э}_2 = k_{2Э}\dot{E}_2 \\ \sqrt{2}\dot{I}_1 &= k_{1I}\dot{H}_1 \quad \sqrt{2}\dot{I}_2 = k_{2I}\dot{H}_2 \end{aligned} \tag{12.13}$$

where $k_{1Э} = 2w_1 k_{w1} l_0; k_{2Э} = 2w_2 k_{w2} l_0; k_{1I} = \frac{\tau_0 p}{m_1 w_1 k_{w1}}; k_{2I} = \frac{\tau_0 p}{m_2 w_2 k_{w2}}$.

For the squirrel-cage rotor, from (12.7) and (12.12), we obtain

$$\begin{aligned} \sqrt{2}\dot{Э}_2 &= k_{2Э}\dot{E}_2 \\ \sqrt{2}\dot{I}_2 &= k_{2I}\dot{H}_2 \end{aligned} \tag{12.14}$$

where $k_{2Э} = l_0$; $k_{2I} = t_{02}$.

The expressions given in (12.13) and (12.14) connect the electric field intensities with the emf induced in the stator and rotor windings and also the magnetic field intensities with the currents flowing in the stator and rotor windings.

12.2 Generalized emf Equations of Air Gap

We proceed from the air gap emf equations represented in (6.19). These equations are written out in relation to the stator winding emf $\dot{Э}_1$ and the rotor winding emf $\dot{Э}_2$, and they contain the inductively coupled elements determined in the phase system of units $(\dot{Э}, \dot{I})$. Here, we express the air gap emf Eq. (6.19) through the specific quantities $(\dot{E}, \dot{H})$. As shown in the Chap. 7, the Eq. (6.19) reflect the magnetic field existing in the machine air gap. Therefore, the elements used in the emf Eq. (6.19) can be determined based on an analysis of the magnetic field in the machine air gap.

The air gap model can be represented as the flat layer having smooth surfaces. On the smooth surfaces of the air gap layer, the currents of the stator and rotor windings are distributed as the linear densities of the surface currents. For the stator and rotor steel, we have that $\mu_c = \infty$. The air gap magnetic permeability is μ_0. We accept that in the air gap there is a single radial component of the magnetic flux density the value of which does not change along the air gap radial length. The presence on the stator and rotor surfaces of the axial teeth is taken into account by introducing the calculated air gap length δ', where $\delta' = \delta k_\delta$. Here, k_δ represents the air gap factor.

To determine the generalized air gap equations, it is necessary to set the magnetic field in the air gap and to find the components of the electric field intensities of the emf induced in the stator and rotor windings. Then, the components of the electric field intensities obtained in this way should be expressed through the corresponding reactances. Now, the air gap equations can be obtained based on the superposition of the components of the electric field intensities of the emf induced in the stator and rotor windings. Below, we will consider the features of the implementation of this approach under the condition of the use of the specific system of units $(\dot{E}, \dot{H})$.

The components of the electric field intensities of the emf induced in the stator and rotor windings. We will consider the case of supplying the current to the stator winding. In the presence of the current in the stator winding ($\dot{H}_{x1} \neq 0$) and in the absence of the current in the rotor winding ($\dot{H}_{x2} = 0$), the magnetic flux density created by the linear density of the surface current of the stator winding $\varkappa_1$ ($\varkappa_1 = H_{x1}$) is determined to proceed from (4.10) and with considering the conditions of kind $\dot{B}_1 = \lambda_1 \dot{F}_1$ and $\dot{F}_1 = (\tau_{01}/\pi) H_{x1}$ as.

$$\dot{B}_1 = \frac{\mu_0 \tau_{01}}{\pi \delta'} \dot{H}_{x1} \tag{12.15}$$

where τ_{01} is the pole pitch defined on the stator bore surface.

We consider the flat air gap obtained in relation to the surface $0 - 0$ passing through the middle of the radial length of the air gap and having the axial length equal to l_0 and also the pole pitch equal to τ_0. In this regard, the value of $\dot{H}_{x1}$ used in (12.15) should be reduced to this air gap surface $0 - 0$ for what the expression of kind $\dot{H}_1 = \dot{H}_{x1}(\tau_{01}/\tau_0)$ can be applied. Now, the expression (12.15) acquires

$$\dot{B}_1 = \frac{\mu_0 \tau_0}{\pi \delta'} \dot{H}_1 \tag{12.16}$$

The magnetic flux density $\dot{B}_1$ causes the electric field intensity $\dot{E}_{11}$ of the self-induction emf induced in the stator winding and also the electric field intensity $\dot{E}_{12}$ of the mutual induction emf induced in the rotor winding. We will determine the values of $\dot{E}_{11}$ and $\dot{E}_{12}$. For this purpose, the following expression can be used

$$\dot{E} = \upsilon \dot{B} \tag{12.17}$$

where $\upsilon = \dfrac{\omega_1 \tau_0}{\pi}$.

Then, for the $\dot{E}_{11}$ from (12.16) and (12.17), we have

$$\dot{E}_{11} = \frac{\omega_1 \mu_0 \tau_0^2}{\pi^2 \delta'} \dot{H}_1 \tag{12.18}$$

The electric field intensity $\dot{E}_{12}$ of the mutual induction emf induced in the rotor winding is determined as

$$\dot{E}_{12} = \frac{\omega_1 \mu_0 \tau_0^2}{\pi^2 \delta'} \dot{H}_1 \tag{12.19}$$

Now, we will consider the case of supplying the current to the rotor winding, i.e. $\dot{H}_{x2} \neq 0$, and of no the current in the stator winding, i.e. $\dot{H}_{x1} = 0$. By analogy with (12.18) the magnetic flux density created by the linear density of the surface current of the rotor winding $\dot{\varkappa}_2$ $\left(\dot{\varkappa}_2 = \dot{H}_{x2}\right)$ can be defined as

$$\dot{B}_2 = \frac{\mu_0 \tau_{02}}{\pi \delta'} \dot{H}_{x2} \tag{12.20}$$

where τ_{02} is the pole pitch defined on the rotor surface.

In (12.20), the value of $\dot{H}_{x2}$ should be reduced to the air gap surface $0-0$. Using in (12.20) the expression of kind $\dot{H}_2 = \dot{H}_{x2}(\tau_{02}/\tau_0)$, we can obtain

$$\dot{B}_2 = \frac{\mu_0 \tau_0}{\pi \delta'} \dot{H}_2 \tag{12.21}$$

The magnetic flux density $\dot{B}_2$ causes the electric field intensity $\dot{E}_{22}$ of the self-induction emf induced in the rotor winding and also the electric field intensity $\dot{E}_{21}$ of the mutual induction emf induced in the stator winding. For $\dot{E}_{22}$, from (12.17) and (12.21), it arises

$$\dot{E}_{22} = \frac{\omega_1 \mu_0 \tau_0^2}{\pi^2 \delta'} \dot{H}_2 \tag{12.22}$$

The value of the $\dot{E}_{21}$ follows from the expression

$$\dot{E}_{21} = \frac{\omega_1 \mu_0 \tau_0^2}{\pi^2 \delta'} \dot{H}_2 \tag{12.23}$$

The equations for the electric field intensities of the emf induced in the stator and rotor windings. The conditions obtained in (12.18), (12.19) and (12.22), (12.23) can be represented as

$$\begin{aligned} \dot{E}_{11} &= \frac{\omega_1 \mu_0 \tau_0^2}{\pi^2 \delta'} \dot{H}_1 \quad \dot{E}_{21} = \frac{\omega_1 \mu_0 \tau_0^2}{\pi^2 \delta'} \dot{H}_2 \\ \dot{E}_{12} &= \frac{\omega_1 \mu_0 \tau_0^2}{\pi^2 \delta'} \dot{H}_1 \quad \dot{E}_{22} = \frac{\omega_1 \mu_0 \tau_0^2}{\pi^2 \delta'} \dot{H}_2 \end{aligned} \tag{12.24}$$

We derive the equations for the electric field intensities of the emf induced in the stator and rotor windings. For this purpose, the superposition principle can be used. Then, with considering (12.24), we can receive

$$\begin{aligned} \dot{E}_1 &= \dot{E}_{11} + \dot{E}_{21} \\ \dot{E}_2 &= \dot{E}_{12} + \dot{E}_{22} \end{aligned} \tag{12.25}$$

Here, $\dot{E}_1$ represents the total value of the electric field intensity of the emf induced in the stator winding. The quantity of $\dot{E}_2$ reflects the total value of the electric field intensity of the emf induced in the rotor winding.

Now, from (12.24) and (12.25), for $\dot{E}_1$ and $\dot{E}_2$, we obtain

$$\begin{aligned} \dot{E}_1 &= \frac{\omega_1 \mu_0 \tau_0^2}{\pi^2 \delta'} \dot{H}_1 + \frac{\omega_1 \mu_0 \tau_0^2}{\pi^2 \delta'} \dot{H}_2 \\ \dot{E}_2 &= \frac{\omega_1 \mu_0 \tau_0^2}{\pi^2 \delta'} \dot{H}_1 + \frac{\omega_1 \mu_0 \tau_0^2}{\pi^2 \delta'} \dot{H}_2 \end{aligned} \tag{12.26}$$

In the Eq. (12.26), we introduce the reactances $x_{11}^*, x_{21}^*, x_{12}^*$ and x_{22}^* determined as

$$x_{11}^* = x_{21}^* = x_{12}^* = x_{22}^* = \frac{\omega_1 \mu_0 \tau_0^2}{\pi^2 \delta'} \tag{12.27}$$

Then, it follows

$$\begin{aligned}\dot{E}_1 &= x_{11}^*\dot{H}_1 + x_{21}^*\dot{H}_2 \\ \dot{E}_2 &= x_{12}^*\dot{H}_1 + x_{22}^*\dot{H}_2\end{aligned} \tag{12.28}$$

In (12.28), the values of $x_{11}^*, x_{21}^*, x_{12}^*$ and x_{22}^* represent the self-and mutual induction reactances of the air gap. They are expressed in a specific system of units. The system of the Eq. (12.28) can be considered as the generalized air gap emf equations obtained in the specific system of units.

The reactances $x_{11}^*, x_{21}^*, x_{12}^*$ and x_{22}^* used in (12.28) can be also determined based on the emf Eqs (6.19) and the expressions received for the reactances of these emf equations in the Chap. 7 of this work. In this case, the reactances of the emf Eq. (6.19) should be divided into the coefficients given in the Chap. 7 as

$$\begin{aligned}k_1 &= \frac{2m_1(w_1k_{w1})^2 l_0}{p\tau_0} \qquad k_{21} = \frac{2m_2w_1k_{w1}w_2k_{w2}l_0}{p\tau_0} \\ k_{12} &= \frac{2m_1w_1k_{w1}w_2k_{w2}l_0}{p\tau_0} \qquad k_2 = \frac{2m_2(w_2k_{w2})^2 l_0}{p\tau_0}\end{aligned} \tag{12.29}$$

As a result, the reactances used in the emf Eq. (6.19) can be reduced to the form shown in (12.27) in the specific system of units.

12.3 Equivalent Circuit of Air Gap: "No-Load and Under-Load" Concept

We transform the air gap emf Eq. (12.28) for what the "no-load and under-load" concept can be used. By an analogy with the condition (8.4) obtained for the stator winding current as a result of applying the "no-load and under-load" concept, the magnetic field intensity of the stator winding $\dot{H}_1$ can be represented as the sum of the magnetic field intensity of the stator winding $\dot{H}_0$ determined in the no-load mode (when $\dot{H}_2 = 0$) and the magnetic field intensity of the stator winding $\dot{H}_{1н}$ taking place in the machine in the under-load mode, i.e.

$$\dot{H}_1 = \dot{H}_0 + \dot{H}_{1\text{H}} \tag{12.30}$$

In (12.30), we determine the values of $\dot{H}_0$ and $\dot{H}_{1н}$. Note that in the Eq. (12.28), the condition of the no-load mode is implemented when $\dot{H}_2 = 0$, and the condition of the under-load mode is obtained when $\dot{E}_1 = 0$, since in the under-load mode the currents of the stator and rotor windings do not create the mutual induction magnetic fields. Below, these conditions ($\dot{E}_1 = 0$ and $\dot{H}_2 = 0$) will be used when transforming the Eq. (12.28).

No-load mode. When $H_2 = 0$ (no-load mode) from the Eq. (12.28) and with considering the conditions given for the reactances shown in (12.27), we can have

$$\begin{aligned} \dot{E}_1 &= x_{11}^* \dot{H}_0 = x_0^* \dot{H}_0 \\ \dot{E}_2 &= x_{12}^* \dot{H}_0 = x_0^* \dot{H}_0 \end{aligned} \tag{12.31}$$

where $x_{11}^* = x_{12}^* = x_0^*$.

Here, we exclude the value of $\dot{H}_0$. Then, it follows

$$\dot{E}_1 = \dot{E}_2 \tag{12.32}$$

As arises from (12.32), in this case (in the specific system of units), the need of reducing the electric field intensity $\dot{E}_2$ of the emf induced in the rotor winding to the value of the electric field intensity $\dot{E}_1$ of the emf induced in the stator winding is disappeared.

Based on the conditions (12.27), the reactance x_0^* used in (12.31) can be reduced to the form

$$x_0^* = \frac{\omega_1 \mu_0 \tau_0^2}{\pi^2 \delta'} \tag{12.33}$$

Here, the value x_0^* represents the flat air gap magnetizing reactance expressed in the specific system of units.

Under-load mode. When $\dot{E}_1 = 0$, from the first equation of the system (12.28), the following condition arises

$$\dot{H}_{1\text{н}} = (-\dot{H}_2) \tag{12.34}$$

From (12.34), we can note that in the specific system of units the need of reducing the magnetic field intensity $\dot{H}_2$ of the rotor winding to the value of the magnetic field intensity of the stator winding is disappeared. In addition, from (12.34) it follows that in the under-load mode the currents of the stator and rotor windings do not create the mutual induction magnetic field, since under the condition $\dot{H}_{1н} = (-\dot{H}_2)$, the magnetic fields created by the currents of the stator and rotor windings are mutually compensated. Consequently, the magnetizing field taking place in the machine air gap in the no-load mode remains unchanged in the operating mode of the machine.

Operating mode. For the machine operating mode, the magnetic field intensity $\dot{H}_1$ of the stator winding obtains from (12.30) and (12.34) the form

$$\dot{H}_1 = \dot{H}_0 + (-\dot{H}_2) \tag{12.35}$$

From here, for the $\dot{H}_0$, we have

$$\dot{H}_0 = \dot{H}_1 + \dot{H}_2 \tag{12.36}$$

Then, the system of the Eq. (12.31) with considering (12.36) acquires

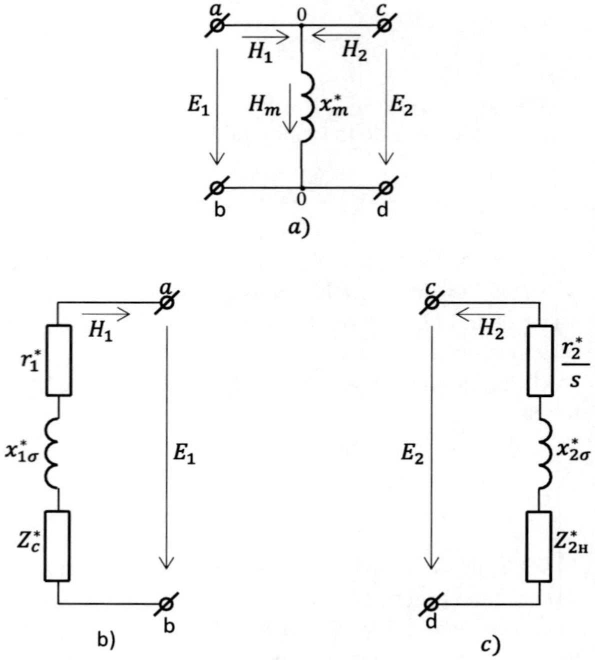

Fig. 12.1 Fragments of machine equivalent circuit in the specific system of units: (**a**) machine magnetizing equivalent circuit; (**b**) stator slot leakage equivalent circuit; (**c**) rotor slot leakage equivalent circuit

$$\begin{aligned} \dot{E}_1 &= x_0^* \dot{H}_0 \\ \dot{E}_2 &= x_0^* \dot{H}_0 \\ \dot{H}_0 &= \dot{H}_1 + \dot{H}_2 \end{aligned} \tag{12.37a}$$

The system of the Eq. (12.31) reflects the no-load mode, and the Eq. (12.37a) represent the operating mode of the machine. In these equations, the reactance x_0^* determined by the expression (12.33) has the same value. Consequently, the magnetic field taking place in the machine air gap in the no-load mode is also kept in the operating mode of the machine, although in this mode the value $\dot{H}_0$ is calculated as $\dot{H}_0 = \dot{H}_1 + \dot{H}_2$.

The system of the Eq. (12.37a) can be visualized by the equivalent circuit shown in Fig. 12.1a. In practice, in the equivalent circuit in Fig. 12.1a, instead of the air gap magnetizing reactance x_0^* the machine magnetizing reactance x_m^* is used. The reactance x_m^* takes into account the reluctance of the stator and rotor magnetic circuit sections that are usually implemented by introducing the machine saturation factor, for example, [1–15, 17]. In this case, the expression for the reactance x_m^* can be obtained from (12.33), if in it the value of δ' to define as $\delta' = \delta k_\delta k_н$, where $k_н$ is the saturation factor of the machine. Now, the equivalent circuit shown in Fig. 12.1a can be represented as the single-element equivalent circuit with the input terminals $a - b$ and with the output terminals $c - d$. Based on the equivalent circuit in Fig. 12.1a, the Eq. (12.37a) acquire the form

$$\dot{E}_1 = x_m^* \dot{H}_m$$
$$\dot{E}_2 = x_m^* \dot{H}_m \qquad (12.37b)$$
$$\dot{H}_m = \dot{H}_1 + \dot{H}_2$$

where x_m^* is the machine magnetizing reactance in the specific system of units. The reactance x_m^* takes into account the reluctance of the layers of the stator and rotor magnetic circuits.

12.4 Slot Leakage Equivalent Circuits of Stator and Rotor

Now, we turn to the consideration of the stator and rotor slot leakage equivalent circuits. For this purpose, we proceed from the voltage equation for the stator and rotor windings obtained in (6.15). In relation to the stator winding emf $\dot{Э}_1$ and the rotor winding emf $\dot{Э}_2$, these equations acquires the form

$$\dot{Э}_1 = (r_1 + jx_{1\sigma})\dot{I}_1 - \dot{U}_1$$
$$\dot{Э}_2 = \left(\frac{r_2}{s} + jx_{2\sigma}\right)\dot{I}_2 + \frac{\dot{U}_2}{s} \qquad (12.38)$$

The Eq. (12.38) can be expressed in specific system of units. For this, the expressions obtained in (12.13) can be used.

12.4.1 Slot Leakage Equivalent Circuit of Stator

We consider the first equation of the system (12.38). With considering (12.13), the stator slot leakage emf equation can be reduced to the form

$$\frac{\dot{Э}_1}{k_{1Э}} = \left(\frac{r_1}{k_1} + j\frac{x_{1\sigma}}{k_1}\right)\frac{\dot{I}_1}{k_{1I}} - \frac{\dot{U}_1}{k_{1Э}} \qquad (12.39)$$

In (12.39), we introduce the ohmic resistance and the stator slot leakage reactance expressed in the specific system of units and determined as

$$r_1^* = \frac{r_1}{k_1} \quad x_{1\sigma}^* = \frac{x_{1\sigma}}{k_1} \qquad (12.40)$$

where $k_1 = \frac{k_{1Э}}{k_{1I}} = \frac{2m_1(w_1k_{w1})^2 l_0}{p\tau_0}$; $k_{1I} = \frac{p\tau_0}{m_1 w_1 k_{w1}}$; $k_{1Э} = 2w_1k_{w1}l_0$.

Here, r_1^* and $x_{1\sigma}^*$ represent the ohmic resistance and stator slot leakage reactance in the specific system of units.

In (12.39), the following conditions can be introduced

$$\dot{E}_1 = \frac{\sqrt{2}\dot{Э}_1}{k_{1Э}} \quad \dot{U}_1^* = \frac{\dot{U}_1}{k_{1Э}} \quad \dot{H}_1 = \frac{\sqrt{2}\dot{I}_1}{k_{1I}} \tag{12.41}$$

Then, from (12.39)–(12.41), we can receive

$$\dot{E}_1 = \left(r_1^* + jx_{1\sigma}^*\right)\dot{H}_1 - \dot{U}_1^* \tag{12.42a}$$

Here, the value $\dot{U}_1^*$ reflects the voltage on the stator winding terminals, and it is represented in the specific system of units.

In the Eq. (12.42a), the value of the voltage on the stator winding terminals $\dot{U}_1^*$ can be replaced by the voltage drop across the power supply network impedance Z_c^*, i.e. $\left(-\dot{U}_1^*\right) = Z_c^*\dot{H}_1$. As a result, the Eq. (12.42a) acquires

$$\dot{E}_1 = \left(r_1^* + jx_{1\sigma}^* + Z_c^*\right)\dot{H}_1 \tag{12.42b}$$

The Eq. (12.42b) can be visualized by the stator slot leakage equivalent circuit shown in Fig. 12.1b.

12.4.2 Slot Leakage Equivalent Circuit of Rotor

We consider the second equation of the system (12.38). It can be represented as

$$\frac{\dot{Э}_2}{k_{2Э}} = \left(\frac{r_2}{k_2 s} + j\frac{x_{2\sigma}}{k_2}\right)\frac{\dot{I}_2}{k_{2I}} + \frac{\dot{U}_2}{k_{2Э}s} \tag{12.43}$$

In (12.43), we introduce the rotor winding ohmic resistance and rotor slot leakage reactance expressed in the specific system of units and determined as

$$r_2^* = \frac{r_2}{k_2} \quad x_{2\sigma}^* = \frac{x_{2\sigma}}{k_2} \tag{12.44}$$

where $k_2 = \frac{k_{2Э}}{k_{2I}} = \frac{2m_2(w_2k_{w2})^2 l_0}{p\tau_0}$; $k_{2I} = \frac{p\tau_0}{m_2w_2k_{w2}}$; $k_{2Э} = 2w_2k_{w2}l_0$.

Here, r_2^* and $x_{2\sigma}^*$ represent the rotor winding ohmic resistance and rotor slot leakage reactance in the specific system of units.

In (12.43), we use the following conditions

$$\dot{E}_2 = \frac{\sqrt{2}\dot{Э}_2}{k_{2Э}} \quad \dot{U}_2^* = \frac{\dot{U}_2}{k_{2Э}} \quad \dot{H}_2 = \frac{\sqrt{2}\dot{I}_2}{k_{2I}} \tag{12.45}$$

Then, from (12.43)–(12.45), it is received as

$$\dot{E}_2 = \left(\frac{r_2^*}{s} + jx_{2\sigma}^*\right)\dot{H}_2 + \frac{\dot{U}_2^*}{s} \tag{12.46a}$$

In (12.46a), the value $\dot{U}_2^*$ reflects the voltage on the rotor winding terminals, and it is represented in the specific system of units.

In the Eq. (12.46a), the value of the voltage on the rotor winding terminals $\dot{U}_2^*/s$ can be replaced by the voltage drop across the load impedance $Z_{2н}^*$, i.e. $\left(\dot{U}_2^*/s\right) = Z_{2н}^*\dot{H}_2$. Now, the equation shown in Eq. (12.46a) can be represented as

$$\dot{E}_2 = \left(r_2^*/s + jx_{2\sigma}^* + Z_{2н}^*\right)\dot{H}_2 \tag{12.46b}$$

The Eq. (12.46b) can be represented as the rotor slot leakage equivalent circuit given in Fig. 12.1c.

Thus, in accordance with the Eqs. (12.37a), (12.42a), and (12.46a) compiled in relation to the electric field intensities ($\dot{E}_1$and $\dot{E}_2$) caused by the induced in the stator and rotor windings emf, we have the stator slot leakage equivalent circuit (Fig. 12.1b), the rotor slot leakage equivalent circuit (Fig. 12.1c) and the one-element machine magnetizing equivalent circuit (Fig. 12.1a). Below, we use these equivalent circuits to construct the machine equivalent circuit.

12.5 Equivalent Circuit of Machine: Voltage Equations

The equivalent circuits shown in Fig. 12.1 can be used to construct the machine equivalent circuit for what the cascade way of connecting is applied.

12.5.1 Cascade Way of Connecting: Application

In this case, the construction of the machine equivalent circuit is reduced to connecting the input terminals $c - d$ of the rotor slot leakage equivalent circuit (Fig. 12.1c) to the output terminals $c - d$ of the machine magnetizing equivalent circuit (Fig. 12.1a), and also to connecting the input terminals $a - b$ of the machine magnetizing equivalent circuit (Fig. 12.1a) to the input terminals $a - b$ of the stator slot leakage equivalent circuit (Fig. 12.1b). As noted earlier, such connection is known as the cascade way of connecting. Thus, in this case, the machine equivalent

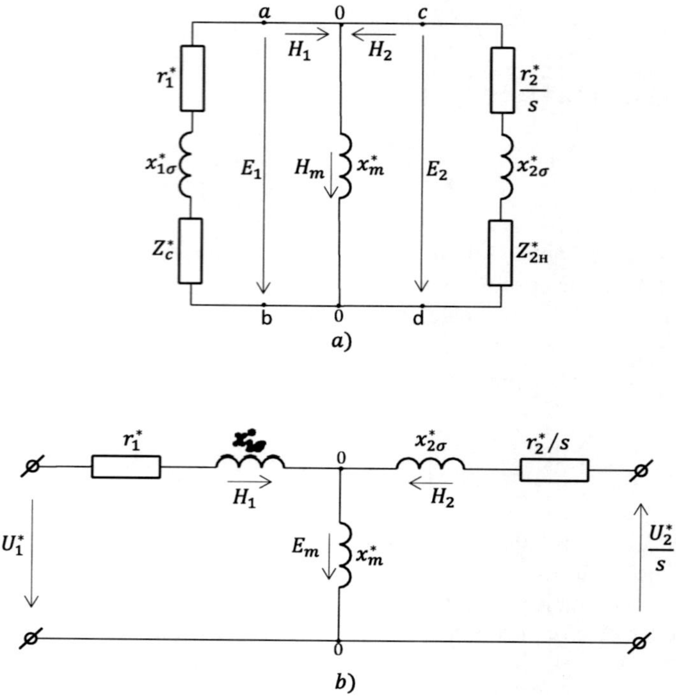

Fig. 12.2 Construction of machine equivalent circuit: (**a**) machine equivalent circuit built in relation to E_1 and E_2 of the emf induced in the stator and rotor windings; (**b**) machine equivalent circuit

circuit arises as a result of the cascade way of connecting the stator slot leakage equivalent circuit (Fig. 12.1b), the machine magnetizing equivalent circuit (Fig. 12.1a), and the rotor slot leakage equivalent circuit (Fig. 12.1c). Now, the machine equivalent circuit acquires the form shown in Fig. 12.2a. If in the equivalent circuit in Fig. 12.2a the voltage drop across the power supply network impedance Z_c^* to represent as $\left(-\dot{U}_1^*\right) = Z_c^* \dot{H}_1$ and also the voltage drop across the load impedance Z_{2H}^* to express as $\left(\dot{U}_2^*/s\right) = Z_{2H}^* \dot{H}_2$, then, the machine equivalent circuit can be given in Fig. 12.2b.

As shown above (Chap. 10), when using the phase quantities, the construction of the machine equivalent circuit is implemented by applying the referred rotor quantities. In the specific system of units, the need for the reduction of the rotor quantities to the number of the turns of the stator winding is eliminated. Therefore, the construction of the machine equivalent circuit by the use of the cascade way of connecting implemented in the specific system of units is simplified. Based on the equivalent circuit in Fig. 12.2b, the voltage equations for the machine windings acquire the following form

$$
\begin{aligned}
\dot{U}_1^* &= (r_1^* + jx_{1\sigma}^*)\dot{H}_1 + (-\dot{E}_m) \\
-\frac{\dot{U}_2^*}{s} &= \left(\frac{r_2^*}{s} + jx_{2\sigma}^*\right)\dot{H}_2 + (-\dot{E}_m) \\
(-\dot{E}_m) &= jx_m^*\dot{H}_m \quad \dot{H}_m = \dot{H}_1 + \dot{H}_2
\end{aligned}
\tag{12.47}
$$

Thus, in this case, the voltage equations for the stator and rotor windings are written out by the use of the machine equivalent circuit (Fig. 12.2b).

12.5.2 *Application of "Folded" Equivalent Circuits of Stator and Rotor*

In Fig. 12.1, the fragment equivalent circuits of the machine equivalent circuit are shown. We can use the equivalent circuits in Fig. 12.1 to construct the machine equivalent circuit in a different way. For this purpose, the machine magnetizing equivalent circuit in Fig. 12.1a can be decomposed in relation to the emf $\dot{E}_1 = \dot{E}_m = \dot{E}_2$ into the separate fragments. For this, the reactance x_m^* used in the parallel branch $0 - 0$ of the machine magnetizing equivalent circuit shown in Fig. 12.1a can be expressed through the parallel connected two reactances x_{m1}^* and x_{m2}^*. Then, the equivalent circuit shown in Fig. 12.1a can be reduced to the form shown in Fig. 12.3a. In Fig. 12.3a, it is shown the stator magnetizing equivalent circuit. In Fig. 12.3b, it is given the rotor magnetizing equivalent circuit. The equivalent circuits in Fig. 12.3a, b will be represented as the "folded" stator and rotor magnetizing equivalent circuits, respectively. In Fig. 12.3c, d, the "folded" stator and rotor slot leakage equivalent circuits are shown. The equivalent circuits in Fig. 12.3 can be used to construct the machine equivalent circuit.

"Folded" stator equivalent circuit. The stator equivalent circuit can be constructed by use of the fragment equivalent circuits given in Fig. 12.3a, c. For this, the terminals $a - b$ of the equivalent circuits in Fig. 12.3a, ccan be connected in the cascade way. As a result, the "folded" stator equivalent circuit takes the form in Fig. 12.4a.

"Folded" rotor equivalent circuit. The rotor equivalent circuit can be constructed by use of the fragment equivalent circuits given in Fig. 12.3b, d. For this purpose, the terminals $c - d$ of the equivalent circuits in Fig. 12.3b, d can be connected in the cascade way. Then, the "folded" rotor equivalent circuit acquires the form in Fig. 12.4b.

The equivalent circuits obtained in this way in Fig. 12.4 contain the reactances x_{m1}^* and x_{m2}^* which are taken into account the saturation of the stator and rotor magnetic circuits.

Machine equivalent circuit. In the presence of the stator and rotor equivalent circuits, the construction of the machine equivalent circuit is simplified. For this, in the stator and rotor equivalent circuits (Fig. 12.4), the parallel branches $0 - 0$

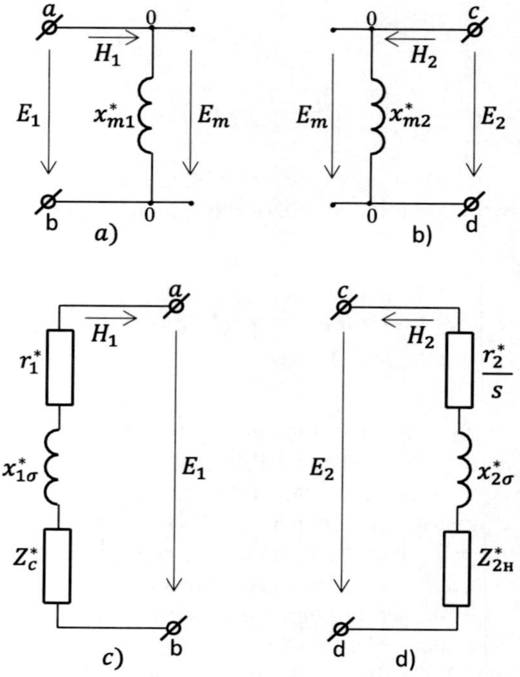

Fig. 12.3 Fragments of machine equivalent circuit: (**a**) stator magnetizing equivalent circuit; (**b**) rotor magnetizing equivalent circuit; (**c**) stator slot leakage equivalent circuit; (**d**) rotor slot leakage equivalent circuit

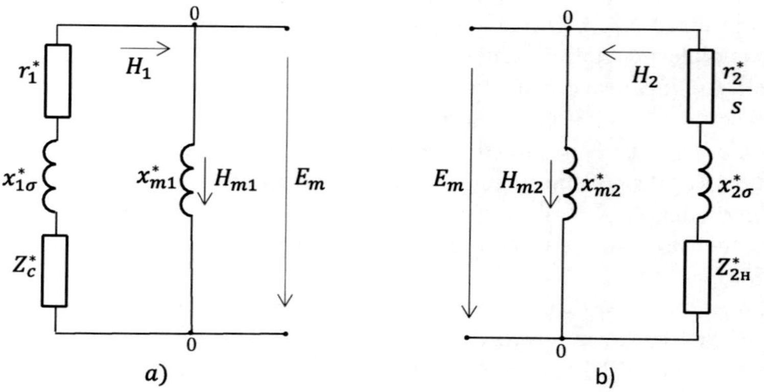

Fig. 12.4 Stator and rotor equivalent circuits in the specific system of units: (**a**) stator equivalent circuit; (**b**) rotor equivalent circuit

containing the reactances x^*_{m1} and x^*_{m2} should be combined into the one common branch with the total reactance x^*_m determined as $x^*_m = x^*_{m1}x^*_{m2}/\left(x^*_{m1}+x^*_{m2}\right)$. In addition, in them, it is necessary to use the conditions of kind $\left(-\dot{U}^*_1\right) = Z^*_c\dot{H}_1$ and $\dot{U}^*_2/s = Z^*_{2н}\dot{H}_2$. Now, the machine equivalent circuit can be received in the form in Fig. 12.2b.

Below, we will use this way of constructing the machine equivalent circuit. In accordance with this way, first, the stator and rotor equivalent circuits should be built, and then, using them, the machine equivalent circuit can be obtained.

12.5.3 Application of "Unfolded" Equivalent Circuits of Stator and Rotor

Here, the construction of the machine equivalent circuit is implemented by the use of the "unfolded" stator and rotor equivalent circuits. In this case, to obtain such stator and rotor equivalent circuits, the "unfolded" stator and rotor magnetizing equivalent circuits and the "unfolded" stator and rotor slot leakage equivalent circuits are applied.

"Unfolded" magnetizing equivalent circuits of stator and rotor. In Fig. 12.3a, b, the stator and rotor magnetizing equivalent circuits are shown. In the equivalent circuit in Fig. 12.3a, the stator magnetizing equivalent circuit is represented by the one parallel branch 0 – 0 including the one lumped reactance x^*_{m1}. In the equivalent circuit in Fig. 12.3b, the rotor magnetizing equivalent circuit is represented by the one parallel branch 0 – 0 containing the one lumped reactance x^*_{m2}. We will assume that in Fig. 12.3a, b, the "folded" stator and rotor magnetizing equivalent circuits are represented. Here, we will consider the "unfolded" stator and rotor magnetizing equivalent circuits which contain the magnetizing reactances reflecting the layers belonging to the models of the stator and rotor magnetizing circuits (Fig. F.1). Such, the stator and rotor magnetizing equivalent circuits corresponding to the machine model in Fig. F.1 are obtained in Appendix F (Fig. F.2). Using Fig. F.2a, the "unfolded" stator magnetizing equivalent circuit is represented in Fig. 12.5a. Applying Fig. F.2a, the "unfolded" rotor magnetizing equivalent circuit is shown in Fig. 12.6a. In the equivalent circuits in Figs. 12.5a and 12.6a, the reactances x^*_{01} and x^*_{02} represent the magnetizing reactances of the air gap sublayer adjoining to the stator bore surface and the air gap sublayers adjoining to the rotor surface, respectively. The equivalent circuits shown in Figs. 12.5a and 12.6a contain the reactances $x^*_{01}, x^*_{ks}, x^*_{zs}$ and $x^*_{02}, x^*_{kR}, x^*_{zR}$ representing the longitudinal magnetizing circuit loops of the layers reflecting the stator and the rotor magnetic circuits. The reactances x^*_{as} and x^*_{aR} used in the equivalent circuits in Figs. 12.5a and 12.6a represent the transverse magnetizing circuit loops of the layers reflecting the stator and rotor yoke areas.

The expressions for the reactances applied in the equivalent circuits in Figs. 12.5a and 12.6a are obtained in Appendix F. These expressions can be also received by the formulas given in the Chap. 11. From (F.9), for the reactances used in the longitudinal magnetizing circuit loops (Fig. 12.5a) of the layers representing the air gap sublayer adjoining to the stator bore surface, the stator slot-wedge area, and the stator winding location area, we can have

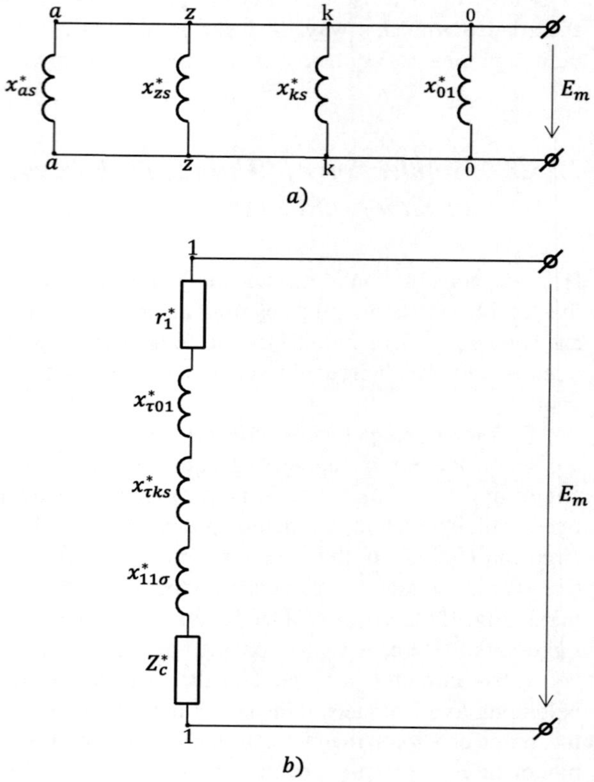

Fig. 12.5 "Unfolded" stator magnetizing equivalent circuit and the "unfolded" stator slot leakage equivalent circuit: (**a**) "unfolded" stator magnetizing equivalent circuit; (**b**) "unfolded" stator slot leakage equivalent circuit

$$x^*_{01} = \frac{\omega_1 \mu_0 \tau_0^2}{\pi^2 \delta_1}$$
$$x^*_{ks} = \frac{\omega_1 \mu_{yks} \tau_{ks}^2}{\pi^2 h_{ks}} \frac{\xi_{Lks}}{\xi_{\tau ks}} \qquad (12.48)$$
$$x^*_{zs} = \frac{\omega_1 \mu_{yzs} \tau_{zs}^2}{\pi^2 h_{zs}} \frac{\xi_{Lzs}}{\xi_{\tau zs}}$$

The reactance used in the transverse magnetizing circuit loop (Fig. 12.5a) representing the stator yoke layer is determined from (F.10) as

$$x^*_{as} = \omega_1 \mu_{as} t_0 \frac{h_{as}}{t_{as}} \xi_{Las} \qquad (12.49)$$

where $t_0 = 2p\tau_0/Z_1$; $t_{as} = 2p\tau_{as}/Z_1$.

Here, t_{as} represents the conditional tooth pitch determined on the inner stator yoke surface in relation to the stator teeth number.

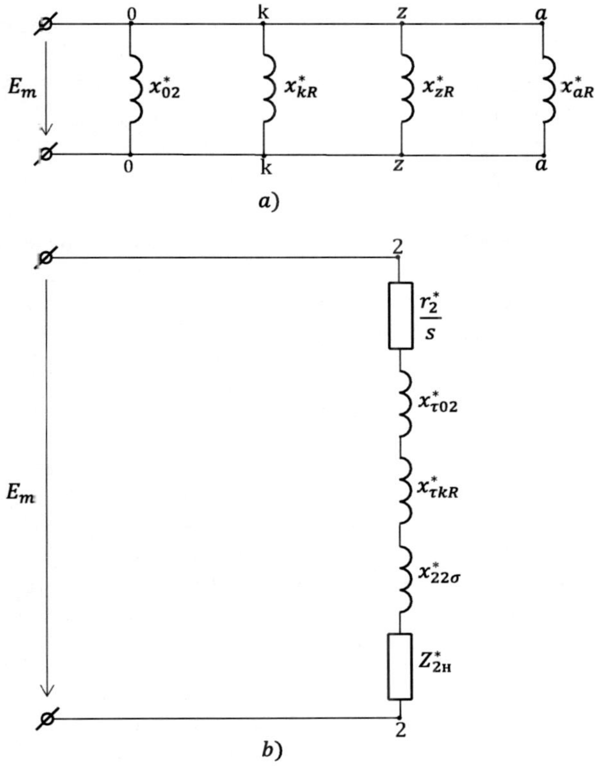

Fig. 12.6 "Unfolded" rotor magnetizing equivalent circuit and the "unfolded" rotor slot leakage equivalent circuit: (**a**) "unfolded" rotor magnetizing equivalent circuit; (**b**) "unfolded" rotor slot leakage equivalent circuit

From (F.11), for the reactances used in the longitudinal magnetizing circuit loops (Fig. 12.6a) of the layers reflecting the air gap sublayer adjoining to the rotor surface, the rotor slot-wedge area, and the rotor winding location area, we can receive

$$x^*_{02} = \frac{\omega_1 \mu_0 \tau_0^2}{\pi^2 \delta_2}$$
$$x^*_{kR} = \frac{\omega_1 \mu_{ykR} \tau_{kR}^2}{\pi^2 h_{kR}} \frac{\xi_{LkR}}{\xi_{\tau kR}} \quad (12.50)$$
$$x^*_{zR} = \frac{\omega_1 \mu_{yzR} \tau_{zR}^2}{\pi^2 h_{zR}} \frac{\xi_{LzR}}{\xi_{\tau zR}}$$

The reactance used in the transverse magnetizing circuit loop (Fig. 12.6a) reflecting the rotor yoke layer is determined from (F.12) as

$$x^*_{aR} = \omega_1 \mu_{aR} t_0 \frac{h_{aR}}{t_{aR}} k_R \xi_{LaR} \quad (12.51)$$

where $t_{aR} = 2p\tau_{aR}/Z_2$; $k_R = Z_1/Z_2$.

Here, t_{aR} represents the conditional tooth pitch determined on the outer rotor yoke surface in relation to the rotor teeth number.

The "unfolded" equivalent circuits obtained in Fig. 12.5a and Fig. 12.6a contain the distributed elements and they can be used to form the "folded" the stator and rotor magnetizing equivalent circuits.

"Unfolded" slot leakage equivalent circuits of stator and rotor. Here, we consider the slot leakage reactance of the stator winding $x^*_{1\sigma}$ and the slot leakage reactance of the rotor winding $x^*_{2\sigma}$ used in the equivalent circuits obtained in Fig. 12.3c, d. In these equivalent circuits, the slot leakage reactances $x^*_{1\sigma}$ and $x^*_{2\sigma}$ are determined in the "folded" (lumped) form. In Appendix F, the description of the way of determining the reactances $x^*_{1\sigma}$ and $x^*_{2\sigma}$ shown in the "unfolded" form is given. From (F.25), the rotor slot leakage reactance $x^*_{2\sigma}$ can be expressed as

$$x^*_{2\sigma} = x^*_{\tau 02} + x^*_{\tau kR} + x^*_{22\sigma} \tag{12.52}$$

The slot leakage reactance of the rotor winding $x^*_{2\sigma}$ can be considered as the sum of the three reactances: the leakage reactance $x^*_{\tau 02}$ of the air gap sublayer adjoining to the rotor surface, the leakage reactance $x^*_{\tau kR}$ of the layer of the rotor slot-wedge area, and the leakage reactance $x^*_{22\sigma}$ of the layer of the rotor winding location area. With considering the expressions (F.26) obtained in Appendix F, the components of the reactance $x^*_{2\sigma}$ can be represented as

$$\begin{aligned} x^*_{\tau 02} &= \omega_1 \mu_0 t_0 \frac{\delta}{2t_{02}} k_{2\sigma} k_R \\ x^*_{\tau kR} &= \omega_1 \mu_0 t_0 \frac{h_{kR}}{b_{kR}} k_R \xi_{LkR} \\ x^*_{22\sigma} &= \omega_1 \mu_0 t_0 \frac{h_{zR}}{3b_{nR}} k_R \xi_{LzR} \end{aligned} \tag{12.53}$$

The analogous expressions can be obtained for the components of the slot leakage reactance of the stator winding $x^*_{1\sigma}$. According to (F .27), the slot leakage reactance of the stator winding $x^*_{1\sigma}$ is represented as the sum of the three reactances: the leakage reactance $x^*_{\tau 01}$ of the air gap sublayer adjoining to the stator bore surface, the leakage reactance $x^*_{\tau ks}$ of the layer of the stator slot-wedge area, and the leakage reactance $x^*_{11\sigma}$ of the layer of the stator winding location area. In this regard, we have

$$x^*_{1\sigma} = x^*_{\tau 01} + x^*_{\tau ks} + x^*_{11\sigma} \tag{12.54}$$

For the reactances $x^*_{\tau 01}$, $x^*_{\tau ks}$ and $x^*_{11\sigma}$ on the basis of the expressions (F.28) obtained in Appendix F, we can receive

$$
\begin{aligned}
x^*_{\tau 01} &= \omega_1 \mu_0 t_0 \frac{\delta}{2t_0} k_{1\sigma} \\
x^*_{\tau ks} &= \omega_1 \mu_0 t_0 \frac{h_{ks}}{b_{ks}} \xi_{Lks} \\
x^*_{11\sigma} &= \omega_1 \mu_0 t_0 \frac{h_{zs}}{3b_{ns}} \xi_{Lzs}
\end{aligned}
\tag{12.55}
$$

The features of determining the slot leakage reactances of the stator winding $x^*_{1\sigma}$ and the rotor winding $x^*_{2\sigma}$ are described in Appendix F. As follows from Appendix F, the components of the reactances $x^*_{1\sigma}$ and $x^*_{2\sigma}$ are determined on the basis of representing stator and rotor slot leakage magnetic circuits subdivided into the layers (Fig. F.5b). The stator and rotor slot leakage magnetic circuits usually contain the air gap sublayers forming in relation to the surface $0 - 0$ passing through the middle of the radial length of the air gap, as well as the layers of the stator and rotor slot-wedge areas, and the layers of the stator and rotor winding location areas.

The equation for $\dot{E}_m$ of the emf induced in the rotor winding from (12.46b) and (12.52) can be expressed as

$$
\begin{aligned}
\dot{E}_m &= \left(r^*_2/s + jx^*_{2\sigma} + Z^*_{2H}\right)\dot{H}_2 \\
&= \left[j\left(x^*_{\tau 02} + x^*_{\tau kR} + x^*_{22\sigma}\right) + r^*_2/s + Z^*_{2H}\right]\dot{H}_2
\end{aligned}
\tag{12.56}
$$

where $x^*_{2\sigma} = x^*_{\tau 02} + x^*_{\tau kR} + x^*_{22\sigma}$.

In (12.56), the components of the reactance $x^*_{2\sigma}$ are determined by the expressions shown in (12.53).

The equation for $\dot{E}_m$ of the emf induced in the stator winding from (12.42b) and (12.54) can be represented as

$$
\begin{aligned}
\dot{E}_m &= \left(r^*_1 + jx^*_{1\sigma} + Z^*_c\right)\dot{H}_1 \\
&= \left[j\left(x^*_{\tau 01} - x^*_{\tau ks} + x^*_{11\sigma}\right) + r^*_1 + Z^*_c\right]\dot{H}_1
\end{aligned}
\tag{12.57}
$$

where $x^*_{1\sigma} = x^*_{\tau 01} + x^*_{\tau ks} + x^*_{11c}$.

In (12.57), the components of the reactance $x^*_{1\sigma}$ are determined by the expressions shown in (12.55).

The Eq. (12.56) describes the rotor slot leakage circuit loop. The Eq. (12.57) represents the stator slot leakage circuit loop. In Appendix F, the stator and rotor slot leakage equivalent circuits are obtained (Fig. F.6). With considering Fig. F.6b, the Eq. (12.56) can be visualized by the rotor slot leakage equivalent circuit in Fig. 12.6b. According to Fig. F.6a, the Eq. (12.57) can be represented as the stator slot leakage equivalent circuit in Fig. 12.5b.

The equivalent circuits shown in Fig. 12.5 reflect the "unfolded" stator magnetizing equivalent circuit and the "unfolded" stator slot leakage equivalent circuit. The equivalent circuits obtained in Fig. 12.6 represent the "unfolded" rotor magnetizing equivalent circuit and the "unfolded" rotor slot leakage equivalent circuit. In accordance with the "no-load and under-load" concept, these equivalent circuits can be used to construct the stator and rotor equivalent circuits reflecting the resulting magnetic fields taking place in the stator and rotor of the machine. In this case, the stator equivalent circuit corresponding to the stator resulting magnetic field can be obtained as a result of the superposition of the magnetizing equivalent circuit in Fig. 12.5a and the slot leakage equivalent circuit in Fig. 12.5b. The equivalent circuit obtained in this way represents the "unfolded" stator equivalent circuit. The equivalent circuit reflecting the rotor resulting magnetic field can be constructed as a result of the superposition of the magnetizing equivalent circuit in Fig. 12.6a and the slot leakage equivalent circuit in Fig. 12.6b. The equivalent circuit constructed in this way represents the "unfolded" rotor equivalent circuit. Below, we will show it.

"Unfolded" stator equivalent circuit. We consider the stator magnetizing equivalent circuit in Fig. 12.5a and the stator slot leakage equivalent circuit in Fig. 12.5b. The stator magnetizing equivalent circuit in Fig. 12.5a reflects the machine no-load mode. The stator slot leakage equivalent circuit in Fig. 12.5b corresponds to the machine under-load mode. In these conditions, to obtain the stator equivalent circuit corresponding to the resulting magnetic field, it is enough to connect the terminals of the parallel branch $z - z$ of the equivalent circuit in Fig. 12.5a to the terminals $1 - 1$ of the equivalent circuit in Fig. 12.5b. As a result, the stator equivalent circuit can be given in the form in Fig. 12.7. This equivalent circuit is the "unfolded" stator equivalent circuit, and it represents the multi-loop equivalent circuit with the parallel connection of the elements. In the equivalent circuit in Fig. 12.7, its transverse branches $0 - 0$, $k - k$, $z - z$, and $a - a$ represent the longitudinal magnetizing circuit loops of the layers reflecting the air gap sublayer adjoining to the stator bore surface, the stator slot-wedge area, and the stator winding location area and also the

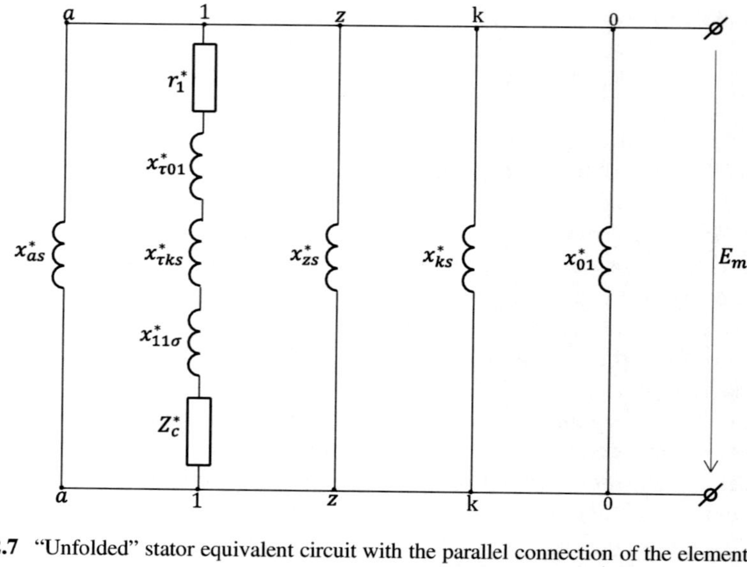

Fig. 12.7 "Unfolded" stator equivalent circuit with the parallel connection of the elements

transverse magnetizing circuit loop of the layer reflecting the stator yoke area. The transverse branch $1 - 1$ containing the impedance $\left[r_1^* + j\left(x_{\tau 01}^* + x_{\tau ks}^* + x_{11\sigma}^*\right) + Z_c^*\right]$ represents in the equivalent circuit in Fig. 12.7 the stator slot leakage circuit loop.

As it was noted above, the "unfolded" stator equivalent circuit represents the multi-loop equivalent circuit with the parallel connection of the elements. Such structure of the stator equivalent circuit is associated with the fact that the one-dimensional stator magnetizing field and the one-dimensional stator slot leakage magnetic field are considered. Thus, the stator equivalent circuit corresponding to the resulting magnetic field has a rather complex structure, and therefore, it is not convenient for practical use. In this connection, the equivalent circuit in Fig. 12.7 needs to be transformed which will be discussed below.

"Unfolded" rotor equivalent circuit. We proceed from the rotor magnetizing equivalent circuit in Fig. 12.6a and the rotor slot leakage equivalent circuit in Fig. 12.6b. The rotor magnetizing equivalent circuit in Fig. 12.6a reflects the machine no-load mode. The rotor slot leakage equivalent circuit in Fig. 12.6b corresponds to the machine under-load mode. In this regard, to receive the rotor equivalent circuit corresponding to the resulting magnetic field, it is enough to connect the terminals of the parallel branch $z - z$ of the equivalent circuit in Fig. 12.6a to the terminals $2 - 2$ of the equivalent circuit in Fig. 12.6b. Now, the rotor equivalent circuit can be shown in the form in Fig. 12.8. This equivalent circuit is the "unfolded" rotor equivalent circuit, and it represents the multi-loop equivalent circuit with the parallel connection of the elements. In the equivalent circuit in Fig. 12.8, its transverse branches $0 - 0$, $k - k$, $z - z$, and $a - a$ represent the longitudinal magnetizing circuit loops of the layers representing the air gap sublayer adjoining to the rotor surface, the rotor slot-wedge area, and the rotor winding location area and also the transverse magnetizing circuit loop of the layer reflecting

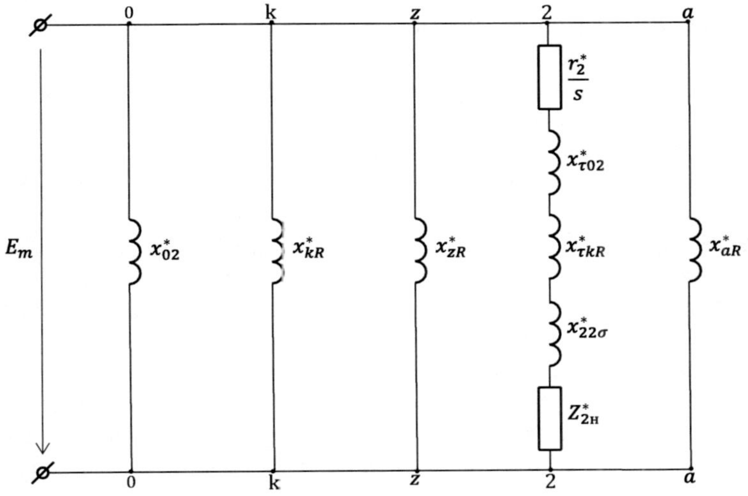

Fig. 12.8 "Unfolded" rotor equivalent circuit with the parallel connection of the elements

the rotor yoke area. The transverse branch 2 – 2 containing the impedance $\left[r_2^*/s + j\left(x_{\tau 02}^* + x_{\tau kR}^* + x_{22\sigma}^*\right) + Z_{2н}^*\right]$ reflects in the equivalent circuit in Fig. 12.8 the rotor slot leakage circuit loop.

As it was noted above, the "unfolded" rotor equivalent circuit represents the multi-loop equivalent circuit with the parallel connection of the elements. Such structure of the rotor equivalent circuit is associated with the fact that the one-dimensional rotor magnetizing field and the one-dimensional rotor slot leakage magnetic field are considered. Thus, the rotor equivalent circuit corresponding to the resulting magnetic field has a rather complex structure, and therefore, it is not convenient for practical use. In this regard, the equivalent circuit in Fig. 12.8 needs to be transformed which will be discussed below.

Magnetizing and slot leakage circuit loops. "Folded" stator equivalent circuit. We proceed from the equivalent circuit obtained in Fig. 12.7. In the equivalent circuit in Fig. 12.7, the parallel branches 0 – 0, $k - k$, $z - z$ and $a - a$ represent the magnetizing circuit loops of the layers reflecting the stator magnetic circuit. In this regard, they can be combined into the one common circuit loop that will represent the stator magnetizing circuit loop. The reactance used in the stator magnetizing circuit loop reflects the stator magnetizing reactance. Based on the equivalent circuit in Fig. 12.7, the stator magnetizing reactance can be expressed as

$$x_{m1}^* = \frac{1}{\frac{1}{x_{01}^*} + \frac{1}{x_{ks}^*} + \frac{1}{x_{zs}^*} + \frac{1}{x_{as}^*}} = \frac{x_{01}^*}{1 + x_{01}^*\left(\frac{1}{x_{ks}^*} + \frac{1}{x_{zs}^*} + \frac{1}{x_{as}^*}\right)} = \frac{x_{01}^*}{k_{н1}} \tag{12.58}$$

where $k_{н1} = 1 + x_{01}^*\left(\frac{1}{x_{ks}} + \frac{1}{x_{zs}^*} + \frac{1}{x_{as}^*}\right)$.

Here, the reactance x_{m1}^* represents the stator magnetizing reactance, and the value of $k_{н1}$ reflects the saturation factor of the stator magnetic circuit.

In the equivalent circuit in Fig. 12.7, the parallel branch 1 – 1 represents the stator slot leakage circuit loop. In accordance with (12.54), the sum of the reactances $x_{\tau 01}^*, x_{\tau ks}^*$ and $x_{11\sigma}^*$ used in the circuit loop 1 – 1 reflects the stator slot leakage reactance for which we have that $x_{1\sigma}^* = x_{\tau 01}^* + x_{\tau ks}^* + x_{11\sigma}^*$. Applying in the parallel branch 1 – 1 of the equivalent circuit in Fig. 12.7 the condition (12.54) for the reactance $x_{1\sigma}^*$ and taking into account the equation for the reactance x_{m1}^* given in (12.58), the equivalent circuit in Fig. 12.7 can be reduced to the form in Fig. 12.4a. In this case, the equivalent circuit in Fig. 12.4a represents the "folded" stator equivalent circuit that arises from the "unfolded" stator equivalent circuit with the parallel connection of the elements shown in Fig. 12.7. As follows from the equivalent circuit in Fig. 12.4a, in relation to the $\dot{E}_m$, the stator is represented as the two circuit loops: the magnetizing circuit loop containing the magnetizing reactance x_{m1}^* and the slot leakage circuit loop with the slot leakage reactance $x_{1\sigma}^*$. The "folded" stator equivalent circuit in Fig. 12.4a has a simple structure, and therefore, it is convenient for practical use.

In the "folded" stator equivalent circuit in Fig. 12.4a, the reactance $x_{1\sigma}^*$ which represents the stator slot leakage reactance is used. This reactance consists of three components: the leakage reactance $x^*_{\tau 01}$ of the air gap sublayer adjoining to the stator bore surface, the leakage reactance $x^*_{\tau ks}$ of the layer of the stator slot-wedge area, and leakage reactance $x^*_{11\sigma}$ of the layer of the stator winding location area. As follows from (12.55), the reactances $x^*_{\tau 01}, x^*_{\tau ks}$ and $x^*_{11\sigma}$ are determined as

$$
\begin{aligned}
x^*_{\tau 01} &= \omega_1 \mu_0 t_0 \frac{\delta}{2t_0} k_{1\sigma} \\
x^*_{\tau ks} &= \omega_1 \mu_{xks} t_0 \frac{h_{ks}}{t_{ks}} \xi_{Lks} \\
x^*_{11\sigma} &= \omega_1 \mu_{xzs} t_0 \frac{h_{zs}}{3t_{zs}} \xi_{Lzs}
\end{aligned}
\tag{12.59}
$$

In (12.59), we can use the conditions of kind following from (F.14) as

$$
\mu_{xks} = \mu_0 \frac{t_{ks}}{b_{ks}} \quad \mu_{xzs} = \mu_0 \frac{t_{zs}}{b_{ns}} \tag{12.60}
$$

Now, from (12.59) and (12.60), the reactances $x^*_{\tau 01}, x^*_{\tau ks}$ and $x^*_{11\sigma}$ can be expressed in the form

$$
\begin{aligned}
x^*_{\tau 01} &= \omega_1 \mu_0 t_0 \frac{\delta}{2t_0} k_{1\sigma} = \omega_1 \mu_0 t_0 \lambda_{\tau 01} \\
x^*_{\tau ks} &= \omega_1 \mu_0 t_0 \frac{h_{ks}}{b_{ks}} \xi_{Lks} = \omega_1 \mu_0 t_0 \lambda_{\tau ks} \\
x^*_{11\sigma} &= \omega_1 \mu_0 t_0 \frac{h_{zs}}{3b_{ns}} \xi_{Lzs} = \omega_1 \mu_0 t_0 \lambda_{11\sigma}
\end{aligned}
\tag{12.61}
$$

where $\lambda_{\tau 01} = \frac{\delta}{2t_0} k_{1\sigma}; \lambda_{\tau ks} = \frac{h_{ks}}{b_{ks}} \xi_{Lks}; \lambda_{11\sigma} = \frac{h_{zs}}{3b_{ns}} \xi_{Lzs}$.

In (12.61), the dimensionless quantities $\lambda_{\tau 01}$, $\lambda_{\tau ks}$, and $\lambda_{11\sigma}$ represent the relative magnetic conductivities for the tangential components of the magnetic fluxes in the layers reflecting the air gap sublayer adjoining to the stator bore surface, the stator slot-wedge area, and the stator winding location area, respectively.

Now, from (12.54) and (12.61), the reactance $x^*_{1\sigma}$ can be expressed in the form

$$
\begin{aligned}
x^*_{1\sigma} &= x^*_{\tau 01} + x^*_{\tau ks} + x^*_{11\sigma} \\
&= \omega_1 \mu_0 t_0 (\lambda_{\tau 01} + \lambda_{\tau ks} + \lambda_{11\sigma}) = \omega_1 \mu_0 t_0 \lambda_{1\sigma}
\end{aligned}
\tag{12.62}
$$

where $\lambda_{1\sigma} = \lambda_{\tau 01} + \lambda_{\tau ks} + \lambda_{11\sigma}$.

Here, the quantity $\lambda_{1\sigma}$ reflects the relative magnetic conductivity for the slot leakage magnetic flux of the stator winding.

Magnetizing and slot leakage circuit loops. "Folded" rotor equivalent circuit. We use the equivalent circuit obtained in Fig. 12.8. In the equivalent circuit in Fig. 12.8, the parallel branches $0-0$, $k-k$, $z-z$ and $a-a$ represent the magnetizing circuit loops of the layers reflecting the rotor magnetic circuit. In this regard, they can be combined into the one common circuit loop that will represent the rotor magnetizing circuit loop. The reactance used in the rotor magnetizing circuit loop reflects the rotor magnetizing reactance. Based on the equivalent circuit in Fig. 12.8, the rotor magnetizing reactance can be expressed as

$$x_{m2}^{*}=\frac{1}{\frac{1}{x_{02}^{*}}+\frac{1}{x_{kR}^{*}}+\frac{1}{x_{zR}^{*}}+\frac{1}{x_{aR}^{*}}}=\frac{x_{02}^{*}}{1+x_{02}^{*}\left(\frac{1}{x_{kR}^{*}}+\frac{1}{x_{zR}^{*}}+\frac{1}{x_{aR}^{*}}\right)}=\frac{x_{02}^{*}}{k_{н2}} \tag{12.63}$$

where $k_{\text{н}2}=1+x_{02}^{*}\left(\frac{1}{x_{kR}^{*}}+\frac{1}{x_{zR}^{*}}+\frac{1}{x_{aR}^{*}}\right)$.

Here, the reactance x_{m2}^{*} represents the rotor magnetizing reactance, and the value of $k_{н2}$ reflects the saturation factor of the rotor magnetic circuit.

In the equivalent circuit in Fig. 12.8, the parallel branch 2 – 2 represents the rotor slot leakage circuit loop. In accordance with (12.52), the sum of the reactances $x_{\tau 02}^{*}, x_{\tau kP}^{*}$ and $x_{22\sigma}^{*}$ used in the circuit loop 2 – 2 reflects the rotor slot leakage reactance $x_{2\sigma}^{*}$ for which we have that $x_{2\sigma}^{*}=x_{\tau 02}^{*}+x_{\tau kR}^{*}+x_{22\sigma}^{*}$. Applying in the parallel branch 2 – 2 of the equivalent circuit in Fig. 12.8 the condition (12.52) for the reactance $x_{2\sigma}^{*}$ and taking into account the equation for the reactance x_{m2}^{*} given in (12.63), the equivalent circuit in Fig. 12.8 can be reduced to the form in Fig. 12.4b. In this case, the equivalent circuit in Fig. 12.4b represents the "folded" rotor equivalent circuit that arises from the "unfolded" rotor equivalent circuit with the parallel connection of the elements shown in Fig. 12.8. As follows from the equivalent circuit in Fig. 12.4b, in relation to the $\dot{E}_m$ the machine rotor is represented as the two circuit loops: the magnetizing circuit loop containing the magnetizing reactance x_{m2}^{*} and the slot leakage circuit loop with the slot leakage reactance $x_{2\sigma}^{*}$. The "folded" rotor equivalent circuit in Fig. 12.4b has a simple structure, and therefore, it is convenient for practical use.

In the "folded" rotor equivalent circuit in Fig. 12.4b, the reactance $x_{2\sigma}^{*}$ which represents the rotor slot leakage reactance is used. This reactance consists of three components: the leakage reactance $x_{\tau 02}^{*}$ of the air gap sublayer adjoining to the rotor surface, the leakage reactance $x_{\tau kR}^{*}$ of the layer of the rotor slot-wedge area, and leakage reactance $x_{22\sigma}^{*}$ of the layer of the rotor winding location area. As follows from (12.53), the reactances $x_{\tau 02}^{*}, x_{\tau kR}^{*}$ and $x_{22\sigma}^{*}$ are determined as

$$\begin{aligned}
x_{\tau 02}^{*}&=\omega_1\mu_0 t_0\frac{\delta}{2t_{02}}k_{2\sigma}k_R\\
x_{\tau kR}^{*}&=\omega_1\mu_{xkR}t_0\frac{h_{kR}}{t_{kR}}k_R\xi_{LkR}\\
x_{22\sigma}^{*}&=\omega_1\mu_{xzR}t_0\frac{h_{zR}}{3t_{zR}}k_R\xi_{LzR}
\end{aligned} \tag{12.64}$$

In (12.64), we can use the conditions following from (F.13) as

$$\mu_{xkR} = \mu_0 \frac{t_{kR}}{b_{kR}} \quad \mu_{xzR} = \mu_0 \frac{t_{zR}}{b_{nR}} \tag{12.65}$$

Now, from (12.64) and (12.65), the reactances $x^*_{\tau 02}, x^*_{\tau kR}$ and $x^*_{22\sigma}$ can be represented in the form

$$\begin{aligned} x^*_{\tau 02} &= \omega_1 \mu_0 t_0 \frac{\delta}{2t_{02}} k_{2\sigma} k_R = \omega_1 \mu_0 t_0 \lambda_{\tau 02} \\ x^*_{\tau kR} &= \omega_1 \mu_0 t_0 \frac{h_{kR}}{b_{kR}} k_R \xi_{LkR} = \omega_1 \mu_0 t_0 \lambda_{\tau kR} \\ x^*_{22\sigma} &= \omega_1 \mu_0 t_0 \frac{h_{zR}}{3b_{nR}} k_R \xi_{LzR} = \omega_1 \mu_0 t_0 \lambda_{22\sigma} \end{aligned} \tag{12.66}$$

where $\lambda_{\tau 02} = \frac{\delta}{2t_{02}} k_{2\sigma} k_R; \lambda_{\tau kR} = \frac{h_{kR}}{b_{kR}} k_R \xi_{LkR}; \lambda_{22\sigma} = \frac{h_{zR}}{3b_{\text{п}R}} k_R \xi_{LzR}$.

In (12.66), the dimensionless quantities $\lambda_{\tau 02}$, $\lambda_{\tau kR}$, and $\lambda_{22\sigma}$ represent the relative magnetic conductivities for the tangential components of the magnetic fluxes in the layers reflecting the air gap sublayer adjoining to the rotor surface, the rotor slot-wedge area, and the rotor winding location area, respectively.

Now, from (12.52) and (12.66), the reactance $x^*_{2\sigma}$ can be expressed in the form

$$\begin{aligned} x^*_{2\sigma} &= x^*_{\tau 02} + x^*_{\tau kR} + x^*_{22\sigma} \\ &= \omega_1 \mu_0 t_0 (\lambda_{\tau 02} + \lambda_{\tau kR} + \lambda_{22\sigma}) = \omega_1 \mu_0 t_0 \lambda_{2\sigma} \end{aligned} \tag{12.67}$$

where $\lambda_{2\sigma} = \lambda_{\tau 02} + \lambda_{\tau kR} + \lambda_{22\sigma}$.

Here, the quantity $\lambda_{2\sigma}$ reflects the relative magnetic conductivity for the slot leakage magnetic flux of the rotor winding.

Machine equivalent circuit. As it was shown above, the "unfolded" stator and rotor equivalent circuits with the parallel connection of the elements can be represented as the "folded" stator and rotor equivalent circuits given in Fig. 12.4. Here, we use the "folded" stator and rotor equivalent circuits to construct the machine equivalent circuit. For this, based on the equivalent circuits in Fig. 12.4, it is necessary to introduce the machine magnetizing reactance x^*_m determined as $x^*_m = x^*_{m1} x^*_{m2} / (x^*_{m1} + x^*_{m2})$ and for the impedances Z^*_c and Z^*_{2H} to use the conditions of a kind $(-\dot{U}^*_1) = Z^*_c \dot{H}_1$ and $\dot{U}^*_2 / s = Z^*_{2H} \dot{H}_2$. Then, the machine equivalent circuit acquires the form shown in Fig. 12.2b.

12.6 L-Shaped Equivalent Circuit of Machine: Specific System of Units

The equivalent circuit given in Fig. 12.2b is known as the T-shaped equivalent circuit represented for the asynchronous machine in the specific system of units. By the equivalent circuit in Fig. 12.2b, we can obtain the L-shaped machine equivalent

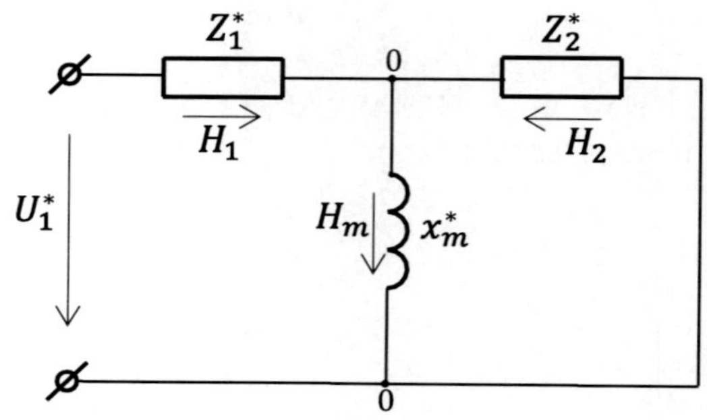

Fig. 12.9 T-shaped machine equivalent circuit

circuit for what in the equivalent circuit in Fig. 12.2b its parallel branch 0 – 0 (magnetizing branch) can be reduced to the terminals of the voltage $\dot{U}_1^*$, for example, [10]. In addition, in the equivalent circuit in Fig. 12.2b, the conditions of kind $Z_1^* = r_1^* + jx_{1\sigma}^*, Z_2^* = r_2^*/s + jx_{2\sigma}^*$ and $\dot{U}_2^*/s = 0$ should be used. Now, the equivalent circuit in Fig. 12.2b acquires the form in Fig. 12.9. In this equivalent circuit, we can implement the transferring the parallel branch 0 – 0 to the terminals of the voltage $\dot{U}_1^*$ for what the transfer technique described in the Chap. 9 can be used. In this considered case, the transfer technique is based on the representation of the quantity $\dot{H}_1$ used in the equivalent circuit in Fig. 12.9 in the form

$$\dot{H}_1 = \dot{H}_m' + (-\dot{H}_2') \tag{12.68}$$

In (12.68), the quantity $\dot{H}_m'$ is determined by the equivalent circuit in Fig. 12.9 when in it the condition of kind $\dot{H}_2 = 0$ is implemented (machine no-load mode). Then, for the value of $(-\dot{H}_2')$, we can have

$$(-\dot{H}_2') = \dot{H}_1 - \dot{H}_m' \tag{12.69}$$

Here, the value of $\dot{H}_1$ is determined by the equivalent circuit in Fig. 12.9 and by the expression obtained in (10.10), and it can be represented as

$$\dot{H}_1 = \dot{U}_1^* \frac{Z_2^* + jx_m^*}{jx_m^*\left(Z_1^* + Z_2^*\dot{c}_1\right)} \tag{12.70}$$

where $\dot{c}_1 = 1 + Z_1^*/jx_m^*$.

We will find the quantities $\dot{H}_m'$ and $(-\dot{H}_2')$. Based on the equivalent circuit in Fig. 12.9, the quantity $\dot{H}_m'$ acquires the form (when $\dot{H}_2 = 0$)

$$\dot{H}_m' = \frac{\dot{U}_1^*}{Z_1^* + jx_m^*} = \frac{\dot{U}_1^*}{jx_m^*\dot{c}_1} \tag{12.71}$$

The quantity $(-\dot{H}_2')$ follows from the Eq. (12.69), and it is determined as $(-\dot{H}_2') = \dot{H}_1 - \dot{H}_m'$. Therefore, from (12.70) and (12.71), we can obtain

$$(-\dot{H}_2') = \dot{H}_1 - \dot{H}_m' = \frac{\dot{U}_1^*}{Z_1^*\dot{c}_1 + Z_2^*\dot{c}_1^2} \tag{12.72}$$

From the Eqs. (12.68), (12.71), and (12.72), the quantity $\dot{H}_1$ can be expressed as

$$\dot{H}_1 = \dot{H}_m' + (-\dot{H}_2') = \frac{\dot{U}_1^*}{Z_1^* + jx_m^*} + \frac{\dot{U}_1^*}{Z_1^*\dot{c}_1 + Z_2^*\dot{c}_1^2} \tag{12.73}$$

The Eq. (12.73) corresponds to the L-shaped equivalent circuit shown in Fig. 12.10. It can be shown that by an analogy with the condition obtained in (10.15) for the quantities $\dot{H}_2$ and $\dot{H}_2'$ used in the equivalent circuits in Figs. 12.9 and 12.10, we can receive

$$\dot{H}_2 = \dot{c}_1\dot{H}_2' \tag{12.74}$$

In practice, for the complex transfer factor $\dot{c}_1$, the approximate condition is often used. Then, we can have

$$c_1 \approx 1 + \frac{x_{1\sigma}^*}{x_m^*} \tag{12.75}$$

Based on (12.72) and (12.74) and also keeping in mind that $Z_1^* = r_1^* + jx_{1\sigma}^*$ and $Z_2^* = r_2^*/s + jx_{2\sigma}^*$, the quantity $\dot{H}_2$ can be represented as

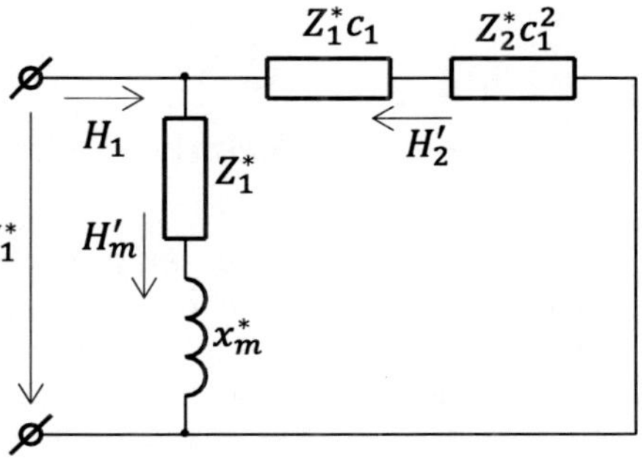

Fig. 12.10 L-shaped machine equivalent circuit

$$\dot{H}_2 = c_1\dot{H}_2' = \frac{\dot{U}_1^*}{\sqrt{\left(r_1^* + c_1\frac{r_2^*}{s}\right)^2 + \left(x_{1\sigma}^* + c_1 x_{2\sigma}^*\right)^2}} \tag{12.76}$$

As follows from (12.76), the calculation results obtained by the L-shaped equivalent circuit shown in Fig. 12.10 can be used to define the value of $\dot{H}_2$ applied in the equivalent circuit in Fig. 12.9.

12.7 Voltage Equations in Specific System of Units: "Inductively Coupled Circuits" Concept

In the Chap. 10 of this work, the voltage equations obtained on the basis of the "inductively coupled circuits" concept and expressed through the referred rotor quantities were considered. In this case, the voltage equations were represented in the phase system of units $(\dot{Э}, \dot{I})$. The machine voltage equations obtained in this way are used in the practice. Below, the voltage equations, we represent in the specific system of units $(\dot{E}, \dot{H})$. For this purpose, we proceed from the system of the equations given in (12.47). In the Eq. (12.47), for the quantity $\dot{H}_m$, the condition of kind $\dot{H}_m = \dot{H}_1 + \dot{H}_2$ can be used. Then, the Eq. (12.47) can be reduced to the form

$$\begin{aligned} \dot{U}_1^* &= r_1^*\dot{H}_1 + j(x_{1\sigma}^* + x_m^*)\dot{H}_1 + jx_m^*\dot{H}_2 = r_1^*\dot{H}_1 + jx_1^*\dot{H}_1 + jx_{21}^*\dot{H}_2 \\ -\frac{\dot{U}_2^*}{s} &= \frac{r_2^*}{s}\dot{H}_2 + j(x_{2\sigma}^* + x_m^*)\dot{H}_2 + jx_m^*\dot{H}_2 = \frac{r_2^*}{s}\dot{H}_2 + jx_2^*\dot{H}_2 + jx_{12}^*\dot{H}_2 \end{aligned} \tag{12.77}$$

where $x_1^* = x_{1\sigma}^* + x_m^*; x_2^* = x_{2\sigma}^* + x_m^*; x_{21}^* = x_{12}^* = x_m^*$.

The system of the Eq. (12.77) corresponds to the "inductively coupled circuits" concept. In this case, the reactances used in them are determined based on the "no-load and under-load" concept, and they are expressed in the specific system of units $(\dot{E}, \dot{H})$. As follows from (12.77), in the specific system of units, the voltage equations do not contain the referred rotor quantities.

12.8 System of Relative Units

The machine voltage equations can be written in the system of physical units and also they can be expressed in the system of relative units represented as the fractions of some quantities called the base quantities. Here, the base quantities are supplied with the index σ and have the same physical dimension. In this case, the machine voltage equations are said to be written in relative units.

At present, the system of physical units is represented by the international SI system. Using this system of units allows you to express the quantities in the named units which makes it easier to understand the physical processes taking place in the machine. However, in practice, the system of relative units has become widespread. This system of units facilitates the calculations and makes it easy to compare the results of the calculations performed for the different types of electric machines.

In this work, the values expressed both in the phase system of units and in the specific system of units are used. In the case of using the phase system of units, the rated values are usually taken as the base quantities of the asynchronous machine. For the stator winding, the base quantities are the rated phase current, i.e. $\dot{I}_\sigma = \dot{I}_r$, the rated phase voltage, i.e. $\dot{U}_\sigma = \dot{U}_r$, the rated angular frequency, i.e. $\omega_\sigma = \omega_r = 2\pi f_r$, where f_r is the rated power supply network frequency, the rated machine complex power, i.e. $S_\sigma = S_r = m_1 \dot{U}_r \dot{I}_r$, the rated stator impedance, i.e. $z_\sigma = z_r = \dot{U}_r / \dot{I}_r$ and the rated machine torque, i.e. $M_\sigma = M_r = P_r p/\omega_r$. From the point of view of the practice of the calculations, for the rotor windings, it is convenient to use the base quantities applied for the stator winding. In this case, the rotor values should be reduced to the number of turns and phases of the stator winding. As it was noted earlier, the system of relative units described above is used in the case when the voltage equations of the machine are represented in the phase system of units.

Further, we consider the system of relative units in the case of using the specific system of units. For the stator winding, the base-specific quantities are the rated magnetic field intensity, i.e. $\dot{H}_\sigma = \dot{H}_r$, where $\dot{H}_r = \dot{\varkappa}_r$, the rated electric field intensity, i.e. $\dot{E}_\sigma = \dot{E}_r$, where $\dot{E}_r = \dot{B}_r(\omega_r \tau_0 / \pi)$, the rated machine-specific complex power, i.e. $S^*_\sigma = S^*_r = 0.5 \dot{E}_r \dot{H}_r$, the rated specific impedance, $z^*_\sigma = z^*_r = \dot{E}_r / \dot{H}_r$. The value of $\dot{\varkappa}_r$ used above represents the linear density of the surface current of the stator winding.

In the specific system of units, for the rotor winding, the base-specific quantities used for the stator winding are convenient to apply. In this case, the rotor-specific quantities should not be reduced to the number of turns and phases of the stator winding.

12.9 Summary

1. In this chapter, the three techniques for constructing the machine equivalent circuit are described. In this case, the specific system of units is used. The characteristic feature of the proposed techniques for constructing the machine equivalent circuit is the use of the fragment equivalent circuits into which the machine equivalent circuit is decomposed. The fragment equivalent circuits are determined in relation to the electric and magnetic field intensities (E, H) given on the "smooth" surfaces of the layers reflecting the stator and rotor models, and they are formed by the use of the "no-load and under-load" concept. In these conditions, to obtain the machine equivalent circuit, the fragment equivalent

circuits can be connected in a cascade. Consequently, proposed in this chapter the techniques for constructing the machine equivalent circuit are based on the use of the superposition principle, the "fragmentation" concept, the "no-load and under-load" concept and also applying the cascade way of connecting.

2. When using the first technique, as the fragment equivalent circuits are considered the machine magnetizing equivalent circuit, the stator slot leakage equivalent circuit, and the rotor slot leakage equivalent circuit. In this case, the machine magnetizing equivalent circuit and the rotor slot leakage equivalent circuit are built on the basis of the use of the "no-load and under-load" concept. The construction of the machine equivalent circuit is reduced to the cascade way of connecting the machine magnetizing equivalent circuit, the stator slot leakage equivalent circuit, and the rotor slot leakage equivalent circuit.
3. When using the second technique, as the fragment equivalent circuits are considered the stator magnetizing equivalent circuit, and the stator slot leakage equivalent circuit, and also the rotor magnetizing equivalent circuit, and the rotor slot leakage equivalent circuit. In this case, the magnetizing equivalent circuits of the stator and rotor are defined as a result of the decomposition of the machine magnetizing equivalent circuit. The construction of the machine equivalent circuit is implemented in the two stages. At the first stage, the stator and rotor equivalent circuits are built. The stator equivalent circuit is built as a result of a cascade way of connecting the stator magnetizing equivalent circuit and the stator slot leakage equivalent circuit. The rotor equivalent circuit is built as a result of the cascade way of connecting the rotor magnetizing equivalent circuit and the rotor slot leakage equivalent circuit. The equivalent circuits obtained in this way represent the "folded" equivalent circuits of the stator and rotor. Further, the "folded" equivalent circuits of the stator and rotor are combined to introduce the machine magnetizing circuit loop and determine the machine magnetizing reactance. As a result, the machine equivalent circuit is followed.
4. The third technique is the further development of the second technique used to construct the machine equivalent circuit. Here, the construction of the "folded" equivalent circuits of the stator and rotor is implemented in the two stages. At the first stage, the "unfolded" equivalent circuits of the stator and rotor are obtained. In this chapter, such equivalent circuits are determined as a result of the superposition of the "unfolded" magnetizing equivalent circuits of the stator and rotor and also the "unfolded" slot leakage equivalent circuits of the stator and rotor. The equivalent circuits obtained in this way represent the "unfolded" stator and rotor equivalent circuits with the parallel connection of the elements. At the second stage, based on considering the "unfolded" stator and rotor equivalent circuits with the parallel connection of the elements, the stator and rotor magnetizing circuit loops and the stator and rotor slot leakage circuit loops are introduced. As a result of this procedure, the "folded" stator and rotor equivalent circuits are followed and they can be used to construct the machine equivalent circuit.
5. The "folded" stator and rotor equivalent circuits are the basis for building the machine equivalent circuit. By the "folded" stator and rotor equivalent circuits,

the machine magnetizing circuit loop is introduced, and the machine magnetizing reactance is determined as a result of which the machine equivalent circuit is followed. In relation to $\dot{E}_m$ representing the electric field intensity on the air gap surface $0 - 0$, the machine equivalent circuit consists of the three circuit loops: the machine magnetizing circuit loop, the stator slot leakage circuit loop, and the rotor slot leakage circuit loop.

6. In this chapter, the one-dimensional magnetizing field and one-dimensional slot leakage magnetic fields are considered, and the specific system of units is applied. In these conditions, the stator and rotor magnetizing circuits and the stator and rotor slot leakage magnetic circuits are represented subdivided into the corresponding layers. The multi layer models significantly simplify the process of calculating the magnetic circuits of the electric machine. We note that when using the phase system of units, the machine magnetic circuits are subdivided into the corresponding sections.
7. The "unfolded" magnetizing equivalent circuits of the stator and rotor correspond to the no-load mode. The "unfolded" slot leakage equivalent circuits of the stator and rotor reflect the under-load mode. In accordance with "no-load and under-load" concept, for these equivalent circuits, the superposition principle can be applied. As a result of which, the "unfolded" stator and rotor equivalent circuits representing the machine resulting magnetic field can be received. In our case, the equivalent circuits obtained in this way represent the "unfolded" stator and rotor equivalent circuits with the parallel connection of the elements.
8. When constructing the machine equivalent circuit based on the use of the machine resulting magnetic field, first the machine equivalent circuit is obtained, and then, the voltage equations are drawn up using this equivalent circuit. Thus, in this case, the voltage equations are compiled by the machine equivalent circuit.
9. In the specific system of units, the determination of the machine basic calculation elements based on the consideration of the machine resulting magnetic field can be reduced to the applying the following provisions:

   - decomposition of the machine magnetizing circuit into the stator and rotor magnetic circuits;
   - decomposition of the stator and rotor magnetizing circuits and also the stator and rotor slot leakage magnetic circuits into the layers and representation of the stator and rotor magnetic circuit as the multi layer system;
   - building the "unfolded" stator and rotor magnetizing equivalent circuits and the "unfolded" stator and rotor slot leakage equivalent circuits;
   - building the "unfolded" stator and rotor equivalent circuits corresponding to the machine resulting magnetic field;
   - introduction of the stator and rotor magnetizing circuit loops and the stator and rotor slot leakage circuit loops, and the determination of the stator and rotor magnetizing reactances and the slot leakage reactances of the stator and rotor windings;
   - building the "folded" stator and rotor equivalent circuits;

- introduction of the machine magnetizing circuit loop and the determination of the machine magnetizing reactance;
- building the machine equivalent circuit by the use of the "folded" stator and rotor equivalent circuits;
- compiling the machine voltage equations.

These provisions correspond to the case when the machine resulting magnetic field is determined as a result of the superposition of the one-dimensional machine magnetizing field and the one-dimensional stator and rotor slot leakage magnetic fields.

10. In this chapter, the machine equivalent circuit is represented through the specific quantities. In this case, the specific quantities mean the electric and magnetic field intensities (E, H) determined on the "smooth" surfaces of the layers reflecting the stator and rotor models. In this work, the expressions are proposed that reflect the mutual relationship between the quantities presented in the phase and specific systems of units. Applying such expressions helps to expand the capabilities of the methods developed based on using the phase and specific systems of units.
11. In this work, to determine the machine calculation elements, the "fragmentation" concept, the superposition principle, the "no-load and under-load" concept, the cascade way of connecting, and the system of the reduction of the "unfolded" equivalent circuits to the "folded" one are applied. In accordance with the "no-load and under-load" concept, the machine calculation elements are expressed through the referred rotor quantities in the phase system of units. It is shown that in the specific system of units the rotor quantities used in the machine calculation elements are not reduced to the number of the turns and the phase of the stator winding. In this case, the procedure of applying the "no-load and under-load" concept is simplified.

References

1. Richter, R.: Electric Machines, vol. 1-5. United Scientific and Technical Publishing, Moscow (1934-1939)
2. Livshits, A.: Electric Machines, vol. 1-3.United Scientific and Technical Publishing, Moscow (1936)
3. Kostenko, M.: Electric Machines. Special Part, Gosenergoizdat, Moscow (1949)
4. Alger, P.: Induction Machines, Their Behavior and Uses, Gordon and Beach, New York (1970)
5. Adking, B.: General Theory of Electrical Machines, John Wiley and Sons, Inc., New York (1957)
6. Krause, P.: Analysis of Electric Machinery, McGraw-Hill, New York (1986)
7. Fitzgerald, A., Kingsley, C., Umans, S.: Electric Machinery, McGraw-Hill, Inc., 5th ed., New York (1990)
8. Say, M.: Alternating Current Machines, John Wiley and Sons, Inc., New York (1983)
9. Shenfer, K.: Asynchronous Machines, State United Scientific and Technical Publishing, Moscow/ Leningrad (1938)
10. Kostenko, M, Piotrovskij, L.: Electric Machines, Publishing Energiya, Leningrad (1973)

11. Voldek. A.: Electric Machines, Energiya, Leningrad (1974)
12. Ivanov-Smolensky, A.: Electric Machines, Energiya, Moscow (1980)
13. Vazhnov, A.: Electric Machines, Energiya, Leningrad (1968)
14. Piotrovskij, L.: Electric Machines, Publishing Energiya, Leningrad (1975)
15. Kopylov, I.: Electric Machines, Publishing Energoatomizdat, Moscow (1986)
16. Seely, S.: Electromechanical Energy Conversion, McGraw-Hill Book Company, Inc., New York (1952)
17. Asanbayev, V.: Alternating Current Multi-Circuit Electric Machines: A New Approach to Steady-State Parameter Determination, Springer Cham Heidelberg, New York, Dordrecht London (2015)

Chapter 13
Basic Calculation Elements Using Resulting Magnetic Field: Specific System of Units

13.1 Magnetizing Circuit of Machine as Layered System

In this chapter, we proceed from the fact that in the machine magnetizing circuit there is a two-dimensional magnetic field. In these conditions, the stator and rotor slot-teeth areas can be represented as the corresponding magneto-anisotropic layers having the magnetic permeabilities μ_{xi} and μ_{yi}, for example [1–9]. In Fig. 13.1, the flat model for the machine magnetizing circuit is given. In this model, the sections of the stator and rotor slot-teeth areas are represented as the homogeneous layers having "smooth" surfaces.

Average values of the magnetic permeabilities. For the slot-teeth layers, the magnetic permeabilities having the average values are used [1–9]. We will demonstrate this provision on the example of considering the slot-tooth fragment shown in Fig. 13.2. In accordance with Fig. 13.2a, the radial magnetic flux Φ_i per single tooth pitch passes through the single tooth which can be taken when $B_z \leq 1.8T$. For the radial magnetic flux Φ_i, the reluctance of the single tooth having the width equal to b_z and the height h_i (Fig. 13.2a) is determined by the geometry and by the permeability of the single tooth. Now, for the reluctance of the single tooth R_{by}, we can obtain that $R_{by} = \frac{h_i}{\mu_z b_z l_i}$. When in relation to the magnetic flux Φ_i the slot-tooth area is represented as the homogeneous layer with the permeability μ_{yi}, the reluctance R_{ty} of this layer having the width equal to t_i is expressed in the form $R_{ty} = \frac{h_i}{\mu_{yi} t_i l_i}$. We proceed from the fact that the reluctance of the single tooth R_{by} and the reluctance of the slot-tooth area layer R_{ty} have the same value, i.e., $R_{by} = R_{ty}$. We use this condition to determine the average value of the permeability μ_{yi} for the slot-tooth layer. Taking into account that $R_{by} = \frac{h_i}{\mu_z b_z l_i}$ and $R_{ty} = \frac{h_i}{\mu_{yi} t_i l_i}$ and also $R_{by} = R_{ty}$, the layer permeability μ_{yi} is expressed as $\mu_{yi} = \mu_z \frac{b_z}{t_i}$.

In accordance with Fig. 13.2a, the flux Φ_i is determined as $\Phi_i = B_i b_z l_i$, where B_i is the magnetic flux density in the single tooth. From Fig. 13.2b, we can obtain for the flux Φ_i the expression of kind $\Phi_i = B_{icp} t_i l_i$, where B_{icp} is the average value of the

V. Asanbayev, *Asynchronous Machines*,
https://doi.org/10.1007/978-3-030-92284-9_13

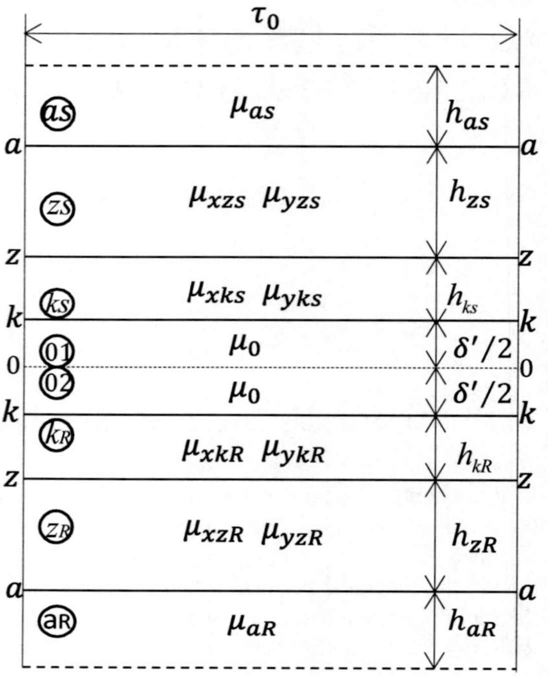

Fig. 13.1 Flat-layered model of the machine magnetizing circuit

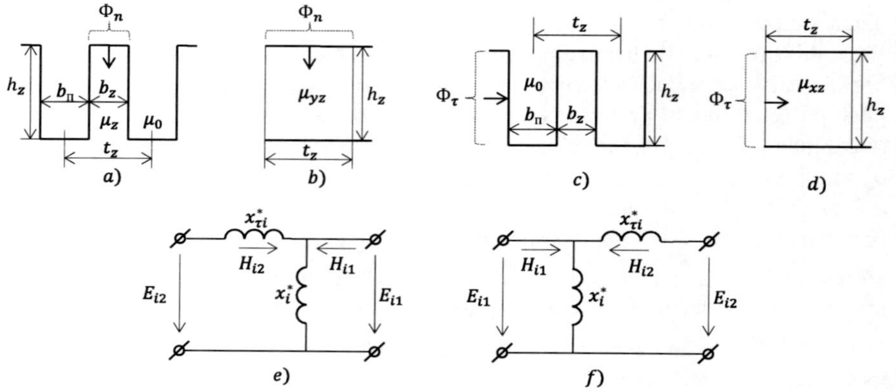

Fig. 13.2 To the calculation of the average values of the radial and the tangential components of the magnetic permeabilities of the tooth area layer. L-shaped equivalent circuits for the stator and the rotor i^{th} layers: (**a**) fragment of the tooth area in relation to the radial magnetic flux; (**b**) layered model of the tooth area in relation to the radial magnetic flux; (**c**) fragment of the tooth area in relation to the tangential magnetic flux; (**d**) layered model of the tooth area in relation to the tangential magnetic flux; (**e**) L-shaped equivalent circuit for the stator i^{th} layer; (**f**) L-shaped equivalent circuit for the rotor i^{th} layer

magnetic flux density in the slot-tooth layer. For the average value of the magnetic flux density in the slot-tooth layer B_{icp}, we can have that $B_{icp} = B_i \frac{b_z}{t_i}$. Consequently, the introduction of the slot-tooth layer is associated with the use of the average value of the magnetic flux density B_{icp} and the average value of the magnetic permeability μ_{yi} (Fig. 13.2b). The expression for μ_{yi} determined as $\mu_{yi} = \mu_z \frac{b_z}{t_i}$, can be used when $B_z \leq 1.8T$. When $B_z \geq 1.8T$, the part of the radial magnetic flux Φ_i passes through the tooth and another part of this flux is branched off into the slot. The magnetic permeability μ_{yi} corresponding to the condition $B_z \geq 1.8T$ is obtained in [1]. Thus, the slot-tooth area can be also represented as the layer when $B_z \geq 1.8T$.

Now, we consider the determination of the magnetic permeability μ_{xi}. In Fig. 13.2c, it is shown the fragment of the slot-tooth area that can be considered in relation to the tangential magnetic flux $\Phi_{\tau i}$. In this case, the reluctance of the slot-tooth area R_n is determined by the slot width equal to b_n (Fig. 13.2c). Therefore, the reluctance R_n can be obtained as $R_n = \frac{b_n}{\mu_0 h_i l_i}$.

Now, in relation to the tangential magnetic flux $\Phi_{\tau i}$, the slot-tooth area can be represented as the homogeneous layer with the magnetic permeability equal to μ_{xi}. Then, the reluctance of the slot-tooth layer (Fig. 13.2d) is determined in the form $R_t = \frac{t_i}{\mu_{xi} h_i l_i}$. We proceed from the fact that the condition of kind $R_n = R_t$ is true. Taking into account that $R_n = \frac{b_n}{\mu_0 h_i l_i}$ and $R_t = \frac{t_i}{\mu_{xi} h_i l_i}$ and also $R_n = R_t$, the value of μ_{xi} can be obtained as $\mu_{xi} = \mu_0 \frac{t_i}{b_n}$. Now, the fragments of the stator and rotor slot-tooth areas reflecting the winding location sections can be represented as the magneto-anisotropic layers having the magnetic permeabilities

$$\mu_{zi} = \mu_0 \frac{t_i}{b_\text{п}} \quad \mu_{yi} = \mu_i \frac{b_z}{t_i} \tag{13.1}$$

where $i = ks, zs, kR, zR$.

Acting in a similar way, the sections of the stator and rotor slot-tooth areas can be represented as the magneto-anisotropic layers with the "smooth" surfaces. Then, the machine magnetizing circuit can be reflected consisting of the seven layers (Fig. 13.1). In [1], the layered model of kind Fig. 13.1 is used to calculate the machine magnetizing circuit. The layered model in Fig. 13.1 allows us to take into account the two-dimensional magnetic field in the machine magnetizing circuit.

L-shaped equivalent circuit for the ith layer and reactances used in it. As shown in [1], the layers of the model in Fig. 13.1 can be represented as the L-shaped equivalent circuits (with the exception of the layers reflecting the stator and rotor yoke areas). In Fig. 13.2e, the L-shaped equivalent circuit for the ith layer of the stator magnetizing circuit is shown. In Fig. 13.2f, the L-shaped equivalent circuit for the ith layer of the rotor magnetizing circuit is given. We will determine the reactances used in these equivalent circuits. When considering the reactance of the ith layer x_i, where in accordance with Fig. 13.1 $i = 01, ks, zs, 02, kR, zR$, we can use the total current law equation obtained in this case the form $\dot{H}_i h_i = \dot{\varkappa}_{im} \frac{\tau_0}{\pi}$, where $\dot{H}_i$ and h_i are the radial component of the magnetic field intensity and the thickness of the ith layer of the magnetizing circuit. Here, the value $\dot{\varkappa}_{im}$ is the linear density of the stator winding surface current corresponding to the ith layer of the magnetizing circuit.

On the basis of Fig. 13.2b, the magnetic field intensity $\dot{H}_i$ can be represented as $\dot{H}_i = \dot{B}_i/\mu_{yi}$. Here, $\dot{B}_i$ is the magnetic flux density in the ith layer of the magnetic circuit; μ_{yi} is the magnetic permeability of the ith layer of the magnetic circuit. Now, with considering the condition for $\dot{H}_i$, the total current law equation can be reduced to the form $\dot{B}_i \frac{\pi h_i}{\mu_{yi}\tau_0} = \dot{\varkappa}_{im}$. Here, the value of $\dot{B}_i$ can be expressed through the magnetic flux density $\dot{B}_m$ determined on the air gap surface 0 – 0 for that the condition of kind $\dot{B}_i = (\tau_0/\tau_i)(\dot{B}_m/\xi_{Li})$, where $\xi_{Li} = l_i/l_0$), is used. Then, we can obtain the equation of the form $\dot{B}_m \frac{\pi h_i}{\mu_{yi}\tau_i\xi_{Li}} = \dot{\varkappa}_{im}$. Taking into account that $\dot{E}_m = \upsilon_1 \dot{B}_m = \frac{\omega_1\tau_0}{\pi}\dot{B}_m$, where $\upsilon = \frac{\omega_1\tau_0}{\pi}$, this equation acquires the form $\dot{E}_m \frac{\pi^2 h_i}{\omega_1\mu_{yi}\tau_i^2}\frac{\xi_{\tau i}}{\xi_{Li}} = \dot{\varkappa}_{im}$. From here, the longitudinal magnetizing reactance of the ith layer of the magnetic circuit can be expressed as $x_i = \frac{\omega_1\mu_{yi}\tau_i^2}{\pi^2 h_i}\frac{\xi_{Li}}{\xi_{\tau i}}$.

Now, we move on to the definition of the transverse magnetizing reactance of the ith layer of the magnetic circuit. In the equivalent circuits shown in Fig. 13.2e, f, this reactance is represented as $x_{\tau i}$. To determine the reactance $x_{\tau i}$, we can use the total current law equation of kind $\frac{1}{2}\dot{H}_{\tau i} l_i = \dot{\varkappa}_{im}\frac{\tau_0}{\pi}$, where $\dot{H}_{\tau i}$ and l_i are the tangential component of the magnetic field intensity and the length of the average value of its line in the ith layer of the magnetic circuit. Here, to determine the value of the length l_i, the condition of kind $l_i = (2/\pi)\tau_i$ can be used [1]. The value of $\dot{H}_{\tau i}$ can be expressed through the normal component of the magnetic flux density $\dot{B}_i$ determining on the surface of the ith layer. Then, we can obtain the condition of the form $\dot{H}_{\tau i} = \dot{B}_i \frac{1}{\pi}\frac{\tau_i}{\mu_{xi}h_i}$. Now, for this case, the total current law equation acquires the form $\dot{B}_i \frac{\tau_i l_i}{2\mu_{xi}h_i\tau_0} = \dot{\varkappa}_{im}$. The value of $\dot{B}_i$ can be expressed through the magnetic flux density $\dot{B}_m$ for that the condition of kind $\dot{B}_i = (\tau_0/\tau_i)(\dot{B}_m/\xi_{Li})$, where $\xi_{Li} = l_i/l_0$) is used. As a result, the total current law equation can be reduced to the form $\dot{B}_m \frac{l_i}{2\mu_{xi}h_i\xi_{Li}} = \dot{\varkappa}_{im}$. Taking into account that $\dot{E}_m = \upsilon_1\dot{B}_m = \frac{\omega_1\tau_0}{\pi}\dot{B}_m$, where $\upsilon = \frac{\omega_1\tau_0}{\pi}$, this equation acquires $\dot{E}_m \frac{\pi l_i}{2\omega_1\mu_{xi}h_i\tau_0\xi_{Li}} = \dot{\varkappa}_{im}$. Here, the length l_i is determined as $l_i = (2/\pi)\tau_i$. Therefore, we have that $\dot{E}_m \frac{\tau_i}{\omega_1\mu_{xi}h_i\tau_0\xi_{Li}} = \dot{\varkappa}_{im}$. From here, the transverse magnetizing reactance of the ith layer $x_{\tau i}$ can be represented in the form $x_{\tau i} = \omega_1\mu_{xi}\frac{\tau_0}{\tau_i}h_i\xi_{Li}$. If we bear in mind that for the ith stator layer $\tau_0 = t_0\frac{Z_1}{2p}$, $\tau_i = t_i\frac{Z_1}{2p}$ and for the ith rotor layer $\tau_0 = t_0\frac{Z_1}{2p}$, $\tau_i = t_i\frac{Z_2}{2p}$, then, the expression for the reactance $x_{\tau i}$ acquires the form for the ith stator layer $x_{\tau is} = \omega_1\mu_{xi}\frac{t_0}{t_i}h_i\xi_{Li}$ and for the ith rotor layer $x_{\tau iR} = \omega_1\mu_{xi}\frac{t_0}{t_i}h_i k_R\xi_{Li}$, where $k_R = Z_1/Z_2$. On the basis of the expressions obtained above for the reactances used in the equivalent circuits shown in Fig. 13.2e and f, we can receive

$$x_i^* = \frac{\omega_1\mu_{yi}\tau_i^2}{\pi^2 h_i}\frac{\xi_{Li}}{\xi_{\tau i}} \quad x_{\tau i}^* = \omega_1\mu_{xi}t_0\frac{h_i}{t_i}k_R\xi_{Li} \tag{13.2}$$

where $i = ks, zs, kR, zR$; $k_R = Z_1/Z_2$.

When determining the reactances used in the L-shaped equivalent circuit of the stator layer, the value of the factor k_R is taken equal to $k_R = 1.0$.
In this case, the stator and rotor magnetizing equivalent circuits can be constructed as a result of the cascade way of connecting the L-shaped equivalent circuits representing the corresponding layers of the model shown in Fig. 13.1 [1]. In this work, the equivalent circuits obtained in this way are considered as the "unfolded" stator and rotor magnetizing equivalent circuits.

13.2 "Unfolded" Magnetizing Equivalent Circuits of Stator and Rotor

The construction of the "unfolded" stator and rotor magnetizing equivalent circuits are based on the representation of the machine air gap consisting of the two sub layers having the calculated thicknesses equal to $\delta_1 = (\delta/2)k_{1\delta}$ and $\delta_1 = (\delta/2)$ $k_{2\delta}$, where $k_{1\delta}$ and $k_{2\delta}$ are the air gap factors [1]. Therefore, in accordance with [1], the machine air gap can be represented by the two L-shaped equivalent circuits having the forms shown in Fig. 12.3. As follows from Appendix B, the decomposition of the machine air gap into the two sub layers can be implemented in relation to the surface $0 - 0$ passing through the middle of the radial length of the air gap. The upper sub layer with the thickness $(\delta/2)$ is adjoined to the stator bore surface and the lower sub layer with the thickness $(\delta/2)$ is adjoined to the rotor surface. In this case, the stator magnetizing circuit includes the air gap upper sub-layer, the layer of the slot-wedge area, the layer of the winding location area, and the layer of the stator yoke area (Fig. 13.1). The rotor magnetizing circuit contains the air gap lower sub layer, the layer of the slot-wedge area, the layer of the winding location area, and the layer of the rotor yoke area (Fig. 13.1). The "unfolded" stator magnetizing equivalent circuit given in Fig. 13.3a corresponds to this nature of the decomposition into the layers of the stator magnetic circuit [1]. The "unfolded" rotor magnetizing equivalent circuit is shown in Fig. 13.4a. In the equivalent circuits in Figs. 13.3a and 13.4a, the reactances x^*_{01} and x^*_{02} are determined as a result of the decomposition of the air gap magnetizing reactance x^*_0 into the parallel connected two reactances [1]. The value of x^*_{01} represents the magnetizing reactance of the air gap upper sub layer and x^*_{02} reflects the magnetizing reactance of the air gap lower sub layer. The equivalent circuits in Figs. 13.3a and 13.4a contain the transverse reactances $x^*_{01}, x^*_{ks}, x^*_{zs}$, and $x^*_{02}, x^*_{kR}, x^*_{zR}$, representing the longitudinal magnetizing own circuit loops of the layers and also include the longitudinal reactances $x^*_{\tau 01}, x^*_{\tau ks}, x^*_{\tau zs}, x^*_{as}$, and $x^*_{\tau 02}, x^*_{\tau kR}, x^*_{\tau zR}, x^*_{aR}$. The transverse and longitudinal reactances used in the equivalent circuits in indicate the presence of the two-dimensional magnetizing field in the layers of the model shown in Fig. 13.1. In the "unfolded" equivalent circuits in Figs. 13.3a and 13.4a, the longitudinal reactances $x^*_{\tau 01}, x^*_{\tau ks}, x^*_{\tau zs}$ and $x^*_{\tau 02}, x^*_{\tau kR}, x^*_{\tau zR}$ can be considered representing the longitudinal branches connecting the "own"

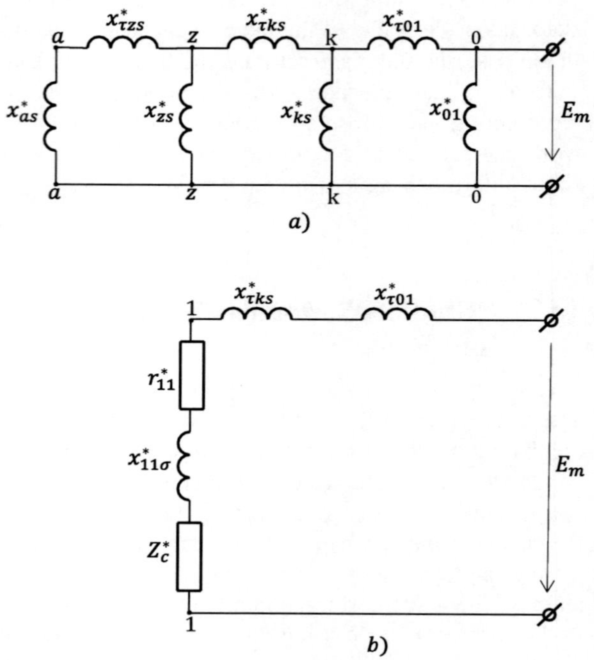

Fig. 13.3 "Unfolded" stator magnetizing equivalent circuit and the "unfolded" stator slot leakage equivalent: (**a**) "unfolded" stator magnetizing equivalent circuit; (**b**) "unfolded" stator slot leakage equivalent circuit

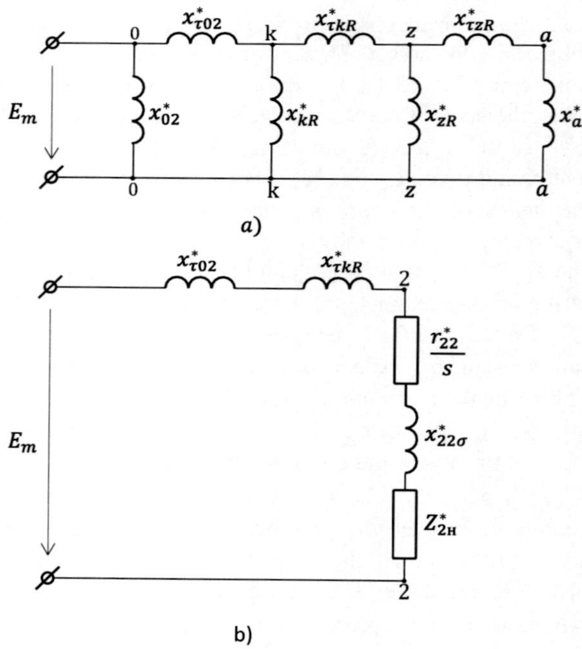

Fig. 13.4 "Unfolded" rotor magnetizing equivalent circuit and the "unfolded" rotor slot leakage equivalent circuit: (**a**) "unfolded" rotor magnetizing equivalent circuit; (**b**) "unfolded" rotor slot leakage equivalent circuit

circuit loops of the adjacent layers of the models reflecting the stator and rotor magnetizing circuits.

The reactances used in the equivalent circuits in Figs. 13.3a and 13.4a can be obtained on the basis of the expressions derived in [1]. In the "unfolded" equivalent circuit in Fig. 13.3a, the reactances of the longitudinal magnetizing "own" circuit loops representing the air gap sub layer, the layer of the stator slot-wedge area, the layer of the stator winding location area are calculated by the expressions used in the twelfth chapter and having from (12.48) the form

$$
\begin{aligned}
x_{01}^{*} &= \frac{\omega_1 \mu_0 \tau_0^2}{\pi^2 \delta_1} \\
x_{ks}^{*} &= \frac{\omega_1 \mu_{yks} \tau_{ks}^2}{\pi^2 h_{ks}} \frac{\xi_{Lks}}{\xi_{\tau ks}} \\
x_{zs}^{*} &= \frac{\omega_1 \mu_{yzs} \tau_{zs}^2}{\pi^2 h_{zs}} \frac{\xi_{Lzs}}{\xi_{\tau zs}}
\end{aligned}
\tag{13.3}
$$

In the "unfolded" equivalent circuit in Fig. 13.3a, the longitudinal reactances representing the air gap upper sub layer, the layer of the stator slot-wedge area, the layer of the stator winding location area, and the layer of the stator yoke area can be received according to [1] as

$$
\begin{aligned}
x_{\tau 01}^{*} &= \omega_1 \mu_0 t_0 \frac{\delta}{2t_0} k_{1\sigma} & x_{\tau ks}^{*} &= \omega_1 \mu_{xks} t_0 \frac{h_{ks}}{t_{ks}} \xi_{Lks} \\
x_{\tau zs}^{*} &= \omega_1 \mu_{xzs} t_0 \frac{h_{zs}}{t_{zs}} \xi_{Lzs} & x_{as}^{*} &= \omega_1 \mu_{as} t_0 \frac{h_{as}}{t_{as}} \xi_{Las}
\end{aligned}
\tag{13.4}
$$

where $t_{as} = 2p\tau_{as}/Z_1$.

Here, t_{as} represents the conditional tooth pitch determined on the inner surface of the stator yoke area in relation to the number of the stator teeth. The factor $k_{1\sigma}$ is used to take into account the distortion of the magnetic flux density lines in the air gap sub layer adjoining to the stator bore surface. The value of this factor can be determined by the expressions proposed, for example, in [10–13].

Now, we consider the reactances used in the "unfolded" rotor magnetizing equivalent circuit. Such rotor equivalent circuit is shown in Fig. 13.4a. In the "unfolded" rotor equivalent circuit in Fig. 13.4a, the reactances of the longitudinal magnetizing "own" circuit loops representing the air gap lower sub layer, the layer of the rotor slot-wedge area, the layer of the rotor winding location area are calculated by the expressions used in Chap. 12 and having from (12.50) the form

$$
\begin{aligned}
x_{02}^{*} &= \frac{\omega_1 \mu_0 \tau_0^2}{\pi^2 \delta_2} \\
x_{kR}^{*} &= \frac{\omega_1 \mu_{ykR} \tau_{kR}^2}{\pi^2 h_{kR}} \frac{\xi_{LkR}}{\xi_{\tau kR}} \\
x_{zR}^{*} &= \frac{\omega_1 \mu_{yzR} \tau_{zR}^2}{\pi^2 h_{zR}} \frac{\xi_{LzR}}{\xi_{\tau zR}}
\end{aligned}
\tag{13.5}
$$

In the "unfolded" equivalent circuit in Fig. 13.4a, the longitudinal reactances representing the air gap lower sub layer, the layer of the rotor slot-wedge area, the layer of the rotor winding location area, and the layer of the rotor yoke area can be received by the expressions shown in [1]. Then, we have

$$
\begin{aligned}
x_{\tau 02}^{*} &= \omega_1 \mu_0 t_0 \frac{\delta}{2 t_{02}} k_{2\sigma} k_R \qquad & x_{\tau kR}^{*} &= \omega_1 \mu_{xkR} t_0 \frac{h_{kR}}{t_{kR}} k_R \xi_{LkR} \\
x_{\tau zR}^{*} &= \omega_1 \mu_{xzR} t_0 \frac{h_{zR}}{t_{zR}} k_R \xi_{LzR} \qquad & x_{aR}^{*} &= \omega_1 \mu_{aR} t_0 \frac{h_{aR}}{t_{aR}} k_R \xi_{LaR}
\end{aligned}
\tag{13.6}
$$

where $k_R = Z_1/Z_2; t_{aR} = 2p\tau_{aR}/Z_2$.

Here, t_{aR} represents the conditional tooth pitch determined on the outer surface of the rotor yoke area in relation to the number of the rotor teeth. The factor $k_{2\sigma}$ is used to take into account the distortion of the magnetic flux density lines in the air gap sub layer adjoining to the rotor surface. The value of this factor can be determined by the expressions proposed, for example, in [10–13].

In further, we use the "unfolded" equivalent circuits obtained for the stator and rotor magnetizing circuits in Figs. 13.3a and 13.4a and reflected more strictly the two-dimensional magnetizing field distribution in the machine.

In the "unfolded" equivalent circuits in Figs. 13.3a and 13.4a, the layers of the machine magnetizing circuit model are represented as the L-shaped equivalent circuits (with the exception of the layers of the stator and rotor yoke areas) having the transverse reactances x_n^* and the longitudinal reactances $x_{\tau n}^*$, where $i=$ 01, *ks*, *zs*, 02, *kR*, and *zR*. In addition, in the "unfolded" stator and rotor magnetizing equivalent circuits (Figs. 13.3a and 13.4a), the terminals $k - k$, $z - z$, and $a - a$ correspond to the surfaces $k - k$, $z - z$ and $a - a$ formed by the corresponding adjacent layers of the machine magnetizing model (Fig. 13.1). On the surfaces $k - k$, $z - z$, and $a - a$ of the machine magnetizing model (Fig. 13.1), the known boundary conditions are met for the components of the electric and magnetic field intensities. These conditions are also met on the terminals $k - k$, $z - z$ and $a - a$ of the "unfolded" stator and rotor magnetizing equivalent circuits (Figs. 13.3a and 13.4a) due to which in these equivalent circuits the input terminals of the subsequent L-shaped equivalent circuits are connected to the output terminals of the previous L-shaped equivalent circuits forming in our case the cascade way of connecting the L-shaped equivalent circuits. Consequently, when using the specific system of units $(\dot{E}, \dot{H})$, the "unfolded" stator and rotor magnetizing equivalent circuits (Figs. 13.3a and 13.4a) can be obtained as a result of the cascade way of connecting the L-shaped equivalent circuits play scng reflecting corresponding layers of the machine magnetizing circuit model (Fig. 13.1).

13.3 "Unfolded" Slot Leakage Equivalent Circuits of Stator and Rotor

Stator and rotor slot leakage magnetic circuits as the layered systems. In practice, the slot leakage calculations are implemented for the single slot. In this regard, the single slot can be considered as the layered system consisting of the homogeneous layers having the length of the single tooth pitch. Such representation of the slot is demonstrated in Appendix F and shown in Fig. F.5b. For example, we consider the rotor slot section occupied by the winding and represent this slot section as the homogeneous layer with the height equal to h_{zR}, and with the magnetic permeability equal to μ_{xzR} (Fig. F.5b). This homogeneous slot layer has the width equal to the single tooth pitch t_{zR}. We define the value of μ_{xzR}. For this purpose, the reluctance of the single slot with the width b_{nR} and the height h_{zR} (Fig. F.5a) and also the reluctance of the homogeneous slot layer with the width t_{zR} and the height h_{zR} (Fig. F.5b) can be considered. In this case, we will take into account that the magnetic field intensity created by the slot current increases linearly from the base of the slot. Here, the value of the tooth pitch t_{zR} is determined on the upper surface of the slot section of the rotor winding location area. The slot leakage magnetic flux can be obtained as $\Phi_{nR} = \mu_0 \frac{h_{zR} l_{zR}}{2b_{nR}} S_n \sqrt{2}\dot{I}_2$. Then, the slot leakage reluctance acquires the form $R_n = \frac{2b_{nR}}{\mu_0 h_{zR} l_{zR}}$. The leakage magnetic flux of the slot layer is calculated by the expression of kind $\Phi_t = \mu_{xzR} \frac{h_{zR} l_{zR}}{2t_{zR}} S_n \sqrt{2}\dot{I}_2$. From here, the reluctance of the slot layer takes the form $R_t = \frac{2t_{zR}}{\mu_{xzR} h_{zR} l_{zR}}$. We proceed from the fact that $R_n = R_t$. Using this condition and also keeping in mind that $R_n = \frac{2b_{nR}}{\mu_0 h_{zR} l_{zR}}$ and $R_t = \frac{2t_{zR}}{\mu_{xzR} h_{zR} l_{zR}}$, the magnetic permeability of the slot layer corresponding to the rotor winding location area can be expressed as $\mu_{xzR} = \mu_0 \frac{t_{zR}}{b_{nR}}$. The expression for μ_{xzR} obtained in this way can be used to determine the magnetic permeability of the layer representing the rotor slot-wedge section. As a result, we can have

$$\mu_{xkP} = \mu_0 \frac{t_{kR}}{b_{kR}} \quad \mu_{xzR} = \mu_0 \frac{t_{zR}}{b_{пR}} \tag{13.7}$$

Acting in a similar way, the magnetic permeabilities of the layers reflecting the slot areas reflecting the stator winding location section and the stator slot-wedge region can be determined (Fig. F.5b). Now, we obtain

$$\mu_{xks} = \mu_0 \frac{t_{ks}}{b_{ks}} \quad \mu_{xzs} = \mu_0 \frac{t_{zs}}{b_{\text{п}s}} \tag{13.8}$$

The magnetic permeabilities μ_{xkn} and μ_{xzn} determined from (13.1) and (13.2) represent the slot leakages caused by the magnetizing fields existing in the stator and rotor layers reflecting the winding location areas and the slot-wedge areas. The magnetic permeabilities μ_{xki} and μ_{xzi} obtained in (13.7) and (13.8) reflect the slot leakages created by the volume of the slot currents in stator and rotor layers representing the winding location areas and the slot-wedge areas. We note that despite the differences in the definitions, the calculations of μ_{xki} and μ_{xzi} are carried out using the same expressions.

Thus, in relation to the transverse slot leakage magnetic flux, the sections of the stator and rotor slot-teeth areas can be represented as multilayer systems. In Fig. F.5, the multilayer models for the stator and rotor slot leakage magnetic circuits are shown. In the further, we will come from the models of the stator and rotor slot leakage magnetic circuits given in Fig. F.5.

"Unfolded" stator slot leakage equivalent circuit. The stator and rotor slot leakage emf equations are obtained in (12.56) and (12.57). Below, we will use these equations. The equations shown in (12.57) can be expressed as

$$\dot{E}_m = \left[j\left(x^*_{\tau 01} + x^*_{\tau ks} + x^*_{11\sigma}\right) + r^*_1 + Z^*_c \right] \dot{H}_1 \tag{13.9}$$

The "unfolded" stator slot leakage equivalent circuit corresponding to this equation is obtained in Appendix F and it is given in Fig. F.6a. In this chapter, Fig. F.6a is shown in Fig. 13.3b. This equivalent circuit has the form convenient for an analysis. For the reactances $x^*_{\tau 01}, x^*_{\tau ks}$, and $x^*_{11\sigma}$ used in the equivalent circuit in Fig. 13.3b, the expressions obtained in (12.55) can be applied. Then, we receive

$$\begin{aligned} x^*_{\tau 01} &= \omega_1 \mu_0 t_0 \frac{\delta}{2t_0} k_{1\sigma} \\ x^*_{\tau ks} &= \omega_1 \mu_{xks} t_0 \frac{h_{ks}}{t_{ks}} \xi_{Lks} \\ x^*_{11\sigma} &= \omega_1 \mu_{xzs} t_0 \frac{h_{zs}}{3t_{zs}} \xi_{Lzs} \end{aligned} \tag{13.10}$$

"Unfolded" rotor slot leakage equivalent circuit. On the basis of the equation given in (12.56), the rotor slot leakage emf equation acquires the form

$$\dot{E}_m = \left[j\left(x^*_{\tau 02} + x^*_{\tau kR} + x^*_{22\sigma}\right) + r^*_2/s + Z^*_{2\text{н}} \right] \dot{H}_2 \tag{13.11}$$

The "unfolded" rotor slot leakage equivalent circuit arising from this equation is received in Appendix F and it is given in Fig. F.6b. In this chapter, the "unfolded" rotor slot leakage equivalent circuit in Fig. A.12.6b is shown in Fig. 13.4b. The equivalent circuit given in Fig. 13.4b has the form convenient for an analysis. The reactances $x^*_{\tau 02}, x^*_{\tau kR}$, and $x^*_{22\sigma}$ applied in the equivalent circuit in Fig. 13.4b can be determined by the expressions obtained in (12.53). Therefore, we can have

$$\begin{aligned} x^*_{\tau 02} &= \omega_1 \mu_0 t_0 \frac{\delta}{2t_{02}} k_{2\sigma} k_R \\ x^*_{\tau kR} &= \omega_1 \mu_{xkR} t_0 \frac{h_{kR}}{t_{kR}} k_R \xi_{LkR} \\ x^*_{22\sigma} &= \omega_1 \mu_{xzR} t_0 \frac{h_{zR}}{3t_{zR}} k_R \xi_{LzR} \end{aligned} \qquad (13.12)$$

The "unfolded" stator magnetizing equivalent circuit shown in Fig. 13.3a corresponds to the no-load mode, and the "unfolded" stator slot leakage equivalent circuit given in Fig. 13.3b reflects the under-load mode. The "unfolded" rotor magnetizing equivalent circuit shown in Fig. 13.4a corresponds to the no-load mode, and the "unfolded" rotor slot leakage equivalent circuit given in Fig. 13.4b reflects the under-load mode. Consequently, in accordance with the "no-load and under-load" concept, these equivalent circuits can be used to build the "unfolded" stator and rotor equivalent circuits reflecting the machine resulting magnetic field. In this case, to build the "unfolded" equivalent circuit corresponding to the stator resulting magnetic field, the equivalent circuits obtained in Fig. 13.3 can be used. The "unfolded" equivalent circuit reflecting the rotor resulting magnetic field can be derived from the equivalent circuit shown in Fig. 13.4. The equivalent circuits obtained in this way represent the "unfolded" stator and rotor equivalent circuits with the mixed connection of the elements. Below, we will show this.

13.4 "Unfolded" Equivalent Circuit with Mixed Connection of Elements of Stator

We consider the "unfolded" stator magnetizing equivalent circuit (Fig. 13.3a) and the "unfolded" stator slot leakage equivalent circuit (Fig. 13.3b). The reactances $x^*_{\tau 01}$ and $x^*_{\tau ks}$ used in the equivalent circuit in Fig. 13.3a represent the tangential components of the magnetizing flux taking place in the air gap sub layer adjoining to the stator bore surface and in the layer reflecting the stator slot-wedge area. Consequently, the reactances $x^*_{\tau 01}$ and $x^*_{\tau ks}$ (Fig. 13.3a) represent the leakage fluxes caused by the machine magnetizing flux in the air gap upper sub layer and in the layer of the stator slot-wedge area. In the equivalent circuit in Fig. 13.3b, the reactances $x^*_{\tau 01}$ and $x^*_{\tau ks}$ represent the leakage fluxes created by the linear density of the stator winding surface current in the air gap sub layer adjoining to the stator bore surface and in the layer of the stator slot-wedge area. Thus, the reactances $x^*_{\tau 01}$ and $x^*_{\tau ks}$ used in the equivalent circuits in Fig. 13.3a, b have a different physical nature of the formation. However, as follows from the expressions obtained in (13.4) and (13.10), the reactance $x^*_{\tau 01}$ used in the equivalent circuits in Fig. 13.3a, b, and also the reactance

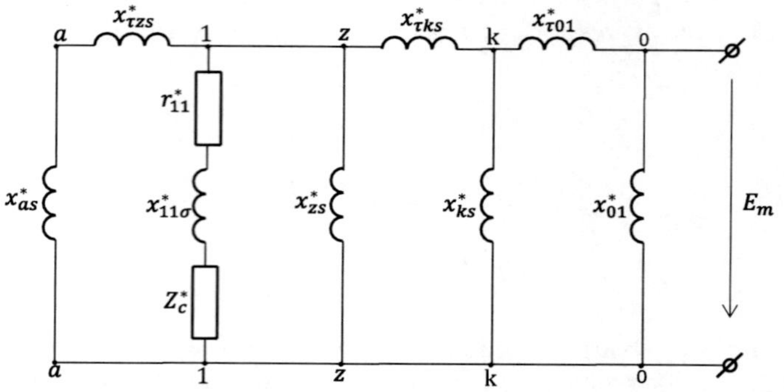

Fig. 13.5 "Unfolded" stator equivalent circuit with the mixed connection of the elements

$x^*_{\tau ks}$ applied in the equivalent circuits in Fig. 13.3a, b are described by the formulas having a similar structure. In these conditions, to obtain "unfolded" stator equivalent circuit corresponding to the resulting magnetic field, it is necessary to connect the terminals of the branch $1 - 1$ of the equivalent circuit in Fig. 13.3b to the terminals of the parallel branch $z - z$ of the equivalent circuit in Fig. 13.3a. As a result, the "unfolded" stator equivalent circuit acquires the form given in Fig. 13.5.

As follows from the equivalent circuit in Fig. 13.5, the air gap sub layer adjoining to the stator bore surface, the layer of the stator slot-wedge area, and the layer of the stator winding location area is represented by the L-shaped equivalent circuits. In this case, in the equivalent circuit in Fig. 13.5, its transverse branches $0 - 0$, $k - k$ and $z - z$ reflect the longitudinal magnetizing "own" circuit loops representing the air gap sub layer, the layer of the stator slot-wedge area and the layer of the stator winding location area. In the equivalent circuit in Fig. 13.5, the layer of the stator winding location area is represented by the two parallel branches. The first transverse branch $z - z$ having the reactance x^*_{zs} reflects the longitudinal magnetizing "own" circuit loop of the considered stator layer, and the second transverse branch $1 - 1$ containing the complex impedance $\left(jx^*_{11\sigma} + r^*_{11} + Z^*_c\right)$ represents the slot leakage "own" circuit loop of the stator winding. In the equivalent circuit in Fig. 13.5, the transverse branch $a - a$ reflects the transverse magnetizing "own" circuit loop of the stator yoke layer.

The equivalent circuit in Fig. 13.5 represents the "unfolded" stator equivalent circuit with the mixed connection of the elements. Thus, the equivalent circuit (Fig. 13.5) corresponding to the stator resulting magnetic field has a rather complex structure, and therefore, it is not convenient for practical use. In order to make it look convenient for practical use, the "unfolded" equivalent circuit in Fig. 13.5 is reduced to the "folded" form for what the multiple transfer technique can be used. Below, we will show this.

13.5 "Unfolded" Equivalent Circuit with Mixed Connection of Elements of Rotor

We consider the "unfolded" rotor magnetizing equivalent circuit (Fig. 13.4a) and the "unfolded" rotor slot leakage equivalent circuit (Fig. 13.4b). The reactances $x^*_{\tau 02}$ and $x^*_{\tau kR}$ used in the equivalent circuit in Fig. 13.4a represent the tangential components of the magnetizing flux taking place in the air gap sub layer adjoining to the rotor surface and in the layer reflecting the rotor slot-wedge area. Consequently, the reactances $x^*_{\tau 02}$ and $x^*_{\tau kR}$ (Fig. 13.4a) represent the leakage fluxes caused by the machine magnetizing flux in the air gap lower sub layer and in the layer of the rotor slot-wedge area. In the equivalent circuit in Fig. 13.4b, the reactances $x^*_{\tau 02}$ and $x^*_{\tau kR}$ represent the leakage fluxes created by the linear density of the rotor winding surface current in the air gap sub layer adjoining to the rotor surface and in the layer of the rotor slot-wedge area. Thus, the reactances $x^*_{\tau 02}$ and $x^*_{\tau kR}$ used in the equivalent circuits in Fig. 13.4a, b have a different physical nature of the formation. However, as follows from the expressions obtained in (13.6) and (13.12), the reactance $x^*_{\tau 02}$ used in the equivalent circuits in Fig. 13.4a, b and also the reactance $x^*_{\tau kR}$ applied in the equivalent circuits in Fig. 13.4a, b are described by the formulas having a similar structure. In these conditions, to obtain the "unfolded" rotor equivalent circuit corresponding to the machine resulting magnetic field, it is necessary to connect the terminals of the branch $2-2$ of the equivalent circuit in Fig. 13.4b to the terminals of the parallel branch $z-z$ of the equivalent circuit in Fig. 13.4a. As a result, the "unfolded" rotor equivalent circuit acquires the form given in Fig. 13.6. This equivalent circuit corresponds to the rotor resulting magnetic field.

As follows from the equivalent circuit in Fig. 13.6, the air gap sub layer adjoining to the rotor surface, the layer of the rotor slot-wedge area, and the layer of the rotor winding location area is represented by the L-shaped equivalent circuits. In this case, in the equivalent circuit in Fig. 13.6, its transverse branches $0-0$, $k-k$ and $z-z$ reflect the longitudinal magnetizing "own" circuit loops representing the air gap

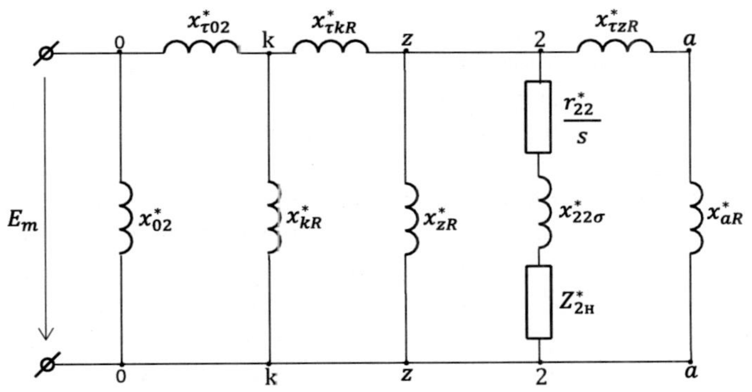

Fig. 13.6 "Unfolded" rotor equivalent circuit with the mixed connection of the elements

sub layer, the layer of the rotor slot-wedge area, and the layer of the rotor winding location area. In the equivalent circuit in Fig. 13.6, the layer of the rotor winding location area is represented by the two parallel branches. The first transverse branch $z - z$ having the reactance x^{*}_{zR} reflects the longitudinal magnetizing "own" circuit loop of the considered rotor layer, and the second transverse branch $2 - 2$ containing the complex impedance $\left(jx^{*}_{22\sigma} + r^{*}_{22}/s + Z^{*}_{2H}\right)$ represents the slot leakage "own" circuit loop of the rotor winding. In the equivalent circuit in Fig. 13.6, the transverse branch $a - a$ reflects the transverse magnetizing "own" circuit loop of the rotor yoke layer.

The equivalent circuit in Fig. 13.6 represents the "unfolded" rotor equivalent circuit with the mixed connection of the elements. Thus, the equivalent circuit (Fig. 13.6) corresponding to the rotor resulting magnetic field has a rather complex structure, and therefore, it is not convenient for practical use. In order to make it look convenient for practical use, the "unfolded" equivalent circuit in Fig. 13.6 is reduced to the "folded" form for what the multiple transfer technique can be used. Below, we will show this.

The "unfolded" equivalent circuits given in Figs. 13.5 and 13.6 represent the result of the cascade way of connecting the L-shaped equivalent circuits reflecting the corresponding layers of the stator and rotor magnetic circuit models (with the exception of the layers of the stator and rotor yoke areas). These equivalent circuits reflect the spatial distribution of the machine's two-dimensional magnetizing field and the one-dimensional slot leakage magnetic fields of the stator and rotor windings. Therefore, the "unfolded" equivalent circuits given in Figs. 13.5 and 13.6 have a mixed connection of the elements. Taking into account the two-dimensionality of the machine magnetic field leads to the complication of the structure of the "unfolded" stator and rotor equivalent circuits.

13.6 Features of Reducing "Unfolded" Equivalent Circuits to "Folded" Form

In Chap. 12, under the conditions of considering the one-dimensional magnetizing fields and the one-dimensional slot leakage magnetic fields, the "unfolded" stator and rotor equivalent circuits with the parallel connection of the elements were obtained. In this case, the reducing process of the "unfolded" stator and rotor equivalent circuits with the parallel connection of the elements to the "folded" form was described by introducing for the stator and rotor the magnetizing and the slot leakage circuit loops. The "folded" stator and rotor equivalent circuits obtained in this way are used in constructing the machine equivalent circuit.

In this chapter, in accordance with Figs. 13.5 and 13.6, we are dealing with the "unfolded" stator and rotor equivalent circuits with the mixed connection of the elements which are the consequence of considering the machine two-dimensional magnetizing field. In this case, the process of constructing the machine equivalent circuit can be considered consisting of several stages. At the first stage, the "unfolded" stator and rotor equivalent circuits with the mixed connection of the elements are reduced to the "unfolded" stator and rotor equivalent circuits with the parallel connection of the elements. This process of reduction can be carried out by the use of the multiple transfer technique [1]. At the second stage, on the basis of considering the obtained "unfolded" stator and rotor equivalent circuits with the parallel connection of the elements can be introduced for the stator and rotor the magnetizing and slot leakage circuit loops, and then, the "folded" stator and rotor equivalent circuits are constructed. At the third stage, the "folded" stator and rotor equivalent circuits received in this way are used to construct the machine equivalent circuit.

13.7 "Unfolded" Equivalent Circuit with Parallel Connection of Elements of Stator

The "unfolded" stator equivalent circuit with the parallel connection of the elements can be obtained by the use of the "unfolded" stator equivalent circuit with the mixed connection of the elements (Fig. 13.5). In the equivalent circuit in Fig. 13.5, its parallel branches $k-k$, $z-z$, $1-1$, and $a-a$ can be transferred to the terminals of the transverse branch $0-0$. In this work, the process of transferring the parallel branches $k-k$, $z-z$, $1-1$, and $a-a$ is carried out, as was noted, by the use of the multiple transfer technique [1]. The characteristic feature of this technique is the use of transfer factors. The implementation of the process of transferring the parallel branches of the equivalent circuit in Fig. 13.5, it is convenient to start with transferring its parallel branch $k-k$ to the terminals $0-0$ (Appendix F). For the parallel branch $k-k$, the transfer factor is

$$c_{\tau ks} = 1 + \frac{x^*_{\tau 01}}{x^*_{ks}} \tag{13.13}$$

The reactance of the transferred parallel branch $k-k$ is determined as

$$\overset{\circ}{x}_{ks} = x^*_{ks} c_{\tau ks} = x^*_{ks} + x^*_{\tau 01} \tag{13.14}$$

The transfer factor for the parallel branch $z-z$ is

$$c_{\tau zs} = 1 + \frac{x^*_{\tau ks} c^2_{\tau ks} + x^*_{\tau 01} c_{\tau ks}}{x^*_{zs} c^2_{\tau ks}} \tag{13.15}$$

The reactance of the transferred parallel branch $z-z$ is derived as

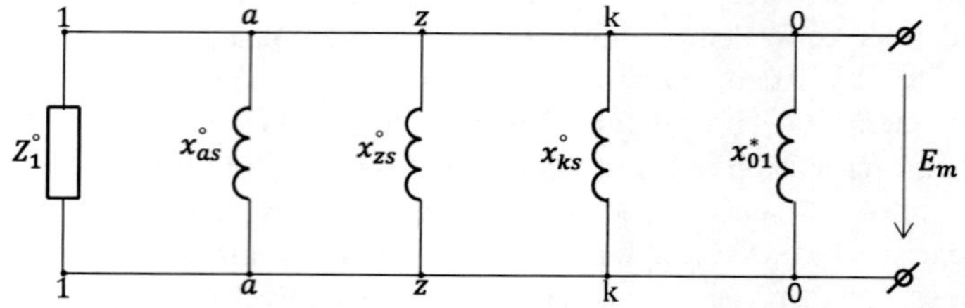

Fig. 13.7 "Unfolded" stator equivalent circuit with the parallel connection of the elements

$$\overset{\circ}{x}_{zs} = x^*_{zs} c_{\tau zs} c^2_{\tau ks} = x^*_{zs} c^2_{\tau ks} + x^*_{\tau ks} c^2_{\tau ks} + x^*_{\tau 01} c_{\tau ks} \tag{13.16}$$

The transfer factor for the parallel branch $1 - 1$ is

$$c_{\tau 1} = 1 + \frac{x^*_{\tau ks} c^2_{\tau ks} c_{\tau zs} + x^*_{\tau 01} c_{\tau ks} c_{\tau zs}}{\left(r^*_{11} + jx^*_{11\sigma} + Z^*_c\right) c^2_{\tau ks} c^2_{\tau zs}} \tag{13.17}$$

The impedance of the transferred parallel branch $1 - 1$ is expressed as

$$\begin{aligned} \overset{\circ}{Z}_1 &= r^*_{11} c^2_{\tau ks} c^2_{\tau zs} + j\left(x^*_{\tau 01} c_{\tau ks} c_{\tau zs} + x^*_{\tau ks} c^2_{\tau ks} c_{\tau zs} + x^*_{11\sigma} c^2_{\tau ks} c^2_{\tau zs}\right) + Z^*_c c^2_{\tau ks} c^2_{\tau zs} \\ &= r^*_{11} c^2_{\tau ks} c^2_{\tau zs} + jx^*_{1\sigma} + Z^*_c c^2_{\tau ks} c^2_{\tau zs} \end{aligned} \tag{13.18a}$$

where $x^*_{1\sigma} = x^*_{\tau 01} c_{\tau ks} c_{\tau zs} + x^*_{\tau ks} c^2_{\tau ks} c_{\tau zs} + x^*_{11\sigma} c^2_{\tau ks} c^2_{\tau zs}$.

Or

$$\begin{aligned} \overset{\circ}{Z}_1 &= r^*_{11} c^2_{\tau ks} c^2_{\tau zs} \\ &+ j\left[x^*_{\tau 01}(1/c_{\tau ks} c_{\tau zs}) + x^*_{\tau ks}(1/c_{\tau zs}) + x^*_{11\sigma}\right] c^2_{\tau ks} c^2_{\tau zs} + Z^*_c c^2_{\tau ks} c^2_{\tau zs} \\ &= \left(r^*_{11} + j\widetilde{x}^*_{11\sigma} + Z^*_c\right) c^2_{\tau ks} c^2_{\tau zs} \end{aligned} \tag{13.18b}$$

where $\widetilde{x}^*_{11\sigma} = x^*_{\tau 01}(1/c_{\tau ks} c_{\tau zs}) + x^*_{\tau ks}(1/c_{\tau zs}) + x^*_{11\sigma}; x^*_{1\sigma} = \widetilde{x}^*_{11\sigma} c^2_{\tau ks} c^2_{\tau zs}$.

When $c_{\tau 1} \approx 1.0$, the reactance of the transferred parallel branch $a - a$ is calculated by the expression

$$\overset{\circ}{x}_{as} = x^*_{as} c^2_{\tau ks} c^2_{\tau zs} + x^*_{\tau zs} c^2_{\tau ks} c^2_{\tau zs} + x^*_{\tau ks} c^2_{\tau ks} c_{\tau zs} + x^*_{\tau 01} c_{\tau ks} c_{\tau zs} \tag{13.19}$$

When using the reactances obtained in (13.14), (13.16), (13.18a), and (13.19), the "unfolded" stator equivalent circuit with the mixed connection of the elements (Fig. 13.5) acquires the form of the "unfolded" stator equivalent circuit with the parallel connection of the elements (Fig. 13.7).

13.8 Magnetizing and Slot Leakage Circuit Loops: “Folded” Equivalent Circuit of Stator

Stator magnetizing circuit loop and stator magnetizing reactance. In the equivalent circuit in Fig. 13.7, the parallel branches $0 - 0$, $k - k$, $z - z$, and $a - a$ represent the magnetizing circuit loops reflecting the layers of the stator model (Fig. 13.1). In this regard, they can be combined into the one common circuit loop that will represent the stator magnetizing circuit loop. The reactance used in this common circuit loop reflects the stator magnetizing reactance. On the basis of the equivalent circuit in Fig. 13.7, the stator magnetizing reactance is determined as

$$x^*_{m1} = \frac{1}{\frac{1}{x^*_{01}} + \frac{1}{\overset{\circ}{x}_{ks}} + \frac{1}{\overset{\circ}{x}_{zs}} + \frac{1}{\overset{\circ}{x}_{as}}} = \frac{x^*_{01}}{1 + x^*_{01}\left(\frac{1}{\overset{\circ}{x}_{ks}} + \frac{1}{\overset{\circ}{x}_{zs}} + \frac{1}{\overset{\circ}{x}_{as}}\right)} = \frac{x^*_{01}}{k_{H1}} \tag{13.20}$$

where $k_{H1} = 1 + x^*_{01}\left(\frac{1}{\overset{\circ}{x}_{ks}} + \frac{1}{\overset{\circ}{x}_{zs}} + \frac{1}{\overset{\circ}{x}_{as}}\right)$.

Here, the reactance x^*_{m1} represents the stator magnetizing reactance, and the value of $k_{н1}$ reflects the saturation factor of the stator magnetizing circuit.

Stator slot leakage circuit loop and stator slot leakage reactance. In the equivalent circuit in Fig. 13.7, the parallel branch $1 - 1$ represents the stator slot leakage circuit loop. The reactance used in the stator slot leakage circuit loop is determined by the expression (13.18). In (13.18), we use the following factors

$$k^2_{11} = c^2_{\tau ks} c^2_{\tau zs}; k_{1k} = c^2_{\tau ks} c_{\tau zs}; k_{10} = c_{\tau ks} c_{\tau zs} \tag{13.21}$$

Now, for the impedance $\overset{\circ}{Z}_1$ used in (13.18), we can obtain with considering (13.21)

$$\overset{\circ}{Z}_1 = r^*_{11} k^2_{11} + j\left(x^*_{\tau 01} k_{10} + x^*_{\tau ks} k_{1k} + x^*_{11\sigma} k^2_{11}\right) + Z^*_c k^2_{11} \tag{13.22}$$

In accordance with (13.22), the sum of the reactances $x^*_{\tau 01} k_{10}, x^*_{\tau ks} k_{1k}$ and $x^*_{11\sigma} k^2_{11}$, used in the circuit loop $1 - 1$ reflects the slot leakage reactance of the stator winding $x^*_{1\sigma}$ for which we have that $x^*_{1\sigma} = \widetilde{x}^*_{11\sigma} c^2_{\tau ks} c^2_{\tau zs} = x^*_{\tau 01} k_{10} + x^*_{\tau ks} k_{1k} + x^*_{11\sigma} k^2_{11}$. Using in (13.22) the condition for $x^*_{1\sigma}$ and the condition for the impedance $Z^*_c k^2_{11}$ in the form $\left(-\widetilde{U}^*_1\right) = Z^*_c k^2_{11} \widetilde{H}_1$, where $\widetilde{U}^*_1 = \dot{U}^*_1 k_{11}$ and $\widetilde{H}_1 = \dot{H}_1 / k_{11}$ (here, we assume that $k_{11} \approx 1.0$), and also taking into account the equation for the reactance x^*_{m1} obtained in (13.20), the equivalent circuit in Fig. 13.7 can be reduced to the form in Fig. 13.8. In this case, Fig. 13.8 represents the “folded” stator equivalent circuit which is determined from the “unfolded” stator equivalent circuit with the parallel connection of the elements (Fig. 13.7). As follows from the equivalent circuit in Fig. 13.8, in relation to $\dot{E}_m$ the “folded” stator equivalent circuit is represented as the two parallel circuit loops: the magnetizing circuit loop containing the reactance x^*_{m1} and the slot

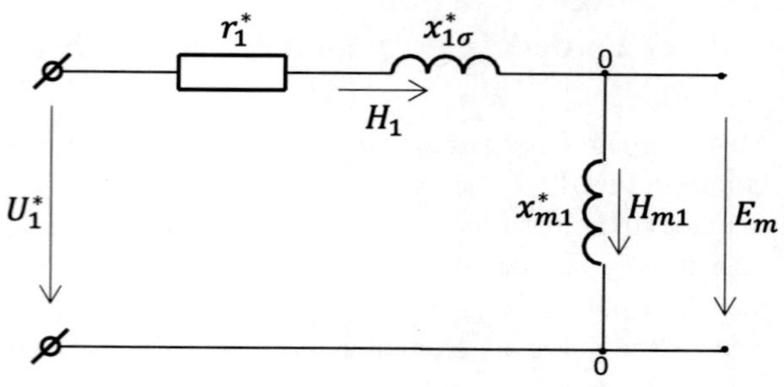

Fig. 13.8 "Folded" stator equivalent circuit

leakage circuit loop having the reactance $x_{1\sigma}^*$. The equivalent circuit in Fig. 13.8 has a simple structure, and therefore, it is convenient for practical use.

In the equivalent circuit in Fig. 13.8, the ohmic resistance of the stator winding is used. For this resistance, we have

$$r_1^* = r_{11}^* k_{11}^2 \tag{13.23}$$

In Fig. 13.8, the reactance $x_{1\sigma}^*$ reflects the slot leakage reactance of the stator winding. For this reactance, from (13.22), we have

$$x_{1\sigma}^* = x_{\tau 01}^* k_{10} + x_{\tau ks}^* k_{1k} + x_{11\sigma}^* k_{11}^2 \tag{13.24}$$

This reactance consists of three components: the leakage reactance of the air gap sub layer adjoining to the stator bore surface $\left(x_{\tau 01}^* k_{10}\right)$, the leakage reactance of the layer reflecting the slot-wedge area $\left(x_{\tau ks}^* k_{1k}\right)$, and the leakage reactance of the layer representing the stator winding location area $(x_{11\sigma}^* k_{11}^2)$. As follows from (13.24), these reactances are expressed through the reactances $x_{\tau 01}^*, x_{\tau ks}^*$ and $x_{11\sigma}^*$ determined in accordance with (13.10) as

$$\begin{aligned} x_{\tau 01}^* &= \omega_1 \mu_0 t_0 \frac{\delta}{2t_0} k_{1\sigma} \\ x_{\tau ks}^* &= \omega_1 \mu_{xks} t_0 \frac{h_{ks}}{t_{ks}} \xi_{Lks} \\ x_{11\sigma}^* &= \omega_1 \mu_{xzs} t_0 \frac{h_{zs}}{3t_{zs}} \xi_{Lzs} \end{aligned} \tag{13.25}$$

In (13.25), we use the conditions of kind following from (13.8)

$$\mu_{xks} = \mu_0 \frac{t_{ks}}{b_{ks}} \quad \mu_{xzs} = \mu_0 \frac{t_{zs}}{b_{\text{пs}}} \tag{13.26}$$

Then, from (13.24), (13.25), and (13.26), the reactances $x_{\tau 01}^* k_{10}, x_{\tau ks}^* k_{1k}$, and $x_{11\sigma}^* k_{11}^2$ used in (13.24) can be expressed as

$$
\begin{aligned}
x^*_{\tau 01}k_{10} &= \omega_1\mu_0 t_0 \frac{\delta}{2t_0}k_{1\sigma}k_{10} = \omega_1\mu_0 t_0 \lambda_{\tau 01} \\
x^*_{\tau ks}k_{1k} &= \omega_1\mu_0 t_0 \frac{h_{ks}}{b_{ks}}\xi_{Lks}k_{1k} = \omega_1\mu_0 t_0 \lambda_{\tau ks} \\
x^*_{11\sigma}k^2_{11} &= \omega_1\mu_0 t_0 \frac{h_{zs}}{3b_{\text{пs}}}\xi_{Lzs}k^2_{11} = \omega_1\mu_0 t_0 \lambda_{11\sigma}
\end{aligned}
\tag{13.27}
$$

where $\lambda_{\tau 01} = \frac{\delta}{2t_0}k_{1\sigma}k_{10}; \lambda_{\tau ks} = \frac{h_{ks}}{b_{ks}}\xi_{Lks}k_{1k}; \lambda_{11\sigma} = \frac{h_{zs}}{3b_{\text{пs}}}\xi_{Lzs}k^2_{11}$.

In (13.27), the dimensionless quantities $\lambda_{\tau 01}$, $\lambda_{\tau ks}$ and $\lambda_{11\sigma}$ represent the relative magnetic conductivities for the tangential components of the magnetic fluxes in the layers reflecting the air gap sub layer adjoining to the stator bore surface, the stator slot-wedge area, and the stator winding location area, respectively.

Now, from (13.24) and (13.27), the reactance $x^*_{1\sigma}$ can be obtained

$$
\begin{aligned}
x^*_{1\sigma} &= x^*_{\tau 01}k_{10} + x^*_{\tau ks}k_{1k} + x^*_{11\sigma}k^2_{11} \\
&= \omega_1\mu_0 t_0(\lambda_{\tau 01} + \lambda_{\tau ks} + \lambda_{11\sigma}) = \omega_1\mu_0 t_0 \lambda_{1\sigma}
\end{aligned}
\tag{13.28}
$$

where $\lambda_{1\sigma} = \lambda_{\tau 01} + \lambda_{\tau ks} + \lambda_{11\sigma}$.

Here, the quantity $\lambda_{1\sigma}$ reflects the relative magnetic conductivity for the slot leakage magnetic flux of the stator winding.

13.9 "Unfolded" Equivalent Circuit with Parallel Connection of Elements of Rotor

We consider the "unfolded" rotor equivalent circuit with the mixed connection of the elements (Fig. 13.6). By this equivalent circuit, the "unfolded" rotor equivalent circuit with the parallel connection of the elements can be obtained. For this purpose, in the equivalent circuit in Fig. 13.6, its parallel branches $k - k$, $z - z$, $2 - 2$, and $a - a$ can be transferred to the terminals of the transverse branch $0 - 0$. The process of transferring the parallel branches $k - k$, $z - z$, $2 - 2$, and $a - a$ is carried out by the use of the multiple transfer technique realized by introducing the transfer factors [1]. The implementation of the process of transferring the parallel branches of the equivalent circuit in Fig. 13.6. it is convenient to start with transferring its parallel branch $k - k$ to the terminals $0 - 0$ (Appendix F). For the parallel branch $k - k$, the transfer factor is

$$
c_{\tau kR} = 1 + \frac{x^*_{\tau 02}}{x^*_{kR}} \tag{13.29}
$$

The reactance of the transferred parallel branch $k - k$ is determined as

$$\overset{\circ}{x}_{kR} = x^*_{kR} c_{\tau kR} = x^*_{kR} + x^*_{\tau 02} \tag{13.30}$$

The transfer factor for the parallel branch $z - z$ is.

$$c_{\tau zR} = 1 + \frac{x^*_{\tau kR} c^2_{\tau kR} + x^*_{\tau 02} c_{\tau kR}}{x^*_{zR} c^2_{\tau kR}} \tag{13.31}$$

The reactance of the transferred parallel branch $z - z$ is derived as

$$\overset{\circ}{x}_{zR} = x^*_{zR} c_{\tau zR} c^2_{\tau kR} = x^*_{zR} c^2_{\tau kR} + x^*_{\tau kR} c^2_{\tau kR} + x^*_{\tau 02} c_{\tau kR} \tag{13.32}$$

The transfer factor for the parallel branch $2 - 2$ is

$$c_{\tau 2} = 1 + \frac{x^*_{\tau kR} c^2_{\tau kR} c_{\tau zR} + x^*_{\tau 02} c_{\tau kR} c_{\tau zR}}{\left(r^*_{22}/s + jx^*_{22\sigma} + Z^*_{2\text{H}}\right) c^2_{\tau kR} c^2_{\tau zR}} \tag{13.33}$$

The reactance of the transferred parallel branch $2 - 2$ is expressed as

$$\begin{aligned}\overset{\circ}{Z}_2 &= \left(r^*_{22}/s\right) c^2_{\tau kR} c^2_{\tau zR} + j\left(x^*_{22\sigma} c^2_{\tau kR} c^2_{\tau zR} + x^*_{\tau kR} c^2_{\tau kR} c_{\tau zR} + x^*_{\tau 02} c_{\tau kR} c_{\tau zR}\right) \\ &+ Z^*_{2H} c^2_{\tau kR} c^2_{\tau zR} = \left(r^*_{22}/s\right) c^2_{\tau kR} c^2_{\tau zR} + jx^*_{2\sigma} + Z^*_{2H} c^2_{\tau kR} c^2_{\tau zR}\end{aligned} \tag{13.34a}$$

where $x^*_{2\sigma} = x^*_{22\sigma} c^2_{\tau kR} c^2_{\tau zR} + x^*_{\tau kR} c^2_{\tau kR} c_{\tau zR} + x^*_{\tau 02} c_{\tau kR} c_{\tau zR}$.

In another form

$$\begin{aligned}\overset{\circ}{Z}_2 &= \left(r^*_{22}/s\right) c^2_{\tau kR} c^2_{\tau zR} \\ &+ j\left[x^*_{\tau 02}\left(1/c_{\tau kR} c_{\tau zR}\right) + x^*_{\tau kR}\left(1/c_{\tau zR}\right) + x^*_{22\sigma}\right] c^2_{\tau kR} c^2_{\tau zR} + Z^*_{2H} c^2_{\tau kR} c^2_{\tau zR} \\ &= \left(r^*_{22}/s + \widetilde{x}^*_{22\sigma} + Z^*_{2H}\right) c^2_{\tau kR} c^2_{\tau zR}\end{aligned} \tag{13.34b}$$

where $\widetilde{x}^*_{22\sigma} = x^*_{\tau 02}\left(1/c_{\tau kR} c_{\tau zR}\right) + x^*_{\tau kR}\left(1/c_{\tau zR}\right) + x^*_{22\sigma}; x^*_{2\sigma} = \widetilde{x}^*_{22\sigma} c^2_{\tau kR} c^2_{\tau zR}$.

When $c_{\tau 2} \approx 1.0$, the reactance of the transferred parallel branch $a - a$ is calculated by the expression

$$\overset{\circ}{x}_{aR} = x^*_{aR} c^2_{\tau kR} c^2_{\tau zR} + x^*_{\tau zR} c^2_{\tau kR} c^2_{\tau zR} + x^*_{\tau kR} c^2_{\tau kR} c_{\tau zR} + x^*_{\tau 02} c_{\tau kR} c_{\tau zR} \tag{13.35}$$

When using the reactances obtained in (13.30), (13.32), (13.34a), and (13.35), the "unfolded" rotor equivalent circuit with the mixed connection of the elements (Fig. 13.6) acquires the form of the "unfolded" rotor equivalent circuit with the parallel connection of the elements (Fig. 13.9).

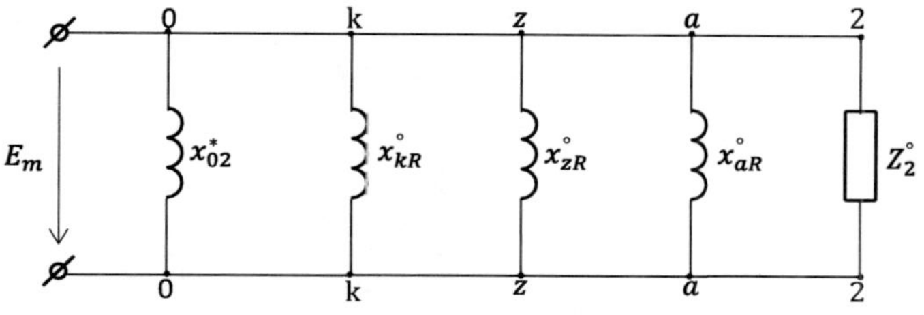

Fig. 13.9 "Unfolded" rotor equivalent circuit with the parallel connection of the elements

13.10 Magnetizing and Slot Leakage Circuit Loops: "Folded" Equivalent Circuit of Rotor

Magnetizing circuit loop and rotor magnetizing reactance. In the equivalent circuit in Fig. 13.9, the parallel branches $0-0$, $k-k$, $z-z$, and $a-a$ represent the magnetizing circuit loops reflecting the layers of the rotor model (Fig. 13.1). In this regard, they can be combined into the one common circuit loop that will represent the rotor magnetizing circuit loop. The reactance used in this common circuit loop reflects the rotor magnetizing reactance. On the basis of the equivalent circuit in Fig. 13.9, the rotor magnetizing reactance is determined as

$$x^*_{m2} = \frac{1}{\dfrac{1}{x^*_{02}} + \dfrac{1}{\overset{\circ}{x}_{kR}} + \dfrac{1}{\overset{\circ}{x}_{zR}} + \dfrac{1}{\overset{\circ}{x}_{aR}}} = \frac{x^*_{02}}{1 + x^*_{02}\left(\dfrac{1}{\overset{\circ}{x}_{kR}} + \dfrac{1}{\overset{\circ}{x}_{zR}} + \dfrac{1}{\overset{\circ}{x}_{aR}}\right)} = \frac{x^*_{02}}{k_{н2}} \tag{13.36}$$

where $k_{н2} = 1 + x^*_{02}\left(\dfrac{1}{\overset{\circ}{x}_{kR}} + \dfrac{1}{\overset{\circ}{x}_{zR}} + \dfrac{1}{\overset{\circ}{x}_{aR}}\right)$.

Here, the reactance x^*_{m2} represents the rotor magnetizing reactance, and the value of $k_{н2}$ reflects the saturation factor of the rotor magnetizing circuit.

Slot leakage circuit loop and rotor slot leakage reactance. In the equivalent circuit in Fig. 13.9, the parallel branch $2-2$ represents the rotor slot leakage circuit loop. The reactance used in the rotor slot leakage circuit loop is determined by the expression (13.34). In (13.34), we use the following factors

$$k^2_{22} = c^2_{\tau kR} c^2_{\tau zR};\, k_{2k} = c^2_{\tau kR} c_{\tau zR};\, k_{20} = c_{\tau kR} c_{\tau zR} \tag{13.37}$$

Now, for the impedance $\overset{\circ}{Z}_2$ used in (13.34), we can obtain with considering (13.37)

$$\overset{\circ}{Z}_2 = \frac{r^*_{22}}{s} k^2_{22} + j\left(x^*_{\tau 02} k_{20} + x^*_{\tau kR} k_{2k} + x^*_{22\sigma} k^2_{22}\right) + Z^*_{2н} k^2_{22} \tag{13.38}$$

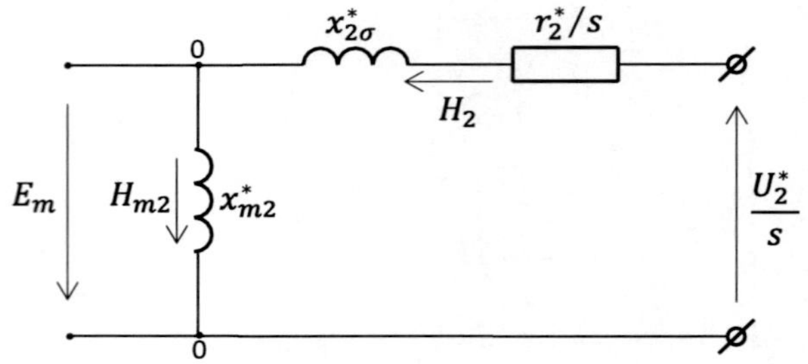

Fig. 13.10 "Folded" rotor equivalent circuit

In accordance with (13.38), the sum of the reactances $x_{\tau 02}^* k_{20}, x_{\tau kR}^* k_{2k}$ and $x_{22\sigma}^* k_{22}^2$ used in the circuit loop 2 – 2 reflects the slot leakage reactance of the rotor winding $x_{2\sigma}^*$ for which we have that $x_{2\sigma}^* = \widetilde{x}_{22\sigma}^* c_{\tau kR}^2 c_{\tau zR}^2 = x_{\tau 02}^* k_{20} + x_{\tau kR}^* k_{2k} + x_{22\sigma}^* k_{22}^2$. Using in (13.38) the condition for $x_{2\sigma}^*$ and the condition for the impedance $Z_{2H}^* k_{22}^2$ in the form $\left(\widetilde{U}_2^*/s\right) = Z_{2H}^* k_{22}^2 \widetilde{H}_2$, where $\widetilde{U}_2^*/s = \left(\dot{U}_2^*/s\right) k_{22}$ and $\widetilde{H}_2 = \dot{H}_2/k_{22}$, and also taking into account the equation for the reactance x_{m2}^* obtained in (13.36), the equivalent circuit in Fig. 13.9 can be reduced to the form in Fig. 13.10. In this case, Fig. 13.10 represents the "folded" rotor equivalent circuit which is determined from the "unfolded" rotor equivalent circuit with the parallel connection of the elements (Fig. 13.9). As follows from the equivalent circuit in Fig. 13.10, in relation to $\dot{E}_m$ the "folded" rotor equivalent circuit is represented as the two parallel circuitloops: the magnetizing circuit loop containing the reactance x_{m2}^* and the slot leakage circuit loop having the reactance $x_{2\sigma}^*$. The equivalent circuit in Fig. 13.10 has a simple structure, and therefore, it is convenient for practical use.

In the equivalent circuit in Fig. 13.10, the ohmic resistance of the rotor winding r_2^*/s is used. For this resistance, we have

$$\frac{r_2^*}{s} = \frac{r_{22}^*}{s} \tag{13.39}$$

In Fig. 13.10, the reactance $x_{2\sigma}^*$ reflects the slot leakage reactance of the rotor winding. For this reactance, from (13.38), we have

$$x_{2\sigma}^* = x_{\tau 02}^* k_{20} + x_{\tau kR}^* k_{2k} + x_{22\sigma}^* k_{22}^2 \tag{13.40}$$

This reactance consists of three components: the leakage reactance of the air gap sub layer adjoining to the rotor surface $\left(x_{\tau 02}^* k_{20}\right)$, the leakage reactance of the layer reflecting the rotor slot-wedge area $\left(x_{\tau kR}^* k_{2k}\right)$, and the leakage reactance of the layer representing the rotor winding location area $(x_{22\sigma}^* k_{22}^2)$. As follows from (13.40), these reactances are expressed through the reactances $x_{\tau 02}^*, x_{\tau kR}^*$, and $x_{22\sigma}^*$ determined in accordance with (13.12) as

$$
\begin{aligned}
x^*_{\tau 02} &= \omega_1 \mu_0 t_0 \frac{\delta}{2t_{02}} k_{2\sigma} k_R \\
x^*_{\tau kR} &= \omega_1 \mu_{xkR} t_0 \frac{h_{kR}}{t_{kR}} k_R \xi_{LkR} \\
x^*_{22\sigma} &= \omega_1 \mu_{xzR} t_0 \frac{h_{zR}}{3t_{zR}} k_R \xi_{LzR}
\end{aligned}
\tag{13.41}
$$

In (13.41), for μ_{xkR} and μ_{xzR}, we use the conditions of kind arising from (13.7)

$$
\mu_{xkR} = \mu_0 \frac{t_{kR}}{b_{kR}} \quad \mu_{xzR} = \mu_0 \frac{t_{zR}}{b_{nR}} \tag{13.42}
$$

Then, from (13.40) to (13.42), the reactances $x^*_{\tau 02} k_{20}, x^*_{\tau kR} k_{2k}$, and $x^*_{22\sigma} k^2_{22}$ can be expressed in the form

$$
\begin{aligned}
x^*_{\tau 02} k_{20} &= \omega_1 \mu_0 t_0 \frac{\delta}{2t_{02}} k_{2\sigma} k_R k_{20} = \omega_1 \mu_0 t_0 \lambda_{\tau 02} \\
x^*_{\tau kR} k_{2k} &= \omega_1 \mu_0 t_0 \frac{h_{kR}}{b_{kR}} k_R \xi_{LkR} k_{2k} = \omega_1 \mu_0 t_0 \lambda_{\tau kR} \\
x^*_{22\sigma} k^2_{22} &= \omega_1 \mu_0 t_0 \frac{h_{zR}}{3b_{nR}} k_R \xi_{LzR} k^2_{22} = \omega_1 \mu_0 t_0 \lambda_{22\sigma}
\end{aligned}
\tag{13.43}
$$

where $\lambda_{\tau 02} = \frac{\delta}{2t_{02}} k_{2\sigma} k_R k_{20}$; $\lambda_{\tau kR} = \frac{h_{kR}}{b_{kR}} k_R \xi_{LkR} k_{2k}$; $\lambda_{22\sigma} = \frac{h_{zR}}{3b_{\Pi R}} k_R \xi_{LzR} k^2_{22}$.

In (13.43), the dimensionless quantities $\lambda_{\tau 02}$, $\lambda_{\tau kR}$, and $\lambda_{22\sigma}$ represent the relative magnetic conductivities for the leakage magnetic fluxes in the layers reflecting the air gap sub layer adjoining to the rotor surface, the rotor slot-wedge area, and the rotor winding location area, respectively.

Now, from (13.40) and (13.43), the reactance $x^*_{2\sigma}$ can be obtained

$$
\begin{aligned}
x^*_{2\sigma} &= x^*_{\tau 02} k_{20} + x^*_{\tau kR} k_{2k} + x^*_{22\sigma} k^2_{22} \\
&= \omega_1 \mu_0 t_0 (\lambda_{\tau 02} + \lambda_{\tau kR} + \lambda_{22\sigma}) = \omega_1 \mu_0 t_0 \lambda_{2\sigma}
\end{aligned}
\tag{13.44}
$$

where $\lambda_{2\sigma} = \lambda_{\tau 02} + \lambda_{\tau kR} + \lambda_{22\sigma}$.

Here, the quantity $\lambda_{2\sigma}$ reflects the relative magnetic conductivity for the slot leakage magnetic flux of the rotor winding.

13.11 Equivalent Circuit of Machine

On the basis of the "folded" stator and rotor equivalent circuits obtained in Figs. 13.8 and 13.10, the machine equivalent circuit can be built. For this, the terminals $0-0$ of the equivalent circuit in Fig. 13.8 should be connected with the terminals $0-0$ of the equivalent circuit in Fig. 13.10 and also to introduce the machine magnetizing reactance determined as $x_m^* = x_{m1}^* x_{m2}^* / \left(x_{m1}^* + x_{m2}^*\right)$. Then, the equivalent circuits given in Figs. 13.8 and 13.10 can be combined into the one common equivalent circuit, as a result of which the machine equivalent circuit takes the form in Fig. 12.2b. It means that the machine equivalent circuit obtained on the basis of considering the resulting magnetic field and the machine equivalent circuit constructed with the use of the "partial" magnetic fields takes the same structure.

13.12 Basic Calculation Elements: Main Provisions for Determination

This work is devoted to the development of the method allowing us to determine the machine calculation elements on the basis of considering the machine resulting magnetic field obtained by Maxwell's equations. To solve this problem, the methods developed in the technical literature by the use of the "partial" magnetic fields can be analyzed in order to adapt them to the conditions of the application of the resulting magnetic field of the electric machine. Such an analysis is represented in this chapter above for the case when the machine resulting magnetic field is determined using the superposition of the equivalent circuits corresponding to the "partial" magnetic fields (machine magnetizing field and slot leakage magnetic fields of the stator and rotor windings) arising from the application of the "no-load and under-load" concept. Here, we use the results of this analysis to formulate the provisions that can be applied when the calculation elements of the electric machine are determined using the resulting magnetic field found by the superposition principle and by Maxwell's equations.

In this chapter, the machine resulting magnetic field is reflected by the "unfolded" stator and the "unfolded" rotor equivalent circuits. The "unfolded" stator equivalent circuit corresponding to the stator resulting magnetic field follows as a result of the superposition of the "unfolded" stator magnetizing equivalent circuit and the "unfolded" stator slot leakage equivalent circuit. The "unfolded" rotor equivalent circuit corresponding to the rotor resulting magnetic field follows as a result of the superposition of the "unfolded" rotor magnetizing equivalent circuit and the "unfolded" rotor slot leakage equivalent circuit.

In this chapter, the specific system of units $\left(\dot{E}, \dot{H}\right)$ is applied. In relation to the specific system of units $\left(\dot{E}, \dot{H}\right)$, the stator and rotor magnetizing circuits and the stator and rotor slot leakage magnetic circuits can be represented as the systems of the layers, i.e., as the multilayer systems. In this case, for the layers representing the stator and rotor slot-teeth areas, the calculated magnetic permeabilities are used. The surfaces of the layers reflecting these stator and rotor areas are taken conditionally "smooth." In this chapter, by the use of the layered models, the two-dimensional magnetizing fields and the one-dimensional slot leakage magnetic fields taking place in the stator and rotor of the machine are considered. Now, the "unfolded" stator and the "unfolded" rotor magnetizing equivalent circuits follow on the basis of considering the layered models for the stator and rotor magnetizing circuits. The "unfolded" stator and the "unfolded" rotor slot leakage equivalent circuits follow on the basis of considering the layered models for the stator and rotor slot leakage magnetic circuits. The "unfolded" stator and the "unfolded" rotor magnetizing equivalent circuits obtained in this way represent the result of the cascade way of connecting the L-shaped equivalent circuits reflecting the corresponding layers of the stator and rotor magnetizing circuit models (with the exception of the layers of the stator and rotor yoke areas). Therefore, the "unfolded" stator and the "unfolded" rotor equivalent circuits reflecting the machine resulting magnetic field represent the result of the cascade way of connecting the L-shaped equivalent circuits corresponding to the layers of the stator and rotor magnetic circuit models.

The equivalent circuits following in this way represent the "unfolded" stator and rotor equivalent circuits with the mixed connection of the elements. Then, the "unfolded" stator and rotor equivalent circuits with the mixed connection of the elements are reduced to the "folded" stator and rotor equivalent circuits for that the multiple transfer technique is used. Now, the construction of the machine equivalent circuit and the determination of the machine calculation elements are carried out using the "folded" stator and rotor equivalent circuits.

13.12.1 Application of Superposition Principle

When using the specific system of units $(\dot{E}, \dot{H})$ and establishing the machine resulting magnetic field by the application of the superposition principle, the determination of the calculation elements of the machine can be reduced to the use of the following basic provisions:

- introduction of the radial and tangential components of the magnetic permeabilities for the stator and rotor slot-teeth areas and representation of these stator and rotor areas as the magneto-anisotropic layers

- representation of the machine magnetizing circuit model as the multilayer system,
- decomposition of the machine magnetizing circuit model into the stator and rotor magnetizing circuit models,
- representation of the layers of the stator and rotor magnetizing circuit models as the L-shaped equivalent circuits (with the exception of the layers of the stator and rotor yoke areas),
- obtaining the "unfolded" stator and the "unfolded" rotor magnetizing equivalent circuits as the cascade way of connecting the L-shaped equivalent circuits,
- representation of the stator and rotor slot leakage magnetic circuit models as the multilayer systems,
- reduction of the layers of the stator and rotor magnetic circuit models to the calculated axial length equal to l_0 and to the pole pitch τ_0 determined on the surface $0 - 0$ passing through the middle of the radial length of the machine air gap,

- construction of the "unfolded" stator and the "unfolded" rotor slot leakage equivalent circuits,
- construction of the "unfolded" stator equivalent circuit corresponding to the machine resulting magnetic field as a result of the superposition of the "unfolded" stator magnetizing equivalent circuit and the "unfolded" stator slot leakage equivalent circuit and obtaining the "unfolded" stator equivalent circuit with the mixed connection of the elements,
- reducing the "unfolded" stator equivalent circuit with the mixed connection of the elements into the "unfolded" stator equivalent circuit with the parallel connection of the elements,
- introduction of the stator magnetizing and the stator slot leakage circuit loops and,
- determination of the magnetizing and slot leakage reactances used in them,
- obtaining the "folded" stator equivalent circuit,
- construction of the "unfolded" rotor equivalent circuit corresponding to the machine resulting magnetic field as a result of the superposition of the "unfolded" rotor magnetizing equivalent circuit and the "unfolded" rotor slot leakage equivalent circuit and obtaining the "unfolded" rotor equivalent circuit with the mixed connection of the elements,
- reducing the "unfolded" rotor equivalent circuit with the mixed connection of the elements into the "unfolded" rotor equivalent circuit with the parallel connection of the elements,
- introduction of the rotor magnetizing and the rotor slot leakage circuit loops and,
- determination of the magnetizing and slot leakage reactances used in them,
- obtaining the "folded" rotor equivalent circuit,
- combining the "folded" stator and the "folded" rotor equivalent circuits by the introduction of the machine magnetizing circuit loop and the determination of the machine magnetizing reactance,
- obtaining the equivalent circuit of the machine,
- compiling the machine voltage equations using the equivalent circuit of the machine.

These provisions correspond to the case when the machine resulting magnetic field is determined as a result of the superposition of the two-dimensional machine magnetizing field and the one-dimensional stator and rotor slot leakage magnetic fields for what the corresponding equivalent circuits are used.

13.12.2 Application of "Field" Concept

Above, the provisions corresponding to the case, when the machine resulting magnetic field is implemented using the "unfolded" stator and the "unfolded" rotor equivalent circuits obtained as a result of the superposition of the "unfolded" stator and the "unfolded" rotor magnetizing equivalent circuits and the "unfolded" stator and the "unfolded" rotor slot leakage equivalent circuits were described. Let us generalize the provisions given above for the case of constructing the equivalent circuit of the machine using the resulting magnetic field determined applying Maxwell's equations ("field" concept). We note that the construction of the machine equivalent circuit is reduced, in this case, to the use of the techniques and procedures aimed at decomposing the resulting magnetic field obtained by applying Maxwell's equations into the machine magnetizing field and the stator and rotor slot leakage magnetic fields. Now, when applying the "field"concept the determination of the calculation elements of the electric machine can be characterized by using the following provisions:

- obtaining the models for the stator and rotor of the machine. In this case, it is assumed that there is a two-dimensional magnetic field in the cross section of the machine and there is a one-dimensional electric field in the axial direction of the machine. In this regard, we can talk about building the layered models for the stator and rotor. In these models, the stator and rotor slot-teeth areas can be represented as the homogeneous layers for which the calculated values of the radial and tangential components of the magnetic permeabilities can be obtained. In general case, the stator and rotor layered models can include the conductive, nonconductive, isotropic, anisotropic, passive, and active layers. In this work, the layered models representing the stator and rotor of the machine are determined in relation to the surface $0-0$ passing through the middle of the radial length of the air gap. In this case, the machine air gap can be considered subdivided into two sub layers. The air gap upper sub layer adjoining to the stator bore surface belongs to the stator model, and the air gap lower sub layer adjoining to the rotor surface belongs to the rotor model.

- solving the field equations, obtaining the expressions for the tangential components of the electric and magnetic field intensities. The equations derived in this way can be expressed through the tangential components of the electric and magnetic field intensities given on the outer and inner surfaces of the ith layer of the stator and rotor models. The equations obtained in relation to the electric and magnetic field intensities defined on the outer and inner surfaces of the ith layer, represent the generalized field equations for the ith layer of the stator and rotor models.

- visualization of the field equations for the ith layer by the L-shaped equivalent circuit. In our case, using the L-shaped equivalent circuit means representing the ith layer as the two circuit loops: the "own" circuit loop and the "connecting" circuit loop. The "connecting" circuit loop connects the "own" circuit loops representing the adjacent layers of the stator and rotor models. In the case of considering the ith conductive layer at the weak skin effect, its "own" circuit loop is represented consisting of the two parallel circuit loops: the "own" magnetizing circuit loop of the ith layer and the "own" circuit loop of the eddy currents induced in the ith layer.

- construction of the stator and rotor equivalent circuits by the cascade way of connecting the L-shaped equivalent circuits reflecting the corresponding layers of the stator and rotor models. The equivalent circuits obtained in this way represent the "unfolded" stator and the "unfolded" rotor equivalent circuits with the mixed connection of the elements.

Further, the transformation of the "unfolded" stator and the "unfolded" rotor equivalent circuits and determination of the calculation elements of the electric machine are carried out according to the similar scheme described above for the case of determining the machine resulting magnetic field using the superposition principle.

With the above way of defining the machine calculation elements, the machine resulting magnetic field determined by Maxwell's equations can be represented decomposed into the machine magnetizing field and the slot leakage magnetic fields of the stator and rotor windings.

As follows from the above, the basic provisions following from the application of the machine resulting magnetic field determined as a result of the superposition of the "partial" magnetic fields and described using Maxwell's equations are not fundamentally different. Below, we will use these provisions when defining the machine calculation elements is based on considering the machine resulting magnetic field found by Maxwell's equations.

13.13 Summary

1. In this chapter, the specific system of units $\left(\dot{E}, \dot{H}\right)$ is used. In relation to the specific quantities, the surfaces of the sections representing the stator and rotor magnetizing circuits and the surfaces of the sections reflecting the stator and rotor slot leakage magnetic circuits can be considered as conditionally "smooth." Consequently, the stator and rotor magnetizing circuits and stator and rotor slot leakage magnetic circuits can be represented as layered systems. On this basis, for the stator and rotor magnetizing circuits and for the stator and rotor slot leakage magnetic circuits, the layered models are proposed.
2. The "unfolded" stator and the "unfolded" rotor magnetizing equivalent circuits are obtained. The "unfolded" stator and the "unfolded" rotor magnetizing equivalent circuits reflect the two-dimensional magnetizing field taking place in the machine, and they are obtained as a result of the cascade way of connecting the L-shaped equivalent circuits representing the corresponding layers of the stator and the rotor magnetizing circuit models.
3. The "unfolded" stator and the "unfolded" rotor slot leakage equivalent circuits are received. The "unfolded" stator and the "unfolded" rotor slot leakage equivalent circuits correspond to the one-dimensional slot leakage magnetic fields and they are constructed as a result of series connecting the one-element branches reflecting the corresponding layers of the stator and the rotor slot leakage magnetic circuit models.
4. The "unfolded" stator and the "unfolded" rotor equivalent circuits with the mixed connection of the elements are proposed. The "unfolded" stator equivalent circuit represents the result of the superposition of the "unfolded" stator magnetizing equivalent circuit and the "unfolded" stator slot leakage equivalent circuit. The "unfolded" rotor equivalent circuit represents the result of the superposition of the "unfolded" rotor magnetizing equivalent circuit and the

"unfolded" rotor slot leakage equivalent circuit. The "unfolded" stator and the rotor equivalent circuits obtained in this way reflect the resulting magnetic field taking place in the electric machine.

5. The "unfolded" stator and the "unfolded" rotor equivalent circuits with the mixed connection of the elements represent the result of the cascade way of connecting the L-shaped equivalent circuits reflecting the corresponding layers of the stator and rotor magnetic circuit models. The use of the L-shaped equivalent circuits indicates the presence of the radial and the tangential magnetic field components in the layers of the stator and the rotor magnetic circuit models.
6. The "unfolded" stator and the "unfolded" rotor equivalent circuits are reduced to the "folded" stator and the "folded" rotor equivalent circuits for what the multiple transfer technique is applied. The "folded" stator equivalent circuit is due to the introduction of the stator magnetizing circuit loop and the stator slot leakage circuit loop. The "folded" rotor equivalent circuit is due to the introduction of the rotor magnetizing circuit loop and the rotor slot leakage circuit loop. The "folded" stator and the "folded" rotor equivalent circuits are convenient for practical use.
7. The construction of the "folded" stator and the "folded" rotor equivalent circuits can be carried out in the two stages. At the first stage, the "unfolded" stator and the "unfolded" rotor equivalent circuits with the mixed connection of the elements are reduced to the "unfolded" stator and the "folded" rotor equivalent circuits with the parallel connection of the elements. At the second stage, the "unfolded" stator and the "unfolded" rotor equivalent circuits with the parallel connection of the elements are reduced to the "folded" stator and the "folded" rotor equivalent circuits.
8. The "folded" stator and the "folded" rotor equivalent circuits can be used to construct the machine equivalent circuit for which the machine magnetizing circuit loop and machine magnetizing reactance can be introduced. The machine equivalent circuit can be considered consisting of the three circuit loops: the machine magnetizing circuit loop, the stator slot leakage circuit loop, and the rotor slot leakage circuit loop.
9. The machine voltage equations are derived by the use of the machine equivalent circuit. Consequently, when considering the machine resulting magnetic field, first, the machine equivalent circuit is built, and then, using this equivalent circuit, the machine voltage equations are obtained. Consequently, when using the resulting magnetic field, the role of the equivalent circuit in the theory of the electric machine increases.
10. In this chapter, the machine calculation elements are received on the basis of considering the machine resulting magnetic field determined as a result of the superposition of the machine magnetizing field and the stator and the rotor slot leakage magnetic fields. As the generalization of this approach, the basic provisions are formulated for the case when the machine resulting magnetic field is determined by Maxwell's equations. Below, to define the machine calculation elements applying the "field" concept, we will use these provisions.

References

1. Asanbayev, V.: Alternating Current Multi-Circuit Electric Machines: A New Approach to Steady-State Parameter Determination, Springer Charm Heidelberg, New-York, Dordrecht London (2015)
2. Mishkin, E.: Theory of Squirrel-Cage Induction Motor Derived Directly from Maxwell's Field Equations, Quart. J. Mech. Math., 7, pt. 4, p. 472-487 (1954)
3. Veselovsky, O., Konyaev, A., Sarapulov, F.: Linear Asynchronous Motor, Publishing Energoatomizdat, Moscow (1991)
4. Postnikov, I., Bezusyi, L.: Calculation of Traveling Electromagnetic Field in Multilayer Medium, Izvestiya ANSSSR. Energetika i Transport, n.2, (1970)
5. Freeman, E.: Traveling Waves in Induction Machines, Proc. IEEE, vol. 114, n. 11, 1968, pp. 1772-1776
6. Freeman, E.: Equivalent Circuits from Electromagnetic Theory. Low Frequency Induction Devices, Proc. IEEE, vol. 121, n. 10, 1974, pp. 1117-1121
7. Boldea, I., Bobescu, M.: Multilayer Approach to the Analysis of Single-Sided Linear Induction Motors, Proc. IEEE, vol. 125, n. 4, 1978, pp. 280-287
8. Williamson, S.: The Anisotropic Layer Theory of Induction Machines and Devices, Inst. Math. And Appl., 1976, Vol. 17, n. 1, pp. 69-84
9. Inkin, A., Litvinov, B.: Synthesis of Cascade Equivalent Circuits for Induction Electric Machines, Elektrotechnika, 1977, n. 1, pp. 29-34
10. Voldek. A.: Electric Machines, Energiya, Leningrad (1974)
11. Ivanov-Smolensky, A.: Electric Machines, Energiya, Moscow (1980)
12. Richter, R.: Electric Machines, vol. 1-5. United Scientific and Technical Publishing, Moscow (1934-1939)
13. Titov, V., Khutoretskiy, G., Zagarodnaya, G., Vartanyan, G., Zaslavsky, D., Smotrov, I.: Turbo Generators. Calculation and Design, Publishing Energiya, Leningrad (1967)

Chapter 14
L-Shaped Equivalent Circuits for Design Fragments of Machine

14.1 Layered Model of Machine: Prerequisites for Application

The electric machine is a rather complex design created on the basis of the need for the rational combination of the electric and magnetic circuits in it. The currents flowing in the windings cause the resulting magnetic field in the electric machine. Therefore, we can proceed from the fact that the electromagnetic processes taking place in the machine electric and magnetic circuits represent the result of the action of the resulting magnetic field. Due to this, for an analysis of the electromagnetic processes in the electric machine, it is necessary to have information about the complete picture of the resulting magnetic field distribution determined with considering the real design features of the electric and magnetic circuits. However, obtaining the complete picture of the electromagnetic field that accurately takes into account the real design of the electric machine is associated with huge costs, and in some cases, it seems practically impossible.

The electric machine is the device designed to convert an electromagnetic energy. As a consequence of this, the general theory for calculating the electric machines is focused on the description of the electromechanical processes associated with the conversion of an electromagnetic energy in the electric machine. The electromagnetic field aspects that do not significantly affect the energy processes in the electric machine may not be considered in determining the electromagnetic field, or the field aspects can be taken into account in the final expressions using the coefficients obtained as a result of the special studies. Thus, from the point of view of the energy processes (the processes associated with the conversion of an electromagnetic energy), the field picture in the electric machine can be simplified, and the simplified machine model can be used to calculate this simplified field picture.

The simplified machine model assumes some approximation of the real structure of the electric machine design. The replacement of the real machine windings with the surface current layers can be considered as such important approximation, for

V. Asanbayev, *Asynchronous Machines*,
https://doi.org/10.1007/978-3-030-92284-9_14

example [1–12, 25, 26]. The introduction of the surface currents allows the fairly complex toothed surfaces of the stator and rotor to be presented as smooth, which greatly contributes to the simplification of the machine model. The surface current layers are determined based on the analysis of the current distribution caused by the real machine windings. Although the current sheets represent the fictitious (calculated) current sheets, their introduction opens up wide opportunities for using the various analytical methods that simplify the analysis. So, for example, the surface current layers can be decomposed into the series of spatial harmonics with the sinusoidal spatial distribution, which makes it possible to obtain fairly simple solutions for the vectors of electromagnetic field intensities.

The next technique for simplifying the machine design is the neglect of the end effects caused by the frontal parts of the windings (finiteness of the axial length) of the electric machine. The processes associated with the end effects are usually considered separately [1–12]. In this regard, they are taken into account in the final expressions using the appropriate coefficients obtained as a result of the special analysis. Under these conditions, the electromagnetic field can be represented as localized in the active zone of the electric machine, where the main process of converting an electromagnetic energy takes place. It is assumed here that the values of the electromagnetic parameters ρ and μ of the materials used in the active zone of the machine do not depend on the field intensity, i.e., they have constant values. This provision allows the application of the superposition method. In addition, it is assumed that the magnetic field does not change in the direction of the axial length of the machine active zone. In this regard, the magnetic field in the machine becomes plane parallel and exists in the cross section of the electric machine. In accordance with these assumptions, the structure of the magnetic field is determined by the tangential and radial components of the magnetic field intensities $\left(\dot{H}_{\varphi} \text{ and } \dot{H}_R\right)$, located in the plane of the cross section of the machine, and by the one axial component of the current density vector (or electric field intensity $\dot{E} = \dot{E}_z$).

When developing methods for calculating and analysis, the resulting magnetic field of the machine created by the currents of the stator and rotor windings in its active zone is considered decomposed into the magnetizing field and the leakage magnetic fields of the machine windings [1–12]. The frontal and differential leakage magnetic fields are not considered in this work. In this regard, the calculation of the resulting magnetic field of the machine is reduced, in this case, to the calculation of the simple in the structure magnetizing field and the slot leakage magnetic fields of the stator and rotor windings.

In order to simplify the calculations, the magnetizing circuits and the slot leakage magnetic circuits of the stator and rotor are subdivided into sections [1–12]. It is assumed that the magnetic flux of the magnetizing field is closed mainly passing in the radial direction through the successive sections of the stator and rotor magnetizing circuits (in our case, the tangential components of the magnetizing flux of the machine are taken into account), and the magnetic fluxes of the slot leakage fields are closed passing in the tangential direction through the parallel sections of the slots leakage magnetic circuits. The representation of the magnetic circuits of the magnetizing fluxes and the magnetic circuits of the slot leakage fluxes in the form of the systems of the corresponding sections allows us to use the total current law, which simplifies the analysis. In this case, the phase system of units $(\dot{Э}, \dot{I})$ is usually used.

For example, in accordance with [13], for the sections of the magnetic circuits of the magnetizing fluxes and the magnetic circuits of the slot leakage fluxes, the average values of the magnetic permeabilities μ can be used. In this case, for the sections of the magnetic circuits of the magnetizing fluxes, the average value of the radial component of the magnetic permeability μ_R is usually determined (in our case, the radial and tangential components of the magnetic permeabilities μ_φ and μ_R are determined), and for the sections of the magnetic circuits of the slot leakage fluxes, the average value of the tangential component of the magnetic permeability μ_φ (with the exception of the machine air gap sublayers) is found. Under these conditions, the sections of the magnetic circuits of the magnetizing and the slot leakage fluxes can be represented in the form of homogeneous layers with smooth surfaces. Then, the models of the magnetic circuits of the magnetizing and the slot leakage fluxes are simplified, and they can be considered in the form of the layered systems [13]. In the layers obtained in this way for the magnetic circuits of the magnetizing fluxes, there is usually the radial (one-dimensional) magnetizing field (in our case, the two-dimensional magnetizing field is considered), and in the layers of the magnetic circuits of the slot leakage fluxes, there is the tangential (one-dimensional) slot leakage magnetic field, which can be calculated using the total current law. As shown in [13], when using the specific system of units $(\dot{E}, \dot{H})$, the layered models greatly simplify the analysis of the magnetic circuits of the magnetizing and the slot leakage magnetic fluxes.

As noted above, the resulting magnetic field that occurs in the active zone of the electric machine can be considered as a result of the superposition of the magnetizing field and the slot leakage fields of the machine windings. In this regard, the model corresponding to the resulting magnetic field of the machine can be obtained by the superposition of the layered model of the magnetic circuit of the magnetizing flux and the layered models of the magnetic circuits of slot leakage magnetic fluxes of the stator and rotor windings. As a result, we can have the layered model that allows us to calculate the resulting magnetic field in the electric machine. In this model, the layers, representing the corresponding sections of the slot-tooth areas of the stator and rotor magnetic circuits, are characterized by the radial and tangential components of the magnetic permeabilities, and they can be represented as the magneto-anisotropic layers. The presence of the windings in the corresponding layers of the slot-tooth areas can be taken into account by introducing the average value of the resistivity of the conductors of the stator and rotor windings. Then, such layers are usually considered as passive or active conducting magneto-anisotropic layers.

The air gap of the electric machine is represented as the nonconductive isotropic layer with the magnetic permeability equal to μ_0. Thus, the layered model intended for calculating the resulting magnetic field in the machine will contain the layers having the different electromagnetic parameters ρ and μ and reflecting the corresponding design areas of the electric machine active zone, for example [13–21].

In the model, the phase winding is represented in the form of the infinitely thin current filaments uniformly distributed in the layer reflecting the winding location section. When the phase winding is powered from the external source, the area of the phase winding is usually considered as the active layer. Consequently, in the general case, the model of the electric machine can contain the passive and active layers, as well as the isotropic and magneto-anisotropic, conducting and nonconducting layers. The models obtained in this way can be applied to describe using the field equations the electromagnetic processes occurring in the electric machine (Appendix H).

The layered models described above are more preferable for solving the problems related to the determination of the electric machine calculation elements (the circuit loops, parameters, equivalent circuits, and voltage equations). As shown in [13], when considering such problems, the components of the electric and magnetic fields intensities E and H, which represent the corresponding layers of the machine model, are reduced to some preselected surface. As such surface, in this work, the surface $0-0$ passing through the middle of the radial length of the air gap of the machine and having the axial calculated length equal to l_0 and the radius equal to R_0 is taken. When using the flat layered model, instead of R_0, the pole pitch τ_0 defined as $\tau_0 = (\pi/p)R_0$ should be used. Then, this "reduction" technique is reduced to multiplying the tangential components of the magnetic field intensities of the ith layer by the coefficient $\xi_{\tau i}$, equal to $\xi_{\tau i} = \tau_{i1}/\tau_0$, i.e. $\dot{H}_i = \dot{H}_{\varphi i}\xi_{\tau i}$. Here, τ_{i1} is the pole pitch of the outer (in the case of the rotor) and the inner (in the case of the stator) surface of the ith layer. In this case, the component of the electric field intensity of the ith layer is multiplied by the coefficient ξ_{Li} defined as $\xi_{Li} = l_i/l_0$, i.e. $\dot{E}_i = \dot{E}_{zi}\xi_{Li}$. The value l_i represents the axial length of the ith layer. Here $\dot{E}_i$ and $\dot{H}_i$ are the components of the electric and magnetic fields intensities of the ith layer reduced to the surface $0-0$ having the axial length l_0 and the pole pitch τ_0. $\dot{E}_{zi}$ and $\dot{H}_{\varphi i}$ are the magnitudes of the components of the electric and magnetic fields intensities on the outer (in the case of the rotor) and the inner (in the case of the stator) surface of the ith layer. The described "reduction" procedure is implemented with respect to the values $\dot{E}_{zi}$ and $\dot{H}_{\varphi i}$, which represent the layers used in the model of the electric machine. Therefore, the determination of the machine calculation elements is carried out, in this case, in relation to the components of the electric and magnetic field intensities $\dot{E}_i$ and $\dot{H}_i$ calculated as

$$\dot{E}_i = \dot{E}_{zi}\xi_{Li} \; \dot{H}_i = \dot{H}_{\varphi i}\xi_{\tau i} \tag{14.1}$$

As follows from these conditions, the impedances of the ith layer will be corrected, in this case, by the dimensionless coefficient equal to $\xi_{Li}/\xi_{\tau i}$ since in accordance with (14.1), we have that $Z_i = \dot{E}_i/\dot{H}_i = Z(\xi_{Li}/\xi_{\tau i})$, where $Z = \dot{E}_{zi}/\dot{H}_{\varphi i}$. The use of the calculated values $\dot{E}_i = \dot{E}_{zi}\xi_{Li}$ and $\dot{H}_i = \dot{H}_{\varphi i}\xi_{\tau i}$ means that the components of the electric and magnetic field intensities, representing the corresponding layers of the machine model, are reduced, as noted above, to the one surface $0 - 0$ having the axial length l_0 and the pole pitch equal to τ_0. Consequently, in the specific system of units $(\dot{E}_i, \dot{H}_i)$, the conditional layers of the electric machine model acquire the same axial length and the pole pitch (in our case, l_0 and τ_0).

The machine model in the form of the system of the flat layers is more convenient for an analysis. In this regard, the question of the transition from the system of the cylindrical layers to the system of the conditional flat layers can be considered. In this case, the cylindrical layers of the model can be reduced to the conditional flat layers with the axial length l_0 and the pole pitch τ_0 [13]. This reduction process can be represented as follows. At the first stage, the ith cylindrical layer of the machine is conventionally considered "elongated" into the ith flat layer, which has the axial length l_i and the pole pitch τ_i, where $\tau_i = (\pi/p)R_{i1}$. Here R_{i1} is the radius of the outer (in the case of the rotor) and the inner (in the case of the stator) surface of the considered i^{th} cylindrical layer. Then, for this i^{th} conditional flat layer, having the pole pitch τ_i, the analytical expressions for the components of the electric and magnetic fields intensities are written (in the Cartesian coordinate system). Further, these relations for the components of the electric and magnetic fields intensities of the ith conditional flat layer are reduced to the surface $0 - 0$ with the calculated axial length l_0 and the pole pitch equal to τ_0. Now, the formulas for $\dot{E}_i$ and $\dot{H}_i$ obtained for the ith cylindrical layer can be expressed through the expressions for $\dot{E}_i$ and $\dot{H}_i$ written for the ith conditional flat layer. The expressions obtained in this way for $\dot{E}_i$ and $\dot{H}_i$ will contain the hyperbolic and trigonometric functions characteristic of the plane model, and, in addition, will include dimensionless coefficients that take into account the curvature of the surfaces of the ith layers. Thus, in relation to these values $\dot{E}_i$ and $\dot{H}_i$, the cylindrical model of the electric machine can be represented as the conditional flat layered system. The described procedure is implemented for the layers used in the cylindrical model of the electric machine. As a result, we come to the machine model which represents the system of the conditional flat layers having the same axial length and the pole pitch, which are equal to l_0 and to τ_0, respectively. The machine planar layered model corresponding to the provisions given above is shown in Fig. 14.1.

The machine model obtained in this way (Fig. 14.1) consists, in our case, of the seven layers, reflecting the machine air gap area, the stator and rotor slot-wedge areas, and the stator and rotor winding location areas, as well as the stator and rotor yoke areas. On the surfaces of the adjacent layers of this machine model, the values of the electric and magnetic field intensities reduced to the length l_0 and the pole pitch τ_0 are used. For these field intensities, we can have

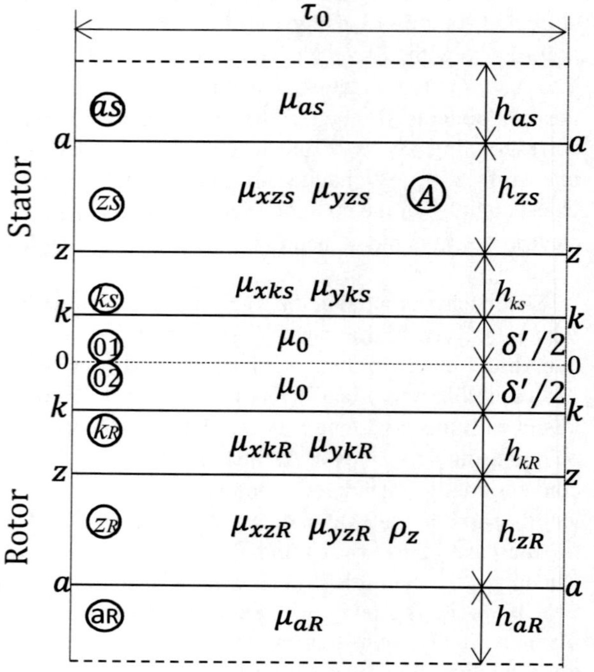

Fig. 14.1 Flat multilayer model of the electric machine

1. on the inner surfaces of the stator layers

$$
\begin{aligned}
\dot{E}_{01} &= \dot{E}_{ks1}; \dot{H}_{01} = \dot{H}_{ks1} \\
\dot{E}_{ks2} &= \dot{E}_{zs1}; \dot{H}_{ks2} = \dot{H}_{zs1} \\
\dot{E}_{zs2} &= \dot{E}_{as1}; \dot{H}_{zs2} = \dot{H}_{as1}
\end{aligned}
\tag{14.2a}
$$

2. on the outer surfaces of the rotor layers

$$
\begin{aligned}
\dot{E}_{01} &= \dot{E}_{kR1}; \dot{H}_{01} = \dot{H}_{kR1} \\
\dot{E}_{kR2} &= \dot{E}_{zR1}; \dot{H}_{kR2} = \dot{H}_{zR1} \\
\dot{E}_{zR2} &= \dot{E}_{aR1}; \dot{H}_{zR2} = \dot{H}_{aR1}
\end{aligned}
\tag{14.2b}
$$

The quantities $\dot{E}_{i1}, \dot{H}_{i1}$, and $\dot{E}_{i2}, \dot{H}_{i2}$, where $i = ks, zs, as$ and kR, zR, aR, obeying the conditions given in (14.2a) are determined using the expressions obtained in (14.1).

On the stator and rotor surfaces, the axial slots are milled, in which the windings and the slot wedges are placed. The distributed phase winding is usually placed in the slots of the stator, and the short-circuited winding in the form of the squirrel cage is often placed in the slots of the rotor. In this regard, in this work, the areas of the slot-teeth zones of the stator and rotor consist of the two layers. In the layered model in Fig. 14.1, the first layers of the stator and rotor represent the slot-wedge areas, and the second layers of the stator and rotor represent the winding location areas. For the layers representing in the machine model in Fig. 14.1 the stator and rotor slot-wedge areas and the stator and rotor winding location areas, the average values of the electromagnetic parameters ρ and μ can be used, as indicated above. Based on the works, for example [13, 14, 19–21], it can be obtained

3. for the layer of the stator slot-wedge area and the layer of the stator winding location area

$$\begin{aligned} \mu_{xks} &= \mu_0 \frac{t_{ks}}{b_{\text{п}s}} \quad \mu_{yks} = \mu_{zks} \frac{b_{zs}}{t_{ks}} \\ \mu_{xzs} &= \mu_0 \frac{t_{zs}}{b_{\text{п}s}} \quad \mu_{yzs} = \mu_{zzs} \frac{b_{zs}}{t_{zs}} \end{aligned} \tag{14.3a}$$

for the layer of the rotor slot-wedge area and the layer of the rotor winding location area

$$\begin{aligned} \mu_{xkR} &= \mu_0 \frac{t_{kR}}{b_{\text{п}R}} \qquad \mu_{ykR} = \mu_{zkR} \frac{b_{zR}}{t_{kR}} \\ \mu_{xzR} &= \mu_0 \frac{t_{zR}}{b_{\text{п}R}} \quad \mu_{yzR} = \mu_{zzR} \frac{b_{zR}}{t_{zR}}; \rho_2 = \rho_c \frac{t_2}{b_{\text{п}}} \end{aligned} \tag{14.3b}$$

where ρ_c is the resistivity of the rotor winding conductor.

14.1.1 Layered Models of Stator and Rotor

For the convenience of an analysis, the model of the electric machine shown in Fig. 14.1 is divided into two parts with respect to the surface $0 - 0$, passing through the middle of the radial length of the air gap. In this case, in relation to the surface $0 - 0$, the upper part of the model in Fig. 14.1 is attributed to the model of the stator, and its lower part is added to the model of the rotor. Thus, as follows from Fig. 14.1, the air gap sublayer adjacent to the stator bore surface and the layers reflecting the corresponding stator regions will represent the model of the stator. In Fig. 14.1, the model of the rotor includes the air gap sublayer adjacent to the rotor surface and the layers representing the corresponding areas of the rotor. In accordance with Fig. 14.1, the layered models of the stator and rotor represent the symmetric system and have a similar designation system. On the surfaces of the adjacent layers of the stator and rotor models, there are the electric and magnetic field intensities $\dot{E}_i$ and $\dot{H}_i$ reduced to the surface $0 - 0$. For them, the conditions given in (14.2a) are satisfied.

14.2 Generalized Equations of Layer: Circuit Visualization

In the multilayer model of the electric machine in Fig. 14.1, we will consider the ith layers related to the stator and rotor of the machine. A schematic outline of these layers is shown in Fig. 14.2. Figure 14.2a shows the ith stator layer and Fig. 14.2b shows the ith rotor conducting layer. Below, we will consider the ith rotor layer in accordance with Fig. 14.2b. In Fig. 14.2b, the values $\dot{E}_{i1}$ and $\dot{H}_{i1}$ represent the outer surface of the layer, and the values $\dot{E}_{i2}$ and $\dot{H}_{i2}$ refer to the inner surface of the ith rotor layer. Suppose that Fig. 14.2b shows the magneto-anisotropic conducting layer corresponding to the general case. For such i^{th} layer, we obtain generalized equations compiled with respect to the values of the tangential components of the electric and magnetic field intensities given on the outer and inner surfaces of the considered layer, and then, these equations can be visualized by the equivalent circuit. To solve these problems, it is necessary, first of all, to establish the expressions for the values of the components of the electric and magnetic field intensities on the outer and inner surfaces of the ith layer, i.e., in other words, to get the generalized equations for the ith layer.

14.2.1 T-Shaped Equivalent Circuit of Layer

To derive the generalized equations for the ith layer and to visualize these equations by the equivalent circuit, we will use the results of the analytical and numerical methods of solving the field equations in the considered layer.

Generalized equations for the layer. The generalized equations for the ith layer are obtained in Appendix H. Below, we will use these equations.

Analytical solution of the field equations. The analytical solution of the field equations can be represented as (Appendix H)

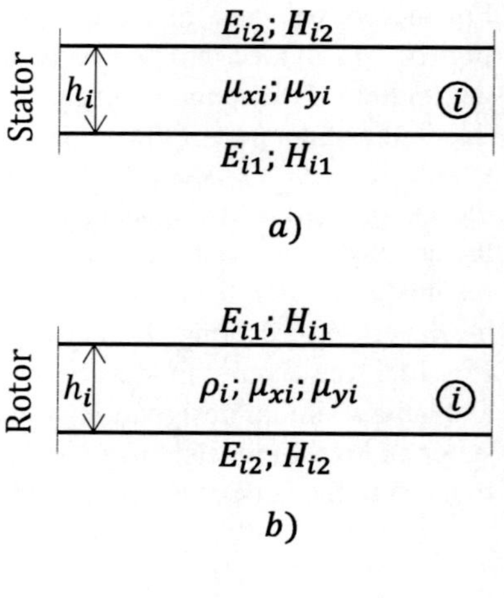

Fig. 14.2 Sketches of the ith flat layers of the stator and rotor: (**a**) stator i^{th} layer; (**b**) rotor i^{th} layer

$$\begin{aligned} \dot{E}_{i1} &= Z_{11}^i \dot{H}_{i1} + Z_{21}^i \dot{H}_{i2} \\ \dot{E}_{i2} &= Z_{12}^i \dot{H}_{i1} + Z_{22}^i \dot{H}_{i2} \end{aligned} \tag{14.4}$$

Here, for the coefficients $Z_{11}^i, Z_{21}^i, Z_{12}^i$ and Z_{22}^i obtained in Appendix H, we have

$$\begin{aligned} Z_{11}^i &= \frac{j\omega_1 \mu_{xi}}{\beta_i \tanh \beta_i h_i} \frac{\xi_{Li}}{\xi_{\tau i}}; Z_{21}^i = \frac{j\omega_1 \mu_{xi}}{\beta_i \sinh \beta_i h_i} \frac{\xi_{Li}}{\xi_{\tau i}} \\ Z_{12}^i &= \frac{j\omega_1 \mu_{xi}}{\beta_i \sinh \beta_i h_i} \frac{\xi_{Li}}{\xi_{\tau i}}; Z_{22}^i = \frac{j\omega_1 \mu_{xi}}{\beta_i \tanh \beta_i h_i} \frac{\xi_{Li}}{\xi_{\tau i}} \end{aligned} \tag{14.5}$$

where $\beta_i = \sqrt{\left(k_{\mu i} \frac{\pi}{\tau_i}\right)^2 + \frac{j\omega_1 s \mu_{xi}}{\rho_i}}; \xi_{\tau i} = \frac{\tau_i}{\tau_0}; \xi_{Li} = \frac{l_i}{l_0}$.

The system of the Eq. (14.4) represents the generalized equations of the ith layer, expressed in the form of Z.

T-shaped equivalent circuit of the layer. We get the T-shaped equivalent circuit for the ith layer. To do this, we will use the equations given in (14.4). To the first equation of the system (14.4), we add and subtract from it the term $Z_{21}^i \dot{H}_{i1}$. To the second equation of this system, we add and subtract from it the term $Z_{12}^i \dot{H}_{i2}$. Then, for the first equation of the system (14.4), we can have

$$\begin{aligned} \dot{E}_{i1} &= \left(Z_{11}^i - Z_{21}^i\right)\dot{H}_{i1} + Z_{21}^i\left(\dot{H}_{i1} + \dot{H}_{i2}\right) = Z_{\tau i1}^* \dot{H}_{i1} + Z_{i0}^* \dot{H}_{i0} \\ &= Z_{\tau i1}^* \dot{H}_{i1} + E_{i0} \end{aligned} \tag{14.6a}$$

The second equation of the system (14.4) takes the form

$$\begin{aligned} \dot{E}_{i2} &= \left(Z_{22}^i - Z_{12}^i\right)\dot{H}_{i2} + Z_{12}^i\left(\dot{H}_{i1} + \dot{H}_{i2}\right) = Z_{\tau i2}^* \dot{H}_{i2} + Z_{i0}^* \dot{H}_{i0} \\ &= Z_{\tau i2}^* \dot{H}_{i2} + E_{i0} \end{aligned} \tag{14.6b}$$

In the Eqs. (14.6a) and (14.6b), the following conditions are satisfied

$$\begin{aligned} & Z_{i0}^* = Z_{21}^i = Z_{12}^i \quad Z_{\tau i1}^* = Z_{11}^i - Z_{21}^i \quad Z_{\tau i2}^* = Z_{22}^i - Z_{12}^i \\ & \dot{H}_{i0} = \dot{H}_{i1} + \dot{H}_{i2} \qquad Z_{\tau i1}^* = Z_{\tau i2}^* \qquad E_{i0} = Z_{i0}^* \dot{H}_{i0} \end{aligned} \tag{14.7}$$

Based on the equations obtained in (14.6a) and (14.6b), for the ith layer, the T-shaped equivalent circuit shown in Fig. 14.3 follows. The parameters $Z_{\tau i1}^*, Z_{\tau i2}^*$ and Z_{i0}^* used in the equivalent circuit shown in Fig. 14.3 for the ith layer are obtained from (14.5) to (14.7) in the following form

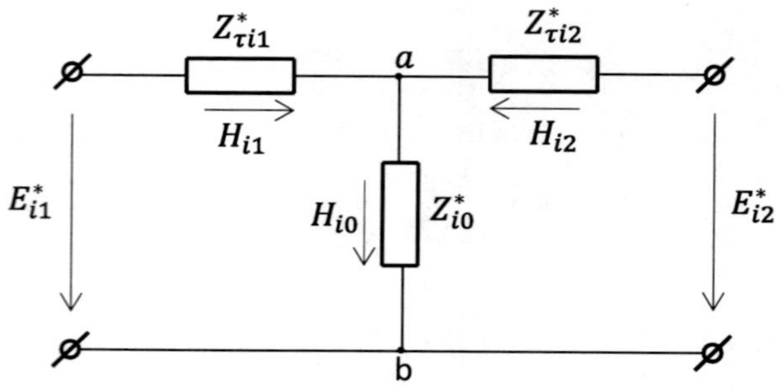

Fig. 14.3 T-shaped equivalent circuit of the ith layer

$$\begin{aligned} Z^*_{i0} &= \frac{j\omega_1\mu_{xi}}{\beta_i \sinh \beta_i h_i} \frac{\xi_{Li}}{\xi_{\tau i}} \\ Z^*_{\tau i1} = Z^*_{\tau i2} &= \frac{j\omega_1\mu_{xi}}{\beta_i} \tanh \beta_i \frac{h_i}{2} \frac{\xi_{Li}}{\xi_{\tau i}} \end{aligned} \tag{14.8}$$

As follows from (14.8), in the equivalent circuit in Fig. 14.3, the impedances presented in the general form are expressed through the hyperbolic functions.

Numerical solution of the field equations. In the numerical solution of the field equations, the numerical values of the components of the electric and magnetic field intensities are determined on the outer and inner surfaces of the ith layer. In this regard, in order to obtain the generalized equations for the ith layer, we use, in this case, the Poynting vector, for example [23–25]. An energy dissipated in the ith layer can be calculated using the Poynting vector as the difference in an energy penetrating into the layer through the one (e.g., outer) surface and exiting through the other (inner) surface of it. Then, for the specific power dissipated in the ith layer of the finite thickness, we can obtain

$$S_i = 0.5\left(\dot{E}_{i1}, \dot{H}^*_{i1} - \dot{E}_{i2}, \dot{H}^*_{i2}\right) \tag{14.9}$$

Here, the vectors $\dot{E}_{i1}, \dot{H}^*_{i1}$, and $\dot{E}_{i2}, \dot{H}^*_{i2}$ represent the values of the components of the electric and magnetic field intensities on the outer and inner surfaces of the ith layer, respectively (Fig. 14.2b). It should be noted that in (14.9) the components of the magnetic field intensities have the conjugate values.

We consider the equation given in (14.9). In (14.9), we introduce the undefined quantity $\dot{E}_{i0}$ for which we add and subtract in the right side of the Eq. (14.9) the terms represented as $0.5\dot{E}_{i0}\dot{H}_{i1}$ and $0.5\dot{E}_{i0}\dot{H}_{i2}$ [22]. Then, we get

$$\begin{aligned} S_i &= 0.5\left(\dot{E}_{i1}\dot{H}^*_{i1} - \dot{E}_{i2}\dot{H}^*_{i2} + \dot{E}_{i0}\dot{H}^*_{i1} - \dot{E}_{i0}\dot{H}^*_{i1} + \dot{E}_{i0}\dot{H}^*_{i2} - \dot{E}_{i0}\dot{H}^*_{i2}\right) \\ &= 0.5\left[\dot{H}^*_{i1}\left(\dot{E}_{i1} - \dot{E}_{i0}\right) + \dot{E}_{i0}\left(\dot{H}^*_{i1} - \dot{H}^*_{i2}\right) + \dot{H}^*_{i2}\left(\dot{E}_{i0} - \dot{E}_{i2}\right)\right] \\ &= S_{\tau i1} + S_{i0} + S_{\tau i2} \end{aligned} \tag{14.10}$$

where $S_{\tau i1} = 0.5\dot{H}^*_{i1}\left(\dot{E}_{i1} - \dot{E}_{i0}\right)$; $S_{i0} = 0.5\dot{E}_{i0}\left(\dot{H}^*_{i1} - \dot{H}^*_{i2}\right)$; $S_{\tau i2} = 0.5\dot{H}^*_{i2}\left(\dot{E}_{i0} - \dot{E}_{i2}\right)$.

We consider the components of the specific power $S_{\tau i1}$, S_{i0}, and $S_{\tau i2}$ dissipated in the ith layer. The component of the power $S_{\tau i1}$ can be reduced to the form

$$S_{\tau i1} = \frac{1}{2}\dot{H}_{i1}^{*}\left(\dot{E}_{i1} - \dot{E}_{i0}\right) = \frac{1}{2}\frac{\dot{E}_{i1} - \dot{E}_{i0}}{\dot{H}_{i1}}[H_{i1}]^2 = Z_{\tau i1}^{*}\frac{[H_{i1}]^2}{2} \tag{14.11}$$

where $Z_{\tau i1}^{*} = \dfrac{\dot{E}_{i1} - \dot{E}_{i0}}{\dot{H}_{i1}}$.

The power component of the ith layer $S_{\tau i2}$ can be represented as

$$S_{\tau i2} = \frac{1}{2}\dot{H}_{i2}^{*}\left(\dot{E}_{i0} - \dot{E}_{i2}\right) = \frac{1}{2}\frac{\dot{E}_{i0} - \dot{E}_{i2}}{\dot{H}_{i2}}[H_{i2}]^2 = Z_{\tau i2}^{*}\frac{[H_{i2}]^2}{2} \tag{14.12}$$

where $Z_{\tau i2}^{*} = \dfrac{\dot{E}_{i0} - \dot{E}_{i2}}{\dot{H}_{i2}}$.

The power component of the ith layer S_{i0} can be transformed as follows

$$S_{i0} = \frac{1}{2}\dot{E}_{i0}\left(\dot{H}_{i1}^{*} - \dot{H}_{i2}^{*}\right) = \frac{\dot{E}_{i0}}{\dot{H}_{i1} - \dot{H}_{i2}}\frac{[H_{i1} - H_{i2}]^2}{2} = Z_{i0}^{*}\frac{[H_{i1} - H_{i2}]^2}{2} \tag{14.13}$$

where $Z_{i0}^{*} = \dfrac{\dot{E}_{i0}}{\dot{H}_{i1} - \dot{H}_{i2}}$.

For the electric field intensity on the outer surface of the ith layer $\dot{E}_{i1}$, from (14.11), it follows

$$\dot{E}_{i1} = Z_{\tau i1}^{*}\dot{H}_{i1} + \dot{E}_{i0} \tag{14.14}$$

The electric field intensity on the inner surface of the ith layer $\dot{E}_{i2}$, follows from (14.12). Moreover, in (14.12), instead of $\left(-\dot{H}_{i2}\right)$, it is necessary to use $\dot{H}_{i2}$. Then, we can have

$$\dot{E}_{i2} = Z_{\tau i2}^{*}\dot{H}_{i2} + \dot{E}_{i0} \tag{14.15}$$

Using the expression (14.13), in which instead of $\left(-\dot{H}_{i2}\right)$ it is necessary in this case to use $\dot{H}_{i2}$, and taking into account the condition

$$\dot{E}_{i0} = Z_{i0}^{*}\dot{H}_{i0} \tag{14.16}$$

where $\dot{H}_{i0} = \dot{H}_{i1} + \dot{H}_{i2}$.

The equations (14.14) and (14.15) are transformed

$$\begin{aligned} \dot{E}_{i1} &= Z_{\tau i1}^{*}\dot{H}_{i1} + Z_{i0}^{*}\dot{H}_{i0} \\ \dot{E}_{i2} &= Z_{\tau i2}^{*}\dot{H}_{i2} + Z_{i0}^{*}\dot{H}_{i0} \end{aligned} \tag{14.17}$$

From (14.17), it follows that the ith layer can be visualized by the T-shaped equivalent circuit shown in Fig. 14.3. As follows from this equivalent circuit, the ith

layer is represented in relation to the quantities $\dot{E}_{i1}$ and $\dot{E}_{i2}$ by three circuit loops. The first two circuit loops, described by the system of Eq. (14.17), can be represented in relation to the quantities $\dot{E}_{i1}$ and $\dot{E}_{i2}$ determined on the outer and inner surfaces of the ith layer. In this regard, the circuit loops described by the system of the Eq. (14.17) and visualized by the equivalent circuit in Fig. 14.3 will be called the "own" circuit loops of the ith layer. The third circuit loop of the ith layer determined from the equivalent circuit in Fig. 14.3 and described in the form

$$\dot{E}_{i1} = Z^*_{\tau i1}\dot{H}_{i1} + Z^*_{\tau i2}\dot{H}_{i2} + \left(-\dot{E}_{i2}\right)$$

connects the "own" circuit loops of the adjacent layers.

Thus, in accordance with the equivalent circuit in Fig. 14.3, the ith layer is characterized by the "own" circuit loops and the "connecting" circuit loop. These circuit loops arising from the equivalent circuit in Fig. 14.3 are interconnected. As will be shown below, this feature of the T-shaped equivalent circuit causes serious difficulties in solving the problems associated with determining the calculation elements of the electric machine [13]. In this regard, in this work, T-shaped equivalent circuits, as the tool for visualizing the layers of the machine model (with exception of the machine air gap layer), are not used.

The impedances used in the equivalent circuit in Fig. 14.3 are determined by the expressions given in (14.11)–(14.13) and are presented through the unknown value $\dot{E}_{i0}$. Let us determine the value $\dot{E}_{i0}$ for which we use the relations shown in (14.11) and (14.12). The considered i^{th} layer (Fig. 14.2) represents the symmetric system. In this regard, for the impedances $Z^*_{\tau i1}$ and $Z^*_{\tau i2}$ used in the Eqs. (14.11) and (14.12), the condition of the form $Z^*_{\tau i1} = Z^*_{\tau i2}$ is implemented. Taking this condition into account and using relations (14.11) and (14.12), for the quantity $\dot{E}_{i0}$, we have

$$\dot{E}_{i0} = \frac{\dot{E}_{i1}\dot{H}_{i2} + \dot{E}_{i2}\dot{H}_{i1}}{\dot{H}_{i1} + \dot{H}_{i2}} \tag{14.18}$$

Now, for the impedance of the transverse branch $a - b$ of the equivalent circuit in Fig. 14.3, from (14.13) and (14.18), we can obtain

$$Z^*_{i0} = \frac{\dot{E}_{i1}\dot{H}_{i2} + \dot{E}_{i2}\dot{H}_{i1}}{\dot{H}^2_{i1} - \dot{H}^2_{i2}} \tag{14.19}$$

For the longitudinal impedances $Z^*_{\tau i1}$ and $Z^*_{\tau i2}$ used in Eqs. (14.11) and (14.12), it follows taking into account the relation (14.18)

$$Z^*_{\tau i1} = Z^*_{\tau i2} = \frac{\dot{E}_{i1} - \dot{E}_{i2}}{\dot{H}_{i1} + \dot{H}_{i2}} \tag{14.20}$$

As follows from (14.19) and (14.20), the impedances of the equivalent circuit in Fig. 14.3 can be determined using the fairly simple relations if the numerical values of the components of the magnetic and electric field intensities on the outer and inner surfaces of the ith layer are given.

Note also, that in accordance with (14.20), the impedances $Z^*_{\tau i1}$ and $Z^*_{\tau i2}$ are determined as the difference in the values of the components of the electric field intensities found on the outer and inner surfaces of the ith layer, i.e., $\dot{E}_{i1} - \dot{E}_{i2}$. The implementation of this condition with the required accuracy puts forward the serious requirements for the level of reliability in determining the values of $\dot{E}_{i1}$ and $\dot{E}_{i2}$.

14.2.2 L-Shaped Equivalent Circuit of Layer

The T-shaped equivalent circuit of the ith layer, shown in Fig. 14.3, consists, as noted above, of the three circuit loops, mutually connected to each other. This circumstance, as noted above, seriously complicates the effective use of the T-shaped equivalent circuits when solving the problem associated with the determination of machine parameters, for example [13]. In this regard, the T-shaped equivalent circuit in Fig. 14.3 needs to be transformed in order to simplify its structure and to make it look convenient for practical use.

In Chaps. 11 and 13 of this work, it was shown that when determining the parameters of the machine, the sections (layers) of the stator and rotor magnetic circuits are represented in the form of the L-shaped equivalent circuits. We will use this provision, and for the ith layer, we will construct the L-shaped equivalent circuit. To this purpose, we can, in our case, proceed from the T-shaped equivalent circuit of the ith layer (Fig. 14.3) in order to transform it into the L-shaped equivalent circuit. To convert the T-shaped equivalent circuit of the ith layer into the L-shaped circuit, we will use the "no-load and load" concept applied above when constructing the air gap magnetizing equivalent circuit and obtaining the machine equivalent circuit. According to the "no-load and load" concept, the stator winding current $\dot{I}_1$ corresponding to the operating mode of the machine is represented as the sum of the no-load current $\dot{I}_0$ that occurs in the stator winding in the absence of the current in the rotor winding, i.e. $\dot{I}_2 = 0$, and the load current $(-\dot{I}'_2)$ caused by the emf induced in the rotor winding in no-load mode, i.e. $\dot{I}_1 = \dot{I}_0 + (-\dot{I}'_2)$. We consider the component of the magnetic field intensity $\dot{H}_{i1}$ defined on the outer surface of the ith layer (Fig. 14.2) through which the energy of the electromagnetic field penetrates into the ith layer. By analogy with the condition for the stator winding current, defined as $\dot{I}_1 = \dot{I}_0 + (-\dot{I}'_2)$, the magnitude of the component of the magnetic field intensity $\dot{H}_{i1}$ corresponding to the conditional "operating mode" of the ith layer can be represented as

$$\dot{H}_{i1} = \dot{H}_i + (-\dot{H}'_{i2}) \tag{14.21}$$

Here, the value $\dot{H}_i$ reflects the component of the magnetic field intensity on the outer surface of the ith layer when the component of the magnetic field intensity is absent on the inner surface of the ith layer, i.e. $\dot{H}_{i2} = 0$ (conditional "no-load mode" of the ith layer). The value $(-\dot{H}'_{i2})$ is the component of the magnetic field intensity on the outer surface of the ith layer, which takes place in the conditional "load mode" of the ith layer and is determined in accordance with (14.21) by the relation of the form $(-\dot{H}'_{i2}) = \dot{H}_{i1} - \dot{H}_i$. Let us define the values $\dot{H}_i$ and $(-\dot{H}'_{i2})$ used in the Eq. (14.21), for which we can use the equivalent circuit shown in Fig. 14.3.

"No-load mode" of the ith layer. The value $\dot{H}_i$ is determined when the condition $\dot{H}_{i2} = 0$ is satisfied in the equivalent circuit in Fig. 14.3. Then, for $\dot{H}_i$, we can have

$$\dot{H}_i = \frac{\dot{E}_{i1}}{Z^*_{i0} + Z^*_{\tau i1}} = \frac{\dot{E}_{i1}}{Z^*_i} \tag{14.22}$$

where $Z^*_i = Z^*_{i0} + Z^*_{\tau i1}$.

When $\dot{H}_{i2} = 0$, for $\dot{E}_{i2}$, we can get according to the equivalent circuit in Fig. 14.3

$$\dot{E}_{i2} = Z^*_{i0} H_i \tag{14.23}$$

Using (14.22) and (14.23), for the emf $\dot{E}_{i1}$ and $\dot{E}_{i2}$, we have

$$\begin{aligned} \dot{E}_{i1} &= Z^*_i \dot{H}_i \\ \dot{E}_{i2} &= Z^*_{i0} H_i \end{aligned} \tag{14.24}$$

In (14.24), we remove the value of H_i. Then, it follows that

$$\dot{E}_{i1} = \frac{Z^*_i}{Z^*_{i0}} \dot{E}_{i2} = c_i \dot{E}_{i2} = \dot{E}'_{i2} \tag{14.25}$$

where $c_i = Z^*_i / Z^*_{i0}$.

Hence, for $\dot{E}'_{i2}$, it follows

$$\dot{E}'_{i2} = c_i \dot{E}_{i2} \tag{14.26}$$

In (14.26) the quantity $\dot{E}'_{i2}$ represents referred to the outer surface of the ith layer the value of the electric field intensity on its inner surface.

From (14.22) and (14.25), for the factor c_i, we obtain

$$c_i = \frac{Z_i^*}{Z_{i0}^*} = 1 + \frac{Z_{\tau i1}^*}{Z_{i0}^*} \tag{14.27}$$

"Under-Load mode" of the ith layer. Let's define the value $(-\dot{H}_{i2}^{'})$ corresponding to the conditional "load mode" of the ith layer. For $(-\dot{H}_{i2}^{'})$ from (14.21), we have that $(-\dot{H}_{i2}^{'}) = \dot{H}_{i1} - \dot{H}_i$. It follows that to determine the value $(-\dot{H}_{i2}^{'})$ it is necessary to set the values of $\dot{H}_{i1}$ and $\dot{H}_i$. Here $\dot{H}_i$ follows from relation (14.22). To determine $\dot{H}_{i1}$, we use the equivalent circuit shown in Fig. 14.3. In this case, in the circuit in Fig. 14.3, the value $\dot{E}_{i2}$ can be replaced by the voltage drop across the impedance Z_{iH}^*, i.e. $\dot{E}_{i2} = Z_{iH}^* \dot{H}_{i2}$. Here, the quantity of Z_{iH}^* represents the "load" impedance and reflects the input impedance of the subsequent layers of the model of the electric machine. The equivalent circuit of the ith layer corresponding to this condition follows from the equivalent circuit in Fig. 14.3 and takes the form in Fig. 14.4a. Based on the equivalent circuit in Fig. 14.4a, for $\dot{H}_{i1}$, we can get

$$\dot{H}_{i1} = \dot{E}_{i1} \frac{(Z_{iH}^* + Z_{\tau i2}^*) + Z_{i0}^*}{Z_{\tau i1}^* (Z_{iH}^* + Z_{\tau i2}^*) + Z_{\tau i1}^* Z_{i0}^* + Z_{i0}^* (Z_{iH}^* + Z_{\tau i2}^*)} \tag{14.28}$$

Now, for the equation $(-\dot{H}_{i2}^{'}) = \dot{H}_{i1} - \dot{H}_i$ from (14.22) and (14.28), it follows

$$\begin{aligned}(-\dot{H}_{i2}^{'}) &= \dot{E}_{i1} \left(\frac{(Z_{iH}^* + Z_{\tau i2}^*) + Z_{i0}^*}{Z_{\tau i1}^* (Z_{iH}^* + Z_{\tau i2}^*) + Z_{\tau i1}^* Z_{i0}^* + Z_{i0}^* (Z_{iH}^* + Z_{\tau i2}^*)} - \frac{1}{Z_{\tau i1}^* + Z_{i0}^*} \right) \\ &= \frac{\dot{E}_{i1}}{Z_{\tau i1}^* c_i + Z_{\tau i2}^* c_i^2 + Z_{iH}^* c_i^2}\end{aligned} \tag{14.29}$$

where $c_i = 1 + Z_{\tau i1}^* / Z_{i0}^*$.

It can be shown that on the basis of the equivalent circuit in Fig. 14.4a, the magnitude of the magnetic field intensity on the inner surface of the ith layer $(-\dot{H}_{i2})$ can be calculated by the relation of the form

$$\begin{aligned}(-\dot{H}_{i2}) &= \frac{\dot{E}_{i1} - Z_{\tau i1}^* \dot{H}_{i1}}{(Z_{iH}^* + Z_{\tau i2}^*)} \\ &= \frac{\dot{E}_{i1}}{(Z_{iH}^* + Z_{\tau i2}^*)} \left[1 - \frac{(Z_{iH}^* + Z_{\tau i2}^* + Z_{i0}^*) Z_{\tau i1}^*}{Z_{\tau i1}^* (Z_{iH}^* + Z_{\tau i2}^*) + Z_{\tau i1}^* Z_{i0}^* + (Z_{iH}^* + Z_{\tau i2}^*) Z_{i0}^*} \right] \\ &= \frac{\dot{E}_{i1}}{Z_{\tau i1}^* + Z_{\tau i2}^* c_i + Z_{iH}^* c_i}\end{aligned} \tag{14.30}$$

where $c_i = 1 + Z_{\tau i1}^* / Z_{i0}^*$.

From (14.29) and (14.30), the ratio $\dot{H}_{i2} / \dot{H}_{i2}^{'}$ will be

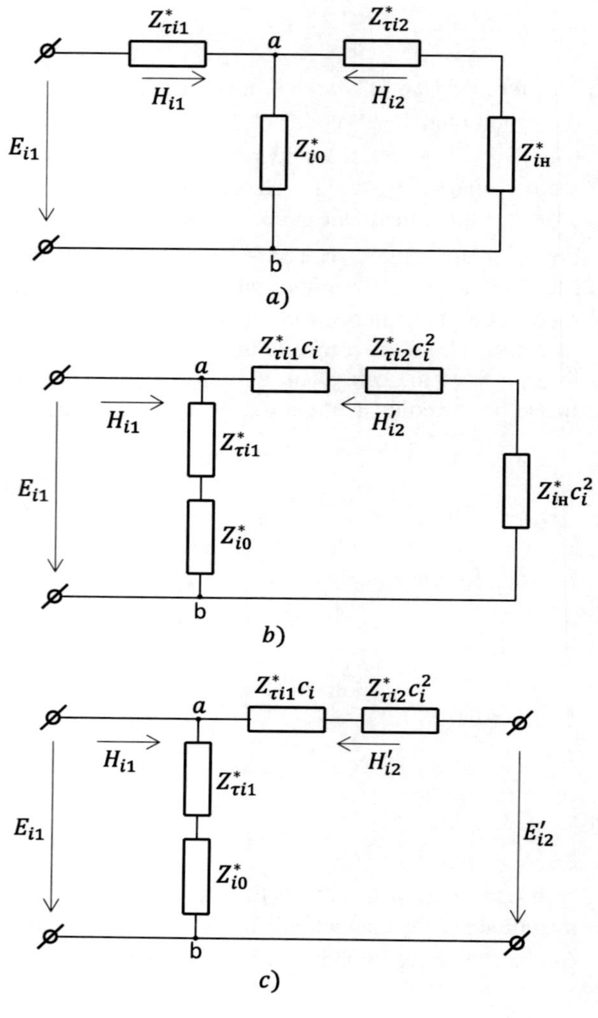

Fig. 14.4 Converting T-shaped equivalent circuit of the ith layer to the L-shaped one: (**a**) T-shaped equivalent circuit of the ith layer with the closed secondary terminals; (**b**) L-shaped equivalent circuit of the ith layer with the closed secondary terminals; (**c**) L-shaped equivalent circuit of the ith layer with the open secondary terminals

$$\frac{\dot{H}_{i2}}{\dot{H}_{i2}^{'}} = c_i \quad \dot{H}_{i2}^{'} = \frac{\dot{H}_{i2}}{c_i} \tag{14.31}$$

In (14.31), $\dot{H}_{i2}^{'}$ is the referred to the outer surface of the ith layer value of the magnetic field intensity on the inner surface of the ith layer. The value c_i used in (14.27), (14.30), and (14.31), represents, as noted above, the transfer factor of the parallel branch $a - b$ to the terminals $\dot{E}_{i1}$ of the equivalent circuit in Fig. 14.3. In the technical literature, this factor is known under various names, such as the transfer factor, the reduction factor, and the transformation factor [13].

When transforming the equations of the ith layer, the value c_i is considered as the reduction factor or the transformation factor. When transforming the equivalent circuit of the ith layer, the value c_i is represented as the transfer factor. In this work, the value c_i will be called the transfer factor, since below we will mainly deal with the transformation of the equivalent circuits of the electric machines.

"Operating mode" of the ith layer. In the "operating mode" of the ith layer, the magnetic field intensity $\dot{H}_{i1}$ on the outer surface of the ith layer is determined using the relation (14.21) as $\dot{H}_{i1} = \dot{H}_i + (-\dot{H}'_{i2})$. Taking into account the expressions obtained in (14.22) and (14.29), this equation can be represented in the form

$$\dot{H}_{i1} = \dot{H}_i + (-\dot{H}'_{i2}) = \dot{E}_{i1}\left(\frac{1}{Z^*_{i0} + Z^*_{\tau i1}} + \frac{1}{Z^*_{\tau i1}c_i + Z^*_{\tau i2}c_i^2 + Z^*_{iH}c_i^2}\right) \tag{14.32}$$

This equation corresponds to the L-shaped equivalent circuit shown in Fig. 14.4b. As follows from Fig. 14.4b, in relation to the terminals of $\dot{E}_{i1}$ belonging to the outer surface of the ith layer, the L-shaped equivalent circuit of the ith layer consists of the two independent circuit loops. In comparison with the original T-shaped equivalent circuit (Fig. 14.4a) in the L-shaped equivalent circuit (Fig. 14.4b), the impedances of the second circuit loop $Z^*_{\tau i2}$ and Z^*_{iH} are multiplied by the square of the transfer factor c_i^2, i.e. $Z^*_{\tau i2}c_i^2$ and $Z^*_{iH}c_i^2$. In this case, the impedance $Z^*_{\tau i1}$ of the first circuit loop arising from the T-shaped equivalent circuit (Fig. 14.4a) goes to the second circuit loop of the L-shaped equivalent circuit (Fig. 14.4b), and the value of this impedance increased by the factor c_i, i.e. $Z^*_{\tau i1}c_i$, is connected in series with the impedances $Z^*_{\tau i2}c_i^2$ and $Z^*_{iH}c_i^2$, i.e., the impedance of the second circuit loop of the L-shaped equivalent circuit will be $Z^*_{\tau i1}c_i + Z^*_{\tau i2}c_i^2 + Z^*_{iH}c_i^2$ (Fig. 14.4b).

The impedance of the parallel branch $a - b$ of the T-shaped equivalent circuit Z^*_{i0} (Fig. 14.4a) is used in the first circuit loop of the L-shaped equivalent circuit of the ith layer (Fig. 14.4b). In this case, its value, as follows from the L-shaped equivalent circuit in Fig. 14.4b, increases by the value of the impedance $Z^*_{\tau i1}$ used in the first circuit loop of the T-shaped equivalent circuit (Fig. 14.4a). As a result, the total impedance of the first circuit loop of the L-shaped equivalent circuit in Fig. 14.4b is $Z^*_{i0} + Z^*_{\tau i1}$. As noted above, the value of the impedance Z^*_{iH} used in the equivalent circuits in Fig. 14.4a represents the input impedance of the subsequent layers of the machine model. Therefore, the multiplication of the impedance Z^*_{iH} by the square of the transfer factor c_i^2, i.e. $Z^*_{iH}c_i^2$, means that when using the L-shaped equivalent circuit for the ith layer, the impedances of the subsequent layers of the machine model are multiplied by the square of the transfer factor of the ith layer, i.e., by c_i^2. In accordance with Fig. 14.4b, the ith layer is represented by the two independent circuit loops. In this work, these circuit loops will be understood as the "own" circuit loop of the ith layer, determined with respect to the electric field intensity $\dot{E}_{i1}$ given on the outer surface of the ith layer and the "conducting" circuit loop which unites the "own" circuit loop of the adjacent layers. Let us obtain the equations for these circuit loops of the ith layer. For this purpose, we will use the equivalent circuit of the ith

layer shown in Fig. 14.4b. Based on Fig. 14.4b, for the equation of the "own" circuit loop of the ith layer, we have

$$\dot{E}_{i1} = (Z^*_{i0} + Z^*_{\tau i1})\dot{H}_i \tag{14.33}$$

The equation for the "conducting" circuit loop takes the form from Fig. 14.4b

$$\dot{E}_{i1} = (Z^*_{\tau i1}c_i + Z^*_{\tau i2}c_i^2 + Z^*_{i\text{н}}c_i^2)\dot{H}'_{i2} \tag{14.34}$$

The equation $\dot{E}_{i2} = Z^*_{i\text{н}}\dot{H}_{i2}$ arising from the equivalent circuit in Fig. 14.3, can be expressed through the referred values of $\dot{E}'_{i2}$ and $\dot{H}'_{i2}$. Then, bearing in mind that $\dot{E}'_{i2} = c_i\dot{E}_{i2}$ and $\dot{H}'_{i2} = \dot{H}_{i2}/c_i$, we obtain that $\dot{E}'_{i2} = Z^*_{i\text{н}}c_i^2\dot{H}'_{i2}$. Using this condition, the Eq. (14.34) can be reduced to the form

$$\dot{E}_{i1} = (Z^*_{\tau i1}c_i + Z^*_{\tau i2}c_i^2)\dot{H}'_{i2} + \dot{E}'_{i2} \tag{14.35}$$

Equations (14.33) and (14.35) correspond to the L-shaped equivalent circuit shown for the ith layer in Fig. 14.4b. From the comparison of the equivalent circuits shown in Figs. 14.3 and 14.4b, it follows that the transformation of the T-shaped equivalent circuit in Fig. 14.3 into the L-shaped equivalent circuit is associated with transferring in the T-shaped equivalent circuit its parallel branch $a - b$ to the terminals of $\dot{E}_{i1}$. In this regard, this transformation technique is known as the transfer technique, for example [3, 4, 13]. Based on the equivalent circuits shown in Figs. 14.3 and 14.4b, we can conclude that the application of the transfer technique allows simplifying the structure of the equivalent circuit of the ith layer. In this work, this transformation technique is widely used in the construction of the equivalent circuits and in determining the parameters of the electric machine [13].

In the equivalent circuit in Fig. 14.4b, the total impedance of the "conducting" circuit loop $(Z^*_{\tau i1}c_i + Z^*_{\tau i2}c_i^2)$ can be represented as

$$Z^*_{\tau i1}c_i + Z^*_{\tau i2}c_i^2 = \left[Z^*_{\tau i1}(1/c_i) + Z^*_{\tau i2}\right]c_i^2 = Z^*_{\tau i}c_i^2 \tag{14.36}$$

Here, the value c_i^2 is the reduction factor of the impedance. From (14.36) for the impedance $Z^*_{\tau i}$, it follows

$$Z^*_{\tau i} = Z^*_{\tau i1}(1/c_i) + Z^*_{\tau i2} \tag{14.37}$$

In the "own" circuit loop of the L-shaped equivalent circuit in Fig. 14.4b, the series-connected impedances $(Z^*_{\tau i1}$ and $Z^*_{i0})$ are used. They can be represented in the form

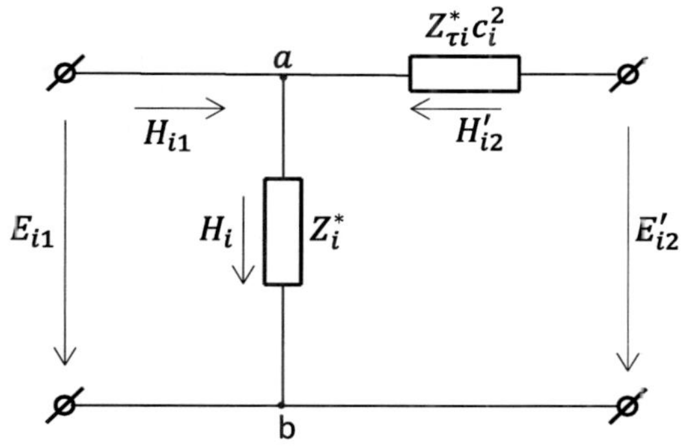

Fig. 14.5 L-shaped equivalent circuit of the ith layer

$$Z_i^* = Z_{i0}^* + Z_{\tau i1}^* = Z_{i0}^*\left(1 + \frac{Z_{\tau i1}^*}{Z_{i0}^*}\right) = Z_{i0}^* c_i \tag{14.38}$$

where $c_i = 1 + Z_{\tau i1}^*/Z_{i0}^*$.

Using the relations (14.36)–(14.38), the equivalent circuit in Fig. 14.4b can be reduced to the form in Fig. 14.5. Figure 14.5 shows the final general form of the L-shaped equivalent circuit of the ith layer. In this equivalent circuit, for the quantities $\dot{E}'_{i2}$ and $\dot{H}'_{i2}$ used in it, we have

$$\dot{E}'_{i2} = \dot{E}_{i2}c_i \quad \dot{H}'_{i2} = \dot{H}_{i2}/c_i \tag{14.39}$$

Based on the L-shaped equivalent circuit in Fig. 14.5, for the ith layer, we can obtain the system of the equations of the form

$$\begin{aligned} \dot{E}_{i1} &= Z_i^*\dot{H}_i \\ \dot{E}'_{i2} &= Z_{\tau i}^* c_i^2 \dot{H}'_{i2} + Z_i^*\dot{H}_i \\ \dot{H}_i &= \dot{H}_{i1} + \dot{H}'_{i2} \end{aligned} \tag{14.40}$$

In (14.40), the first equation reflects the "own" circuit loop of the ith layer and the second equation describes the "conducting" circuit loop.

The impedances used in the L-shaped equivalent circuit in Fig. 14.5 and in the equations shown in (14.40) are expressed in terms of the impedances of the T-shaped equivalent circuit in Fig. 14.3. The analytical expressions for the impedances of the T-shaped equivalent circuit (Fig. 14.3) are obtained in (14.8). We can use these expressions and as a result of which for the impedances of the L-shaped equivalent circuit of the ith layer, we can have

$$Z_i^* = Z_{i0}^* c_i = \frac{j\omega_1 \mu_{xi}}{\beta_i \tanh \beta_i h_i} \frac{\xi_{Li}}{\xi_{\tau i}}$$
$$Z_{\tau i}^* = Z_{\tau i1}^* (1/c_i) + Z_{\tau i2}^* = \frac{j\omega_1 \mu_{xi}}{\beta_i} \frac{\xi_{Li}}{\xi_{\tau i}} \tanh \beta_i h_i \quad (14.41)$$

Based on the relations obtained in (14.8) and (14.22), for the transfer factor of the ith layer c_i, it follows

$$c_i = \cosh \beta_i h_i \quad (14.42)$$

In the equivalent circuit shown in Fig. 14.5, the quantities $\dot{E}'_{i2}$ and $\dot{H}'_{i2}$ are determined by the relations given in (14.39).

For the impedances used in the L-shaped equivalent circuit in Fig. 14.5, the expressions are given in (14.41) for the general case. On the basis of the consideration of the L-shaped equivalent circuit shown in Fig. 14.5, the L-shaped equivalent circuits for the ith conducting layer are constructed for the regions of weak and strong skin effect (Appendix H). The parameters used in these equivalent circuits are obtained on the basis of the expressions shown in (14.41).

Impedances of the L-shaped equivalent circuit based on the numerical values of the field intensities given on the outer and inner surfaces of the ith layer. When using the numerical values for the components of the electric and magnetic field intensities given on the outer and inner surfaces of the ith layer, the expressions for the impedances Z_i^* and $Z_{\tau i}^*$ can be obtained using (14.19), (14.20), and (14.40). Then, for the impedance Z_i^*, we have

$$Z_i^* = Z_{i0}^* c_i = \frac{\dot{E}_{i1}\dot{H}_{i1} + \dot{E}_{i2}\dot{H}_{i2}}{\dot{H}_{i1}^2 - \dot{H}_{i2}^2} \quad (14.43)$$

The impedance $Z_{\tau i}^*$ takes the form

$$Z_{\tau i}^* = Z_{\tau i1}^* (1/c_i) + Z_{\tau i2}^* = \frac{\dot{E}_{i1}^2 - \dot{E}_{i2}^2}{\dot{E}_{i1}\dot{H}_{i1} + \dot{E}_{i2}\dot{H}_{i2}} \quad (14.44)$$

For the transfer factor c_i, from (14.19), (14.20), and (14.22), we have

$$c_i = 1 + \frac{Z_{\tau i1}^*}{Z_{i0}^*} = \frac{\dot{E}_{i1}\dot{H}_{i1} + \dot{E}_{i2}\dot{H}_{i2}}{\dot{E}_{i1}\dot{H}_{i2} + \dot{E}_{i2}\dot{H}_{i1}} \quad (14.45)$$

The L-shaped equivalent circuit of the ith layer, in comparison with its T-shaped equivalent circuit, has a simple structure, since the L-shaped equivalent circuit of the ith layer consists of the two independent "own" circuit loops. The use of the L-shaped equivalent circuits makes it possible to organize the fairly simple and logical process of constructing the equivalent circuits and determining the parameters of the electric machine [13]. In this regard, in this work, the L-shaped equivalent circuits form the basis for constructing the equivalent circuits, obtaining the voltage equations, and determining the parameters of the electric machine.

14.3 L-Shaped Equivalent Circuits of Passive Layers of Stator and Rotor

We consider the equivalent circuit of the ith layer shown in Fig. 14.5. We will use this equivalent circuit to construct the L-shaped equivalent circuits for the passive layers of the stator and rotor of the machine.

L-shaped equivalent circuits for the nonconducting layers representing the stator and rotor slot-wedge areas. In accordance with the layered machine model shown in Fig. 14.1, we will consider the L-shaped equivalent circuits for the non conducting layers reflecting the stator and rotor slot-wedge areas. For this purpose, we proceed from the L-shaped equivalent circuit obtained in Fig. 14.5. The stator and rotor of the electric machine are usually laminated. The resistivity of the laminated iron can be taken as the relatively large value, i.e., $\rho_c = \infty$. This means that the propagation coefficient β_i defined for the layers of the stator and rotor slot-wedge areas can be obtained as $\beta_i = k_{\mu i}(\pi/\tau_i)$, where $i = ks, kR$. In this case, the value of the coefficient β_i is the real number. Consequently, the impedances Z_i^* and $Z_{\tau i}^*$ used in the L-shaped equivalent circuit in Fig. 14.5 become inductive, i.e., $Z_i^* \approx jx_i^*; Z_{\tau i}^* \approx jx_{\tau i}^*$, where $i = ks, kR$. Based on the expressions (14.41), for reactances x_i^* and $x_{\tau i}^*$, we can have

$$x_i^* = \frac{\omega_1 \mu_{zi}}{\beta_i \tanh \beta_i h_i} \frac{\xi_{Li}}{\xi_{\tau i}} \quad x_{\tau i}^* = \omega_1 \mu_{xi} h_i k_i \frac{\xi_{Li}}{\xi_{\tau i}} \tag{14.46}$$

where $k_i = \tanh \beta_i h_i / \beta_i h_i$; $\beta_i = k_{\mu i}(\pi/\tau_i)$; $i = ks, kR$.

Now, from the L-shaped equivalent circuit in Fig. 14.5 and keeping in mind the expressions given in (14.46), the L-shaped equivalent circuits for the layers reflecting the stator and rotor slot-wedge areas can be reduced to the form in Fig. 14.6. Figure 14.6a shows the L-shaped equivalent circuit for the layer of the stator slot-wedge area. Figure 14.6b shows the L-shaped equivalent circuit for the layer of the rotor slot-wedge area. For the layer factor c_{ks} used in the equivalent circuit in Fig. 14.6a, we have

$$c_{ks} = \cosh \beta_{ks} h_{ks} = \cosh \left[k_{\mu ks}(\pi/\tau_{ks})\right] h_{ks} \tag{14.47}$$

where $\beta_{ks} = k_{\mu ks}(\pi/\tau_{ks})$.

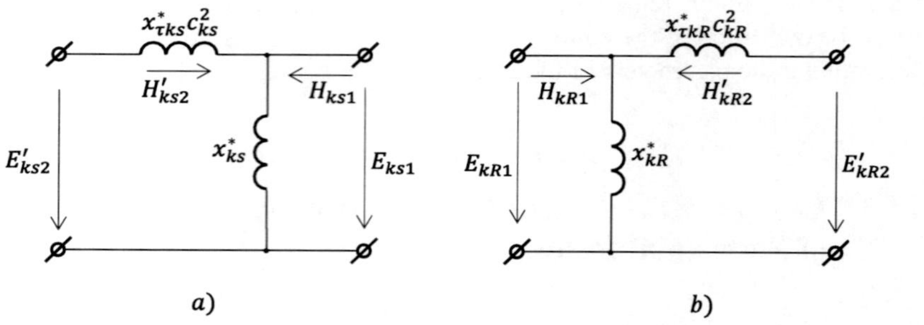

Fig. 14.6 L-shaped equivalent circuits of the layers reflecting the slot-wedge areas of the stator and rotor: (**a**) L-shaped equivalent circuit of the layer representing the stator slot-wedge area; (**b**) L-shaped equivalent circuit of the layer representing the rotor slot-wedge area

In the equivalent circuit in Fig. 14.6a, for the values $\dot{E}'_{ks2}$ and $\dot{H}'_{ks2}$ used in it, it follows

$$\dot{E}'_{ks2} = \dot{E}_{ks2}c_{ks} \quad \dot{H}'_{ks2} = \dot{H}_{ks2}/c_{ks} \tag{14.48}$$

In the equivalent circuit in Fig. 14.6b, the layer factor c_{kR} is used. For this factor, by the analogy with (14.47), we can obtain

$$c_{kR} = \cosh\beta_{kR}h_{kR} = \cosh\left[k_{\mu kR}(\pi/\tau_{kR})\right]h_{kR} \tag{14.49}$$

where $\beta_{kR} = k_{\mu kR}(\pi/\tau_{kR})$.

In the equivalent circuit in Fig. 14.6b, for the values $\dot{E}'_{kR2}$ and $\dot{H}'_{kR2}$ used in it, it follows

$$\dot{E}'_{kR2} = \dot{E}_{kR2}c_{kR} \quad \dot{H}'_{kR2} = \dot{H}_{kR2}/c_{kR} \tag{14.50}$$

Based on the expressions (14.46), for the reactances x^*_{ks} and $x^*_{\tau ks}$ of the equivalent circuit in Fig. 14.6a, we have

$$x^*_{ks} = \frac{\omega_1\mu_{xks}}{\beta_{ks}\tanh\beta_{ks}h_{ks}}\frac{\xi_{Lks}}{\xi_{\tau ks}} \quad x^*_{\tau ks} = \omega_1\mu_{xks}h_{ks}k_{ks}\frac{\xi_{Lks}}{\xi_{\tau ks}} \tag{14.51}$$

where $k_{ks} = \tanh\beta_{ks}h_{ks}/\beta_{ks}h_{ks}$; $\xi_{\tau ks} = \tau_{ks}/\tau_0$; $\xi_{Lks} = l_{ks}/l_0$.

The reactances x^*_{kR} and $x^*_{\tau kR}$ used in the equivalent circuit in Fig. 14.6b with the help of the expressions (14.6a) take the form

$$x^*_{kR} = \frac{\omega_1\mu_{xkR}}{\beta_{kR}\tanh\beta_{kR}h_{kR}}\frac{\xi_{LkR}}{\xi_{\tau kR}} \quad x^*_{\tau kR} = \omega_1\mu_{xkR}h_{kR}k_{kR}\frac{\xi_{LkR}}{\xi_{\tau kR}} \tag{14.52}$$

where $k_{kR} = \tanh\beta_{kR}h_{kR}/\beta_{kR}h_{kR}$; $\xi_{\tau kR} = \tau_{kR}/\tau_0$; $\xi_{LkR} = l_{kR}/l_0$.

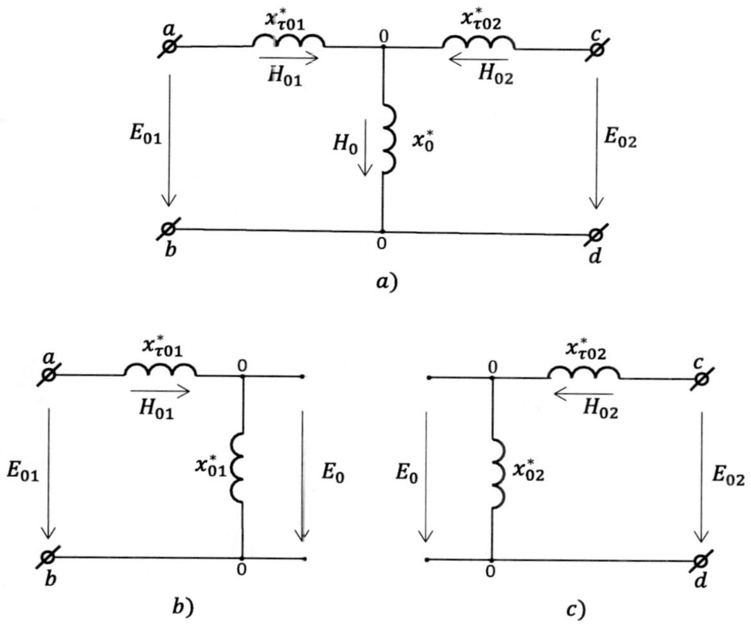

Fig. 14.7 Air gap layer equivalent circuit: (**a**) T-shaped equivalent circuit of the air gap layer; (**b**) L-shaped equivalent circuit of the air gap sublayer adjacent to the stator bore surface; (**c**) L-shaped equivalent circuit of the air gap sublayer adjacent to the rotor surface

L-shaped equivalent circuits for the air gap sublayers of the electric machine. The machine air gap occupies the space between the stator and rotor surfaces and it is considered as the nonconductive layer of the machine model. The T-shaped equivalent circuit is used for the air gap layer [13]. In this regard, based on the equivalent circuit in Fig. 14.3, for the air gap layer, we have the equivalent circuit in Fig. 14.7a. The parameters used in the equivalent circuit in Fig. 14.7a are inductive, since the resistivity of the gap layer is equal to infinity, i.e., $\rho_0 = \infty$. In addition, the magnetic permeability of the air gap layer is μ_0. In this regard, in (14.8), the value β_i can be replaced by β_0 defined as $\beta_0 = \pi/\tau_0$, and the value h_i can be replaced by δ. Here, τ_0 is the pole pitch determined on the surface $0 - 0$ passing through the middle of the radial length of the air gap. As applied to the air gap layer, the expressions obtained in (14.8) can be represented through the value of the pole pitch τ_0. Taking into account these reservations, the parameters used in the T-shaped equivalent circuit for the air gap layer obtain the form from (14.8)

$$x_0^* = \frac{\omega_1 \mu_0}{\beta_0 \sinh \beta_0 \delta'} \quad x_{\tau 01}^* = x_{\tau 02}^* = \frac{\omega_1 \mu_0}{\beta_0} \tanh \beta_0 \frac{\delta}{2} \tag{14.53}$$

where $\beta_0 = \pi/\tau_0$.

The equivalent circuit of the air gap layer shown in Fig. 14.7a can be used to construct the equivalent circuit of the electric machine. However, for the purposes of an analysis, the air gap of the machine can be subdivided with respect to the surface $0-0$ into the two sublayers having the same thickness equal to $\delta/2$. With respect to the surface $0-0$, the upper air gap sublayer (Fig. 14.1) refers to the stator model, and the lower air gap sublayer is adjacent to the rotor model. Let us define the equivalent circuits for the air gap sublayers of the electric machine. For this purpose, we consider the equivalent circuit of the air gap shown in Fig. 14.7a. The terminals $0-0$ of the equivalent circuit in Fig. 14.7a correspond to the surface $0-0$ used in the model of the machine (Fig. 14.1). In this regard, the reactance x_0^* of the parallel branch $0-0$ of the equivalent circuit in Fig. 14.7a can be represented as the total reactance of the two parallel connected reactances x_{01}^* and x_{02}^*, i.e. $x_0^* = x_{01}^* x_{02}^* / \left(x_{01}^* + x_{02}^*\right)$. Then, in relation to the terminals $0-0$, the T-shaped equivalent circuit of the air gap (Fig. 14.7a) can be decomposed into the two L-shaped equivalent circuits shown in Fig. 14.7b, c. The L-shaped equivalent circuit in Fig. 14.7b represents the air gap sublayer adjoined to the stator bore surface and having the calculated radial length equal to $\delta_1 = \delta k_{1\delta}/2$. The L-shaped equivalent circuit in Fig. 14.7c reflects the air gap sublayer adjoined to the rotor surface and having the calculated radial length equal to $\delta_2 = \delta k_{2\delta}/2$. Here, $k_{1\delta}$ and $k_{2\delta}$ are the air gap factors. In this regard, the L-shaped equivalent circuit in Fig. 14.7b is referred to the stator equivalent circuit. The L-shaped equivalent circuit in Fig. 14.7c is attached to the rotor equivalent circuit. Based on the expressions obtained in (14.53), for the parameters of the L-shaped equivalent circuits obtained for the air gap sublayers (Fig. 14.7b, c), we have

$$\begin{aligned} x_{01}^* &= \frac{\omega_1 \mu_0}{\beta_0 \sinh \beta_0 \delta_1} & x_{02}^* &= \frac{\omega_1 \mu_0}{\beta_0 \sinh \beta_0 \delta_2} \\ x_{\tau 01}^* &= \frac{\omega_1 \mu_0}{\beta_0} \tanh \beta_0 \frac{\delta}{2} & x_{\tau 02}^* &= \frac{\omega_1 \mu_0}{\beta_0} \tanh \beta_0 \frac{\delta}{2} \end{aligned} \tag{14.54}$$

The L-shaped equivalent circuits for the air gap sublayers (Fig. 14.7b, c) and the parameters used in them in the form given in (14.54) are used to construct equivalent circuits for the stator and rotor of the electric machine.

Equivalent circuits of the layers representing the stator and rotor yoke areas. In the model shown in Fig. 14.1, the stator and rotor magnetic circuits are represented each by the three layers. The last third layers in Fig. 14.1 represent the stator and the rotor yoke areas. It is assumed that in these stator and rotor layers energy of the electromagnetic field is completely dissipated. To ensure this condition, in this work, it is assumed that on the outer (in the case of the stator) and on the inner (in the case of the rotor) surfaces of the layers reflecting the stator and the rotor yoke areas (Fig. 14.1), the normal component of the magnetic flux density is zero. The equivalent circuits for such layers of the model given in Fig. 14.1 can be obtained using the L-shaped equivalent circuit shown in Fig. 14.5. For this purpose, in the equivalent circuit in Fig. 14.5, it is necessary to use the condition of the form $\dot{E}'_{ai2} = 0$,

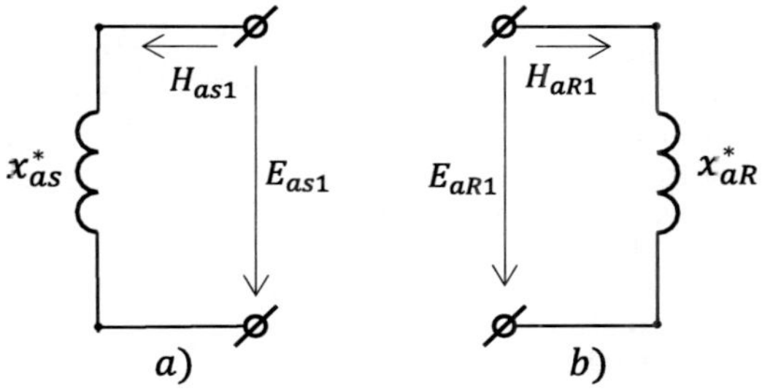

Fig. 14.8 Equivalent circuits for the layers reflecting the stator and rotor yoke areas: (**a**) equivalent circuit of the stator yoke layer; (**b**) equivalent circuit of the rotor yoke layer

where $i = as, aR$, and in addition, to accept that $\rho_{ai} = \infty$ since the stator and rotor of the electric machine are made of the laminated iron. It is shown in [13] that when these conditions are used in the equivalent circuit in Fig. 14.5, the equivalent circuits for the stator and rotor yoke layers take the form of the two-pole equivalent circuit (Fig. 14.8). The equivalent circuit in Fig. 14.8a reflects the layer of the stator yoke area, and the layer of the rotor yoke area is represented by the equivalent circuit in Fig. 14.8b. In this case, the reactance x^*_{as} used in the equivalent circuit reflecting the layer of the stator yoke area (Fig. 14.8a), takes the form

$$x^*_{as} = \omega_1 \mu_{xas} h_{as} k_{as} \frac{\xi_{Las}}{\xi_{\tau as}} \tag{14.55}$$

where $k_{as} = \tanh \beta_{as} h_{as} / \beta_{as} h_{as}$; $\beta_{as} = k_{\mu as}(\pi/\tau_{as})$.

In the equivalent circuit in Fig. 14.8b, the reactance x^*_{aR} is used, and it is determined as

$$x^*_{aR} = \omega_1 \mu_{xaR} h_{aR} k_{aR} \frac{\xi_{LaR}}{\xi_{\tau aR}} \tag{14.56}$$

where $k_{aR} = \tanh \beta_{aR} h_{aR} / \beta_{aR} h_{aR}$; $\beta_{aR} = k_{\mu aR}(\pi/\tau_{aR})$.

14.4 L-Shaped Equivalent Circuits of Winding Location Area Layers of Stator and Rotor

In accordance with the layered model of the electric machine shown in Fig. 14.1, the layers reflecting the stator and rotor windings location areas are considered in the form of the magneto-anisotropic nonconductive layers. In this case, the stator and rotor phase windings are usually represented in the form of the infinitely thin current threads evenly distributed in the considered layers. Under these conditions, the propagation coefficients β_i, where $i = zs, zR$, determined for the layers of the stator and rotor winding location areas as $\beta_i = k_{\mu i}(\pi/\tau_i)$ are characterized, in this case, by the real number. Consequently, the impedances used in the L-shaped equivalent

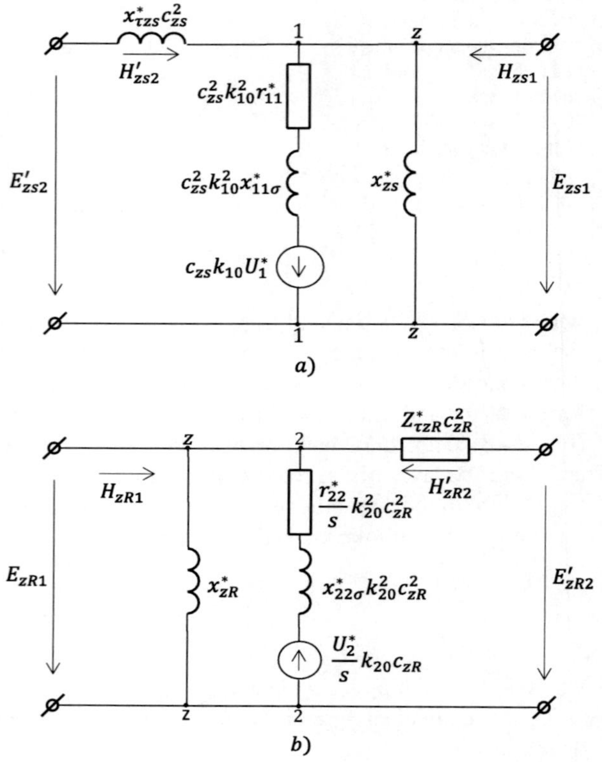

Fig. 14.9 L-shaped equivalent circuits for the layers representing the winding location areas of the stator and rotor: (**a**) L-shaped equivalent circuit for the layer reflecting the winding location area of the stator; (**b**) L-shaped equivalent circuit for the layer reflecting the winding location area of the rotor

circuits for the layers reflecting the stator and rotor winding location areas become inductive. Matching these conditions, the L-shaped equivalent circuits for the layers representing the stator and rotor winding location areas were obtained in [13] and shown in Fig. 14.9. Figure 14.9a shows the L-shaped equivalent circuit obtained for the layer of the stator winding location area. Figure 14.9b shows the L-shaped equivalent circuit for the layer of the rotor winding location area. For the layer factor c_{zs} used in the equivalent circuit in Fig. 14.9a, we have

$$c_{zs} = \cosh \beta_{zs} h_{zs} = \cosh\left[k_{\mu zs}(\pi/\tau_{zs})\right] h_{zs} \tag{14.57}$$

where $\beta_{zs} = k_{\mu zs}(\pi/\tau_{zs})$.

In the equivalent circuit in Fig. 14.9a, for the values $\dot{E}'_{zs2}$ and $\dot{H}'_{zs2}$ used in it, it follows

$$\dot{E}'_{zs2} = \dot{E}_{zs2} c_{zs} \quad \dot{H}'_{zs2} = \dot{H}_{zs2}/c_{zs} \tag{14.58}$$

In the equivalent circuit in Fig. 14.9b, the layer factor c_{zR} is used. For this factor, from (14.47), we obtain

$$c_{zR} = \cosh \beta_{zR} h_{zR} = \cosh \left[k_{\mu zR}(\pi/\tau_{zR})\right] h_{zR} \tag{14.59}$$

where $\beta_{zR} = k_{\mu zR}(\pi/\tau_{zR})$.

In the equivalent circuit in Fig. 14.9b, for the values $\dot{E}'_{zR2}$ and $\dot{H}'_{zR2}$, we can have

$$\dot{E}'_{zR2} = \dot{E}_{zR2} c_{zR} \quad \dot{H}'_{zR2} = \dot{H}_{zR2}/c_{zR} \tag{14.60}$$

The equivalent circuit in Fig. 14.9a uses the reactances x^*_{zs}, $x^*_{\tau zs}$ and $x^*_{11\sigma}$ for which from (14.51) and also from [13], we can obtain

$$\begin{aligned} x^*_{zs} &= \frac{\omega_1 \mu_{xzs}}{\beta_{zs} \tanh \beta_{zs} h_{zs}} \frac{\xi_{Lzs}}{\xi_{\tau zs}} \\ x^*_{\tau zs} &= \omega_1 \mu_{xzs} h_{zs} k_{zs} \frac{\xi_{Lzs}}{\xi_{\tau zs}} \\ x^*_{11\sigma} &= \omega_1 \mu_{xzs} \frac{h_{zs}}{3} \frac{\xi_{Lzs}}{\xi_{\tau zs}} \end{aligned} \tag{14.61}$$

where $k_{zs} = \tanh \beta_{zs} h_{zs} / \beta_{zs} h_{zs}$; $\xi_{\tau zs} = \tau_{zs}/\tau_0$; $\xi_{Lzs} = l_{zs}/l_0$.

In the equivalent circuit in Fig. 14.9b, the reactances x^*_{zR}, $x^*_{\tau zR}$ and $x^*_{22\sigma}$ are used, for which from (14.52) and also from [13], we can obtain

$$\begin{aligned} x^*_{zR} &= \frac{\omega_1 \mu_{xzR}}{\beta_{zR} \tanh \beta_{zR} h_{zR}} \frac{\xi_{LzR}}{\xi_{\tau zR}} \\ x^*_{\tau zR} &= \omega_1 \mu_{xzR} h_{zR} k_{zR} \frac{\xi_{LzR}}{\xi_{\tau zR}} \\ x^*_{22\sigma} &= \omega_1 \mu_{xzR} \frac{h_{zR}}{3} \frac{\xi_{LzR}}{\xi_{\tau zR}} \end{aligned} \tag{14.62}$$

where $k_{zR} = \tanh \beta_{zR} h_{zR} / \beta_{zR} h_{zR}$; $\xi_{\tau zR} = \tau_{zR}/\tau_0$; $\xi_{LzR} = l_{zR}/l_0$.

In the equivalent circuits obtained in Fig. 14.9a, the coefficients k_{10} and k_{20} are determined as

$$k_{10} = \frac{\beta_{zs} h_{zs}}{\sinh \beta_{zs} h_{zs}} \quad k_{20} = \frac{\beta_{zR} h_{zR}}{\sinh \beta_{zR} h_{zR}} \tag{14.63}$$

The quantities r^*_{11} and r^*_{22}/s used in the equivalent circuits in Fig. 14.9, represent the ohmic resistances of the phase windings of the stator and rotor and can be determined using the expressions obtained in [13].

14.5 Summary

1. The selection of the basic fragments of the machine design is the basis for obtaining the electric machine model. In this work, the stator and rotor slot-wedge areas, the stator and rotor winding location areas, the stator and rotor yoke areas, as well as the air gap of the machine are considered as such fragments of the machine design. The pictures of the electromagnetic field distribution in the selected fragments can be simplified. In practice, the simplified pictures of the electric and magnetic field distribution are usually characterized by the radial and tangential components of the magnetic field intensities, which take place in the plane of the cross section of the machine, as well as by the one axial component of the electric field intensity. In order to simplify the analysis, for the stator and rotor slot-tooth areas, the average values of the magnetic permeabilities can be applied, and for the stator and rotor winding conductors, the average values of the resistivities can be used. In addition, the surfaces of the air gap are assumed to be conditionally smooth. Under these conditions, the selected basic fragments of the machine design can be represented in the form of corresponding layers, which leads to the multilayer model of the electric machine. In this work, with respect to the surface $0-0$ passing through the middle of the radial length of the air gap, the multilayer model of the machine is decomposed into the multilayer model of the stator and the multilayer model of the rotor. This decomposition of the machine model simplifies the analysis.
2. Solutions of the field equations are expressed in terms of the components of the electric and magnetic field intensities $(\dot{E}, \dot{H})$, which characterize the electromagnetic field intensity at the point. This work discusses the machine calculation elements used in the analysis of the energy processes occurring in the electric machine. In this regard, this work does not analyze the nature of the field distribution in the electric machine, but only the components of the electric and magnetic field intensities determined on the outer and inner surfaces formed by the adjacent layers of the stator and rotor models are considered. Such representation of the solutions of the field equations helps to simplify the analysis of the energy processes in the electric machine.
3. In accordance with the Poynting vector, the power dissipated in the ith layer of the machine model is determined by the values of the components of the electric and magnetic field intensities specified on the outer and inner surfaces of the considered layer. In this regard, in this work, the solutions of the field equations are represented in terms of these values of the components of the electric and magnetic field intensities. The equations obtained in this way can be considered as the generalized equations for the ith layer of the machine model. Such generalized equations for the ith layer are proposed based on the solution of the field equations obtained using the analytical and numerical methods of the calculation.
4. Generalized equations for the ith layer can be visualized using the T-shaped equivalent circuit. The T-shaped equivalent circuit for the ith layer consists of the three mutually connected circuit loops. Therefore, the use of the T-shaped

equivalent circuits creates great difficulties in solving the issue related to the determination of the calculation elements of the electric machine.

5. In this work, the ith layer of the machine model is replaced by the L-shaped equivalent circuit. The L-shaped equivalent circuit is obtained in relation to the electric and magnetic field intensities $(\dot{E}, \dot{H})$ set on the outer (in the case of the rotor) and on the inner (in the case of the stator) surfaces of the ith layer, and therefore, it (the L-shaped equivalent circuit) contains the two circuit loops independent one from the other. In this regard, the L-shaped equivalent circuit of the ith layer, in comparison with the T-shaped equivalent circuit, has a rather simple structure, convenient for practical use.
6. The L-shaped equivalent circuit of the ith layer follows on the basis of the use of the "no-load and under-load" concept which is the basic concept applied in the theory of electric machines to determine the calculation elements of the machine. In this regard, the parameters applied in the L-shaped equivalent circuit are identical in the structure and the physical essence to the parameters used in the theory of electric machines. This circumstance is a very important factor in a favor of using the L-shaped equivalent circuits. In this work, the L-shaped equivalent circuits form the basis for the methods used to determine the calculation elements of the machine.
7. The L-shaped equivalent circuit of the nonconducting layer consists of the two independent circuit loops: the "own" longitudinal magnetizing circuit loop ("own" circuit loop) of the layer and the "own" transverse magnetizing circuit loop of the layer ("connecting" circuit loop). The "connecting" circuit loop provides the electric connection of the "own" longitudinal magnetizing circuit loops reflecting the adjacent layers of the machine model. In the case of considering the conducting layer in the area of the weak skin effect, the "own" longitudinal magnetizing circuit loop is represented as consisting of the two parallel circuit loops: the "own" longitudinal magnetizing circuit loop of the layer and the "own" leakage circuit loop of the eddy currents induced in the layer. In this case, the L-shaped equivalent circuit of the conducting layer consists of the three independent circuit loops: the "own" longitudinal magnetizing circuit loop of the layer, the "own" leakage circuit loop of the eddy currents induced in the layer, and the "connecting" circuit loop.

References

1. Richter, R.: Electric Machines, vol. 1-5. United Scientific and Technical Publishing, Moscow (1934-1939)
2. Livshits, A.: Electric Machines, vol. 1-3, United Scientific and Technical Publishing, Moscow (1936)
3. Kostenko, M.: Electric Machines. Special Part, Gosenergoizdat, Moscow (1949)
4. Kostenko, M., Piotrovsky, L.: Electric Machines, Energiya, Leningrad (1973)
5. Alger, P.: Induction Machines: Their Behavior and Uses, Gordon and Beach, New York (1970)
6. Say, M.: Alternating Current Machines, John Wiley and Sons, Inc., New York (1983)

7. Adking, B.: The General Theory of Electrical Machines, John Wiley and Sons, Inc., New York (1957)
8. Voldek. A.: Electric Machines, Energiya, Leningrad (1974)
9. Ivanov-Smolensky, A.: Electric Machines, Energiya, Moscow (1980)
10. Vazhnov, A.: Electric Machines, Energiya, Leningrad (1968)
11. Krause, P.: Analysis of Electrical Machines, McGraw-Hill, New York (1986)
12. Fitzgerald, A., Kingsley, C., Umans, S.: Electric Machinery, McGraw-Hill, Inc., 5th ed., New York (1990)
13. Asanbayev, V.: Alternating Current Multi-Circuit Electric Machines: A New Approach to Steady-State Parameter Determination, Springer Cham Heidelberg, New York, Dordrecht London (2015)
14. Mishkin, E.: Theory of the Squirrel-Cage Induction Machine Derived Directly from Maxwell's Field Equations, The Quarterly Journal of Mechanics and Applied Mathematics, Vol.7, Part 4, pp. 472-487, (1954)
15. Freeman, E.: Equivalent Circuits from Electromagnetic Theory: Low Frequency Induction Devices, Procc. IEEE, 12 (115), 1772-1776 (1968)
16. Freeman, E.: Travelling Waves in Induction Machines: Input Impedance and Equivalent Circuits, Procc. IEEE, (115), N 12, pp. 1772-1776 (1968)
17. Postnikov, I., Bezusyi, L.: Calculation of Traveling Electromagnetic Field in Multilayer Medium, Izvestiya ANSSSR. Energetika i Transport, n.2, (1970)
18. Inkin, A., Litvinov, B.: Synthesis of Cascade Equivalent Circuits for Induction Electric Machines, Elektrotechnika, 1977, n.1, pp.29-34
19. Veselovsky, O., Konyaev, A., Sarapulov, F.: Linear Asynchronous Motor, Publishing Energoatomizdat, Moscow (1991)
20. Boldea, I., Babescu, M.: Multi-Layer Approach to the Analysis of Single-Sited Linear Induction Motor, Procc. IEEE, (125), N 4, pp. 280-287 (1978)
21. Williamson, S.: The Anisotropic Layer Theory of Induction Machines and Induction Devices, J. Inst. Maths. And Applies, vol. 17, 1976, pp. 69-84
22. Asanbayev, V., Shumilov, Yu., Strunevich, E., Saratov, V.: Determination of Parameters of AC Electric Machines with the Solid Rotor on the Basis of the Field Calculation by the Finite Element Method. Academy of Science Ukrainian SSR, Institute of Electrodynamics, Preprint, N 358, Kiev (1984)
23. Shimoni, K.: Theoretical Electrical Engineering (Translation from German), Publishing Mir, Moscow (1973)
24. Seely, S.: Electromechanical Energy Conversion, McGraw-Hill Book Company, Inc., (1962)
25. Turovskiy, YA.: Technical Electrodynamics, (Translated from Polish), Publishing Energiya (1974)
26. Voldek, A.: Induction Magneto-hydrodynamic Machines with Liquid-Metal Working Substance, Publishing Energiya, Leningrad (1970)

Chapter 15
Basic Calculation Elements of Machine with Single-Cage Rotor: "Field" Concept

15.1 L-Shaped Equivalent Circuits of Stator Model Layers and Their Parameters

We proceed from the L-shaped equivalent circuits discussed in Chap. 14 and represent the corresponding layers of the stator model (Fig. 14.1). Figure 14.7b shows the L-shaped equivalent circuit for the air gap sublayer adjoined to the stator bore surface. This L-shaped equivalent circuit for the air gap sublayer is given in Fig. 15.1a. The reactances x_{01}^* and $x_{\tau 01}^*$ used in the equivalent circuit in Fig. 15.1a follow from (14.54) in the form

$$x_{01}^* = \frac{\omega_1 \mu_0 \tau_0^2}{\pi^2 \delta_1} \zeta_0 \quad x_{\tau 01}^* = \omega_1 \mu_0 \frac{\delta}{2} k_{1\sigma} \zeta_{\tau 0} \tag{15.1}$$

where
$$\zeta_0 = \frac{4p(\delta/2R_0)\left[1 - (\delta/2R_0)^2\right]^p}{(1 + \delta/2R_0)^{2p} - (1 - \delta/2R_0)^{2p}}$$
$$\zeta_{\tau 0} = \frac{(1 + \delta/2R_0)^p - (1 - \delta/2R_0)^p}{p(\delta/2R_0)[(1 + \delta/2R_0)^p + (1 - \delta/2R_0)^p]}.$$

Here, the coefficients ζ_0 and $\zeta_{\tau 0}$ take into account the curvature of the surfaces of the air gap and they are determined from the relations obtained in [1].

Figure 14.6a shows the L-shaped equivalent circuit for the layer of the stator slot-wedge area. This equivalent circuit is shown in Fig. 15.1b. For the reactances x_{ks}^* and $x_{\tau ks}^*$ used in the equivalent circuit in Fig. 15.1b, we can obtain from (14.51)

$$x_{ks}^* = \frac{\omega_1 \mu_{yks} \tau_{ks}^2}{\pi^2 h_{ks}} \frac{\xi_{Lks}}{\xi_{\tau ks}} \zeta_{ks} \quad x_{\tau ks}^* = \omega_1 \mu_{xks} h_{ks} \frac{\xi_{Lks}}{\xi_{\tau ks}} \zeta_{\tau ks} \tag{15.2}$$

V. Asanbayev, *Asynchronous Machines*,
https://doi.org/10.1007/978-3-030-92284-9_15

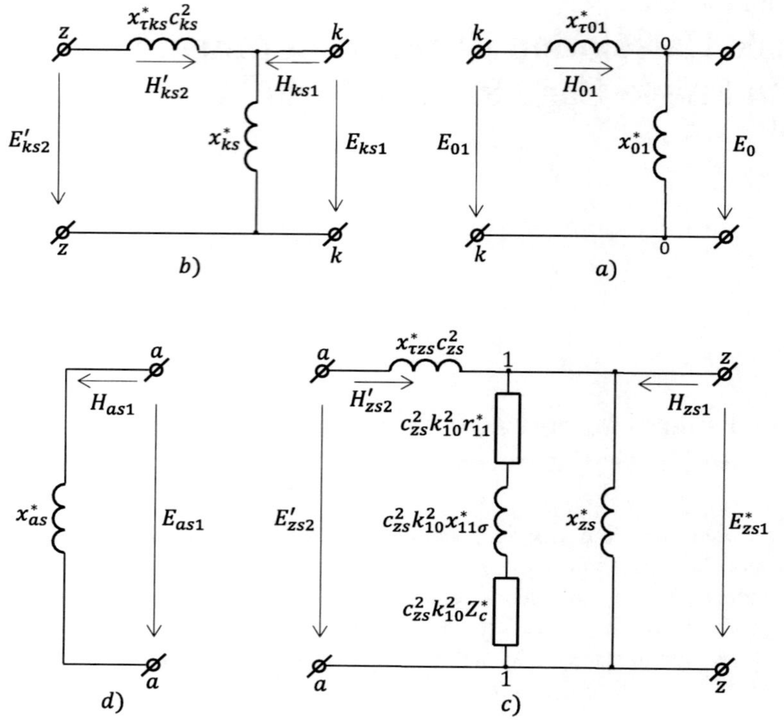

Fig. 15.1 Stator model layer equivalent circuits: (**a**) L-shaped equivalent circuit of the air gap sublayer adjoined to the stator bore surface, (**b**) L-shaped equivalent circuit of the layer of the stator slot-wedge area, (**c**) L-shaped equivalent circuit of the layer of the stator winding location area, (**d**) equivalent circuit of the stator yoke area layer

where

$$\zeta_{ks} = n_{ks}\frac{h_{ks}}{R_{ks1}}\frac{(1 + h_{ks}/R_{ks1})^{2n_{ks}} + 1}{(1 + h_{ks}/R_{ks1})^{2n_{ks}} - 1}$$

$$\zeta_{\tau ks} = \frac{1}{n_{ks}h_{ks}/R_{ks1}}\frac{(1 + h_{ks}/R_{ks1})^{2n_{ks}} - 1}{(1 + h_{ks}/R_{ks1})^{2n_{ks}} + 1}.$$

In (15.2), the coefficients ζ_{ks} and $\zeta_{\tau ks}$ take into account the curvature of the surfaces of the layer representing the stator slot-wedge area and follow from the relations proposed in [1]. The quantity of R_{ks1} used in (15.2) is the radius of the inner surface of the layer of the stator slot-wedge area. The coefficient c_{ks} used in the equivalent circuit in Fig. 15.1b can be calculated using the relationship obtained in (14.47). In accordance with [1], the components of the electric and magnetic field intensities $\dot{E}'_{ks2}$ and $\dot{H}'_{ks2}$ representing the output terminals of the L-shaped equivalent circuit in Fig. 15.1b are determined by the relations given in (14.48).

Figure 14.9a shows the L-shaped equivalent circuit for the layer reflecting the stator winding location area. In this equivalent circuit, the voltage $\dot{U}_1^*$ can be replaced by the voltage drop across the power supply network impedance Z_c^*, i.e. $\left(-\dot{U}_1^*\right) = Z_c^*\dot{H}_1$. Then, the L-shaped equivalent circuit for the layer of the stator winding location area can be represented as the passive equivalent circuit shown in Fig. 15.1c. Based on the relations (14.61), for the reactances x_{zs}^* and $x_{\tau zs}^*$ of the equivalent circuit in Fig. 15.1c, we have

$$x_{zs}^* = \frac{\omega_1 \mu_{yzs} \tau_{zs}^2}{\pi^2 h_{zs}} \frac{\xi_{Lzs}}{\xi_{\tau zs}} \zeta_{zs} \quad x_{\tau zs}^* = \omega_1 \mu_{xzs} h_{zs} \frac{\xi_{Lzs}}{\xi_{\tau zs}} \zeta_{\tau zs} \tag{15.3}$$

where

$$\zeta_{zs} = n_{zs} \frac{h_{zs}}{R_{zs1}} \frac{\left(1 + h_{zs}/R_{zs1}\right)^{2n_{zs}} + 1}{\left(1 + h_{zs}/R_{zs1}\right)^{2n_{zs}} - 1}$$

$$\zeta_{\tau zs} = \frac{1}{n_{zs} h_{zs}/R_{zs1}} \frac{\left(1 + h_{zs}/R_{zs1}\right)^{2n_{zs}} - 1}{\left(1 + h_{zs}/R_{zs1}\right)^{2n_{zs}} + 1}$$

In (15.3), the coefficients ζ_{zs} and $\zeta_{\tau zs}$ are taken into account the curvature of the surfaces of the layer of the stator winding location area and they are obtained in [1]. The quantity of R_{zs1} used in (15.3) is the radius of the inner surface of the layer of the stator winding location area. In the equivalent circuit in Fig. 15.1c, the reactance $x_{11\sigma}^*$ is used. The formula for this reactance can be obtained using the expression given in (14.61). Then, we get

$$x_{11\sigma}^* = \omega_1 \mu_{xzs} \frac{h_{zs}}{3} \frac{\xi_{Lzs}}{\xi_{\tau zs}} \tag{15.4}$$

The coefficient c_{zs} used in the equivalent circuit in Fig. 15.1c is determined by the expression obtained in (14.57). For the components of the electric and magnetic field intensities $\dot{E}_{zs2}'$ and $\dot{H}_{zs2}'$ representing the output terminals of the L-shaped equivalent circuit in Fig. 15.1c, the expressions shown in (14.58) can be used.

The layer of the stator yoke area is replaced by the two-terminal circuit (Fig. 14.8a). This equivalent circuit is shown in Fig. 15.1d. The reactance x_{as}^* used in the equivalent circuit in Fig. 15.1d is determined from (14.55) as

$$x_{as}^* = \omega_1 \mu_{xas} h_{as} \frac{\xi_{Las}}{\xi_{\tau as}} \zeta_{as} \tag{15.5}$$

where $\zeta_{as} = \dfrac{1}{n_{as} h_{as}/R_{as1}} \dfrac{\left(1 + h_{as}/R_{as1}\right)^{2n_{as}} - 1}{\left(1 + h_{as}/R_{as1}\right)^{2n_{as}} + 1}$.

Here, the coefficient ζ_{as} is taken into account the curvature of the surfaces of the stator yoke area layer, R_{as1} reflects the radius of the inner surface of the stator yoke area layer.

15.2 Cascade Equivalent Circuits of Stator

To obtain the cascade equivalent circuit for the stator, it is necessary to use its layered model shown in Fig. 14.1 and the L-shaped equivalent circuits shown in Fig. 15.1. As follows from Fig. 15.1, the construction of the L-shaped equivalent circuits is carried out in relation to the components of the electric and magnetic field intensities $\dot{E}_{i1}, \dot{H}_{i1}$, and $\dot{E}_{i2}, \dot{H}_{i2}$ specified on the inner and outer surfaces of the stator model layers, respectively. If at the boundaries formed by adjacent layers of the stator model (Fig. 14.1), the conditions (14.2a) for the components of the electric and magnetic field intensities are taken into account, then, the L-shaped equivalent circuits representing the corresponding layers of the stator model can be connected in cascade, in which the input terminals of each subsequent L-shaped equivalent circuit are connected to the output terminals of each previous L-shaped equivalent circuit. Now, the stator equivalent circuit follows, in this case, as a result of the cascade way of connecting the L-shaped circuits, representing the corresponding layers of the stator model. Below, the description of this technique for constructing the stator equivalent circuit is given.

According to the layered model of the stator shown in Fig. 14.1, the cascade stator equivalent circuit (in the form of the cascade-connected two-port networks) can be represented in the form in Fig. 15.2a. In this cascade equivalent circuit, the four-terminal network Π_{01} represents the air gap sublayer adjoined to the stator bore surface, the four-terminal network Π_{ks} reflects the layer of the stator slot-wedge area, the four-terminal network Π_{zs} corresponds to the layer of the stator winding location area, and the two-terminal network Π_{as} reflects the stator yoke area layer. The terminals $k-k$, $z-z$ and $a-a$ of the cascade equivalent circuit in Fig. 15.2a correspond to the surfaces $k-k$, $z-z$, and $a-a$ formed by the adjacent layers of the stator model (Fig. 14.1). In this regard, at the terminals $k-k$, $z-z$, and $a-a$ of the cascade equivalent circuit in Fig. 15.2a the conditions given in (14.2a) are fulfilled.

For the schematic visualization of the two-port networks used in the cascade equivalent circuit in Fig. 15.2a, the L-shaped equivalent circuits shown in Fig. 15.1 can be applied. The use of these equivalent circuits means that the four-terminal networks in the equivalent circuit in Fig. 15.2a are replaced by the L-shaped equivalent circuits (with the exception of the stator yoke area layer, which is replaced by the two-terminal equivalent circuit). As follows from the equivalent circuit in Fig. 15.2a, in this case, the conditions (14.2a) are not met at the output terminals of the L-shaped equivalent circuits, which represent the layer of the stator slot-wedge area (four-terminal network Π_{ks}) and the layer of the stator winding location area (four-terminal network Π_{zs}). Therefore, the L-shaped equivalent circuits used in the equivalent circuit in Fig. 15.2a cannot be directly connected in a cascade. Below we will analyze the conditions under which these L-shaped equivalent circuits can be connected in a cascade.

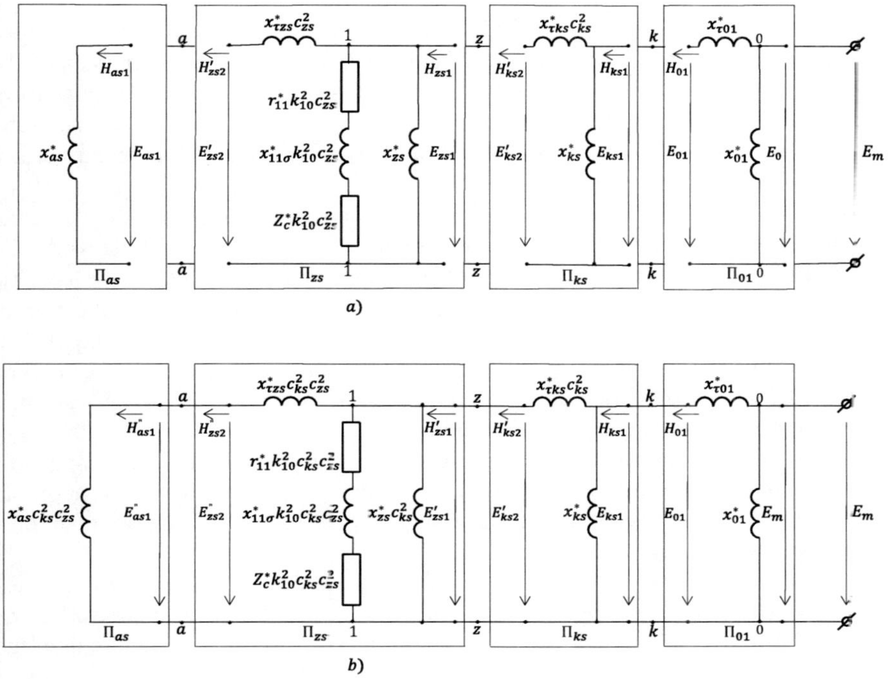

Fig. 15.2 Stator equivalent circuits as the cascade connection of the four-terminal networks: **(a)** schematic visualization of the four-terminal networks; **(b)** cascade connection of the L-shaped equivalent circuits, reflecting the stator model layers

15.3 "Unfolded" Equivalent Circuit with Mixed Connection of Elements of Stator

Let's analyze the stator cascade equivalent circuit in Fig. 15.2a. In this equivalent circuit, we will consider the conditions for connecting the input terminals of the L-shaped equivalent circuits of the subsequent four-terminal networks with the output terminals of the L-shaped equivalent circuits of the previous four-terminal networks. Let's start by considering the conditions for connecting the input terminals of the L-shaped equivalent circuit of the four-terminal network Π_{ks} with the output terminals of the L-shaped equivalent circuit of the four-terminal network Π_{01}. In Fig. 15.2a, the output terminals of the L-shaped equivalent circuit of the four-terminal network Π_{01} correspond to the outer surface of the air gap sublayer adjoined to the stator bore surface. The input terminals of the L-shaped equivalent circuit of the four-terminal network Π_{ks} represent the inner surface of the layer of the stator slot-wedge area. In the stator model (Fig. 14.1), these layers form the adjoined surface $k - k$ on which, in accordance with (14.2a), the conditions of the form $\dot{E}_{01} = \dot{E}_{ks1}$ and $\dot{H}_{01} = H_{ks1}$ are valid. Thus, the components of the electric and magnetic field intensities $\dot{E}_{01}$ and $\dot{H}_{01}$ act on the output terminals of the L-shaped equivalent circuit of the four-terminal network Π_{01}. At the input terminals of the L-shaped equivalent circuit of the four-terminal network Π_{ks}, for the components of the electric and magnetic field intensities, we have that $\dot{E}_{ks1}$ and $\dot{H}_{ks1}$. Therefore, based on the conditions $\dot{E}_{01} = \dot{E}_{ks1}$ and $\dot{H}_{01} = H_{ks1}$, we can note that on the surface $k - k$ of the stator model (Fig. 14.1), as well as on the output terminals of the L-shaped equivalent circuit of the four-terminal network Π_{01} and on the input terminals of the L-shaped equivalent circuit of the four-terminal network Π_{ks} (Fig. 15.2a), the components of the electric and magnetic field intensities have the same value. In this regard, the output terminals of the L-shaped equivalent circuit of the four-terminal network Π_{01} and the input terminals of the L-shaped equivalent circuit of the four-terminal network Π_{ks} can be directly connected in a cascade. The four-terminal network Π_{01} and the four-terminal network Π_{ks} as well as their schematic visualization, are shown in Fig. 15.2b.

Now, in the equivalent circuit in Fig. 15.1a, we will consider the features of connecting the input terminals of the L-shaped equivalent circuit of the four-terminal network Π_{zs} with the output terminals of the L-shaped equivalent circuit of the four-terminal network Π_{ks}.

In Fig. 15.1a, the output terminals of the L-shaped equivalent circuit of the four-terminal network Π_{ks} correspond to the outer surface of the layer representing the stator slot-wedge area. The input terminals of the L-shaped equivalent circuit of the four-terminal network Π_{zs} represent the inner surface of the layer reflecting the stator winding location area. In the stator model (Fig. 14.1), these layers form the adjacent surface $z - z$ on which, in accordance with (14.2a), the conditions of the form $\dot{E}_{ks2} = \dot{E}_{zs1}$ and $\dot{H}_{ks2} = H_{zs1}$ are valid. The components of the electric and magnetic field intensities $\dot{E}'_{ks2}$ and $\dot{H}'_{ks2}$ are on the output terminals of the L-shaped equivalent circuit of the four-terminal network Π_{ks}. At the input terminals of the L-shaped equivalent circuit of the four-terminal network Π_{zs}, for the components of the electric and magnetic field intensities, we have that $\dot{E}_{zs1}$ and $\dot{H}_{zs1}$. When cascading the output terminals of the L-shaped equivalent circuit of the four-terminal network Π_{ks} with the input terminals of the L-shaped equivalent circuit of the four-terminal network Π_{zs}, it should be borne in mind that the values of $\dot{E}'_{ks2}$ and $\dot{H}'_{ks2}$ corresponding to the output terminals of the L-shaped equivalent circuit of the four-terminal network Π_{ks} are defined in accordance with (14.48) as $\dot{E}'_{ks2} = \dot{E}_{ks2}c_{ks}$ and $\dot{H}'_{ks2} = \dot{H}_{ks2}/c_{ks}$. In this regard, and taking into account the conditions at the boundary $z - z$ of the stator model in the form of $\dot{E}_{ks2} = \dot{E}_{zs1}$ and $\dot{H}_{ks2} = H_{zs1}$, the values of $\dot{E}_{zs1}$ and H_{zs1} corresponding to the input terminals of the L-shaped equivalent circuit of the four-terminal network Π_{zs} should be corrected as

$$\dot{E}'_{ks2} = \dot{E}'_{zs1} = \dot{E}_{zs1}c_{ks} \quad \dot{H}'_{ks2} = \dot{H}'_{zs1} = \dot{H}_{zs1}/c_{ks} \tag{15.6}$$

Consequently, when cascading the output terminals of the L-shaped equivalent circuit of the four-terminal network Π_{ks} with the input terminals of the L-shaped equivalent circuit of the four-terminal network Π_{zs} instead of the conditions obtained in (14.2a) in the form $\dot{E}_{ks2} = \dot{E}_{zs1}$ and $\dot{H}_{ks2} = H_{zs1}$, it is necessary to use the new conditions of kind

$$\dot{E}'_{ks2} = \dot{E}'_{zs1} \quad \dot{H}'_{ks2} = \dot{H}'_{zs1} \tag{15.7}$$

where $\dot{E}'_{ks2} = \dot{E}_{ks2}c_{ks}; \dot{E}'_{zs1} = \dot{E}_{zs1}c_{ks}; \dot{H}'_{ks2} = \dot{H}_{ks2}/c_{ks}; \dot{H}'_{zs1} = \dot{H}_{zs1}/c_{ks}$.

With respect to the values $\dot{E}'_{zs1}$ and $\dot{H}'_{zs1}$, the input impedance of the L-shaped equivalent circuit of the four-terminal network Π_{zs} is defined as $Z'_{zs1} = \dot{E}'_{zs1}/\dot{H}'_{zs1} = (\dot{E}_{zs1}/\dot{H}_{zs1})c^2_{ks} = Z_{zs1}c^2_{ks}$, where $Z_{zs1} = \dot{E}_{zs1}/\dot{H}_{zs1}$ and it represents the input impedance of the L-shaped equivalent circuit of the four-terminal network Π_{zs} determined with respect to the values of $\dot{E}_{zs1}$ and $\dot{H}_{zs1}$. Consequently, under the conditions of the transition to the new values $\dot{E}'_{zs1}$ and $\dot{H}'_{zs1}$, the impedances of the L-shaped equivalent circuit of the four-terminal network Π_{zs} in the form $[x^*_{zs}, (r^*_{11}k^2_{10}c^2_{zs}), (x^*_{11\sigma}k^2_{10}c^2_{zs}), (Z^*_c k^2_{10}c^2_{zs}), (x^*_{\tau zs}c^2_{zs})]$ are multiplied by the squared coefficient c^2_{ks} that is $[x^*_{zs}c^2_{ks}, (r^*_{11}k^2_{10}c^2_{ks}c^2_{zs}), (x^*_{11\sigma}k^2_{10}c^2_{ks}c^2_{zs}), (Z^*_c k^2_{10}c^2_{ks}c^2_{zs}), (x^*_{\tau zs}c^2_{ks}c^2_{zs})]$. In addition, when the output terminals of the L-shaped equivalent circuit of the four-terminal network Π_{ks} are cascaded with the input terminals of the L-shaped equivalent circuit of the four-terminal network Π_{zs}, the values $\dot{E}'_{zs2} = \dot{E}_{zs2}c_{zs}$ and $\dot{H}'_{zs2} = \dot{H}_{zs2}/c_{zs}$ corresponding to the output terminals of the L-shaped equivalent circuit of the four-terminal network Π_{zs} and determined by the condition given in (14.58) are corrected by the coefficient c_{ks} and take the form

$$\dot{E}''_{zs2} = \dot{E}'_{zs2}c_{ks} = \dot{E}_{zs2}c_{ks}c_{zs} \quad \dot{H}''_{zs2} = \dot{H}'_{zs2}/c_{ks} = \dot{H}_{zs2}/c_{ks}c_{zs} \tag{15.8}$$

Taking into account the provisions given above, the L-shaped equivalent circuit of the four-terminal network Π_{zs} (Fig. 15.2a) takes the form shown in Fig. 15.2b. In Fig. 15.2b, the output terminals of the L-shaped equivalent circuit of the four-terminal network Π_{ks} and the input terminals of the L-shaped equivalent circuit of the four-terminal network Π_{zs} are connected in a cascade.

Further, in the equivalent circuit in Fig. 15.2a, we will consider the features of the connection of the output terminals of the L-shaped equivalent circuit of the four-terminal network Π_{zs} with the input terminals of the equivalent circuit of the two-terminal network Π_{as}. In Fig. 15.2a, the output terminals of the four-terminal network Π_{zs} correspond to the outer surface of the layer reflecting the stator winding location area. The input terminals of the equivalent circuit of the two-terminal network Π_{as} represent the inner surface of the stator yoke area layer. In the stator model (Fig. 14.1), these layers form the adjacent surface $a - a$ on which, in accordance with (14.2a), the conditions of the form $\dot{E}_{zs2} = \dot{E}_{as1}$ and $\dot{H}_{zs2} = \dot{H}_{as1}$ are valid. According to (15.8), the components of the electric and magnetic field intensities $\dot{E}''_{zs2}$ and $\dot{H}''_{zs2}$ act on the output terminals of the L-shaped equivalent circuit of the four-terminal network Π_{zs}.

At the input terminals of the equivalent circuit of the two-terminal network Π_{as}, for the components of the electric and magnetic field intensities, we have that $\dot{E}_{as1}$ and $\dot{H}_{as1}$. When cascading the output terminals of the L-shaped equivalent circuit of the four-terminal network Π_{zs} with the input terminals of the equivalent circuit of the two-terminal network Π_{as}, it must be borne in mind that the values $\dot{E}''_{zs2}$ and $\dot{H}''_{zs2}$ corresponding to the output terminals of the L-shaped equivalent circuit of the four-terminal network Π_{zs} are determined in accordance with (14.48), (14.58), and (15.8) as $\dot{E}''_{zs2} = \dot{E}_{zs2}c_{ks}c_{zs}$ and $\dot{H}''_{zs2} = \dot{H}_{zs2}/c_{ks}c_{zs}$. In this regard, and taking into account the conditions at the boundary $a - a$ of the stator model in the form $\dot{E}_{zs2} = \dot{E}_{as1}$ and $\dot{H}_{zs2} = H_{as1}$, the quantities $\dot{E}_{as1}$ and H_{as1} corresponding to the input terminals of the equivalent circuit of the two-terminal network Π_{as} should be corrected as

$$\dot{E}''_{zs2} = \dot{E}''_{as1} = \dot{E}_{as1}c_{ks}c_{zs} \quad \dot{H}''_{zs2} = \dot{H}''_{as1} = \dot{H}_{as1}/c_{ks}c_{zs} \tag{15.9}$$

Consequently, when cascading the output terminals of the L-shaped equivalent circuit of the four-terminal network Π_{zs} with the input terminals of the equivalent circuit of the two-terminal network Π_{as}, instead of the conditions obtained in (14.2a) in the form $\dot{E}_{zs2} = \dot{E}_{as1}$ and $\dot{H}_{zs2} = H_{as1}$, it is necessary to use the new conditions of the form

$$\dot{E}''_{zs2} = \dot{E}''_{as1} \quad \dot{H}''_{zs2} = \dot{H}''_{as1} \tag{15.10}$$

where $\dot{E}''_{zs2} = \dot{E}_{zs2}c_{ks}c_{zs}; \dot{E}''_{as1} = \dot{E}_{as1}c_{ks}c_{zs}; \dot{H}''_{zs2} = \dot{H}_{zs2}/c_{ks}c_{zs}; \dot{H}''_{as1} = \dot{H}_{as1}/c_{ks}c_{zs}$.

With respect to the values $\dot{E}''_{as1}$ and $\dot{H}''_{as1}$, for the input impedance of the equivalent circuit of the two-terminal network Π_{as}, we have that $Z'_{as1} = \dot{E}''_{as1}/\dot{H}''_{as1} = (\dot{E}_{as1}/\dot{H}_{as1})c_{ks}^2c_{zs}^2 = Z_{as1}c_{ks}^2c_{zs}^2$, where $Z_{as1} = \dot{E}_{as1}/\dot{H}_{as1}$ and represents the input impedance of the equivalent circuit of the two-terminal network Π_{as} determined with respect to the values of $\dot{E}_{as1}$ and $\dot{H}_{as1}$. Therefore, under the conditions of the transition to the new quantities $\dot{E}''_{as1}$ and $\dot{H}''_{as1}$, the reactance x^*_{as} used in the equivalent circuit of the two-terminal network Π_{as} is multiplied by the product of the square of the coefficients $(c_{ks}^2c_{zs}^2)$, i.e. $x^*_{as}c_{ks}^2c_{zs}^2$. Thus, the reactance of the equivalent circuit of the two-terminal network Π_{as} (Fig. 15.2a) takes the form shown in Fig. 15.2b. In Fig. 15.2b, the output terminals of the L-shaped equivalent circuit of the four-terminal network Π_{zs} and the input terminals of the equivalent circuit of the two-terminal network Π_{as} are connected in a cascade.

At the input and output terminals of the L-shaped equivalent circuits used in the cascade equivalent circuit in Fig. 15.2b, the new conditions are valid. These conditions are determined in accordance with (14.2a), (15.7), and (15.10) in the form

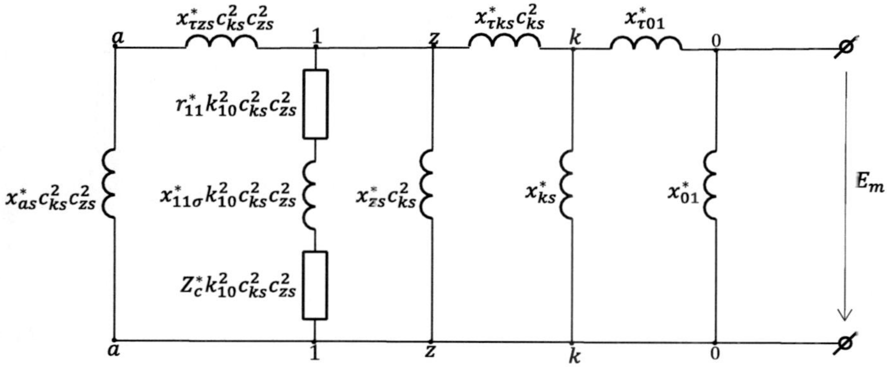

Fig. 15.3 "Unfolded" stator equivalent circuit with the mixed connection of the elements

$$
\begin{aligned}
\dot{E}_{01} &= \dot{E}_{ks1}\ \dot{H}_{01} = H_{ks1} \\
\dot{E}'_{ks2} &= \dot{E}'_{zs1} \quad \dot{H}'_{ks2} = \dot{H}'_{zs1} \\
\dot{E}''_{zs2} &= \dot{E}''_{as1} \quad \dot{H}''_{zs3} = \dot{H}''_{as1}
\end{aligned}
\tag{15.11}
$$

When these conditions are met, the L-shaped equivalent circuits used in Fig. 15.2b can be connected in a cascade. As a result of this connection, the stator equivalent circuit takes the form in Fig. 15.3. This equivalent circuit is the "unfolded" stator equivalent circuit with the mixed connection of the elements.

For the parameters used in the "unfolded" equivalent circuit in Fig. 15.3, we can have

$$
\begin{aligned}
x_{01} = x^*_{01}; x_{ks} = x^*_{ks}; x_{zs} = x^*_{zs}c^2_{ks}; x_{as} = x^*_{as}c^2_{ks}c^2_{zs} \\
x_{\tau 01} = x^*_{\tau 01}; x_{\tau ks} = x^*_{\tau ks}c^2_{ks}; x_{\tau zs} = x^*_{\tau zs}c^2_{ks}c^2_{zs}; \\
r_1 = r^*_{11}k^2_{10}c^2_{ks}c^2_{zs}; x_{11\sigma} = x^*_{11\sigma}k^2_{10}c^2_{ks}c^2_{zs}; Z_c = Z^*_c k^2_{10}c^2_{ks}c^2_{zs}
\end{aligned}
\tag{15.12}
$$

When using the relations given in (15.12), the "unfolded" stator equivalent circuit in Fig. 15.3 takes the form in Fig. 15.4. The elements of this "unfolded" equivalent circuit (Fig. 15.4) have the simplified form, which facilitates an analysis.

As follows from Figs. 13.5 and 15.4, the stator equivalent circuit built on the basis of the resulting magnetic field, obtained as a result of the superposition of the magnetizing field and the stator slot leakage field and, as well as the stator equivalent circuit built using the "field concept," have the same structure. The stator equivalent circuit in Fig. 13.5 is obtained without taking into account the distribution of the magnetic field in the cross section of the stator. The stator equivalent circuit in Fig. 15.4 is built taking into account the distribution of the magnetic field in the cross section of the stator. Therefore, the elements of the equivalent circuit in Fig. 15.4 contain the corresponding coefficients that take into account the distribution of the magnetic field in the cross section of the stator.

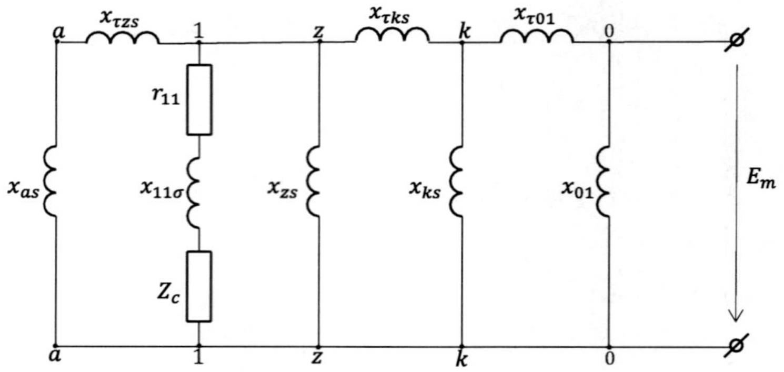

Fig. 15.4 "Unfolded" stator equivalent circuit with the mixed connection of the elements presented in the simplified form

15.4 "Unfolded" Equivalent Circuit with Parallel Connection of Elements of Stator

The "unfolded" equivalent circuit having the mixed connection of the elements and shown in Fig. 15.4 can be reduced to the stator equivalent circuit with the parallel connection of the elements. The essence of the process of reducing the "unfolded" stator equivalent circuit obtained in Fig. 15.4 represents the transfer of the parallel branches $k - k$, $z - z$, $1 - 1$,and $a - a$ to its terminals $0 - 0$. The process of reducing the "unfolded" equivalent circuit in Fig. 15.4 is carried out using the multiple transfer technique described in Appendices G and I. As follows from Appendices G and I, the process of reducing the "unfolded" equivalent circuit in Fig. 15.4 begins with transferring the parallel branch $k - k$ to its terminals $0 - 0$. The transfer factor for this parallel branch is

$$c_{\tau ks} = 1 + \frac{x_{\tau 01}}{x_{ks}} \tag{15.13}$$

The reactance of the transferred parallel branch $k - k$ is defined as

$$\overset{\circ}{x}_{ks} = x_{ks}c_{\tau ks} = x_{ks} + x_{\tau 01} \tag{15.14}$$

The transfer factor for the parallel branch $z - z$ becomes

$$c_{\tau zs} = 1 + \frac{x_{\tau ks}c_{\tau ks}^2 + x_{\tau 01}c_{\tau ks}}{x_{zs}c_{\tau ks}^2} \tag{15.15}$$

The reactance of the transferred parallel branch $z - z$

$$\overset{\circ}{x}_{zs} = x_{zs}c_{\tau ks}^2 + x_{\tau ks}c_{\tau ks}^2 + x_{\tau 01}c_{\tau ks} \tag{15.16}$$

The transfer factor for the parallel branch $1 - 1$

$$c_{\tau 1} = 1 + \frac{x_{\tau ks}c_{\tau ks}^2 c_{\tau zs} + x_{\tau 01}c_{\tau ks}c_{\tau zs}}{(r_{11} + jx_{11\sigma} + Z_c)c_{\tau ks}^2 c_{\tau zs}^2} \tag{15.17}$$

The reactance of the transferred parallel branch $1 - 1$

$$\begin{aligned} \overset{\circ}{Z}_1 &= r_{11}c_{\tau ks}^2 c_{\tau zs}^2 + j\left(x_{11\sigma}c_{\tau ks}^2 c_{\tau zs}^2 + x_{\tau ks}c_{\tau ks}^2 c_{\tau zs} + x_{\tau 01}c_{\tau ks}c_{\tau zs}\right) + Z_c c_{\tau ks}^2 c_{\tau zs}^2 \\ &= r_{11}c_{\tau ks}^2 c_{\tau zs}^2 + jx_{1\sigma} + Z_c c_{\tau ks}^2 c_{\tau zs}^2 \end{aligned} \tag{15.18a}$$

where $x_{1\sigma} = x_{11\sigma}c_{\tau ks}^2 c_{\tau zs}^2 + x_{\tau ks}c_{\tau ks}^2 c_{\tau zs} + x_{\tau 01}c_{\tau ks}c_{\tau zs}$.

Or

$$\begin{aligned} \overset{\circ}{Z}_1 &= \{r_{11} + j[x_{11\sigma} + x_{\tau ks}(1/c_{\tau zs}) + x_{\tau 01}(1/c_{\tau ks}c_{\tau zs})] + Z_c\}c_{\tau ks}^2 c_{\tau zs}^2 \\ &= (r_{11} + j\widetilde{x}_{11\sigma} + Z_c)c_{\tau ks}^2 c_{\tau zs}^2 \end{aligned} \tag{15.18b}$$

where $\widetilde{x}_{11\sigma} = x_{11\sigma} + x_{\tau ks}(1/c_{\tau zs}) + x_{\tau 01}(1/c_{\tau ks}c_{\tau zs}); x_{1\sigma} = \widetilde{x}_{11\sigma}c_{\tau ks}^2 c_{\tau zs}^2$.

The electrical machines used in the power systems are designed with relatively high efficiency. This means that, in this case, the impedance of the power supply network Z_c significantly exceeds the impedance of the machine windings. Under these conditions, the transfer factor of the parallel branch $1 - 1$ determined by (15.17) takes the value close to the unity, i.e., $c_{\tau 1} \approx 1.0$.

Then, for the reactance of the transferred parallel branch $a - a$, we have

$$\overset{\circ}{x}_{as} = x_{as}c_{\tau ks}^2 c_{\tau zs}^2 + x_{\tau zs}c_{\tau ks}^2 c_{\tau zs}^2 + x_{\tau ks}c_{\tau ks}^2 c_{\tau zs} + x_{\tau 01}c_{\tau ks}c_{\tau zs} \tag{15.19}$$

As shown in Appendix I, when using the reactances obtained in (15.14), (15.16), (15.18a), and (15.19), the "unfolded" stator equivalent circuit with the mixed connection of the elements (Fig. 15.4) takes the form of the equivalent circuit with the parallel connection of the elements (Fig. 15.5).

Let us express the impedances $x_{01}, \overset{\circ}{x}_{ks}, \overset{\circ}{x}_{zs}, \overset{\circ}{Z}_1$, and $\overset{\circ}{x}_{as}$ used in the "unfolded" equivalent circuit in Fig. 15.5 through the impedances of the equivalent circuit in Fig. 15.3. For this purpose, we use the relations shown in (15.12). Based on (15.12) and (15.14), for the reactance of the parallel branch $k - k$, we receive

$$\overset{\circ}{x}_{ks} = x_{ks}^* + x_{\tau 01}^* \tag{15.20}$$

The reactance of the parallel branch $z - z$ from (15.12) and (15.16)

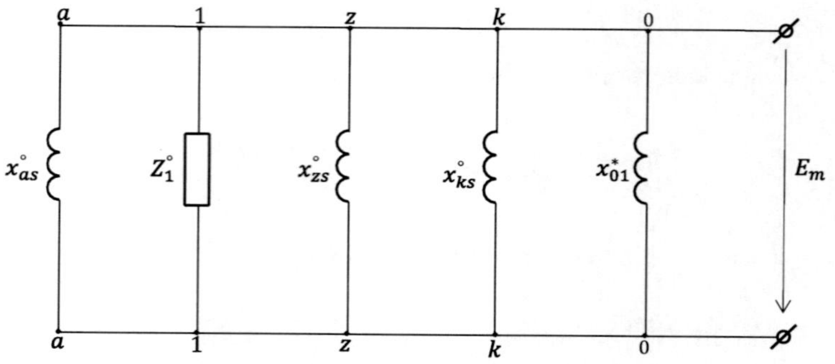

Fig. 15.5 Stator equivalent circuit with the parallel connection of the elements

$$\overset{\circ}{x}_{zs} = x^*_{zs} c^2_{ks} c^2_{\tau ks} + x^*_{\tau ks} c^2_{ks} c^2_{\tau ks} + x^*_{\tau 01} c_{\tau ks} \tag{15.21}$$

The impedance of the parallel branch $1-1$ proceeding from (15.12) and (15.18a)

$$\begin{aligned} \overset{\circ}{Z}_1 &= r^*_{11} k^2_{10} c^2_{ks} c^2_{zs} c^2_{\tau ks} c^2_{\tau zs} + j\big(x^*_{11\sigma} k^2_{10} c^2_{ks} c^2_{zs} c^2_{\tau ks} c^2_{\tau zs} + x^*_{\tau ks} c^2_{ks} c^2_{\tau ks} c_{\tau zs} \\ &\quad + x^*_{\tau 01} c_{\tau ks} c_{\tau zs}\big) + Z^*_c k^2_{10} c^2_{ks} c^2_{zs} c^2_{\tau ks} c^2_{\tau zs} \\ &= r^*_{11} k^2_{10} c^2_{ks} c^2_{zs} c^2_{\tau ks} c^2_{\tau zs} + j x^*_{1\sigma} + Z^*_c k^2_{10} c^2_{ks} c^2_{zs} c^2_{\tau ks} c^2_{\tau zs} \end{aligned} \tag{15.22a}$$

where $x^*_{1\sigma} = x^*_{11\sigma} k^2_{10} c^2_{ks} c^2_{zs} c^2_{\tau ks} c^2_{\tau zs} + x^*_{\tau ks} c^2_{ks} c^2_{\tau ks} c_{\tau zs} + x^*_{\tau 01} c_{\tau ks} c_{\tau zs}$.

In another form

$$\begin{aligned} \overset{\circ}{Z}_1 &= \left\{r^*_{11} + j\left[x^*_{11\sigma} + x^*_{\tau ks}\left(1/k^2_{10} c^2_{zs} c_{\tau zs}\right) + x^*_{\tau 01}\left(1/k^2_{10} c^2_{ks} c^2_{zs} c_{\tau ks} c_{\tau zs}\right)\right]\right. \\ &\quad \left. + Z^*_c\right\} k^2_{10} c^2_{ks} c^2_{zs} c^2_{\tau ks} c^2_{\tau zs} = \left(r^*_{11} + j\widetilde{x}^*_{11\sigma} + Z^*_c\right) k^2_{10} c^2_{ks} c^2_{zs} c^2_{\tau ks} c^2_{\tau zs} \end{aligned} \tag{15.22b}$$

where $\begin{aligned} \widetilde{x}^*_{11\sigma} &= x^*_{11\sigma} + x^*_{\tau ks}\left(1/k^2_{10} c^2_{zs} c_{\tau zs}\right) + x^*_{\tau 01}\left(1/k^2_{10} c^2_{ks} c^2_{zs} c_{\tau ks} c_{\tau zs}\right) \\ x^*_{1\sigma} &= \widetilde{x}^*_{11\sigma} k^2_{10} c^2_{ks} c^2_{zs} c^2_{\tau ks} c^2_{\tau zs} \end{aligned}$

The reactance of the parallel branch $a-a$ by the expressions (15.12) and (15.19)

$$\begin{aligned} \overset{\circ}{x}_{as} &= x^*_{as} c^2_{ks} c^2_{zs} c^2_{\tau ks} c^2_{\tau zs} + \\ &+ x^*_{\tau zs} c^2_{ks} c^2_{zs} c^2_{\tau ks} c^2_{\tau zs} + x^*_{\tau ks} c^2_{ks} c^2_{\tau ks} c_{\tau zs} + x^*_{\tau 01} c_{\tau ks} c_{\tau zs} \end{aligned} \tag{15.23}$$

In accordance with (15.20)–(15.23), the reactances of the parallel branches of the "unfolded" equivalent circuit in Fig. 15.5 are expressed through the reactances used in the L-shaped equivalent circuits for the layers reflecting the stator model. The analytical expressions for these reactances are obtained in (15.1)–(15.5).

15.5 Magnetizing and Slot Leakage Circuit Loops: "Folded" Equivalent Circuit of Stator

As it was shown in Chap. 13, the introduction of the magnetizing and the slot leakage circuit loops can be carried out on the basis of considering the "unfolded" stator equivalent circuit with the parallel connection of the elements. In this case, the "unfolded" stator equivalent circuit with the parallel connection of the elements is obtained in Fig. 15.5. Below we will use this equivalent circuit to introduce the magnetizing and the slot leakage circuit loops and then to obtain the "folded" stator equivalent circuit.

Stator magnetizing circuit loop and magnetizing reactance. In the equivalent circuit in Fig. 15.5, the parallel branches $0-0$, $k-k$, $z-z$ and $a-a$ represent the magnetizing circuit loops of the layers reflecting the stator model (Fig. 14.1). In this regard, they can be combined into one common branch, which will represent the stator magnetizing circuit loop. In this case, the reactance used in the magnetizing circuit loop reflects the stator magnetizing reactance. For the stator magnetizing reactance, based on the equivalent circuit in Fig. 15.5, we can have

$$x^*_{m1} = \frac{1}{\dfrac{1}{x_{01}} + \dfrac{1}{\mathring{x}_{ks}} + \dfrac{1}{\mathring{x}_{zs}} + \dfrac{1}{\mathring{x}_{as}}} = \frac{x_{01}}{1 + x_{01}\left(\dfrac{1}{\mathring{x}_{ks}} + \dfrac{1}{\mathring{x}_{zs}} + \dfrac{1}{\mathring{x}_{as}}\right)} = \frac{x_{01}}{k_{\text{H}1}} \tag{15.24}$$

where $k_{\text{H}1} = 1 + x_{01}\left(\dfrac{1}{\mathring{x}_{ks}} + \dfrac{1}{\mathring{x}_{zs}} + \dfrac{1}{\mathring{x}_{as}}\right)$.

Here, the reactance x^*_{m1} is the stator magnetizing reactance, and the quantity $k_{н1}$ reflects the saturation factor of the stator magnetic circuit.

Stator slot leakage circuit loop and the slot leakage reactance. In the equivalent circuit in Fig. 15.5, the parallel branch $1-1$ represents the stator slot leakage circuit loop. The impedance $\mathring{Z}_1$ used in the stator slot leakage circuit loop is determined by the relation (15.22a). Here, in (15.22a), we use the following coefficients

$$k^2_{11} = k^2_{10} c^2_{ks} c^2_{zs} c^2_{\tau ks} c^2_{\tau zs}; k_{k1} = c^2_{ks} c^2_{\tau ks} c_{\tau zs}; k_{10} = c_{\tau ks} c_{\tau zs} \tag{15.25}$$

Now, for the impedance $\mathring{Z}_1$ used in (15.22a), we can obtain taking into account (15.25)

$$\mathring{Z}_1 = r^*_{11} k^2_{11} + j\left(x^*_{\tau 01} k_{10} + x^*_{\tau ks} k_{k1} + x^*_{11\sigma} k^2_{11}\right) + Z^*_c k^2_{11} \tag{15.26}$$

In (15.26), the ohmic resistance of the stator winding r^*_1 is used. It is defined as

$$r^*_1 = r^*_{11} k^2_{11} \tag{15.27}$$

On the basis of the expression (15.26), we can introduce the stator slot leakage reactance $x^*_{1\sigma}$ for which follows

$$x^*_{1\sigma} = x^*_{\tau 01}k_{10} + x^*_{\tau ks}k_{k1} + x^*_{11\sigma}k^2_{11} \tag{15.28}$$

The slot leakage reactance $x^*_{1\sigma}$ consists of three components: the leakage reactance of the air gap sublayer adjoined to the stator bore surface $\left(x^*_{\tau 01}k_{10}\right)$, the leakage reactance of the layer representing the stator slot-wedge area $\left(x^*_{\tau ks}k_{k1}\right)$, and the leakage reactance of the layer reflecting the stator winding location area $\left(x^*_{\tau ks}k_{k1}\right)$. As follows from (15.28), the components of the slot leakage reactance $x^*_{1\sigma}$ are expressed through the reactances $x^*_{\tau 01}, x^*_{\tau ks}$ and $x^*_{11\sigma}$ determined in accordance with (15.1), (15.2), and (15.4) in the form

$$\begin{aligned} x^*_{\tau 01} &= \omega_1\mu_0 t_0 \frac{\delta}{2t_0} k_{1\sigma}\zeta_{\tau 0} \\ x^*_{\tau ks} &= \omega_1\mu_{xks} t_0 \frac{h_{ks}}{t_{ks}} \xi_{Lks}\zeta_{\tau ks} \\ x^*_{11\sigma} &= \omega_1\mu_{xzs} t_0 \frac{h_{zs}}{3t_{zs}} \xi_{Lzs} \end{aligned} \tag{15.29}$$

Here, for μ_{xks} and μ_{xzs}, we have

$$\mu_{xks} = \mu_0 \frac{t_{ks}}{b_{\text{пs}}} \quad \mu_{xzs} = \mu_0 \frac{t_{zs}}{b_{\text{пs}}} \tag{15.30}$$

Now, the reactances $x^*_{\tau 01}k_{10}, x^*_{\tau ks}k_{k1}$, and $x^*_{11\sigma}k^2_{11}$ can be represented from (15.29) and (15.30) as

$$\begin{aligned} x^*_{\tau 01}k_{10} &= \omega_1\mu_0 t_0 \frac{\delta}{2t_0} k_{1\sigma}\zeta_{\tau 0}k_{10} = \omega_1\mu_0 t_0 \lambda_{\tau 01} \\ x^*_{\tau ks}k_{k1} &= \omega_1\mu_0 t_0 \frac{h_{ks}}{b_{\text{пs}}} \xi_{Lks}\zeta_{\tau ks}k_{k1} = \omega_1\mu_0 t_0 \lambda_{\tau ks} \\ x^*_{11\sigma}k^2_{11} &= \omega_1\mu_0 t_0 \frac{h_{zs}}{3b_{\text{пs}}} \xi_{Lzs}k^2_{11} = \omega_1\mu_0 t_0 \lambda_{11\sigma} \end{aligned} \tag{15.31}$$

where $\lambda_{\tau 01} = \frac{\delta}{2t_0} k_{1\sigma}\zeta_{\tau 0}k_{10}; \lambda_{\tau ks} = \frac{h_{ks}}{b_{\text{пs}}} \xi_{Lks}\zeta_{\tau ks}k_{k1}; \lambda_{11\sigma} = \frac{h_{zs}}{3b_{\text{пs}}} \xi_{Lzs}k^2_{11}$.

In (15.31), the dimensionless quantities $\lambda_{\tau 01}$, $\lambda_{\tau ks}$ and $\lambda_{11\sigma}$ represent the relative magnetic conductivity for the tangential components of the magnetic fluxes in the layers reflecting the air gap sublayer adjoining to the stator bore surface, the stator slot-wedge area, and the stator winding location area, respectively.

Now, from (15.28) and (15.31), for the slot leakage reactance $x^*_{1\sigma}$, we can obtain

$$\begin{aligned} x^*_{1\sigma} &= x^*_{\tau 01}k_{10} + x^*_{\tau ks}k_{k1} + x^*_{11\sigma}k^2_{11} \\ &= \omega_1\mu_0 t_0 (\lambda_{\tau 01} + \lambda_{\tau ks} + \lambda_{11\sigma}) = \omega_1\mu_0 t_0 \lambda_{1\sigma} \end{aligned} \tag{15.32}$$

where $\lambda_{1\sigma} = \lambda_{\tau 01} + \lambda_{\tau ks} + \lambda_{11\sigma}$.

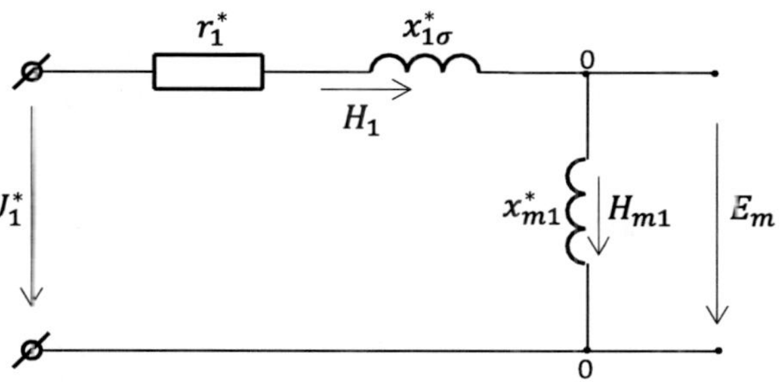

Fig. 15.6 "Folded" stator equivalent circuit

Here, the quantity $\lambda_{1\sigma}$ reflects the relative magnetic conductivity for the slot leakage magnetic flux of the stator winding.

"Folded" stator equivalent circuit. In the equivalent circuit in Fig. 15.5, we use the condition obtained in (15.24). In addition, in relation (15.26), for the impedance $\overset{\circ}{Z}_1$ used in the equivalent circuit in Fig. 15.5, we take into account the conditions given in (15.27) and (15.32). Further, in relation (15.26), for impedance Z_c^*, we apply the equation of the form $\left(-\dot{U}_1^*\right) = Z_c^* \dot{H}_1$. Then, for the impedance $Z_c^* k_{11}^2$ we have that $\left(-\widetilde{U}_1^*\right) = Z_c^* k_{11}^2 \widetilde{H}_1$, where $\widetilde{U}_1^* = \dot{U}_1^* k_{11}$ and $\widetilde{H}_1 = \dot{H}_1 / k_{11}$. Now, taking into account the above, and also bearing in mind that the value of the coefficient k_{11} for the machines of the normal design can be taken equal to the one, i.e., $k_{11} \approx 1.0$, the stator equivalent circuit in Fig. 15.5 can be reduced to the "folded" stator equivalent circuit in Fig. 15.6. As follows from the "folded" equivalent circuit in Fig. 15.6, in relation to $\dot{E}_m$, the stator of the machine is represented in the form of the two independent parallel circuit loops: the magnetizing circuit loop containing the stator magnetizing reactance x_{m1}^* and the slot leakage circuit loop having the stator slot leakage reactance $x_{1\sigma}^*$. The "folded" stator equivalent circuit obtained in this way has a simple structure and it is convenient for practical use.

15.6 L-Shaped Equivalent Circuits of Single-Cage Rotor Model Layers and Their Parameters

We proceed from the layered single-cage rotor model shown in Fig. 14.1. In accordance with the layered model shown in Fig. 14.1, the considered single-cage rotor is represented as consisting of the four layers: the air gap sublayer adjoined to the rotor surface, the slot-wedge area layer, the winding location area layer, and also the rotor yoke area layer. As shown in Chap. 14, the layers of the rotor model are represented in the form of the L-shaped equivalent circuits (with the exception of the rotor yoke area layer, which is replaced by the equivalent circuit of the two-terminal

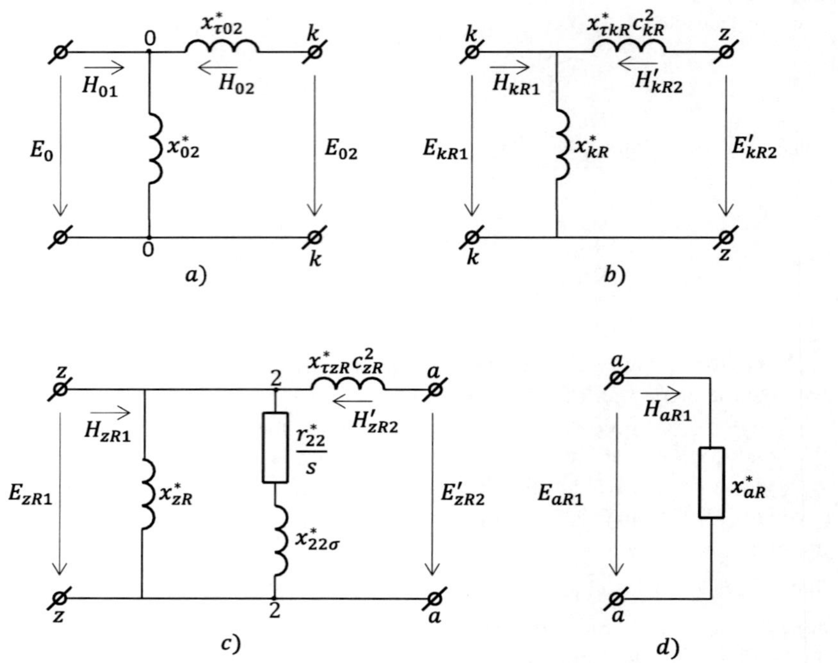

Fig. 15.7 Equivalent circuits for the layers reflecting the single-cage rotor model: (**a**) L-shaped equivalent circuit of the air gap sublayer adjoined to the rotor surface, (**b**) L-shaped equivalent circuit for the layer representing rotor slot-wedge area, (**c**) L-shaped equivalent circuit for the layer reflecting the rotor winding location area, (**d**) equivalent circuit of the rotor yoke area layer

network). Below we will consider the parameters of the L-shaped equivalent circuits reflecting the rotor model layers.

For the air gap sublayer adjoined to the rotor surface, we can use the L-shaped equivalent circuit shown in Fig. 14.7c. Here, this L-shaped equivalent circuit of the air gap sublayer is shown in Fig. 15.7a. The reactances x^*_{02} and $x^*_{\tau 02}$ used in this equivalent circuit are determined from relations (14.54) in the form

$$x^*_{02} = \frac{\omega_1 \mu_0 \tau_0^2}{\pi^2 \delta_2} \zeta_0 \quad x^*_{\tau 02} = \omega_1 \mu_0 t_0 \frac{\delta}{2 t_{02}} k_{2\sigma} k_R \zeta_{\tau 0} \tag{15.33}$$

In (15.33), the coefficients ζ_0 and $\zeta_{\tau 0}$ can be calculated using the relations obtained in (15.1).

The rotor wedge area is represented in the form of the nonconductive magneto-anisotropic layer with the electromagnetic parameters equal to $\rho_{kl} = \infty$ as well as μ_{xkR} and μ_{ykR}. To obtain the L-shaped equivalent circuit for the layer reflecting the rotor slot-wedge area, we will use the equivalent circuit shown in Fig. 14.6b. This L-shaped equivalent circuit is represented in the form in Fig. 15.7b. Based on the relations obtained in (14.52), for the reactances x^*_{kR} and $x^*_{\tau kR}$ used in the equivalent circuit in Fig. 15.7b, we have

$$x^*_{kR} = \frac{\omega_1 \mu_{ykR} \tau^2_{kR}}{\pi^2 h_{kR}} \frac{\xi_{LkR}}{\xi_{\tau kR}} \zeta_{kR} \qquad x^*_{\tau kR} = \omega_1 \mu_{xkR} t_0 \frac{h_{kR}}{t_{kR}} k_R \xi_{LkR} \zeta_{\tau kR} \tag{15.34}$$

where

$$\zeta_{kR} = n_{kR} \frac{h_{kR}}{R_{kR1}} \frac{1 + (1 - h_{kR}/R_{kR1})^{2n_{kR}}}{1 - (1 - h_{kR}/R_{kR1})^{2n_{kR}}}$$

$$\zeta_{\tau kR} = \frac{1}{n_{kR} h_{kR}/R_{kR1}} \frac{1 - (1 - h_{kR}/R_{kR1})^{2n_{kR}}}{1 + (1 - h_{kR}/R_{kR1})^{2n_{kR}}}$$

Here, $n_{kR} = pk_{\mu kR}$ and also R_{kR1} is the radius of the outer surface of the rotor slot-wedge area layer.

The coefficient c_{kR} used in the equivalent circuit in Fig. 15.7b can be calculated using the expression given in (14.49). The values of $\dot{E}'_{kR2}$ and $\dot{H}'_{kR2}$ relating to the output terminals of the equivalent circuit in Fig. 15.7b, are determined by the expressions obtained in (14.50).

The L-shaped equivalent circuit representing the rotor winding location area layer (single-cage rotor) has the form shown in Fig. 14.5. In the rotor model in Fig. 14.1, the winding location area is represented in the form of the conducting magneto-anisotropic layer with the average values of the electromagnetic parameters equal to ρ_2, as well as μ_{xzR} and μ_{yzR}. Therefore, in this case, the active resistance of the rotor winding is determined by introducing the average value of the resistivity ρ_2 representing the rotor winding conductor, for example [1–4]. At the weak skin effect, in the equivalent circuit in Fig. 14.5, the "own" magnetizing circuit loop of the layer reflecting the rotor winding location area and the "own" leakage circuit loop of the eddy currents induced in this layer should be distinguished. For the layer reflecting the rotor winding location area, the L-shaped equivalent circuit containing the "own" magnetizing circuit loop and the "own" leakage circuit loop of the eddy currents induced in the considered layer is obtained in Appendix H and it is shown in Fig. H.2. In this chapter, the equivalent circuit in Fig. H.2 is shown in Fig. 15.7c. For the reactances x^*_{zR} and $x^*_{\tau zR}$ used in the equivalent circuit in Fig. 15.7c, we have from (14.62) the relations of the form

$$x^*_{zR} = \frac{\omega_1 \mu_{yzR} \tau^2_{zR}}{\pi^2 h_{zR}} \frac{\xi_{LzR}}{\xi_{\tau zR}} \zeta_{zR} \qquad x^*_{\tau zR} = \omega_1 \mu_{xzR} t_0 \frac{h_{zR}}{t_{zR}} k_R \xi_{LzR} \zeta_{\tau zR} \tag{15.35}$$

where

$$\zeta_{zR} = n_{zR} \frac{h_{zR}}{R_{zR1}} \frac{1 + (1 - h_{zR}/R_{zR1})^{2n_{zR}}}{1 - (1 - h_{zR}/R_{zR1})^{2n_{zR}}}$$

$$\zeta_{\tau zR} = \frac{1}{n_{zR} h_{zR}/R_{zR1}} \frac{1 - (1 - h_{zR}/R_{zR1})^{2n_{zR}}}{1 + (1 - h_{zR}/R_{zR1})^{2n_{zR}}}$$

Here, $n_{zR} = pk_{\mu zR}$ and also R_{zR1} is the radius of the outer surface of the rotor winding location area layer.

In the equivalent circuit in Fig. 15.7c, the active resistance r_{22}^*/s and the reactance $x_{22\sigma}^*$ represent the rotor winding location area layer. The expression for the active resistance r_{22}^*/s was obtained in [1]. The reactance $x_{22\sigma}^*$ follows from the expression given in (14.62). As a result, for the active resistance r_{22}^*/s and the reactance $x_{22\sigma}^*$, we have

$$\frac{r_{22}^*}{s} = \frac{\rho_2}{h_{zR}s}\frac{\xi_{LRz}}{\xi_{\tau Rz}} \quad x_{22\sigma}^* = \omega_1 \mu_{xRz} t_0 \frac{h_{zR}}{3t_{zR}} k_R \xi_{LzR} \tag{15.36}$$

The coefficient c_{zR} used in the equivalent circuit in Fig. 15.7c can be calculated using the expression given in (14.59). The quantities $\dot{E}'_{zR2}$ and $\dot{H}'_{zR2}$ determined at the output terminals of the equivalent circuit in Fig. 15.7c, follow from the relations (14.60).

In the model in Fig. 14.1, the rotor yoke area is represented as the nonconductive isotropic layer. The equivalent circuit for this rotor area can be obtained on the basis of the L-shaped equivalent circuit built for the ith layer in Fig. 14.5. For this purpose, we can proceed from the fact that the normal component of the magnetic flux density on the inner surface of the rotor yoke area layer (on the surface located in the vicinity of the central axis of the rotor) can be taken equal to zero. This condition means that the electric field intensity on this surface of the rotor yoke area layer acquires the zero value, i.e., $\dot{E}_{aR2} = 0$. The use of this condition in the equivalent circuit in Fig. 14.5 leads to the fact that this equivalent circuit in relation to the rotor yoke area layer is transformed into the equivalent circuit of the two-terminal network [1]. If, in addition, it is borne in mind that the resistivity of the laminated rotor takes the relatively large value, i.e. $\rho_a \approx \infty$, then, the equivalent circuit of the rotor yoke layer takes the form in Fig. 14.8b. In this chapter, this equivalent circuit is shown in Fig. 15.7d. For the reactance of the rotor yoke area layer x_{aR}^* used in the equivalent circuit in Fig. 15.7d, from (14.56), we have

$$x_{aR}^* = \omega_1 \mu_{aR} t_0 \frac{h_{aR}}{t_{aR}} k_R \xi_{LaR} \zeta_{aR} \tag{15.37}$$

where $\zeta_{aR} = \dfrac{1-(1-h_{aR}/R_{aR1})^{2p}}{(ph_{aR}/R_{aR1})\left[1+(1-h_{aR}/R_{aR1})^{2p}\right]}$; $\xi_{LaR} = \dfrac{l_{aR}}{l_0}$; $t_{aR} = \dfrac{2p\tau_{aR}}{Z_2}$.

In (15.37), the coefficient ζ_{aR} takes into account the curvature of the layer surface of the rotor yoke area. In addition, in (15.37), the value R_{aR1} reflects the radius of the outer surface of the rotor yoke area layer, and τ_{aR} represents the pole pitch on this surface.

The equivalent circuits shown in Fig. 15.7 and representing the corresponding layers of the single-cage rotor model (Fig. 14.1) will be used below to construct the cascade equivalent circuit for the single-cage rotor.

15.7 Cascade Equivalent Circuits of Single-Cage Rotor

We proceed from the layered rotor model shown in Fig. 14.1. Here, the rotor model includes the air gap sublayer adjoined to the rotor surface. The rotor body in this model is represented by the three layers: the layer of the slot-wedge area, the layer of the winding location area, and the layer of the rotor yoke area. Let's construct the cascade rotor equivalent circuit using the layered model shown in Fig. 14.1. To do this, you need to have the equivalent circuits that reflect the corresponding layers of the rotor model. For this purpose, we can use the equivalent circuits obtained in Fig. 15.7. In accordance with Fig. 15.7, the layers of the rotor model are replaced by L-shaped equivalent circuits (with the exception of the rotor yoke area layer represented by the equivalent circuit of the two-terminal network). Therefore, based on the model in Fig. 14.1 and the equivalent circuits shown in Fig. 15.7, the rotor equivalent circuit can be obtained as a result of the cascade way of connecting the L-shaped equivalent circuits, reflecting the corresponding layers of the rotor model.

Figure 15.8a shows the equivalent circuit of the rotor in the form of the cascade-connected four-terminal networks. In this cascade equivalent circuit, the four-terminal network Π_{02} represents the air gap sublayer adjoined to the rotor surface, the four-terminal network Π_{kR} reflects the layer of the rotor slot-wedge area, the four-terminal network Π_{zR} corresponds to the layer of the winding location area, and the two-terminal network Π_{aR} shows the rotor yoke area layer.

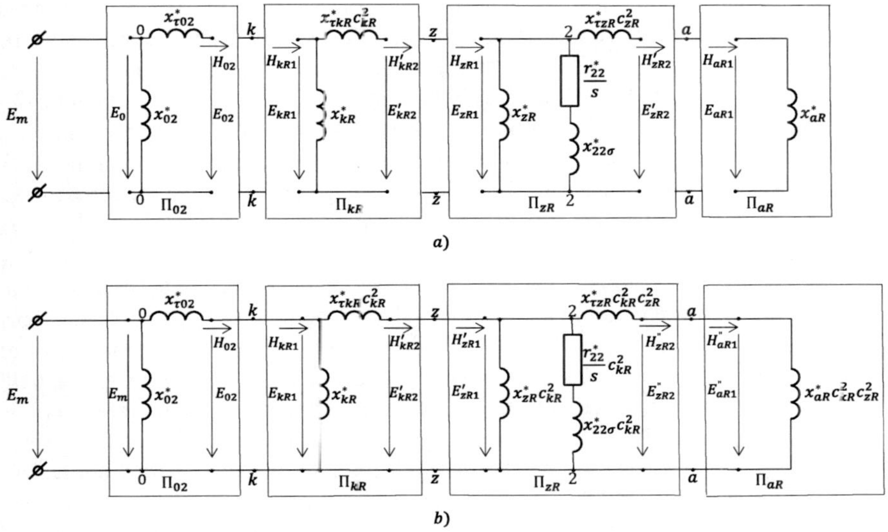

Fig. 15.8 Rotor equivalent circuits as the cascade connection of the four-terminal networks: (**a**) schematic visualization of the four-terminal networks; (**b**) cascade connection of the L-shaped equivalent circuits, reflecting the rotor model layers

For the schematic visualization of the four-terminal networks used in the equivalent circuit in Fig. 15.8a, the L-shaped equivalent circuits obtained in Fig. 15.7 are proposed (with the exception of the rotor yoke area layer).

In the cascade equivalent circuit in Fig. 15.8a, its terminals $k-k$, $z-z$ and $a-a$ correspond to the surfaces $k-k$, $z-z$, and $a-a$ of the layered rotor model shown in Fig. 14.1. In this regard, at the terminals $k-k$, $z-z$, and $a-a$ of the cascade equivalent circuit in Fig. 15.8a, the conditions given in (14.2a) are fulfilled. As follows from the cascade equivalent circuit in Fig. 15.8a, as well as the relations given in (14.50) and (14.60), the conditions (14.2a) are not satisfied at the output terminals of the L-shaped equivalent circuits, representing the layer of the rotor slot-wedge area (four-terminal network Π_{kR}) and the layer of the rotor winding location area (four-terminal network Π_{zR}). In this regard, the L-shaped equivalent circuits used in the equivalent circuit in Fig. 15.8a cannot be directly connected in a cascade. Below we will analyze the conditions under which these L-shaped equivalent circuits can be connected in a cascade.

15.8 "Unfolded" Equivalent Circuit with Mixed Connection of Elements of Single-Cage Rotor

Let's analyze the cascade equivalent circuit of the rotor shown in Fig. 15.8a. The consideration of the cascade equivalent circuit in Fig. 15.8a will begin with an analysis of the conditions for connecting the output terminals of the L-shaped equivalent circuit of the four-terminal network Π_{02} with the input terminals of the L-shaped equivalent circuit of the four-terminal network Π_{kR}. In Fig. 15.8a, the output terminals of the L-shaped equivalent circuit of the four-terminal network Π_{02} correspond to the inner surface of the air gap sublayer adjoined to the rotor surface. The input terminals of the L-shaped equivalent circuit of the four-terminal network Π_{kR} represent the outer surface of the rotor slot-wedge area layer. In the rotor model (Fig. 14.1), these rotor layers form the adjacent surface $k-k$, on which, in accordance with (14.2a), the conditions of the form $\dot{E}_{02}=\dot{E}_{kR1}$ and $\dot{H}_{02}=H_{kR1}$ are valid. Thus, the components of the electric and magnetic field intensities $\dot{E}_{02}$ and $\dot{H}_{02}$ act on the output terminals of the L-shaped equivalent circuit of the four-terminal network Π_{02}. At the input terminals of the L-shaped equivalent circuit of the four-terminal network Π_{kR}, for the components of the electric and magnetic field intensities, we have that $\dot{E}_{kR1}$ and $\dot{H}_{kR1}$. Therefore, proceeding from the conditions that $\dot{E}_{02}=\dot{E}_{kR1}$ and $\dot{H}_{02}=H_{kR1}$, we can note that on the surface $k-k$ of the rotor model (Fig. 14.1), as well as on the output terminals of the L-shaped equivalent circuit of the four-terminal network Π_{02} and on the input terminals of the L-shaped equivalent circuit of the four-terminal network Π_{kR} (Fig. 15.8a), the components of the electric and magnetic field intensities have the same value. In this regard, the output terminals of the L-shaped equivalent circuit of the four-terminal network Π_{02} and the input terminals of the L-shaped equivalent circuit of the four-terminal network Π_{kR} can be directly connected in a cascade. The L-shaped equivalent circuit of the four-terminal network Π_{02} and the L-shaped equivalent circuit of the four-terminal network Π_{kR} connected in a cascade are shown in Fig. 15.8b.

Now, in the equivalent circuit in Fig. 15.8a, we will consider the peculiarities of connecting the output terminals of the L-shaped equivalent circuit of the four-terminal network Π_{kR} with the input terminals of the L-shaped equivalent circuit of the four-terminal network Π_{zR}.

In Fig. 15.8a, the output terminals of the L-shaped equivalent circuit of the four-terminal network Π_{kR} correspond to the inner surface of the rotor slot-wedge area layer. The input terminals of the L-shaped equivalent circuit of the four-terminal network Π_{zR} represent the outer surface of the rotor winding location area layer. In the rotor model (Fig. 14.1), these rotor layers form the adjacent surface $z - z$ on which, in accordance with (14.2a), the conditions of the form $\dot{E}_{kR2} = \dot{E}_{zR1}$ and $\dot{H}_{kR2} = H_{zR1}$ are valid. The components of the electric and magnetic field intensities $\dot{E}'_{kR2}$ and $\dot{H}'_{kR2}$ are on the output terminals of the L-shaped equivalent circuit of the four-terminal network Π_{kR}. At the input terminals of the L-shaped equivalent circuit of the four-terminal network Π_{zR}, for the components of the electric and magnetic field intensities, we have $\dot{E}_{zR1}$ and $\dot{H}_{zR1}$. When cascading the output terminals of the L-shaped equivalent circuit of the four-terminal network Π_{kR} with the input terminals of the L-shaped equivalent circuit of the four-terminal network Π_{zR}, it must be borne in mind that the values $\dot{E}'_{kR2}$ and $\dot{H}'_{kR2}$ corresponding to the output terminals of the L-shaped equivalent circuit of the four-terminal network Π_{kR} are defined in accordance with (14.50) as $\dot{E}'_{kR2} = \dot{E}_{kR2} c_{kR}$ and $\dot{H}'_{kR2} = \dot{H}_{kR2}/c_{kR}$. In this regard, and taking into account the conditions at the boundary $z - z$ of the rotor model in the form $\dot{E}_{kR2} = \dot{E}_{zR1}$ and $\dot{H}_{kR2} = H_{zR1}$, the values of $\dot{E}_{zR1}$ and H_{zR1} corresponding to the input terminals of the L-shaped equivalent circuit of the four-terminal network Π_{zR} should be corrected as

$$\dot{E}'_{kR2} = \dot{E}'_{zR1} = \dot{E}_{zR1} c_{kR} \quad \dot{H}'_{kR2} = \dot{H}'_{zR1} = \dot{H}_{zR1}/c_{kR} \tag{15.38}$$

Therefore, when cascading the output terminals of the L-shaped equivalent circuit of the four-terminal network Π_{kR} with the input terminals of the L-shaped equivalent circuit of the four-terminal network Π_{zR}, instead of the conditions obtained in (14.2a) in the form $\dot{E}_{kR2} = \dot{E}_{zR1}$ and $\dot{H}_{kR2} = H_{zR1}$, it is necessary to use the new conditions of kind

$$\dot{E}'_{kR2} = \dot{E}'_{zR1} \quad \dot{H}'_{kR2} = \dot{H}'_{zR1} \tag{15.39}$$

where $\dot{E}'_{kR2} = \dot{E}_{kR2} c_{kR}; \dot{E}'_{zR1} = \dot{E}_{zR1} c_{kR}; \dot{H}'_{kR2} = \dot{H}_{kR2}/c_{kR}; \dot{H}'_{zR1} = \dot{H}_{zR1}/c_{kR}$.

With respect to the values $\dot{E}'_{zR1}$ and $\dot{H}'_{zR1}$, the input impedance of the L-shaped equivalent circuit of the four-terminal network Π_{zR} is defined as $Z'_{zR1} = \dot{E}'_{zR1}/\dot{H}'_{zR1} = \left(\dot{E}_{zR1}/\dot{H}_{zR1}\right) c_{kR}^2 = Z_{zR1} c_{kR}^2$, where $Z_{zR1} = \dot{E}_{zR1}/\dot{H}_{zR1}$ is the input impedance of the L-shaped equivalent circuit of the four-terminal network Π_{zR}, defined in relation to the values $\dot{E}_{zR1}$ and $\dot{H}_{zR1}$. Therefore, under the conditions of the transition to the new values $\dot{E}'_{zR1}$ and $\dot{H}'_{zR1}$, the impedances of the L-shaped equivalent circuit of the four-terminal network Π_{zR} $\left[x^*_{zR}, (r^*_{22}/s), x^*_{22\sigma}, (x^*_{\tau Rz}c^2_{zR})\right]$ are multiplied by the squared coefficient c^2_{kR}, that is $\left[x^*_{zR}c^2_{kR}, (r^*_{22}/s)c^2_{kR}, x^*_{22\sigma}c^2_{kR}, (x^*_{\tau Rz}c^2_{kR}c^2_{zR})\right]$. In addition, when the output terminals of the L-shaped equivalent circuit of the four-terminal network Π_{kR} are cascaded with the input terminals of the L-shaped equivalent circuit of the four-terminal network Π_{zR}, the values $\dot{E}'_{zR2} = \dot{E}_{zR2}c_{zR}$ and $\dot{H}'_{zR2} = \dot{H}_{zR2}/c_{zR}$ corresponding to the output terminals of the L-shaped equivalent circuit of the four-terminal network Π_{zR} and determined from (14.60) are corrected by the coefficient c_{kR} and take the form

$$\dot{E}''_{zR2} = \dot{E}'_{zR2}c_{kR} = \dot{E}_{zR2}c_{kR}c_{zR} \quad \dot{H}''_{zR2} = \dot{H}'_{zR2}/c_{kR} = \dot{H}_{zR2}/c_{kR}c_{zR} \tag{15.40}$$

Taking into account the provisions shown above, the L-shaped equivalent circuit of the four-terminal network Π_{kR} and the L-shaped equivalent circuit of the four-terminal network Π_{zR}, connected in a cascade take the form shown in Fig. 15.8b.

Further, in the equivalent circuit in Fig. 15.8a, we will consider the peculiarities of connecting the output terminals of the L-shaped equivalent circuit of the four-terminal network Π_{zR} with the input terminals of the equivalent circuit of the two-terminal network Π_{aR}. In Fig. 15.8a, the output terminals of the L-shaped equivalent circuit of the four-terminal network Π_{zR} correspond to the inner surface of the rotor winding location area layer. The input terminals of the equivalent circuit of the two-terminal network Π_{aR} represent the outer surface of the rotor yoke area layer. In the rotor model (Fig. 14.1), these rotor layers form the adjacent surface $a - a$, on which, in accordance with (14.2a), the conditions of the form $\dot{E}_{zR2} = \dot{E}_{aR1}$ and $\dot{H}_{zR2} = H_{aR1}$ are valid. According to (15.40), the components of the electric and magnetic field intensities $\dot{E}''_{zR2}$ and $\dot{H}''_{zR2}$ act on the output terminals of the L-shaped equivalent circuit of the four-terminal network Π_{zR}. At the input terminals of the equivalent circuit of the two-terminal network Π_{aR}, for the components of the electric and magnetic field intensities, we have $\dot{E}_{aR1}$ and $\dot{H}_{aR1}$. When cascading the output terminals of the L-shaped equivalent circuit of the four-terminal network Π_{zR} with the input terminals of the equivalent circuit of the two-terminal network Π_{aR}, it should be borne in mind that the values of $\dot{E}''_{zR2}$ and $\dot{H}''_{zR2}$ corresponding to the output terminals of the L-shaped equivalent circuit of the four-terminal network Π_{zR} are determined in accordance with (15.40) as $\dot{E}''_{zR2} = \dot{E}_{zR2}c_{kR}c_{zR}$ and $\dot{H}''_{zR2} = \dot{H}_{zR2}/c_{kR}c_{zR}$. In this regard, and taking into account the conditions at the boundary $a - a$ of the rotor model in the form $\dot{E}_{zR2} = \dot{E}_{aR1}$ and $\dot{H}_{zR2} = H_{aR1}$, the values of $\dot{E}_{aR1}$ and H_{aR1} corresponding to the input terminals of the equivalent circuit of the two-terminal network Π_{aR} should be corrected as

$$\dot{E}''_{zR2} = \dot{E}''_{aR1} = \dot{E}_{aR1}c_{kR}c_{zR} \quad \dot{H}''_{zR2} = \dot{H}''_{aR1} = \dot{H}_{aR1}/c_{kR}c_{zR} \tag{15.41}$$

Therefore, when cascading the output terminals of the L-shaped equivalent circuit of the four-terminal network Π_{zR} with the input terminals of the equivalent circuit of the two-terminal network Π_{aR}, instead of the conditions obtained in (14.2a) in the form $\dot{E}_{zR2} = \dot{E}_{aR1}$ and $\dot{H}_{aR2} = \dot{H}_{aR1}$, it is necessary to use the new conditions of the form

$$\dot{E}''_{zR2} = \dot{E}''_{aR1} \quad \dot{H}''_{zR2} = \dot{H}''_{aR1} \tag{15.42}$$

where
$$\begin{aligned}\dot{E}''_{zR2} &= \dot{E}_{zR2}c_{kR}c_{zR}; \dot{E}''_{aR1} = \dot{E}_{aR1}c_{kR}c_{zR}\\ \dot{H}''_{zR2} &= \dot{H}_{zR2}/c_{kR}c_{zR}; \dot{H}''_{aR1} = \dot{H}_{aR1}/c_{kR}c_{zR}\end{aligned}$$

With respect to the values $\dot{E}''_{aR1}$ and $\dot{H}''_{aR1}$, the input reactance of the equivalent circuit of the two-terminal network Π_{aR} is defined as $x'_{aR} = \dot{E}''_{aR1}/\dot{H}''_{aR1} = (\dot{E}_{aR1}/\dot{H}_{aR1})c^2_{kR}c^2_{zR} = x_{aR}c^2_{kR}c^2_{zR}$, where $x_{aR} = \dot{E}_{aR1}/\dot{H}_{aR1}$ is the reactance of the equivalent circuit of the two-terminal network Π_{aR} determined with respect to the values $\dot{E}_{aR1}$ and $\dot{H}_{aR1}$. Therefore, under the conditions of the transition to the new values $\dot{E}''_{aR1}$ and $\dot{H}''_{aR1}$, the reactance of the equivalent circuit of the two-terminal network Π_{aR} is multiplied by the product of the square of the coefficients $(c^2_{kR}c^2_{zR})$, i.e. $x^*_{aR}c^2_{kR}c^2_{zR}$. Thus, the reactance of the equivalent circuit of the two-terminal network Π_{aR} (Fig. 15.8a) is determined in the form shown in Fig. 15.8b. Now, taking into account the above, the L-shaped equivalent circuit of the four-terminal network Π_{zR} and the equivalent circuit of the two-terminal network Π_{aR} connected in a cascade, get the form shown in Fig. 15.8b.

In the cascade equivalent circuit in Fig. 15.8b, we passed to the new boundary conditions, determined in accordance with (14.2a), (15.39), and (15.42) in the form

$$\begin{aligned}\dot{E}_{02} &= \dot{E}_{kR1} \quad \dot{H}_{02} = H_{kR1}\\ \dot{E}'_{kR2} &= \dot{E}'_{zR1} \quad \dot{H}'_{kR2} = \dot{H}'_{zR1}\\ \dot{E}''_{zR2} &= \dot{E}''_{aR1} \quad \dot{H}''_{zR2} = \dot{H}''_{aR1}\end{aligned} \tag{15.43}$$

The use of these conditions allows you to cascade the L-shaped equivalent circuits used in Fig. 15.8b. As a result of this connection, the rotor equivalent circuit takes the form in Fig. 15.9. This equivalent circuit is the unfolded equivalent circuit of the single-cage rotor with the mixed connection of the elements.

For the parameters used in the equivalent circuit in Fig. 15.9, for convenience, we can introduce the simplified notations in the form

$$\begin{aligned}x_{02} = x^*_{02}; x_{kR} = x^*_{kR}; x_{zR} = x^*_{zR}c^2_{kR}; x_{aR} = x^*_{aR}c^2_{kR}c^2_{zR}\\ x_{\tau 02} = x^*_{\tau 02}; x_{\tau kR} = x^*_{\tau kR}c^2_{kR}; x_{\tau zR} = x^*_{\tau zR}c^2_{kR}c^2_{zR}\\ r_{22}/s = (r^*_{22}/s)c^2_{kR}; x_{22\sigma} = x^*_{22\sigma}c^2_{kR}\end{aligned} \tag{15.44}$$

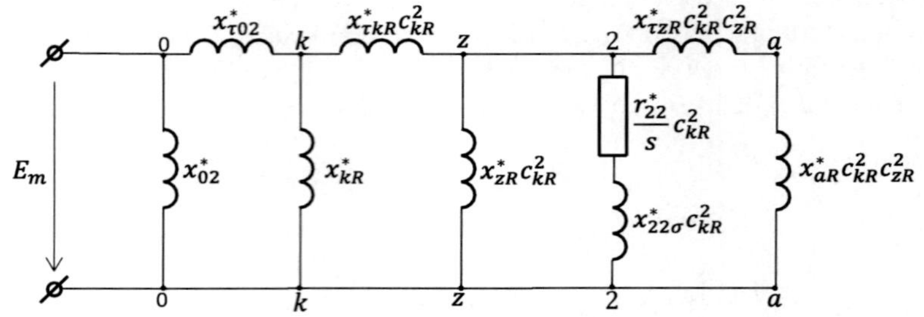

Fig. 15.9 "Unfolded" single-cage rotor equivalent circuit with the mixed connection of the elements

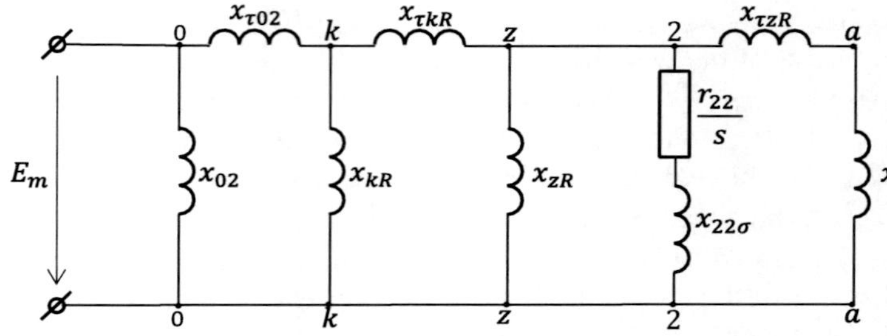

Fig. 15.10 "Unfolded" single-cage rotor equivalent circuit with the mixed connection of the elements presented in the simplified form

When using the simplified notations given in (15.44), the equivalent circuit in Fig. 15.9 takes the form in Fig. 15.10. The elements of the equivalent circuit in Fig. 15.10 help to facilitate an analysis. Below, we bring the transformation of the equivalent circuit in Fig. 15.10 in order to simplify its structure.

As follows from Figs. 13.6 and 15.10, the rotor equivalent circuit built on the basis of the resulting magnetic field, obtained as a result of the superposition of the magnetizing field and the rotor slot leakage field and, as well as the rotor equivalent circuit built using the "field concept," have the same structure. The rotor equivalent circuit in Fig. 13.6 is obtained without taking into account the distribution of the magnetic field in the cross section of the rotor. The rotor equivalent circuit in Fig. 15.10 is built taking into account the distribution of the magnetic field in the cross section of the rotor. Therefore, the elements of the equivalent circuit in Fig. 15.10 contain the corresponding coefficients that take into account the distribution of the magnetic field in the cross section of the rotor.

15.9 "Unfolded" Equivalent Circuit with Parallel Connection of Elements of Single-Cage Rotor

The equivalent circuit in Fig. 15.10 is the "unfolded" equivalent circuit with the mixed connection of the elements in a view of the fact that it is built as a result of the cascade way of connecting the L-shaped equivalent circuits obtained with respect to the values of $\dot{E}$ and $\dot{H}$ specified on the outer surfaces of the layers of the rotor model (surfaces $0-0$, $k-k$, $z-z$, and $a-a$ in Fig. 4.1. The equivalentcircuit in Fig. 15.10 has the complex structure, and therefore, it needs to be transformed in order to make look convenient for practical use. The transformation of the equivalent circuit in Fig. 15.10 is implemented using the multiple transfer technique and involves the introduction of the magnetizing and the slot leakage circuit loops obtained for the rotor in relation to its terminals $0-0$. To do this, the "own" magnetizing circuit loops of the rotor model layers (terminals $k-k$, $z-z$, and $a-a$) and the "own" rotor slot leakage circuit loop (terminals $2-2$) should be transferred to the terminals $0-0$ of the equivalent circuit in Fig. 15.10. As a result, the unfolded equivalent circuit with the mixed connection of the elements (Fig. 15.10) takes the form of the equivalent circuit with the parallel connection of the elements. In this new equivalent circuit, the parallel branches will represent the magnetizing circuit loops of the rotor model layers and the rotor slot leakage circuit loop. We will show this below.

We consider the rotor equivalent circuit in Fig. 15.10. In this equivalent circuit, its parallel branches $k-k$, $z-z$ and $2-2$ as well as $a-a$ having respectively the ohmic and inductive resistances x_{kR}, x_{zR}, $(r_{22}/s+jx_{22\sigma})$ and x_{aR} will be transferred to its terminals $0-0$. Here, we will use the multiple transfer technique described above [1]. Appendix I describes the multiple transfer technique for converting the equivalent circuit shown in Fig. 15.10. It is shown in Appendix I that the transformation of the equivalent circuit in Fig. 15.10 begins with transferring the parallel branch $k-k$ to its terminals $0-0$. The transfer factor for this parallel branch is

$$c_{\tau kR} = 1 + \frac{x_{\tau 02}}{x_{kR}} \tag{15.45}$$

The reactance of the transferred parallel branch $k-k$ is defined as

$$\overset{\circ}{x}_{kR} = x_{kR}c_{\tau kR} = x_{kR} + x_{\tau 02} \tag{15.46}$$

The transfer factor of the parallel branch $z-z$ becomes

$$c_{\tau zR} = 1 + \frac{x_{\tau kR}c_{\tau kR}^2 + x_{\tau 02}c_{\tau kR}}{x_{zR}c_{\tau kR}^2} \tag{15.47}$$

The reactance of the transferred parallel branch $z-z$

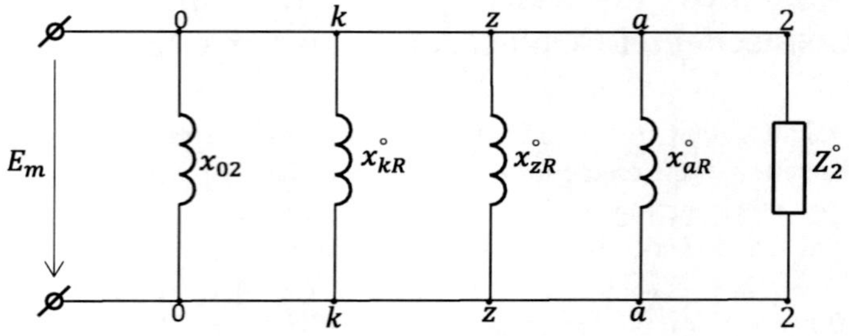

Fig. 15.11 Single-cage rotor equivalent circuit with the parallel connection of the elements

$$\overset{\circ}{x}_{zR} = x_{zR}c_{\tau zR}c_{\tau kR}^2 = x_{zR}c_{\tau kR}^2 + x_{\tau kR}c_{\tau kR}^2 + x_{\tau 02}c_{\tau kR} \tag{15.48}$$

The transfer factor of the parallel branch 2 – 2

$$c_{\tau 2} = 1 + \frac{x_{\tau kR}c_{\tau kR}^2 c_{\tau zR} + x_{\tau 02}c_{\tau kR}c_{\tau zR}}{(r_{22}/s + jx_{22\sigma})c_{\tau kR}^2 c_{\tau zR}^2} \tag{15.49}$$

The reactance of the transferred parallel branch 2 – 2

$$\begin{aligned}\overset{\circ}{Z}_2 &= (r_{22}/s)c_{\tau kR}^2 c_{\tau zR}^2 + j\left(x_{22\sigma}c_{\tau kR}^2 c_{\tau zR}^2 + x_{\tau kR}c_{\tau kR}^2 c_{\tau zR} + x_{\tau 02}c_{\tau kR}c_{\tau zR}\right) \\ &= (r_{22}/s)c_{\tau kR}^2 c_{\tau zR}^2 + jx_{2\sigma}\end{aligned} \tag{15.50a}$$

where $x_{2\sigma} = x_{22\sigma}c_{\tau kR}^2 c_{\tau zR}^2 + x_{\tau kR}c_{\tau kR}^2 c_{\tau zR} + x_{\tau 02}c_{\tau kR}c_{\tau zR}$.

In another form

$$\begin{aligned}\overset{\circ}{Z}_2 &= \{r_{22}/s + j[x_{22\sigma} + x_{\tau kR}(1/c_{\tau zR}) + x_{\tau 02}(1/c_{\tau kR}c_{\tau zR})]\}c_{\tau kR}^2 c_{\tau zR}^2 \\ &= (r_{22}/s + j\widetilde{x}_{22\sigma})c_{\tau kR}^2 c_{\tau zR}^2\end{aligned} \tag{15.50b}$$

where $\widetilde{x}_{22\sigma} = x_{22\sigma} + x_{\tau kR}(1/c_{\tau zR}) + x_{\tau 02}(1/c_{\tau kR}c_{\tau zR})$; $x_{2\sigma} = \widetilde{x}_{22\sigma}c_{\tau kR}^2 c_{\tau zR}^2$.

The reactance of the transferred parallel branch $a - a$

$$\begin{aligned}\overset{\circ}{x}_{aR} &= x_{aR}c_{\tau kR}^2 c_{\tau zR}^2 c_{\tau 2}^2 + x_{\tau zR}c_{\tau kR}^2 c_{\tau zR}^2 c_{\tau 2}^2 \\ &+ x_{\tau kR}c_{\tau kR}^2 c_{\tau zR}c_{\tau 2} + x_{\tau 02}c_{\tau kR}c_{\tau zR}c_{\tau 2}\end{aligned} \tag{15.51}$$

As shown in Appendix I, when using the reactances obtained in (15.46), (15.48), (15.50a), and (15.51), the "unfolded" rotor equivalent circuit with the mixed connection of the elements (Fig. 15.10) takes the form of the "unfolded" equivalent circuits with the parallel connection of the elements (Fig. 15.11).

Let us express the reactances $\overset{\circ}{x}_{02}, \overset{\circ}{x}_{kR}, \overset{\circ}{x}_{z}, \overset{\circ}{x}_{aR}$ and the impedance $\overset{\circ}{Z}_2$ used in the equivalent circuit in Fig. 15.11 through the corresponding elements of the equivalent circuit in Fig. 15.9. Based on (15.44) and (15.46), for the reactance of the parallel branch $k - k$, we have

$$\overset{\circ}{x}_{kR} = x^*_{kR} + x^*_{\tau 02} \tag{15.52}$$

The reactance of the parallel branch $z - z$ from (15.44) and (15.48)

$$\overset{\circ}{x}_{zR} = x^*_{zR} c^2_{kR} c^2_{\tau kR} + x^*_{\tau kR} c^2_{kR} c^2_{\tau kR} + x^*_{\tau 02} c_{\tau kR} \tag{15.53}$$

The impedance of the parallel branch $2 - 2$ based on (15.44) and (15.50a)

$$\begin{aligned}\overset{\circ}{Z}_2 = {} & (r^*_{22}/s) c^2_{kR} c^2_{zR} c^2_{\tau kR} c^2_{\tau zR} + j(x^*_{22\sigma} c^2_{kR} c^2_{zR} c^2_{\tau kR} c^2_{\tau zR} + x^*_{\tau kR} c^2_{kR} c^2_{\tau kR} c_{\tau zR} \\ & + x^*_{\tau 02} c_{\tau kR} c_{\tau zR}) = (r^*_{22}/s) c^2_{kR} c^2_{zR} c^2_{\tau kR} c^2_{\tau zR} + j x^*_{2\sigma}\end{aligned} \tag{15.54a}$$

where $x^*_{2\sigma} = x^*_{22\sigma} c^2_{kR} c^2_{zR} c^2_{\tau kR} c^2_{\tau zR} + x^*_{\tau kR} c^2_{kR} c^2_{\tau kR} c_{\tau zR} + x^*_{\tau 02} c_{\tau kR} c_{\tau zR}$.

Or

$$\begin{aligned}\overset{\circ}{Z}_2 = {} & \left\{ r^*_{22}/s + j\left[x^*_{22\sigma} + x^*_{\tau kR}\left(1/c^2_{zR} c_{\tau zR}\right) + x^*_{\tau 02}\left(1/c^2_{kR} c^2_{zR} c_{\tau kR} c_{\tau zR}\right)\right]\right\} \\ & c^2_{kR} c^2_{zR} c^2_{\tau kR} c^2_{\tau zR} = \left(r^*_{22}/s + j\tilde{x}^*_{22\sigma}\right) c^2_{kR} c^2_{zR} c^2_{\tau kR} c^2_{\tau zR}\end{aligned} \tag{15.54b}$$

where
$$\tilde{x}^*_{22\sigma} = x^*_{22\sigma} + x^*_{\tau kR}\left(1/c^2_{zR} c_{\tau zR}\right) + x^*_{\tau 02}\left(1/c^2_{kR} c^2_{zR} c_{\tau kR} c_{\tau zR}\right)$$
$$x^*_{2\sigma} = \tilde{x}^*_{22\sigma} c^2_{kR} c^2_{zR} c^2_{\tau kR} c^2_{\tau zR}$$

The reactance of the parallel branch $a - a$ from (15.44) and (15.51)

$$\begin{aligned}\overset{\circ}{x}_{aR} = {} & x^*_{aR} c^2_{kR} c^2_{Rz} c^2_{\tau kR} c^2_{\tau zR} c^2_{\tau 2} + x^*_{\tau z} c^2_{kR} c^2_{Rz} c^2_{\tau kR} c^2_{\tau zR} c^2_{\tau 2} \\ & + x^*_{\tau kR} c^2_{kR} c^2_{\tau kR} c_{\tau zR} c_{\tau 2} + x^*_{\tau 02} c_{\tau kR} c_{\tau zR} c_{\tau 2}\end{aligned} \tag{15.55}$$

In accordance with (15.52)–(15.55), the reactances of the parallel branches of the equivalent circuit in Fig. 15.11 are expressed through the reactances of the L-shaped equivalent circuits, representing the corresponding layers of the rotor model (Fig. 14.1). The analytical expressions for these reactances are obtained in (14.52), (14.54), (14.56), and (14.62). Based on the expressions (15.52)–(15.55), the elements of the rotor equivalent circuit (Fig. 15.11) can be calculated using the relations obtained for the reactances of the L-shaped equivalent circuits reflecting the corresponding layers of the rotor model.

15.10 Magnetizing and Slot Leakage Circuit Loops: "Folded" Equivalent Circuit of Single-Cage Rotor

To introduce the magnetizing and the slot leakage circuit loops, as well as to construct the "folded" rotor equivalent circuit, we will use the equivalent circuit obtained in Fig. 15.11.

Magnetizing circuit loop and rotor magnetizing reactance. In the equivalent circuit in Fig. 15.11, the parallel branches $0-0$, $k-k$, $z-z$, and $a-a$ represent the magnetizing circuit loops of the rotor model layers (Fig. 14.1). In this regard, they can be combined into one common branch which will represent the rotor magnetizing circuit loop. In this case, the reactance used in the common branch (magnetizing circuit loop) reflects the rotor magnetizing reactance. For the rotor magnetizing reactance, based on the equivalent circuit in Fig. 15.11, we can have

$$x^*_{m2} = \frac{1}{\dfrac{1}{x_{02}} + \dfrac{1}{\overset{\circ}{x}_{kR}} + \dfrac{1}{\overset{\circ}{x}_{zR}} + \dfrac{1}{\overset{\circ}{x}_{aR}}} = \frac{x_{02}}{1 + x_{02}\left(\dfrac{1}{\overset{\circ}{x}_{kR}} + \dfrac{1}{\overset{\circ}{x}_{zR}} + \dfrac{1}{\overset{\circ}{x}_{aR}}\right)} = \frac{x_{02}}{k_{\text{H}2}} \tag{15.56}$$

where $k_{\text{H}2} = 1 + x_{02}\left(\dfrac{1}{\overset{\circ}{x}_{kR}} + \dfrac{1}{\overset{\circ}{x}_{zR}} + \dfrac{1}{\overset{\circ}{x}_{aR}}\right)$.

Here, the reactance x^*_{m2} represents the rotor magnetizing reactance, and the value $k_{н2}$ reflects the saturation coefficient of the rotor magnetic circuit.

Slot leakage circuit loop and rotor slot leakage reactance. In the equivalent circuit in Fig. 15.11, the parallel branch $2-2$ represents the rotor slot leakage circuit loop. The reactance used in the rotor slot leakage circuit loop (parallel branch $2-2$, Fig. 15.11) is determined by the relation (15.54a). Into this relation, we introduce the following coefficients

$$k^2_{22} = c^2_{kR}c^2_{zR}c^2_{\tau kR}c^2_{\tau zR};\, k_{2k} = c^2_{kR}c^2_{\tau kR}c_{\tau zR};\, k_{20} = c_{\tau kR}c_{\tau zR} \tag{15.57}$$

Then, from relation (15.54a)and (15.57), we can obtain

$$\begin{aligned} \overset{\circ}{Z}_2 &= (r^*_{22}/s)k^2_{22} + j\left(x^*_{\tau 02}k_{20} + x^*_{\tau kR}k_{2k} + x^*_{22\sigma}k^2_{22}\right) \\ &= r^*_{22}/s + jx^*_{2\sigma} \end{aligned} \tag{15.58}$$

where $r^*_2/s = (r^*_{22}/s)k^2_{22}; x^*_{2\sigma} = x^*_{\tau 02}k_{20} + x^*_{\tau kR}k_{2k} + x^*_{22\sigma}k^2_{22}$.

In accordance with the expression (15.58), the ohmic resistance r^*_2/s and the reactance $x^*_{2\sigma}$ are used in the rotor slot leakage circuit loop. In (15.58), the value r^*_2/s reflects the ohmic resistance of the rotor winding, and the value $x^*_{2\sigma}$ is the rotor slot leakage reactance.

From (15.58), for the ohmic resistance of the rotor winding, we can have

$$\frac{r_2^*}{s} = \frac{r_{22}^*}{s} k_{22}^2 \tag{15.59}$$

For the rotor slot leakage reactance $x_{2\sigma}^*$, from (15.58), we have

$$x_{2\sigma}^* = x_{\tau 02}^* k_{20} + x_{\tau kR}^* k_{2k} + x_{22\sigma}^* k_{22}^2 \tag{15.60}$$

As follows from (15.60), this reactance consists of the three components: the leakage reactance of the air gap sublayer adjoined to the rotor surface $\left(x_{\tau 02}^* k_{20}\right)$, the leakage reactance of the layer reflecting the rotor slot-wedge area $\left(x_{\tau kR}^* k_{2k}\right)$, as well as the leakage reactance of the layer representing the rotor winding location area $(x_{22\sigma}^* k_{22}^2)$. These components of the reactance $x_{2\sigma}^*$ are expressed through the reactances $x_{\tau 02}^*, x_{\tau kR}^*$ and $x_{22\sigma}^*$ determined in accordance with (15.33), (15.34), and (15.36) in the form

$$\begin{aligned} x_{\tau 02}^* &= \omega_1 \mu_0 t_0 \frac{\delta}{2t_{02}} k_{2\sigma} k_R \zeta_{\tau 0} \\ x_{\tau kR}^* &= \omega_1 \mu_{xkR} t_0 \frac{h_{kR}}{t_{kR}} k_R \xi_{LkR} \zeta_{\tau kR} \\ x_{22\sigma}^* &= \omega_1 \mu_{xzR} t_0 \frac{h_{zR}}{3t_{zR}} k_R \xi_{LzR} \end{aligned} \tag{15.61}$$

In (15.61), for μ_{xkR} and μ_{xzR}, we can use the following conditions

$$\mu_{xkR} = \mu_0 \frac{t_{kR}}{b_{kR}} \quad \mu_{xzR} = \mu_0 \frac{t_{zR}}{b_{\Pi R}} \tag{15.62}$$

Then, the reactances $x_{\tau 02}^* k_{20}, x_{\tau kR}^* k_{2k}$ and $x_{22\sigma}^* k_{22}^2$ acquire the form from (15.61) and (15.62)

$$\begin{aligned} x_{\tau 02}^* k_{20} &= \omega_1 \mu_0 t_0 \frac{\delta}{2t_{02}} k_{2\sigma} k_R \zeta_{\tau 0} k_{20} = \omega_1 \mu_0 t_0 \lambda_{\tau 02} \\ x_{\tau kR}^* k_{2k} &= \omega_1 \mu_0 t_0 \frac{h_{kR}}{b_{kR}} k_R \xi_{LkR} \zeta_{\tau kR} k_{2k} = \omega_1 \mu_0 t_0 \lambda_{\tau kR} \\ x_{22\sigma}^* k_{22}^2 &= \omega_1 \mu_0 t_0 \frac{h_{zR}}{3b_{\Pi R}} k_R \xi_{LzR} k_{22}^2 = \omega_1 \mu_0 t_0 \lambda_{22\sigma} \end{aligned} \tag{15.63}$$

where $\lambda_{\tau 02} = \frac{\delta}{2t_{02}} k_{2\sigma} k_R \zeta_{\tau 0} k_{20}; \lambda_{\tau kR} = \frac{h_{kR}}{b_{kR}} k_R \xi_{LkR} \zeta_{\tau kR} k_{2k}; \lambda_{22\sigma} = \frac{h_{zR}}{3b_{\Pi R}} k_R \xi_{LzR} k_{22}^2.$

In (15.63), the dimensionless quantities $\lambda_{\tau 02}$, $\lambda_{\tau kR}$ and $\lambda_{22\sigma}$ represent the relative magnetic conductivity for the tangential components of the magnetic fluxes in the layers reflecting the air gap sublayer adjoined to the rotor surface, the rotor slot-wedge area, and the rotor winding location area, respectively.

Now from (15.60) and (15.63), for the rotor slot leakage reactance $x_{2\sigma}^*$, we can obtain

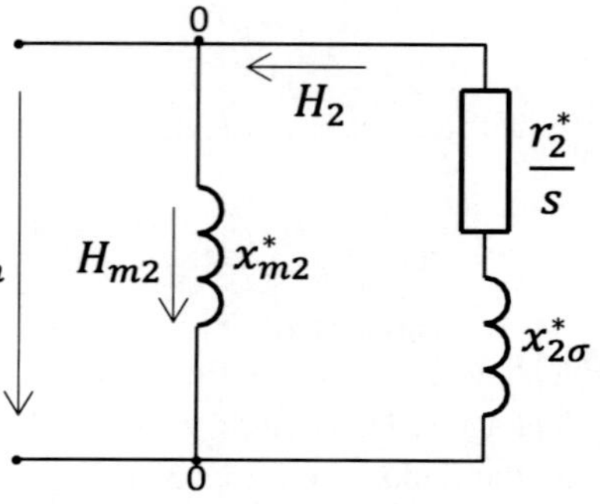

Fig. 15.12 "Folded" single-cage rotor equivalent circuit

$$\begin{aligned} x_{2\sigma}^* &= x_{\tau 02}^* k_{20} + x_{\tau kR}^* k_{2k} + x_{22\sigma}^* k_{22}^2 \\ &= \omega_1 \mu_0 t_0 (\lambda_{\tau 02} + \lambda_{\tau kR} + \lambda_{22\sigma}) = \omega_1 \mu_0 t_0 \lambda_{2\sigma} \end{aligned} \quad (15.64)$$

where $\lambda_{2\sigma} = \lambda_{\tau 02} + \lambda_{\tau kR} + \lambda_{22\sigma}$.

Here, the value $\lambda_{2\sigma}$ reflects the relative magnetic conductivity for the slot leakage magnetic flux of the rotor winding.

"Folded" equivalent circuit of single-cage rotor. Based on the relations (15.56), (15.59), and (15.60), the equivalent circuit in Fig. 15.11 can be reduced to the form given in Fig. 15.11. Figure 15.12 shows the "folded" equivalent circuit of the single-cage rotor. As follows from the equivalent circuit in Fig. 15.12, the single-cage rotor is represented in relation to $\dot{E}_m$ in the form of the two independent circuit loops: the magnetizing circuit loop containing the reactance x_{m2}^* and the rotor slot leakage circuit loop including the ohmic resistance r_2^*/s and the slot leakage reactance $x_{2\sigma}^*$. The "folded" rotor equivalent circuit obtained in this way has a simple structure and it is convenient for practical use.

15.11 Equivalent Circuit of Machine with Single-Cage Rotor: Voltage Equations

The equivalent circuit of the machine with the single-cage rotor can be obtained using the "folded" stator equivalent circuit shown in Fig. 15.6 and the "folded" rotor equivalent circuit given in Fig. 15.12. For this, terminals $0 - 0$ of these equivalent circuits should be connected, and the machine magnetizing reactance defined as $x_m^* = x_{m1}^* x_{m2}^* / x_{m1}^* + x_{m2}^*$ should be introduced. As a result, the equivalent circuit of the machine with the single-cage rotor takes the form in Fig. 15.13.

Using the equivalent circuit in Fig. 15.13, the machine voltage equations can be obtained. Received in this way the voltage equations correspond to the "no-load and under-load" concept. So, we can have

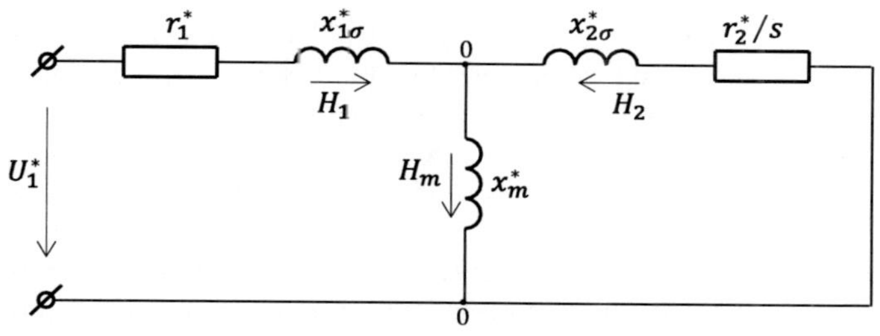

Fig. 15.13 Equivalent circuit of the machine with the single-cage rotor

$$\begin{aligned} U_1^* &= \left(r_1^* + jx_{1\sigma}^*\right)\dot{H}_1 + \left(-\dot{E}_m\right) \\ -\frac{U_2^*}{s} &= \left(\frac{r_2^*}{s} + jx_{2\sigma}^*\right)\dot{H}_2 + \left(-\dot{E}_m\right) \end{aligned} \tag{15.65}$$

Here, $(-\dot{E}_m) = jx_m^*\dot{H}_m$ and $\dot{H}_m = \dot{H}_1 + \dot{H}_2$. We will use these conditions in the system of the Eq. (15.65). Then, we obtain for the voltage equations corresponding to the "inductively coupled circuits" concept

$$\begin{aligned} U_1^* &= r_1^*\dot{H}_1 + j\left(x_{1\sigma}^* + x_m^*\right)\dot{H}_1 + jx_m^*\dot{H}_2 = r_1^*\dot{H}_1 + jx_1^*\dot{H}_1 + jx_{21}^*\dot{H}_2 \\ -\frac{U_2^*}{s} &= \frac{r_2^*}{s}\dot{H}_2 + jx_m^*\dot{H}_1 + j\left(x_{2\sigma}^* + x_m^*\right)\dot{H}_2 = \frac{r_2^*}{s}\dot{H}_2 + jx_{12}^*\dot{H}_1 + jx_2^*\dot{H}_2 \end{aligned} \tag{15.66}$$

where $x_1^* = x_{1\sigma}^* + x_m^*; x_2^* = x_{2\sigma}^* + x_m^*; x_{21}^* = x_m^*; x_{12}^* = x_m^*$.

15.12 Summary

In this chapter, the method for determining the calculation elements of the asynchronous machine is described. The proposed method is based on the use of the resulting electromagnetic field determined in the machine using Maxwell's equations. The peculiarities of the implementation of the method are demonstrated by the example of considering the basic calculation elements of the asynchronous machine with the single-cage rotor.

1. In this chapter of the work, the techniques and procedures for determining the basic calculation elements of the machine with the single-cage rotor have been developed. These techniques and procedures are based on the use of the "field" concept. The proposed techniques and procedures are the result of the implementation of the provisions formulated in Chap. 13 of this work for the case when the resulting magnetic field in the machine is determined as the result of the superposition of the "partial" magnetic fields. The determination of the machine calculation elements is based on the application of the superposition

principle, the "fragmentation" principle, the "no-load and under-load" concept, the cascade way of connecting, and the multiple transfer technique.

2. In this work, the basic fragments of the machine design are highlighted. Further, these fragments of the machine design are represented in the form of the layers with smooth surfaces, as a result of which the layered machine model is followed. The representation of the fragments of the machine design in the form of the layers with the smooth surfaces is associated with the introduction of the average values of the electromagnetic parameters ρ and μ. In this work, to simplify the analysis, the layered models of the stator and rotor are obtained using the layered model of the machine. Consequently, the "fragmentation" principle means the selection of the machine design fragments, the representation of these fragments as the corresponding layers with the smooth surfaces and the representation of these layers as the L-shaped equivalent circuits.
3. The layered model of the machine results from the use of the average values of the magnetic permeabilities μ_{xi} and μ_{yi}, where $i = s, R$, for the slot-tooth area layers and the average values of the resistivity for the winding conductors of the stator and rotor. The machine model consists of the air gap sublayers, the slot-wedge area layers, the winding location area layers, as well as the yoke area layers of the stator and rotor.
4. The layered model of the machine can be used to describe the distribution of the electromagnetic field in the cross section of the machine for that Maxwell's equations can be applied. In this work, the proposed layered model is used to determine the equations arising for the electric and magnetic field intensities determined on the lower and upper surfaces of the machine model layers. The equations obtained in this way can be visualized by the T-shaped equivalent circuits representing the corresponding machine model layers. In this work, T-shaped equivalent circuits are not used due to the fact that their use creates serious difficulties in determining the machine calculation elements.
5. The equations and the T-shaped equivalent circuits corresponding to the machine model layers are transformed. The transformation applied in this work is based on the use of the "no-load and under-load" concept and is implemented using the reduction technique (in the case of transforming the layer equations) or the transfer technique (in the case of transforming the equivalent circuits). As a result, in this work, the layers of the machine model are represented in the form of the L-shaped equivalent circuits.
6. For convenience, the layered machine model is decomposed into the layered stator model and the layered rotor model. This decomposition of the machine model is carried out in relation to the surface $0 - 0$ passing through the middle of the radial length of the air gap of the machine. In this case, the stator model consists of the air gap sublayer adjoined to the stator bore surface, the layer of the slot-wedge area, the layer of the winding location area, and the layer of the stator yoke area. The rotor model includes the air gap sublayer adjoined to the rotor surface, the slot-wedge area layer, the winding location area layer, and the rotor yoke area layer.

7. In this work, the layers of the stator and rotor models are represented in the form of the L-shaped equivalent circuits (with the exception of the layers of the stator and rotor yoke areas, which are replaced by the equivalent circuits of the two-terminal network). The L-shaped equivalent circuits obtained in this way and reflecting the corresponding layers of the stator and rotor models are used to construct their equivalent circuits. In this case, the equivalent circuits of the stator and rotor result from the cascade way of connecting the L-shaped equivalent circuits, reflecting the corresponding layers of their models. The equivalent circuits flowing out in this way are the "unfolded" stator and rotor equivalent circuits with the mixed connection of the elements.
8. The L-shaped equivalent circuit consists of the two independent circuit loops: the "own" circuit loop and the "connecting" circuit loop. The "connecting" circuit loop provides the connection of the "own" circuit loops of the adjacent layers of the stator and rotor models. In the case of the conducting layer (in our case, the layer of the rotor winding location area), the "own" circuit loop is represented in the area of the weak skin effect as consisting of the two circuit loops: the "own" magnetizing circuit loop of the layer and the "own" slot leakage circuit loop of the eddy currents induced in the considered conducting layer. Therefore, the use of the L-shaped equivalent circuits makes it possible to construct the equivalent circuits for the stator and rotor of the simple structure consisting of the magnetizing circuit loops and the slot leakage circuit loops. Such stator and rotor equivalent circuits are convenient for practical use.
9. The stator and rotor equivalent circuits resulting from the cascade way of connecting the L-shaped equivalent circuits are the "unfolded" equivalent circuits with the mixed connection of the elements. The practical use of such stator and rotor equivalent circuits is associated with certain inconveniences. In this regard, the "unfolded" stator and rotor equivalent circuits are reduced, in this work, to the convenient form.
10. The reduction of the "unfolded" stator and rotor equivalent circuits with the mixed connection of the elements is carried out in the two stages. At the first stage, the "unfolded" stator and rotor equivalent circuits with the mixed connection of the elements are reduced to the equivalent circuits with the parallel connection of the elements, for which the multiple transfer technique is applied. At the second stage, the "unfolded" stator and rotor equivalent circuits with the parallel connection of the elements are used to introduce the stator and rotor magnetizing circuit loops and the stator and rotor slot leakage circuit loops, as a result of which the "folded" stator and rotor equivalent circuits follow. The "folded" stator and rotor equivalent circuits have a simple structure and they are convenient for practical use. The "folded" stator and rotor equivalent circuits are used to construct the equivalent circuit of the electric machine.
11. The construction of the machine equivalent circuit is accompanied by the introduction of the magnetizing circuit loop and the determination of the magnetizing reactance of the machine, for which the "folded" stator and rotor equivalent circuits are applied. The machine equivalent circuit obtained in this

way consists of the three circuit loops: the machine magnetizing circuit loop, the stator slot leakage circuit loop, and the rotor slot leakage circuit loop.

12. In this work, the voltage equations for the stator and rotor windings are compiled using the equivalent circuit of the electric machine. In this case, the voltage equations corresponding to the "no-load and under-load" concept can be obtained. These voltage equations are used to compile the voltage equations corresponding to the "inductively coupled circuits" concept.
13. Obtained with the help of the "field" concept, the machine equivalent circuit, the voltage equations, and the parameters used in them have the traditional form and the physical substance. This creates a favorable basis for the analysis and the generalization of the field research results.

References

1. Asanbayev, V.: Alternating Current Multi-Circuit Electric Machines: A New Approach to Steady-State Parameter Determination, Springer Cham Heidelberg, New York, Dordrecht London (2015)
2. Mishkin, E.: Theory of the Squirrel-Cage Induction Machine Derived Directly from Maxwell's Field Equations, The Quarterly Journal of Mechanics and Applied Mathematics, Vol.7, Part 4, pp. 472-487, (1954)
3. Veselovsky, O., Konyaev, A., Sarapulov, F.: Linear Asynchronous Motor, Publishing Energoatomizdat, Moscow (1991)
4. Dombrovsky, V., Zaichik, V.: Asynchronous Machines: Theory, Calculation, Elements of Design, Publishing Energoatomizdat, Leningrad (1990)

Chapter 16
Basic Calculation Elements of Machine with Wound Rotor: “Field” Concept

16.1 L-Shaped Equivalent Circuits of Wound Rotor Model Layers and Their Parameters

In Chap. 14, the layered model of the machine is decomposed into the layered stator model and the layered rotor model with respect to the surface $0-0$ passing through the middle of the radial length of the air gap. The layered stator model is discussed above in Chap. 15. In this regard, the layered rotor model is considered in this chapter. The rotor of the asynchronous machine can be equipped with the distributed phase winding. Let us consider the features of determining the basic calculation elements of the rotor having the distributed phase winding and supplied, in the general case, from the external source. For this purpose, we proceed from the layered model shown for the wound rotor in Fig. 16.1 [1–4]. This rotor model is built in relation to the surface $0-0$ passing through the middle of the radial length of the air gap of the machine. The layered model shown in Fig. 16.1 is similar to the layered model shown for the single-cage rotor in Fig. 14.1 (with the exception of the winding location area layer). The basic difference between the layered model of the wound rotor and the layered model of the single-cage rotor is the presence in the model of the wound rotor of the active layer that reflects the location area of the phase winding, powered by the external source (exciter). In this case, the phase winding of the rotor is usually represented in the form of the infinitely thin current threads evenly distributed in the layer [1]. The wound rotor model in Fig. 16.1 includes the air gap sublayer adjoined to the rotor surface (02), the slot-wedge area layer (kR), the phase winding location area layer (zR), and the rotor yoke area layer (aR). On the surfaces $k-k$, $z-z$ and $a-a$ of the layered model of the wound rotor (Fig. 16.1), the conditions of the following form are valid

V. Asanbayev, *Asynchronous Machines*,
https://doi.org/10.1007/978-3-030-92284-9_16

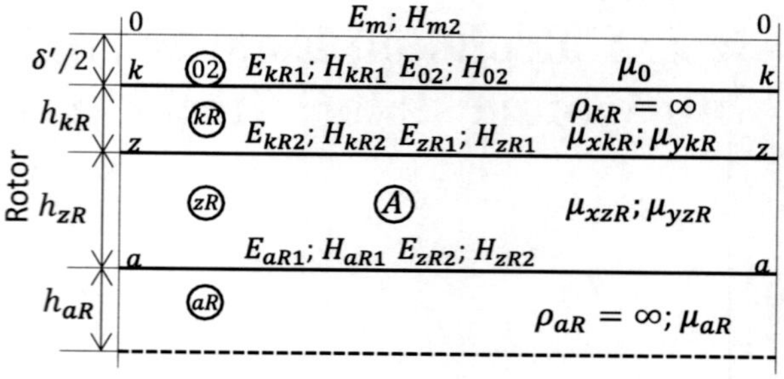

Fig. 16.1 Multilayer wound rotor model

$$
\begin{aligned}
\dot{E}_{02} &= \dot{E}_{kR1} & \dot{H}_{02} &= \dot{H}_{kR1} \\
\dot{E}_{kR2} &= \dot{E}_{zR1} & \dot{H}_{kR2} &= \dot{H}_{zR1} \\
\dot{E}_{zR2} &= \dot{E}_{aR1} & \dot{H}_{zR2} &= \dot{H}_{aR1}
\end{aligned}
\tag{16.1}
$$

In the wound rotor model, the conditions given in (16.1) are typical for the surfaces formed by the corresponding adjacent layers. In this case, the values of the electric and magnetic field intensities used in (16.1) are the quantities reduced to the calculated length l_0 and the pole pitch τ_0 and can be determined applying the expressions obtained in (14.2b). The conditions given in (16.1) will be used when constructing the equivalent circuit for the wound rotor. The layers of the wound rotor model (Fig. 16.1) are represented by the L-shaped equivalent circuits (with the exception of the layer reflecting the rotor yoke area which has the equivalent circuit of the two-terminal network). The air gap sublayer adjoined to the rotor surface is represented by the L-shaped equivalent circuit shown in Fig. 15.7a. The reactances used in the L-shaped equivalent circuit in Fig. 15.7a can be calculated using the expressions obtained in (15.33). In this chapter, the L-shaped equivalent circuit of the air gap sublayer adjoined to the rotor surface is shown in Fig. 16.2a.

The layer representing the rotor slot-wedge area is visualized by the L-shaped equivalent circuit of the form in Fig. 15.7b. The reactances used in this L-shaped equivalent circuit are determined by the expressions given in (15.34). In this chapter, the L-shaped equivalent circuit of the layer reflecting rotor slot-wedge area is shown in Fig. 16.2b. For the components of the electric and magnetic field intensities $\dot{E}'_{kR2}$ and $\dot{H}'_{kR2}$ determined at the output terminals $z - z$ of the equivalent circuit in Fig. 16.2b, the conditions arising from (14.50) are valid

$$
\dot{E}'_{kR2} = \dot{E}_{kR2} c_{kR} \quad \dot{H}'_{kR2} = \dot{H}_{kR2} / c_{kR}
\tag{16.2}
$$

where $c_{kR} = \cosh \beta_{kR} h_{kR}; \beta_{kR} = k_{\mu kR}(\pi/\tau_{kR}); k_{\mu kR} = \sqrt{\mu_{xkR}/\mu_{ykR}}$.

The layer representing the rotor winding location area (active layer) can be reflected by the L-shaped equivalent circuit in Fig. 14.9b. Here, we consider the laminated rotor and the active layer, in this case, can be considered as the nonconducting magneto-anisotropic layer. In this regard, the propagation coefficient for the layer

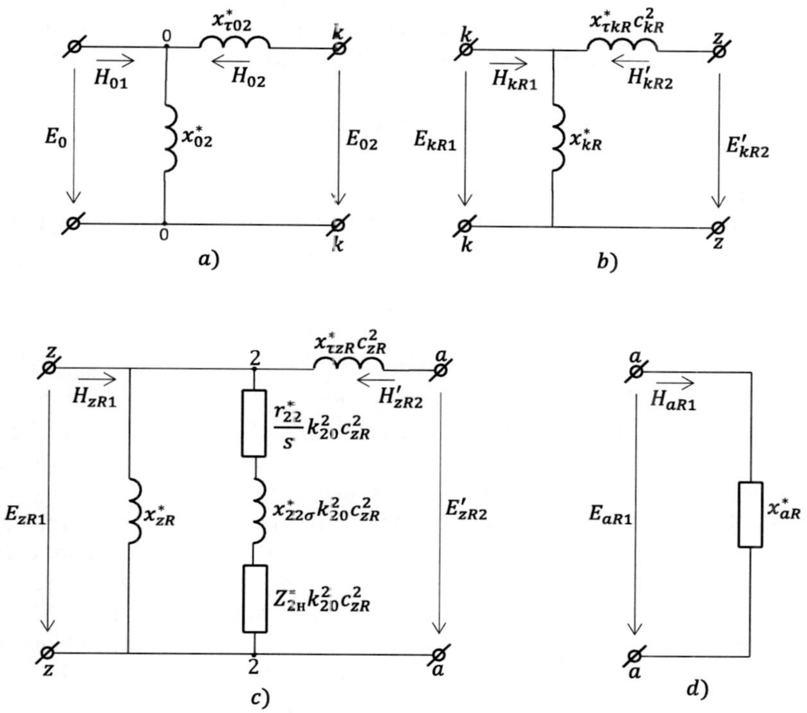

Fig. 16.2 Wound rotor model layer equivalent circuits: (**a**) L-shaped equivalent circuit of the air gap sublayer adjoined to the rotor surface, (**b**) L-shaped equivalent circuit of the layer reflecting the rotor slot-wedge area, (**c**) L-shaped equivalent circuit of the layer representing the rotor winding location area, (**d**) equivalent circuit of the layer of the rotor yoke area

corresponding to the rotor winding location area β_{zR} is the real number, i.e., $\beta_{zR} = k_{\mu zR}(\pi/\tau_{zR})$, where $k_{\mu zR} = \sqrt{\mu_{xzR}/\mu_{yzR}}$ is the coefficient of the magnetic anisotropy, τ_{zR} is the pole pitch defined on the outer surface of the rotor active layer. Now, the L-shaped equivalent circuit of the active layer, shown in Fig. 14.9b, takes the form in Fig. 16.2c. In this case, in the equivalent circuit in Fig. 16.2c, the voltage of the rotor winding $\dot{U}_2^*/s$ is replaced by the voltage drop across the load impedance Z_{2H}^*, i.e., $\dot{U}_2^*/s = Z_{2H}^* \dot{H}_2$. As a result, in Fig. 16.2c, the rotor winding location area layer is represented as the passive equivalent circuit. The reactances $x_{zR}^*, x_{\tau zR}^*$, and $x_{22\sigma}^*$ used in the L-shaped equivalent circuit in Fig. 16.2c can be determined using the expressions obtained in (15.35) and (15.36). The reactances x_{zR}^* and $x_{\tau zR}^*$ follow from the expressions obtained in (15.35). For reactance $x_{22\sigma}^*$, we use the expression given in (15.36). In the L-shaped equivalent circuit in Fig. 16.2c, the rotor winding is represented by the ohmic resistance r_{22}^*/s. To determine this resistance, the expression obtained in [1] can be used. The ohmic resistance r_{22}^*/s

and the reactance $x^*_{22\sigma}$ used in the rotor equivalent circuit (Fig. 16.2c) are multiplied by the factors k^2_{20} and c^2_{zR} determined from (14.59) and (14.63) as

$$k_{20} = \beta_{zR} h_{zR} / \sinh \beta_{zR} h_{zR} \quad c_{zR} = \cosh \beta_{zR} h_{zR} \tag{16.3}$$

The quantities $\dot{E}'_{zR2}$ and $\dot{H}'_{zR2}$ determined at the output terminals $a - a$ of the equivalent circuit in Fig. 16.2c, follow in accordance with (14.60) from conditions of the form

$$\dot{E}'_{zR2} = \dot{E}_{zR2} c_{zR} \quad \dot{H}'_{zR2} = \dot{H}_{zR2} / c_{zR} \tag{16.4}$$

where $c_{zR} = \cosh \beta_{zR} h_{zR}; \beta_{zR} = k_{\mu zR}(\pi/\tau_{zR}); k_{\mu zR} = \sqrt{\mu_{xzR}/\mu_{yzR}}$.

As follows from the L-shaped equivalent circuit in Fig. 16.2c, the "own" circuit loops reflecting the active rotor layer are established in relation to the quantity of $\dot{E}_{zR1}$ determined on the outer surface of this rotor layer.

The rotor yoke area layer is represented in the form of the equivalent circuit of the two-terminal network. Such equivalent circuit is shown in Fig. 15.7d. In this chapter, the equivalent circuit for the rotor yoke area layer is given in Fig. 16.2d. The reactance used in the equivalent circuit in Fig. 16.2d was obtained in (15.37).

The L-shaped equivalent circuits shown in Fig. 16.2 will be used below when constructing the equivalent circuit for the wound rotor.

16.2 Cascade Equivalent Circuits of Wound Rotor

To build the cascade equivalent circuit for the wound rotor, we can use the L-shaped equivalent circuits shown in Fig. 16.2, as well as the layered rotor model obtained in Fig. 16.1. In this case, the cascade equivalent circuit of the rotor will represent the result of the cascade way of connecting the L-shaped equivalent circuits (Fig. 16.2) placed in a space in accordance with the layered rotor model in Fig. 16.1. However, the process of constructing the cascade equivalent circuit for the wound rotor can be significantly simplified if for this purpose we use the cascade equivalent circuit shown in Fig. 15.8a for the single-cage rotor. To do this, in the cascade equivalent circuit in Fig. 15.8a, the L-shaped equivalent circuit, representing the single-cage location area layer (Fig. 15.7c) should be replaced with the L-shaped equivalent circuit, reflecting the winding location area layer (Fig. 16.2c). As a result, for the wound rotor, the cascade equivalent circuit can be represented in the form in Fig. 16.3a. In Fig. 16.3a, the equivalent circuit for the wound rotor takes the form of the passive cascade equivalent circuit.

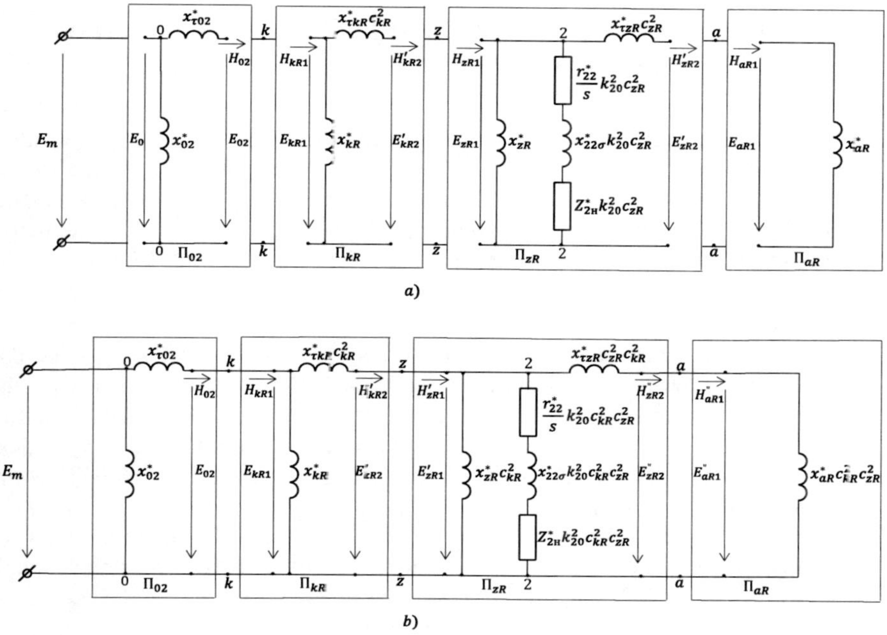

Fig. 16.3 Equivalent circuits of the wound rotor obtained as a result of the cascade way of connecting the four-terminal networks. (**a**) schematic visualization of the four-terminal networks; (**b**) cascade connection of the L-shaped equivalent circuits reflecting the wound rotor model layers

In the cascade equivalent circuit in Fig. 16.3a, its terminals $k - k$, $z - z$, and $a - a$ correspond to the surfaces $k - k$, $z - z$, and $a - a$ formed by the adjacent layers of the wound rotor model (Fig. 16.1). In this regard, at the terminals $k - k$, $z - z$, and $a - a$ of the cascade equivalent circuit in Fig. 16.3a, the conditions given in (16.1) are fulfilled. As follows from the equivalent circuit in Fig. 16.3a, the conditions (16.1) are not fulfilled at the output terminals of the L-shaped equivalent circuit representing the rotor slot-wedge area layer (four-terminal network Π_{kR}) and at the input terminals of the L-shaped equivalent circuit reflecting the rotor winding location area layer (four-terminal network Π_{zR}). In this regard, the L-shaped equivalent circuits used in the equivalent circuit in Fig. 16.3a cannot be directly connected in a cascade. The similar conditions are typical for the output terminals of the L-shaped equivalent circuit reflecting the rotor winding location area layer (four-terminal network Π_{zR}) and for the input terminals of the equivalent circuit representing the rotor yoke area layer (two-terminal network Π_{aR}). Below we will analyze the conditions under which these L-shaped equivalent circuits can be connected in a cascade.

16.3 "Unfolded" Equivalent Circuit with Mixed Connection of Elements of Wound Rotor

Based on the cascade equivalent circuit in Fig. 16.3a, the "unfolded" equivalent circuit with the mixed connection of the elements can be obtained for the wound rotor. The cascade equivalent circuit of the wound rotor (Fig. 16.3a) and the cascade equivalent circuit of the single-cage rotor (Fig. 15.8a) have the same structure. In this regard, to analyze the cascade equivalent circuit of the wound rotor (Fig. 16.3a), the techniques and procedures used in the analysis of the cascade equivalent circuit of the single-cage rotor (Fig. 15.8a) can be applied.

Let's analyze the cascade equivalent circuit of the wound rotor in Fig. 16.3a. The consideration of the equivalent circuit in Fig. 16.3a will begin with an analysis of the conditions for connecting the output terminals of the L-shaped equivalent circuit of the four-terminal network Π_{02} with the input terminals of the L-shaped equivalent circuit of the four-terminal network Π_{kR}. In Fig. 16.3a, the output terminals of the L-shaped equivalent circuit of the four-terminal network Π_{02} correspond to the inner surface of the air gap sublayer adjoined to the rotor surface. The input terminals of the L-shaped equivalent circuit of the four-terminal network Π_{kR} represent the outer surface of the layer reflecting the rotor slot-wedge area. In the wound rotor model (Fig. 16.1), these layers of the rotor form the adjacent surface $k - k$ on which, in accordance with (16.1), the conditions of the form $\dot{E}_{02} = \dot{E}_{kR1}$ and $\dot{H}_{02} = H_{kR1}$ are met. Thus, the components of the electric and magnetic field intensities $\dot{E}_{02}$ and $\dot{H}_{02}$ act on the output terminals of the L-shaped equivalent circuit of the four-terminal network Π_{02}. At the input terminals of the L-shaped equivalent circuit of the four-terminal network Π_{kR}, for the components of the electric and magnetic field intensities, we have $\dot{E}_{kR1}$ and $\dot{H}_{kR1}$. Therefore, based on the conditions $\dot{E}_{02} = \dot{E}_{kR1}$ and $\dot{H}_{02} = H_{kR1}$, we can note that on the surface $k - k$ of the wound rotor model (Fig. 16.1), as well as on the output terminals of the L-shaped equivalent circuit of the four-terminal network Π_{02} and on the input terminals of the L-shaped equivalent circuit of the four-terminal network Π_{kR} (Fig. 16.3a), the components of the electric and magnetic field intensities have the same value. In this regard, the output terminals of the L-shaped equivalent circuit of the four-terminal network Π_{02} and the input terminals of the L-shaped equivalent circuit of the four-terminal network Π_{kR} can be directly connected in a cascade. The L-shaped equivalent circuit of the four-terminal network Π_{02} and the L-shaped equivalent circuit of the four-terminal network Π_{kR} connected in a cascade are shown in Fig. 16.3b.

Now, in the equivalent circuit in Fig. 16.3a, we will consider the peculiarities of connecting the output terminals of the L-shaped equivalent circuit of the four-terminal network Π_{kR} with the input terminals of the L-shaped equivalent circuit of the four-terminal network Π_{zR}. In Fig. 16.3a, the output terminals of the L-shaped equivalent circuit of the four-terminal network Π_{kR} correspond to the inner surface of the layer reflecting the rotor slot-wedge area. The input terminals of the L-shaped equivalent circuit of the four-terminal network Π_{zR} represent the outer surface of the layer representing the rotor winding location area. In the wound rotor model

(Fig. 16.1), these rotor layers form the adjacent surface $z-z$ on which, in accordance with (16.1), the conditions of the form $\dot{E}_{kR2}=\dot{E}_{zR1}$ and $\dot{H}_{kR2}=H_{zR1}$ are met. The components of the electric and magnetic field intensities $\dot{E}'_{kR2}$ and $\dot{H}'_{kR2}$ act on the output terminals of the L-shaped equivalent circuit of the four-terminal network Π_{kR}. At the input terminals of the L-shaped equivalent circuit of the four-terminal network Π_{zR} for the components of the electric and magnetic field intensities, we have that $\dot{E}_{zR1}$ and $\dot{H}_{zR1}$. When cascading the output terminals of the L-shaped equivalent circuit of the four-terminal network Π_{kR} with the input terminals of the L-shaped equivalent circuit of the four-terminal network Π_{zR} it must be borne in mind that the quantities $\dot{E}'_{kR2}$ and $\dot{H}'_{kR2}$ corresponding to the output terminals of the L-shaped equivalent circuit of the four-terminal network Π_{kR} are defined in accordance with (16.2) as $\dot{E}'_{kR2}=\dot{E}_{kR2}c_{kR}$ and $\dot{H}'_{kR2}=\dot{H}_{kR2}/c_{kR}$. In this regard, and taking into account the conditions at the boundary $z-z$ of the wound rotor model in the form $\dot{E}_{kR2}=\dot{E}_{zR1}$ and $\dot{H}_{kR2}=H_{zR1}$, the quantities of $\dot{E}_{zR1}$ and H_{zR1} corresponding to the input terminals of the L-shaped equivalent circuit of the four-terminal network Π_{zR} should be corrected as

$$\dot{E}'_{kR2}=\dot{E}'_{zR1}=\dot{E}_{zR1}c_{kR}\quad \dot{H}'_{kR2}=\dot{H}'_{zR1}=\dot{H}_{zR1}/c_{kR} \tag{16.5}$$

Therefore, when cascading the output terminals of the L-shaped equivalent circuit of the four-terminal network Π_{kR} with the input terminals of the L-shaped equivalent circuit of the four-terminal network Π_{zR}, instead of the conditions obtained in (16.1) in the form $\dot{E}_{kR2}=\dot{E}_{zR1}$ and $\dot{H}_{kR2}=H_{zR1}$, it is necessary to use the new conditions of kind

$$\dot{E}'_{kR2}=\dot{E}'_{zR1}\quad \dot{H}'_{kR2}=\dot{H}'_{zR1} \tag{16.6}$$

where $\dot{E}'_{kR2}=\dot{E}_{kR2}c_{kR};\dot{E}'_{zR1}=\dot{E}_{zR1}c_{kR};\dot{H}'_{kR2}=\dot{H}_{kR2}/c_{kR};\dot{H}'_{zR1}=\dot{H}_{zR1}/c_{kR}$.

In relation to the values $\dot{E}'_{zR1}$ and $\dot{H}'_{zR1}$, the input impedance of the L-shaped equivalent circuit of the four-terminal network Π_{zR} is defined as $Z'_{zR1}=\dot{E}'_{zR1}/\dot{H}'_{zR1}=(\dot{E}_{zR1}/\dot{H}_{zR1})c^2_{Rk}=Z_{zR1}c^2_{kR}$, where $Z_{zR1}=\dot{E}_{zR1}/\dot{H}_{zR1}$ is the input impedance of the L-shaped equivalent circuit of the four-terminal network Π_{zR} determined with respect to the real values of $\dot{E}_{zR1}$ and $\dot{H}_{zR1}$. Therefore, under the conditions of the transition to the new values $\dot{E}'_{zR1}$ and $\dot{H}'_{zR1}$, the impedances of the L-shaped equivalent circuit of the four-terminal network Π_{zR} $\left[x^*_{zR},(r^*_{22}/s)k^2_{20}c^2_{zR},(x^*_{22\sigma}k^2_{20}c^2_{zR}),(x^*_{\tau zR}c^2_{zR}),(Z^*_{2H}k^2_{20}c^2_{zR})\right]$ are multiplied by the squared factor c^2_{kR} that is $\left[x^*_{zR}c^2_{kR},(r^*_{22}/s)k^2_{20}c^2_{kR}c^2_{zR},(x^*_{22\sigma}k^2_{20}c^2_{kR}c^2_{zR}),(x^*_{\tau zR}c^2_{kR}c^2_{zR}),(Z^*_{2H}k^2_{20}c^2_{kR}c^2_{zR})\right]$. Now, the output terminals of the L-shaped circuit of the four-terminal network Π_{kR} and the input terminals of the L-shaped circuit of the four-terminal network Π_{zR} can be directly connected in a cascade. With the cascade way of connecting the output terminals of the L-shaped equivalent circuit of the four-terminal network Π_{kR} and the input terminals of the L-shaped equivalent circuit of the four-terminal network Π_{zR}, the values $\dot{E}'_{zR2}$ and $\dot{H}'_{zR2}$ corresponding to the output terminals of the L-shaped equivalent circuit of the four-terminal network Π_{zR} and determined in accordance with (16.4) as $\dot{E}'_{zR2} = \dot{E}_{zR2}c_{zR}$ and $\dot{H}'_{zR2} = \dot{H}_{zR2}/c_{zR}$ are corrected by the transfer factor c_{kR} and take the form

$$\dot{E}''_{zR2} = \dot{E}'_{zR2}c_{kR} = \dot{E}_{zR2}c_{kR}c_{zR} \quad \dot{H}''_{zR2} = \dot{H}'_{zR2}/c_{kR} = \dot{H}_{zR2}/c_{kR}c_{zR} \tag{16.7}$$

The cascade way of connecting the L-shaped equivalent circuit of the four-terminal network Π_{kR} and the L-shaped equivalent circuit of the four-terminal network Π_{zR} is shown in Fig. 16.3b.

Further, in the equivalent circuit in Fig. 16.3a, we will consider the features of connecting the output terminals of the L-shaped equivalent circuit of the four-terminal network Π_{zR} with the input terminals of the equivalent circuit of the two-terminal network Π_{aR}. In Fig. 16.3a, the output terminals of the L-shaped equivalent circuit of the four-terminal network Π_{zR} correspond to the inner surface of the layer reflecting the rotor winding location area. The input terminals of the equivalent circuit of the two-terminal network Π_{aR} represent the outer surface of the layer representing the rotor yoke area. In the wound rotor model (Fig. 16.1), these rotor layers form the adjacent surface $a - a$ on which, in accordance with (16.1), the conditions of the form $\dot{E}_{zR2} = \dot{E}_{aR1}$ and $\dot{H}_{zR2} = H_{aR1}$ are met. According to (16.7), the components of the electric and magnetic field intensities $\dot{E}''_{zR2}$ and $\dot{H}''_{zR2}$ act on the output terminals of the L-shaped equivalent circuit of the four-terminal network Π_{zR}. At the input terminals of the equivalent circuit of the two-terminal network Π_{aR}, for the components of the electric and magnetic field intensities, we have that $\dot{E}_{aR1}$ and $\dot{H}_{aR1}$. When cascading the output terminals of the L-shaped equivalent circuit of the four-terminal network Π_{zR} with the input terminals of the equivalent circuit of the two-terminal network Π_{aR}, it must be borne in mind that the values of $\dot{E}''_{zR2}$ and $\dot{H}''_{zR2}$ corresponding to the output terminals of the L-shaped equivalent circuit of the four-terminal network Π_{zR} are determined in accordance with (16.7) as $\dot{E}''_{zR2} = \dot{E}_{zR2}c_{kR}c_{zR}$ and $\dot{H}''_{zR2} = \dot{H}_{zR2}/c_{kR}c_{zR}$. In this regard, and taking into account the conditions at the boundary $a - a$ of the wound rotor model in the form $\dot{E}_{zR2} = \dot{E}_{aR1}$ and $\dot{H}_{zR2} = H_{aR1}$, the values of $\dot{E}_{aR1}$ and H_{aR1} corresponding to the input terminals of the equivalent circuit of the two-terminal network Π_{aR} should be corrected as

$$\dot{E}''_{zR2} = \dot{E}''_{aR1} = \dot{E}_{aR1}c_{kR}c_{zR} \quad \dot{H}''_{zR2} = \dot{H}''_{aR1} = \dot{H}_{aR1}/c_{kR}c_{zR} \tag{16.8}$$

Therefore, when cascading the output terminals of the L-shaped equivalent circuit of the four-terminal network Π_{zR} with the input terminals of the equivalent circuit of the two-terminal network Π_{aR}, instead of the conditions obtained in (16.1) as $\dot{E}_{zR2} = \dot{E}_{aR1}$ and $\dot{H}_{aR2} = \dot{H}_{aR1}$, it is necessary to use the new conditions of the form

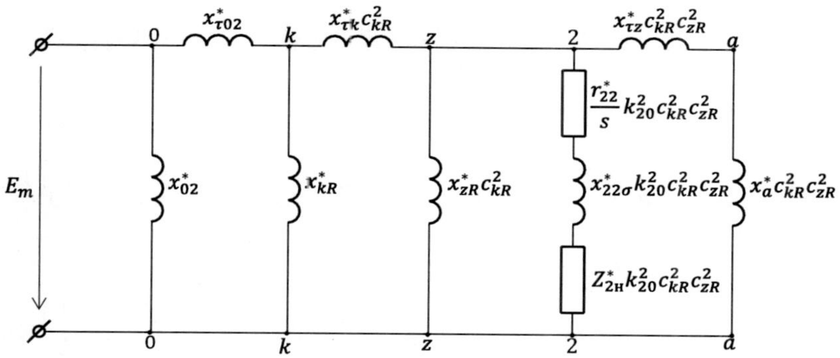

Fig. 16.4 "Unfolded" wound rotor equivalent circuit with the mixed connection of the elements

$$\dot{E}''_{zR2} = \dot{E}''_{aR1} \quad \dot{H}''_{zR2} = \dot{H}''_{aR1} \tag{16.9}$$

where $\dot{E}''_{zR2} = \dot{E}_{zR2}c_{kR}c_{zR}; \dot{E}''_{aR1} = \dot{E}_{aR1}c_{kR}c_{zR}$
$\dot{H}''_{zR2} = \dot{H}_{zR2}/c_{kR}c_{zR}; \dot{H}''_{aR1} = \dot{H}_{aR1}/c_{kR}c_{zR}$.

With respect to the values $\dot{E}''_{aR1}$ and $\dot{H}''_{aR1}$, the input reactance of the equivalent circuit of the two-terminal network Π_{aR} is defined as $x'_{aR} = \dot{E}''_{aR1}/\dot{H}''_{aR1} = (\dot{E}_{aR1}/\dot{H}_{aR1})c^2_{kR}c^2_{zR} = x_{aR}c^2_{kR}c^2_{zR}$, where $x_{aR} = \dot{E}_{aR1}/\dot{H}_{aR1}$ is the reactance of the equivalent circuit of the two-terminal network Π_{aR} determined with respect to the values $\dot{E}_{aR1}$ and $\dot{H}_{aR1}$. Therefore, under the conditions of the transition to the new values $\dot{E}''_{aR1}$ and $\dot{H}''_{aR1}$, the reactance of the equivalent circuit of the two-terminal network Π_{aR} is multiplied by the product of the squared coefficients $(c^2_{kR}c^2_{zR})$, i.e. $x^*_{aR}c^2_{kR}c^2_{zR}$. The L-shaped equivalent circuit of the four-terminal network Π_{zR} and the equivalent circuit of the two-terminal network Π_{aR} connected in a cascade, obtain the form shown in Fig. 16.3b.

In the cascade equivalent circuit in Fig. 16.3b, we passed to the new boundary conditions, determined in accordance with (16.1), (16.6) and (16.9) in the form

$$\begin{aligned} \dot{E}_{02} &= \dot{E}_{kR1} & \dot{H}_{02} &= H_{kR1} \\ \dot{E}'_{kR2} &= \dot{E}'_{zR1} & \dot{H}'_{kR2} &= \dot{H}'_{zR1} \\ \dot{E}''_{zR2} &= \dot{E}''_{aR1} & \dot{H}''_{zR2} &= \dot{H}''_{aR1} \end{aligned} \tag{16.10}$$

The use of these conditions allows you to cascade the L-shaped equivalent circuits used in Fig. 16.3b. As a result of these connections, the equivalent circuit of the wound rotor takes the form in Fig. 16.4. This equivalent circuit is the "unfolded" wound rotor equivalent circuit with the mixed connection of the elements.

For the elements used in the equivalent circuit in Fig. 16.4, we can have

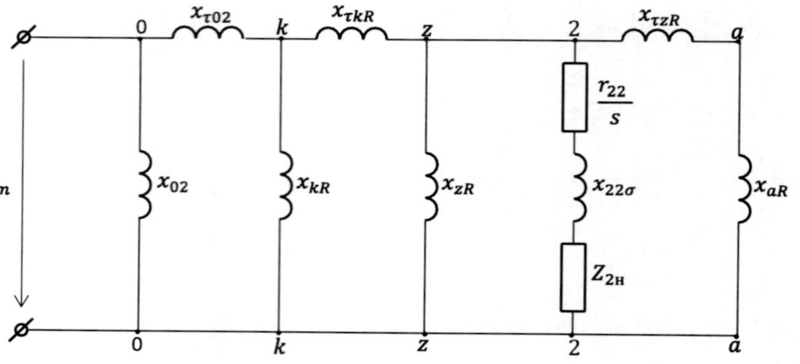

Fig. 16.5 "Unfolded" wound rotor equivalent circuit with the mixed connection of the elements presented in the simplified form

$$\begin{gathered} x_{02} = x_{02}^{*}; x_{kR} = x_{kR}^{*}; x_{zR} = x_{zR}^{*} c_{kR}^{2}; x_{aR} = x_{aR}^{*} c_{kR}^{2} c_{zR}^{2} \\ x_{\tau 02} = x_{\tau 02}^{*}; x_{\tau kR} = x_{\tau kR}^{*} c_{kR}^{2}; x_{\tau zR} = x_{\tau zR}^{*} c_{kR}^{2} c_{zR}^{2} \\ r_{22}/s = \left(r_{22}^{*}/s\right) k_{20}^{2} c_{kR}^{2} c_{zR}^{2}; x_{22\sigma} = x_{22\sigma}^{*} k_{20}^{2} c_{kR}^{2} c_{zR}^{2}; Z_{2\text{H}} = Z_{2\text{H}}^{*} k_{20}^{2} c_{kR}^{2} c_{zR}^{2} \end{gathered} \tag{16.11}$$

When using the expressions given in (16.11), the equivalent circuit in Fig. 16.4 takes the form in Fig. 16.5. The elements of the equivalent circuit in Fig. 16.5 have the simplified view, which facilitates an analysis.

As follows from Figs. 13.6 and 16.5, the wound rotor equivalent circuit built on the basis of the resulting magnetic field, obtained as a result of the superposition of the magnetizing and the slot leakage magnetic fields, as well as the equivalent circuit of the wound rotor built using the "field" concept have the same structure. The equivalent circuit of the wound rotor in Fig. 13.6 was obtained without taking into account the distribution of the resulting magnetic field in the cross section of the rotor. The equivalent circuit of the wound rotor in Fig. 16.5 is built taking into account the distribution of the resulting magnetic field in the cross section of the rotor. Therefore, the elements of the equivalent circuit in Fig. 16.5 contain the corresponding coefficients that take into account the distribution of the magnetic field in the cross section of the rotor.

16.4 "Unfolded" Equivalent Circuit with Parallel Connection of Elements of Wound Rotor

The wound rotor equivalent circuit with the parallel connection of the elements results from the transformation of the "unfolded" wound rotor equivalent circuit with the mixed connection of the elements (Fig. 16.5). The "unfolded" equivalent circuit in Fig. 16.5 completely repeats the structure of the "unfolded" single-cage rotor equivalent circuit shown in Fig. 15.10. Therefore, the transformation of the equivalent circuit in Fig. 16.5 can be carried out using the techniques and procedures applied in the conversion of the single-cage rotor equivalent circuit given in Fig. 15.10 (Appendix I). The transformation of the "unfolded" equivalent circuit in Fig. 16.5 (as well as the "unfolded" equivalent circuit in Fig. 15.10) is reduced to transferring the parallel branches $k - k$, $z - z$, $2 - 2$ and $a - a$ to its terminals $0 - 0$. In accordance with the equivalent circuit in Fig. 16.5, these parallel branches contain the elements shown as x_{kR}, x_{zR}, $(r_{22}/s + jx_{22\sigma} + Z_{2H})$ and x_{aR}. In the equivalent circuit in Fig. 16.5, the transfer process can be started by transferring the parallel branch $k - k$ to its terminals $0 - 0$. The transfer factor for this parallel branch is

$$c_{\tau kR} = 1 + \frac{x_{\tau 02}}{x_{kR}} \tag{16.12}$$

The reactance of the transferred parallel branch $k - k$ is defined as

$$\overset{\circ}{x}_{kR} = x_{kR}c_{\tau kR} = x_{kR} + x_{\tau 02} \tag{16.13}$$

The transfer factor of the parallel branch $z - z$ becomes.

$$c_{\tau zR} = 1 + \frac{x_{\tau kR}c_{\tau kR}^2 + x_{\tau 02}c_{\tau kR}}{x_{zR}c_{\tau kR}^2} \tag{16.14}$$

The reactance of the transferred parallel branch $z - z$.

$$\overset{\circ}{x}_{zR} = x_{zR}c_{\tau zR}c_{\tau kR}^2 = x_{zR}c_{\tau kR}^2 + x_{\tau kR}c_{\tau kR}^2 + x_{\tau 02}c_{\tau kR} \tag{16.15}$$

The transfer factor of the parallel branch $2 - 2$

$$c_{\tau 2} = 1 + \frac{x_{\tau kR}c_{\tau kR}^2c_{\tau zR} + x_{\tau 02}c_{\tau kR}c_{\tau zR}}{(r_{22}/s + jx_{22\sigma} + Z_{2\text{H}})c_{\tau kR}^2c_{\tau zR}^2} \tag{16.16}$$

The impedance of the parallel branch $2 - 2$

$$\begin{aligned} \overset{\circ}{Z}_2 = (r_{22}/s)c_{\tau kR}^2c_{\tau zR}^2 + j\left(x_{22\sigma}c_{\tau kR}^2c_{\tau zR}^2 + x_{\tau kR}c_{\tau kR}^2c_{\tau zR} + x_{\tau 02}c_{\tau kR}c_{\tau zR}\right) \\ + Z_{2\text{H}}c_{\tau kR}^2c_{\tau zR}^2 = (r_{22}/s)c_{\tau kR}^2c_{\tau zR}^2 + jx_{2\sigma} + Z_{2\text{H}}c_{\tau kR}^2c_{\tau zR}^2 \end{aligned} \tag{16.17a}$$

where $x_{2\sigma} = x_{22\sigma}c_{\tau kR}^2c_{\tau zR}^2 + x_{\tau kR}c_{\tau kR}^2c_{\tau zR} + x_{\tau 02}c_{\tau kR}c_{\tau zR}$.

Or in another form

$$\begin{aligned} \overset{\circ}{Z}_2 = \left\{(r_{22}/s)c_{\tau kR}^2c_{\tau zR}^2 + j[x_{22\sigma} + x_{\tau kR}(1/c_{\tau zR}) + x_{\tau 02}(1/c_{\tau kR}c_{\tau zR})] + Z_{2\text{H}}\right\}c_{\tau kR}^2c_{\tau zR}^2 \\ = [(r_{22}/s) + j\widetilde{x}_{22\sigma} + Z_{2\text{H}}]c_{\tau kR}^2c_{\tau zR}^2 \end{aligned} \tag{16.17b}$$

where $\widetilde{x}_{22\sigma} = x_{22\sigma} + x_{\tau kR}(1/c_{\tau zR}) + x_{\tau 02}(1/c_{\tau kR}c_{\tau zR})$; $x_{2\sigma} = \widetilde{x}_{22\sigma}c_{\tau kR}^2c_{\tau zR}^2$.

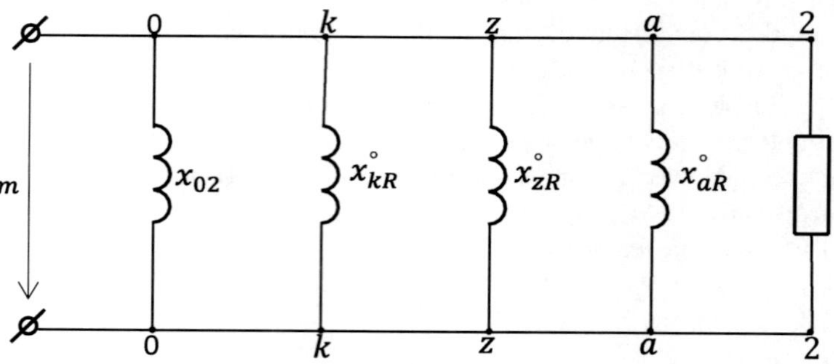

Fig. 16.6 “Unfolded” wound rotor equivalent circuit with the parallel connection of the elements

The reactance of the parallel branch $a - a$

$$\overset{\circ}{x}_{aR} = x_{aR}c^2_{\tau kR}c^2_{\tau zR} + x_{\tau zR}c^2_{\tau kR}c^2_{\tau zR} + x_{\tau kR}c^2_{\tau kR}c_{\tau zR} + x_{\tau 02}c_{\tau kR}c_{\tau zR} \tag{16.18}$$

The expression (16.18), obtained for the reactance $\overset{\circ}{x}_{aR}$ corresponds to the condition when the load impedance value $Z_{2н}$ significantly exceeds the value of the impedance of the rotor winding, i.e., $|Z_{2н}| > |r_{22}/s + jx_{22\sigma}|$. If this condition is met for the transfer factor $c_{\tau 2}$ determined by (16.16), we have that $c_{\tau 2} \approx 1.0$. The expression (16.18) corresponds to this condition.

When using the expressions obtained in (16.13), (16.15), (16.17a), and (16.18), the “unfolded” wound rotor equivalent circuit with the mixed connection of the elements (Fig. 16.5) takes the form of the equivalent circuit with the parallel connection of the elements (Fig. 16.6).

Let us express the reactances $\overset{\circ}{x}_{kR}, \overset{\circ}{x}_{zR}, \overset{\circ}{x}_{aR}$ and the impedance $\overset{\circ}{Z}_2$, , used in the equivalent circuit in Fig. 16.6, through the reactances of the equivalent circuit in Fig. 16.4. Based on (16.11) and (16.13), for the reactance of the parallel branch $k - k$, we have

$$\overset{\circ}{x}_{kR} = x^*_{kR} + x^*_{\tau 02} \tag{16.19}$$

The reactance of the parallel branch $z - z$ from (16.11) and (16.15)

$$\overset{\circ}{x}_{zR} = x^*_{zR}c^2_{kR}c^2_{\tau kR} + x^*_{z\tau R}c^2_{kR}c^2_{\tau kR} + x^*_{\tau kR}c^2_{kR}c^2_{\tau kR} + x^*_{\tau 02}c_{\tau kR} \tag{16.20}$$

The impedance of the parallel branch 2 – 2 proceeding (16.11) and (16.17a)

$$\begin{aligned}\overset{\circ}{Z}_2 &= (r^*_{22}/s)k^2_{20}c^2_{kR}c^2_{zR}c^2_{\tau kR}c^2_{\tau zR} + j\big(x^*_{22\sigma}k^2_{20}c^2_{kR}c^2_{zR}c^2_{\tau kR}c^2_{\tau zR} + x^*_{\tau kR}c^2_{kR}c^2_{\tau kR} \\ &\quad \times c_{\tau zR} + x^*_{\tau 02}c_{\tau kR}c_{\tau zR}\big) + Z^*_{2\text{H}}k^2_{20}c^2_{kR}c^2_{zR}c^2_{\tau kR}c^2_{\tau zR} \\ &= (r^*_{22}/s)k^2_{20}c^2_{kR}c^2_{zR}c^2_{\tau kR}c^2_{\tau zR} + jx^*_{2\sigma} + Z^*_{2\text{H}}k^2_{20}c^2_{kR}c^2_{zR}c^2_{\tau kR}c^2_{\tau zR}\end{aligned} \tag{16.21a}$$

where $x^*_{2\sigma} = x^*_{22\sigma}k^2_{20}c^2_{kR}c^2_{zR}c^2_{\tau kR}c^2_{\tau zR} + x^*_{\tau kR}c^2_{kR}c^2_{\tau kR}c_{\tau zR} + x^*_{\tau 02}c_{\tau kR}c_{\tau zR}$.

In another form

$$\begin{aligned}\overset{\circ}{Z}_2 &= \{r^*_{22}/s + j[x^*_{22\sigma} + x^*_{\tau kR}(1/k^2_{20}c^2_{zR}c_{\tau zR}) \\ &\quad + x^*_{\tau 02}(1/k^2_{20}c^2_{kR}c^2_{zR}c_{\tau kR}c_{\tau zR})] + Z^*_{2\text{H}}\}k^2_{20}c^2_{kR}c^2_{zR}c^2_{\tau kR}c^2_{\tau zR} \\ &= (r^*_{22}/s + \widetilde{x}^*_{22\sigma} + Z^*_{2\text{H}})k^2_{20}c^2_{kR}c^2_{zR}c^2_{\tau kR}c^2_{\tau zR}\end{aligned} \tag{16.21b}$$

where $\widetilde{x}^*_{22\sigma} = x^*_{22\sigma} + x^*_{\tau kR}(1/k^2_{20}c^2_{zR}c_{\tau zR}) + x^*_{\tau 02}(1/k^2_{20}c^2_{kR}c^2_{zR}c_{\tau kR}c_{\tau zR})$

$$x^*_{2\sigma} = \widetilde{x}^*_{22\sigma}k^2_{20}c^2_{kR}c^2_{zR}c^2_{\tau kR}c^2_{\tau zR}$$

The reactance of the parallel branch $a - a$ from (16.11) and (16.18)

$$\begin{aligned}\overset{\circ}{x}_{aR} &= x^*_{aR}c^2_{kR}c^2_{zR}c^2_{\tau kR}c^2_{\tau zR} + x^*_{\tau z}c^2_{kR}c^2_{zR}c^2_{\tau kR}c^2_{\tau zR} \\ &\quad + x^=_{\tau kR}c^2_{kR}c^2_{\tau kR}c_{\tau z} + x^*_{\tau 02}c_{\tau kR}c_{\tau z}\end{aligned} \tag{16.22}$$

In accordance with (16.19)–(16.22), the reactances used in the equivalent circuit in Fig. 16.6 are expressed through the reactances of the L-shaped equivalent circuits representing the layers of the wound rotor model. The analytical expressions for these reactances were obtained above in (15.33)–(15.37).

16.5 Magnetizing and Slot Leakage Circuit Loops: "Folded" Equivalent Circuit of Wound Rotor

Here, to receive the magnetizing and the slot leakage circuit loops, as well as to construct the wound rotor equivalent circuit in the form shown in Fig. 12.4b, we will use the equivalent circuit obtained in Fig. 16.6.

Magnetizing circuit loop and magnetizing reactance of the wound rotor. In the equivalent circuit in Fig. 16.6, the parallel branches $0 - 0$, $k - k$, $z - z$ и $a - a$ represent the magnetizing circuit loops reflecting the layers of the wound rotor model (Fig. 16.1). In this regard, they can be combined into the one common rotor magnetizing circuit loop. The circuit loop obtained in this way will contain the magnetizing reactance of the wound rotor. For the magnetizing reactance of the wound rotor, based on the equivalent circuit in Fig. 16.6, we can have

$$x^*_{m2} = \frac{1}{\frac{1}{x_{02}} + \frac{1}{\overset{\circ}{x}_{kR}} + \frac{1}{\overset{\circ}{x}_{zR}} + \frac{1}{\overset{\circ}{x}_{aR}}} = \frac{x_{02}}{1 + x_{02}\left(\frac{1}{\overset{\circ}{x}_{kR}} + \frac{1}{\overset{\circ}{x}_{zR}} + \frac{1}{\overset{\circ}{x}_{aR}}\right)} = \frac{x_{02}}{k_{H2}} \tag{16.23}$$

where $k_{H2} = 1 + x_{02}\left(\frac{1}{\overset{\circ}{x}_{kR}} + \frac{1}{\overset{\circ}{x}_{zR}} + \frac{1}{\overset{\circ}{x}_{aR}}\right)$.

Here, the reactance x^*_{m2} is the magnetizing reactance of the wound rotor, and the value k_{H2} reflects the saturation factor of the magnetic circuit of the wound rotor.

Slot leakage circuit loop and slot leakage reactance of the wound rotor. The parallel branch 2 − 2 of the equivalent circuit in Fig. 16.6 represents the slot leakage circuit loop of the wound rotor. In this rotor circuit loop, the impedance $\overset{\circ}{Z}_2$ is used, the value of which is determined by the expression obtained in (16.21a). In (16.21a), we introduce the following coefficients

$$k^2_{22} = k^2_{20} c^2_{kR} c^2_{zR} c^2_{\tau kR} c^2_{\tau zR};\, k_{21} = c^2_{kR} c^2_{\tau kR} c_{\tau zR};\, k_{20} = c_{\tau kR} c_{\tau zR} \tag{16.24}$$

From (16.21a)and (16.24), for the parallel branch 2 − 2, we can represent the impedance $\overset{\circ}{Z}_2$ in the form

$$\overset{\circ}{Z}_2 = \frac{r^*_{22}}{s} k^2_{22} + j\left(x^*_{\tau 02} k_{20} + x^*_{\tau kR} k_{21} + x^*_{22\sigma} k^2_{22}\right) + Z^*_{2H} k^2_{22} \tag{16.25}$$

In (16.25), for the ohmic resistance r^*_2/s used in the rotor slot leakage circuit loop, we have

$$\frac{r^*_2}{s} = \frac{r^*_{22}}{s} k^2_{22} \tag{16.26}$$

The slot leakage reactance of the wound rotor $x^*_{2\sigma}$ is determined from (16.21a), (16.24), and (16.25) as

$$x^*_{2\sigma} = x^*_{\tau 02} k_{20} + x^*_{\tau kR} k_{21} + x^*_{22\sigma} k^2_{22} \tag{16.27}$$

From (16.27) it follows that the reactance $x^*_{2\sigma}$ consists of the three components: the leakage reactance of the air gap sublayer adjoined to the rotor surface ($x^*_{\tau 02} k_{20}$), the leakage reactance of the layer reflecting the rotor slot-wedge area ($x^*_{\tau kR} k_{21}$), and the leakage reactance of the layer representing the rotor winding location area ($x^*_{22\sigma} k^2_{22}$). As follows from (16.27), the reactances $x^*_{\tau 02} k_{20}, x^*_{\tau kR} k_{21}$ and $x^*_{22\sigma} k^2_{22}$ are expressed through the reactances $x^*_{\tau 02}, x^*_{\tau kR}$ and $x^*_{22\sigma}$ determined in accordance with (15.33), (15.34), and (15.36) in the form

$$\begin{aligned} x^*_{\tau 02} &= \omega_1 \mu_0 t_0 \frac{\delta_2}{2t_{02}} k_{2\sigma} k_R \zeta_{\tau 0} \\ x^*_{\tau kR} &= \omega_1 \mu_{xkR} t_0 \frac{h_{kR}}{t_{kR}} k_R \xi_{LkR} \zeta_{\tau kR} \\ x^*_{22\sigma} &= \omega_1 \mu_{xzR} t_0 \frac{h_{zR}}{3t_{zR}} k_R \xi_{LzR} \end{aligned} \tag{16.28}$$

In (16.28), the quantities μ_{xkR} and μ_{xzR} can be expressed as

$$\mu_{xkR} = \mu_0 \frac{t_{kR}}{b_{kR}} \quad \mu_{xzR} = \mu_0 \frac{t_{zR}}{b_{\text{п}R}} \tag{16.29}$$

Then, from (16.28) and (16.29), the reactances $x^*_{\tau 02} k_{20}, x^*_{\tau kR} k_{21}$ and $x^*_{22\sigma} k^2_{22}$ acquire the form

$$\begin{aligned} x^*_{\tau 02} k_{20} &= \omega_1 \mu_0 t_0 \frac{\delta_2}{2t_{02}} k_{2\sigma} k_R \zeta_{\tau 0} k_{20} = \omega_1 \mu_0 t_0 \lambda_{\tau 02} \\ x^*_{\tau kR} k_{21} &= \omega_1 \mu_0 t_0 \frac{h_{kR}}{b_{kR}} k_R \xi_{LkR} \zeta_{\tau kR} k_{21} = \omega_1 \mu_0 t_0 \lambda_{\tau kR} \\ x^*_{22\sigma} k^2_{22} &= \omega_1 \mu_0 t_0 \frac{h_{zR}}{3b_{\text{п}R}} k_R \xi_{LzR} k^2_{22} = \omega_1 \mu_0 t_0 \lambda_{22\sigma} \end{aligned} \tag{16.30}$$

where $\lambda_{\tau 02} = \dfrac{\delta_2}{2t_{02}} k_{2\sigma} k_R \zeta_{\tau 0} k_{20}; \lambda_{\tau kR} = \dfrac{h_{kR}}{b_{kR}} k_R \xi_{LkR} \zeta_{\tau kR} k_{21}; \lambda_{22\sigma} = \dfrac{h_{zR}}{3b_{\text{п}R}} k_R \xi_{LzR} k^2_{22}$.

In (16.30), the dimensionless quantities $\lambda_{\tau 02}$, $\lambda_{\tau kR}$ and $\lambda_{22\sigma}$ represent the relative magnetic conductivity for the tangential components of the magnetic fluxes in the layers reflecting the air gap sublayer adjoined to the rotor surface, the rotor slot-wedge area, and the rotor winding location area, respectively.

Now, from (16.27) and (16.30), for the reactance $x^*_{2\sigma}$, we can obtain

$$\begin{aligned} x^*_{2\sigma} &= x^*_{\tau 02} k_{20} + x^*_{\tau kR} k_{21} + x^*_{22\sigma} k^2_{22} \\ &= \omega_1 \mu_0 t_0 (\lambda_{\tau 02} + \lambda_{\tau kR} + \lambda_{22\sigma}) = \omega_1 \mu_0 t_0 \lambda_{2\sigma} \end{aligned} \tag{16.31}$$

where $\lambda_{2\sigma} = \lambda_{\tau 02} + \lambda_{\tau kR} + \lambda_{22\sigma}$.

Here, the quantity $\lambda_{2\sigma}$ reflects the relative magnetic conductivity for the slot leakage magnetic flux of the rotor wound. As follows from (15.22a) and (16.31), the single-cage rotor and the wound rotor have the same value of the coefficient $\lambda_{2\sigma}$ at the weak skin effect.

"Folded" wound rotor equivalent circuit. In the Eq. (16.25) for the impedance $Z^*_{2н}$, we can use the condition of the form $\widetilde{U}^*_2/s = \widetilde{Z}^*_{2н} \widetilde{H}_2$, where $\widetilde{U}^*_2 = \dot{U}_2 k_{22}$ and $\widetilde{H}_2 = \dot{H}_2 / k_{22}$, and also $\widetilde{Z}^*_{2н} = Z^*_{2н} k^2_{22}$. For the electric machines of the normal design, the value of the coefficient k_{22} defined as $k_{22} = k_{20} c_{kR} c_{zR} c_{\tau kR} c_{\tau zR}$ can be taken as the one, i.e., $k_{22} \approx 1.0$. Then, the condition for impedance $Z^*_{2н}$ can be represented in the form $\dot{U}^*_2/s = Z^*_{2н} \dot{H}_2$. Now, taking into account the expressions (16.23), (16.25), (16.26), and (16.31), the equivalent circuit in Fig. 16.6 can be reduced to the

Fig. 16.7 "Folded" wound rotor equivalent circuit

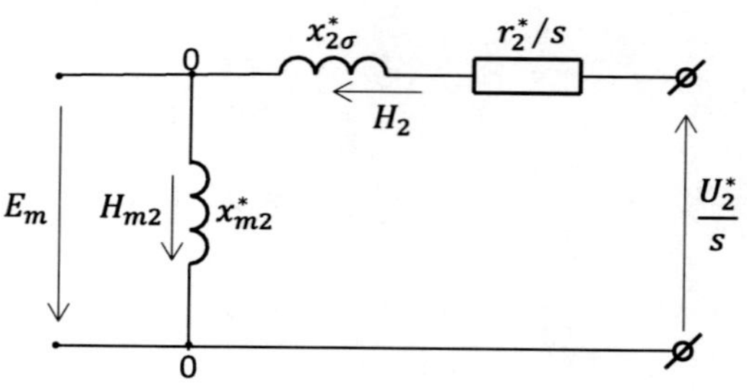

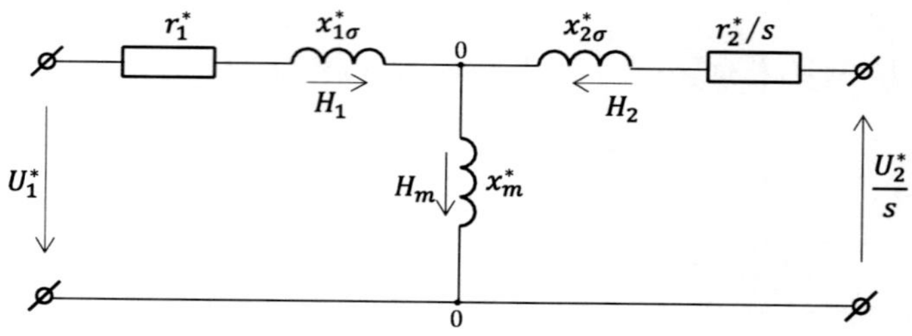

Fig. 16.8 Machine equivalent circuit with the wound rotor

"folded" form in Fig. 16.7. As follows from the "folded" equivalent circuit in Fig. 16.7, the wound rotor of the machine is represented in relation to $\dot{E}_m$ in the form of the two independent circuit loops: the magnetizing circuit loop containing the reactance x^*_{m2} and the slot leakage circuit loop including the reactance $x^*_{2\sigma}$. The "folded" rotor equivalent circuit obtained in this way has the simple structure and is convenient for the practical use.

16.6 Equivalent Circuit of Machine with Wound Rotor: Voltage Equations

The equivalent circuit of the machine with the wound rotor can be obtained using the stator equivalent circuit shown in Fig. 15.6 and the rotor equivalent circuit shown in Fig. 16.7. For this, the terminals 0 – 0 of these equivalent circuits should be connected and the magnetizing reactance of the machine defined as $x^*_m = x^*_{m1}x^*_{m2}/x^*_{m1} + x^*_{m2}$ should be entered. As a result, the equivalent circuit of the machine with the wound rotor takes the form in Fig. 16.8. Using the equivalent circuit in Fig. 16.8, the voltage equations of the machine can be received. Obtained in this way voltage equations correspond to the "no-load and under-load" concept. Then, we can have

$$
\begin{aligned}
\dot{U}_1^* &= \left(r_1^* + jx_{1\sigma}^*\right)\dot{H}_1 + (-\dot{E}_m) \\
-\frac{\dot{U}_2^*}{s} &= \left(\frac{r_2^*}{s} + jx_{2\sigma}^*\right)\dot{H}_2 + (-\dot{E}_m)
\end{aligned} \tag{16.32}
$$

Here, $(-\dot{E}_m) = jx_m^*\dot{H}_m$ and $\dot{H}_m = \dot{H}_1 + \dot{H}_2$. We will use these conditions in the system of the Eq. (16.32). Then, we obtain the voltage equations corresponding to the "inductively coupled circuits" concept

$$
\begin{aligned}
\dot{U}_1^* &= r_1^*\dot{H}_1 + j\left(x_{1\sigma}^* + x_m^*\right)\dot{H}_1 + jx_m^*\dot{H}_2 = r_1^*\dot{H}_1 + jx_1^*\dot{H}_1 + jx_{21}^*\dot{H}_2 \\
-\frac{\dot{U}_2^*}{s} &= \frac{r_2^*}{s}\dot{H}_2 + jx_m^*\dot{H}_1 + j\left(x_{2\sigma}^* + x_m^*\right)\dot{H}_2 = \frac{r_2^*}{s}\dot{H}_2 + jx_{12}^*\dot{H}_1 + jx_2^*\dot{H}_2
\end{aligned} \tag{16.33}
$$

where $x_1^* = x_{1\sigma}^* + x_m^*; x_2^* = x_{2\sigma}^* + x_m^*; x_{21}^* = x_m^*; x_{12}^* = x_m^*$.

16.7 Summary

In this section of the work, the scheme for determining the basic calculation elements of the asynchronous machine with the wound rotor is considered. In this case, the "field" concept is applied. The proposed in this chapter scheme is similar to the scheme developed in Chap. 15 to define the basic calculation elements of the electric machine with the single-cage rotor and is based on the use of the machine resulting magnetic field, determined applying the Maxwell's equations.

The "field" concept is realized using the superposition principle, the "fragmentation" principle, the "no-load and under-load concept, the cascade way of connecting, and the multiple transfer technique. The "fragmentation" principle is used to decompose the cross section of the wound rotor into the fragments (layers) representing the basic elements of the rotor design. The fragments (layers) of the wound rotor design are the air gap sublayer adjoined to the rotor surface, the layer reflecting the slot-wedge area, the layer representing the rotor winding location area, and the layer of the rotor yoke area. In this regard, the wound rotor model is considered as the layered system. In this case, the slot-teeth areas are represented in the form of the layers as a result of the introduction for these rotor areas of the average values of the magnetic permeabilities μ_{xi} and μ_{yi}, where $i = kR, zR$. Since the power supply of the rotor winding from the external source, then, the layer reflecting the rotor winding location area is represented as the active layer.

The layers of the wound rotor model are represented in the form of the L-shaped equivalent circuits, for the construction of which the "no-load and under-load" concept is used. In accordance with the rotor model, the L-shaped equivalent circuits are connected in a cascade, as a result of which the "unfolded" wound rotor equivalent circuit with the mixed connection of the elements follows. Further, the "unfolded" rotor equivalent circuit with the mixed connection of the elements is reduced to the equivalent circuit with the parallel connection of the elements, for which the multiple transfer technique is used. On the basis of the consideration of the rotor equivalent circuit with the parallel connection of the elements, the magnetizing circuit loop and magnetizing reactance and also the slot leakage circuit loop and the slot leakage reactance are introduced. As a result, the "folded" wound rotor equivalent circuit is constructed. The "folded" wound rotor equivalent circuit and the "folded" stator equivalent circuit are used to construct the machine equivalent circuit and to draw up the machine voltage equations. Thus, in this case, the determination of the calculation elements of the machine with the wound rotor is achieved using the "field" concept.

References

1. Asanbayev, V.: Alternating Current Multi-Circuit Electric Machines: A New Approach to Steady-State Parameter Determination, Springer Cham Heidelberg, New York, Dordrecht London (2015)
2. Mishkin, E.: Theory of the Squirrel-Cage Induction Machine Derived Directly from Maxwell's Field Equations, The Quarterly Journal of Mechanics and Applied Mathematics, Vol.7, Part 4, pp. 472-487, (1954)
3. Veselovsky, O., Konyaev, A., Sarapulov, F.: Linear Asynchronous Motor, Publishing Energoatomizdat, Moscow (1991)
4. Dombrovsky, V., Zaichik, V.: Asynchronous Machines: Theory, Calculation, Elements of Design, Publishing Energoatomizdat, Leningrad (1990)

Chapter 17
To Development of Improved Engineering Calculation Methods: Features of Applying Field Equations

17.1 EMF Equations

In practice, the calculation of the asynchronous machine is based on the use of Kirchhoff's equations. Kirchhoff's equations can be represented in the following form

$$\begin{aligned} \dot{U}_1 + \Im_{1p} &= r_1 \dot{I}_1 \\ \left(-\frac{\dot{U}_2}{s}\right) + \dot{\Im}_{2p} &= \frac{r_2}{s} \dot{I}_2 \end{aligned} \tag{17.1}$$

The asynchronous machines are characterized by the property of the emf induction in the stator and rotor windings. Taking this into account, we represent the system of Eq. (17.1) through the values of the emf $\dot{\Im}_{1p}$ and emf $\dot{\Im}_{2p}$ in the form

$$\begin{aligned} (-\Im_{1p}) &= r_1(-\dot{I}_1) + \dot{U}_1 \\ (-\dot{\Im}_{2p}) &= \frac{r_2}{s}(-\dot{I}_2) + \left(-\frac{\dot{U}_2}{s}\right) \end{aligned} \tag{17.2}$$

In the system of Eq. (17.2), the voltage at the terminals of the stator winding $\dot{U}_1$ is replaced by a voltage drop across the impedance of the power supply network Z_c, i.e., $\left(-\dot{U}_1\right) = Z_c \dot{I}_1$, and the voltage at the terminals of the rotor winding $\dot{U}_2/s$ is replaced by a voltage drop across the load impedance $Z_{2н}$, i.e., $\dot{U}_2/s = Z_{2н}\dot{I}_2$. Then, the system of Eq. (17.2) can be represented as

V. Asanbayev, *Asynchronous Machines*,
https://doi.org/10.1007/978-3-030-92284-9_17

$$
\begin{aligned}
(-\dot{\text{Э}}_{1p}) &= (r_1 + Z_c)(-\dot{I}_1) \\
(-\dot{\text{Э}}_{2p}) &= \left(\frac{r_2}{s} + Z_{2\text{н}}\right)(-\dot{I}_2)
\end{aligned}
\tag{17.3}
$$

The ohmic resistances of the stator and rotor windings, as well as the impedance of the power supply circuit, and the load impedance used in the right side of the system of the Eq. (17.3) are assumed to be given, therefore, in this work, these ohmic resistances and the impedances are not specifically considered. In (17.3), the subject of a further analysis is the values of the emf $\dot{\text{Э}}_{1p}$ and emf $\dot{\text{Э}}_{2p}$ induced in the stator and rotor windings by the resulting magnetic field of the machine.

17.2 "Inductively Coupled Circuits" Concept: EMF Equations of Stator and Rotor Windings

In accordance with the "inductively coupled circuits" concept, the emf $\dot{\text{Э}}_{1p}$ and emf $\dot{\text{Э}}_{2p}$ used in the system of the Eq. (17.3) can be represented in the following form

$$
\begin{aligned}
\dot{\text{Э}}_{1p} &= \dot{\text{Э}}_1 + \dot{\text{Э}}_{21} \\
\dot{\text{Э}}_{1p} &= \dot{\text{Э}}_2 + \dot{\text{Э}}_{12}
\end{aligned}
\tag{17.4}
$$

Here, the quantities $\dot{\text{Э}}_1$ and $\dot{\text{Э}}_2$ represent the emfs of the full self-induction of the stator and rotor windings, respectively. In (17.4), the quantities $\dot{\text{Э}}_{21}$ and $\dot{\text{Э}}_{12}$ reflect the emfs of the mutual induction of the stator and rotor windings, respectively. Taking into account the above, the system of the Eq. (17.4) can be represented in the form

$$
\begin{aligned}
(-\dot{\text{Э}}_{1p}) &= jx_1\dot{I}_1 + jx_{21}\dot{I}_2 \\
(-\dot{\text{Э}}_{2p}) &= jx_{12}\dot{I}_1 + jx_2\dot{I}_2
\end{aligned}
\tag{17.5}
$$

The reactances of the full self-induction of the stator and rotor windings x_1 and x_2 given in the system of the Eq. (17.5) can be represented as the sum of the leakage reactances $x_{1\sigma}$ and $x_{2\sigma}$ and the self-induction reactances x_{11} and x_{22} of the stator and rotor windings, respectively, i.e.

$$
\begin{aligned}
x_1 &= x_{1\sigma} + x_{11} \\
x_2 &= x_{2\sigma} + x_{22}
\end{aligned}
\tag{17.6}
$$

Taking into account (17.6), the system of the Eq. (17.5) takes the form

$$\begin{aligned}(-\dot{Э}_{1p}) &= j(x_{1\sigma}+x_{11})\dot{I}_1 + jx_{21}\dot{I}_2 = (-\dot{Э}_{1\sigma}) + (-\dot{Э}_1)\\ (-\dot{Э}_{2p}) &= j(x_{2\sigma}+x_{22})\dot{I}_2 + jx_{12}\dot{I}_1 = (-\dot{Э}_{2\sigma}) + (-\dot{Э}_2)\end{aligned} \quad (17.7)$$

where $\begin{aligned}&(-\dot{Э}_{1\sigma}) = jx_{1\sigma}\dot{I}_1; (-\dot{Э}_1) = jx_{11}\dot{I}_1 + jx_{21}\dot{I}_2\\ &(-\dot{Э}_{2\sigma}) = jx_{2\sigma}\dot{I}_2; (-\dot{Э}_2) = jx_{22}\dot{I}_2 + jx_{12}\dot{I}_1.\end{aligned}$

In (17.7), the quantities $\dot{Э}_{1\sigma}$ and $\dot{Э}_{2\sigma}$ reflect the leakage emfs induced in the stator and rotor windings. The emf $\dot{Э}_{1\sigma}$ and emf $\dot{Э}_{2\sigma}$ are expressed through the leakage reactances of the stator and rotor windings $x_{1\sigma}$ and $x_{2\sigma}$ for which we can have from (17.5) to (17.7)

$$\begin{aligned}x_{1\sigma} &= x_1 - x_{12} = \frac{\omega_1(\Psi_1 - \Psi_{12})}{\dot{I}_1} = \frac{\omega_1\Psi_{1\sigma}}{\dot{I}_1}\\ x_{2\sigma} &= x_2 - x_{21} = \frac{\omega_1(\Psi_2 - \Psi_{21})}{\dot{I}_2} = \frac{\omega_1\Psi_{2\sigma}}{\dot{I}_2}\end{aligned} \quad (17.8)$$

where $\Psi_{1\sigma} = \Psi_1 - \Psi_{12}$; $\Psi_{2\sigma} = \Psi_2 - \Psi_{21}$.

Using the conditions obtained in (17.7), the system of the Eq. (17.3) can be reduced to the form

$$\begin{aligned}(-\dot{Э}_1) &= (jx_{1\sigma} + r_1 + Z_c)(-\dot{I}_1)\\ (-\dot{Э}_2) &= \left(jx_{2\sigma} + \frac{r_2}{s} + Z_{2\text{н}}\right)(-\dot{I}_2)\end{aligned} \quad (17.9)$$

On the other hand, for the emf $\dot{Э}_1$ and emf $\dot{Э}_2$, we can have from (17.7)

$$\begin{aligned}(-\dot{Э}_1) &= jx_{11}\dot{I}_1 + jx_{21}\dot{I}_2\\ (-\dot{Э}_2) &= jx_{22}\dot{I}_2 + jx_{12}\dot{I}_1\end{aligned} \quad (17.10)$$

where $x_{11} = \omega_1\dot{\Psi}_{11}/\dot{I}_1; x_{22} = \omega_1\dot{\Psi}_{22}/\dot{I}_2; x_{21} = \omega_1\dot{\Psi}_{21}/\dot{I}_2; x_{12} = \omega_1\dot{\Psi}_{12}/\dot{I}_1$.

When obtaining the Eq. (17.10), it was assumed that the magnetic permeabilities of the steel of the stator and rotor magnetic circuits are relatively large, i.e., $\mu_c = \infty$. Under these conditions, the magnetic field in the machine exists only in its air gap and is determined as a result of the superposition of the magnetic fields of self-and mutual induction of the stator and rotor windings. In this regard, we can assert that the determination of the reactances x_{11}, x_{12}, x_{21}, and x_{22} used in the Eq. (17.10) is reduced, in this case, to the calculation of the magnetic fields of the self-and mutual induction in the air gap of the machine. In the future, Eq. (17.10) will be considered as a system of the magnetizing emf equations of the air gap of the machine.

17.3 Circuit Visualization of emf Equations

According to (17.9) and (17.10), Kirchhoff's equations given in (17.3) are represented by the two systems of the equations written in relation to the emf $\dot{Э}_1$ and emf $\dot{Э}_2$ induced in the stator and rotor windings. Therefore, the Eqs. (17.9) and (17.10) can be considered as the fragments of Kirchhoff's equations shown in (17.3). In the future, for the convenience of an analysis, the Eqs. (17.9) and (17.10) will be represented in the form of the equivalent circuits. The emf equations of the system (17.9) can be represented in the form of the equivalent circuits shown in Fig. 6.4b, c. These equivalent circuits use the ohmic resistances and impedances (r_1 and Z_c), as well as (r_2/s and $Z_{2н}$). As noted above, these resistances are not specifically considered in this work. Then, the equivalent circuits in Fig. 6.4b, c can be considered representing the processes associated with the leakage of the stator and rotor windings. Because of this, Fig. 6.4b conventionally represents the leakage equivalent circuit of the stator winding. Fig. 6.4c conventionally reflects the leakage equivalent circuit of the rotor winding. Consequently, the equivalent circuits in Fig. 6.4b, c represent the sections of the stator and rotor leakage magnetic circuits.

The air gap magnetizing emf equations shown in (17.10) are represented by the magnetizing equivalent circuit of the air gap in Fig. 6.4a. Thus, the equivalent circuits shown in Fig. 6.4 reflect the corresponding fragments of the machine design.

17.4 Machine Equivalent Circuit: "Inductively Coupled Circuits" Concept

The equivalent circuits in Fig. 6.4 are obtained on the basis of the emf equations presenting in (17.9) and (17.10), and reflecting the corresponding fragments of Kirchhoff's equations shown in (17.3). As is customary in the technical literature, Kirchhoff's equations are the initial equations for constructing the machine equivalent circuit. Therefore, the equivalent circuits in Fig. 6.4 representing the fragments of Kirchhoff's equations can be used to construct the machine equivalent circuit. As follows from Fig. 6.4, the equivalent circuit of the machine can be obtained as a result of connecting the input terminals c–d of the leakage equivalent circuit of the rotor (Fig. 6.4c) with the output terminals c–d of the magnetizing equivalent circuit of the air gap (Fig. 6.4a), and also as a result of connecting the input terminals a–b of the magnetizing equivalent circuit of the air gap (Fig. 6.4a) with the input terminals a–b of the leakage equivalent circuit of the stator (Fig. 6.4b). Such connection is known as the cascade connection. Thus, the equivalent circuit of the machine can be constructed as a result of the cascade way of connecting the magnetizing equivalent circuit of the air gap in Fig. 6.4a, the leakage equivalent circuit of the stator in Fig. 6.4b, and the leakage equivalent circuit of the rotor in Fig. 6.4c. The machine equivalent circuit obtained in this way is shown in Fig. 6.5.

The equivalent circuit shown in Fig. 6.5 has two problems. The first problem is associated with the fact that in it the stator and rotor windings of the machine are presented in the form of two inductively coupled circuits. In accordance with (17.8), the leakage reactances of the windings are determined as the difference between the very close quantities. Therefore, the second problem is associated with the difficulties associated with ensuring the required level of reliability in determining the leakage reactances of the stator and rotor windings of the machine.

In this regard, the equivalent circuit in Fig. 6.5 is not suitable for practical use. In order to eliminate these difficulties, the equivalent circuit of the machine in Fig. 6.5 is transformed. The physical meaning of the transformation used in the theory of the electric machines is reduced to the elimination of the inductively coupled elements that take place in the equivalent circuit of the machine in Fig. 6.5, i.e., is reduced to the construction of the machine equivalent circuit which has a simplified structure and contains the leakage reactances amenable to the sufficiently accurate calculation and experimental determination.

In this work, the equivalent circuit of the machine (Fig. 6.5) follows on the basis of the cascade way of connecting the equivalent circuits of the fragments shown in Fig. 6.4. As follows from Fig. 6.4, the inductively coupled elements are used in the magnetizing equivalent circuit of the air gap (Fig. 6.4a). Therefore, we can restrict ourselves to transforming this equivalent circuit and obtaining the equivalent circuit for the air gap without the inductively coupled elements. Now the construction of the equivalent circuit of the machine can be realized using the cascade way of connecting the transformed magnetizing equivalent circuit of the air gap and the leakage equivalent circuits of the stator and rotor. The machine equivalent circuit obtained in this way will not contain the inductively coupled elements and therefore it will have a simplified structure. Below, the construction of the equivalent circuit of the machine using the cascade way of connecting will be considered in more detail, and we will begin this construction process by converting the magnetizing equivalent circuit of the air gap of the machine (Fig. 6.4a).

17.5 No-Load and Under-Load" Concept: Equations for emf and Current of Stator Winding

The operating condition of the machine can be viewed as a result of the superposition of the no-load and load modes. In this regard, the resulting magnetic field of the machine can be represented as a result of the superposition of the magnetizing field that occurs in the machine in the no-load mode and the leakage magnetic fields created by the currents of the stator and rotor windings in the load mode ("no-load and under-load" concept). In this work, the "no-load and under-load" concept is implemented in relation to the emf $\dot{Э}_1$ of the stator winding. In the technical literature, such scheme for the implementation of the "no-load and under-load" concept has become the most widespread.

According to the "no-load and under-load" concept, the stator winding current can be represented as the sum of the no-load current and the load current. In the no-load mode (in the absence of the current in the rotor winding, i.e. $I_2 = 0$), the no-load current $\dot{I}_0$ flows in the stator winding, which, in accordance with the law of the total current, causes the rotating magnetic field in the machine. This rotating magnetic field crosses the windings of the machine and, according to the law of electromagnetic induction, induces the emf $\dot{\text{Э}}_1$ in the stator winding and the emf $\dot{\text{Э}}_2$ in the rotor winding. When the rotor winding is connected to the load impedance the load current I_2 arises in the rotor winding under the action of the emf $\dot{\text{Э}}_2$ induced in it in the no-load mode. This load current I_2 forms the mmf of the rotor winding F_2 which creates the magnetic field of the rotor. The rotor magnetic field induces the emf in the stator winding under the influence of which the load component of the stator winding current $I_{1\text{н}}$ appears in the stator winding. Provided that $\dot{\text{Э}}_1 = const$, the component of the stator winding current $I_{1\text{н}}$ will have such value that created by it the mmf of the stator $F_{1\text{н}}$ completely balances the value of the rotor mmf F_2, i.e., $F_{1\text{н}} = -F_2$. This means that in the load mode, the currents of the stator and rotor windings do not create the common magnetic field of the mutual induction. Thus, in the load mode, the currents of the stator and rotor windings form the leakage magnetic field covering the stator winding and leakage magnetic field coupled to the rotor winding. Therefore, in this case, the resulting magnetic field of the machine will be presented as a result of the superposition of the magnetizing field that occurs in the machine in the no-load mode and the leakage magnetic fields of the stator and rotor windings existing in the machine in the load mode.

In accordance with the "no-load and under-load" concept, the stator winding current I_1 can be represented as consisting of the two components: the no-load current (magnetizing current) I_0 and the load current $I_{1\text{н}}$, i.e. $I_1 = I_0 + I_{1\text{н}}$. The currents $\dot{I}_0$ and $\dot{I}_{1\text{н}}$ used in the equation $I_1 = I_0 + I_{1\text{н}}$ flow in the stator winding, and therefore these currents should be considered caused by the emf of the stator winding $\dot{\text{Э}}_1$. However, physically, the current $\dot{I}_{1\text{н}}$ arises under the action of the emf $\dot{\text{Э}}_2$ induced in the rotor winding in the no-load mode, and the current $\dot{I}_0$ is created under the action of the emf of the stator winding $\dot{\text{Э}}_1$. In practice, for the emfs of the stator and rotor windings, the condition of the form $\dot{\text{Э}}_1 \neq \dot{\text{Э}}_2$ is usually satisfied. Under these conditions, the equation $I_1 = I_0 + I_{1\text{н}}$ obtained for the stator winding currents has no physical meaning, since the currents $\dot{I}_0$ and $\dot{I}_{1\text{н}}$ are caused by emf having different values. This equation acquires a physical meaning when the condition of the form $\dot{\text{Э}}_1 = \dot{\text{Э}}_2$ is satisfied for the emfs of the stator and rotor windings. To implement this condition, the stator and rotor windings must have the same number of turns. However, in practice, the stator and rotor windings are usually made with different numbers of turns. Under these conditions, it is assumed that the calculated winding is located on the rotor. In this case, the calculated rotor winding and the stator winding acquire the same number of turns. In the technical literature, this technique is known as reducing the rotor winding to the number of the turns of the stator winding. For the reduced rotor winding, the reduced (calculated) value of the emf of the rotor winding $\dot{Э}_2'$ is used for which the condition of the form $\dot{Э}_1 = \dot{Э}_2'$ is satisfied.

To determine the load component of the stator winding current $\dot{I}_{1н}$ used in the equation $I_1 = I_0 + I_{1н}$, we use the condition $F_{1н} = -F_2$. Taking this condition into account for the load component of the stator winding current $\dot{I}_{1н}$, we have that $I_{1н} = (-I_2')$. Now, the emf and the stator winding current equations can be finalized as

$$\begin{aligned} \dot{Э}_1 &= \dot{Э}_2' \\ \dot{I}_1 &= \dot{I}_0 + (-\dot{I}_2') \end{aligned} \quad (17.11)$$

In (17.11) the quantity $\dot{I}_2'$ represents the value of the rotor winding current reduced to the stator winding. According to (17.11), the "no-load and under-load" concept is represented by the equations for the emf and current of the stator winding. In this case, the value of the stator winding current $\dot{I}_1$ is considered as a result of the sum of the currents $\dot{I}_0$ and $(-\dot{I}_2')$ reflecting the components of the stator winding current and occurring in the machine in the no-load mode and in the load mode, respectively. Therefore, based on the Eq. (17.11), we can note that the implementation of the "no-load and under-load" concept is reduced to the establishment of the components of the stator winding currents $\dot{I}_0$ and $(-\dot{I}_2')$ corresponding to these two modes of the machine. As will be shown below, the Eq. (17.11) are very important for the theory of the calculation of asynchronous machines.

17.6 EMF Equations and Equivalent Circuits

We express the system of the emf Eq. (17.10) through the currents $\dot{I}_0$ and $(-\dot{I}_2')$ as a result of which it can be shown that the equivalent circuit in Fig. 6.4a can be reduced to the form in Fig. 10.4a. As follows from the comparison of the equivalent circuits shown in Figs. 6.4a and Fig. 10.4a, the magnetizing equivalent circuit of the air gap corresponding to the "no-load and under-load" concept receives the simplified structure.

In the magnetizing equivalent circuit of the air gap shown in Fig. 10.4a, the reduced rotor values are used $(\dot{Э}_2', \dot{I}_2')$. In this regard, the leakage emf equation for the rotor shown in (17.9) and the leakage equivalent circuit of the rotor obtained in Fig. 6.4c should be expressed through the reduced values of the emf $\dot{Э}_2'$ and the current $\dot{I}_2'$ of the rotor winding. Then, the equivalent circuit in Fig. 6.4c takes the form in Fig. 10.4c.

Based on the leakage equivalent circuit of the stator shown in Fig. 6.4b and the transformed magnetizing equivalent circuit of the air gap shown in Fig. 10.4a, it follows that the same values of the emf $\dot{Э}_1$ and the current $\dot{I}_1$ of the stator winding act on the terminals a–b of these equivalent circuits. Therefore, in our case of applying the "no-load and under-load" concept, the values used in the leakage emf equation of the stator obtained in (17.9) and in the leakage equivalent circuit of the stator shown in Fig. 6.4b remain unchanged. Therefore, the leakage equivalent circuit of the stator (Fig. 6.4b) is shown in Fig. 10.4b.

17.7 Machine Equivalent Circuit: "No-Load and Under-Load" Concept

We proceed from the equivalent circuits shown in Fig. 10.4. The equivalent circuit shown in Fig. 10.4a is the converted magnetizing equivalent circuit of the air gap. The leakage equivalent circuit of the stator is shown in Fig. 10.4b. The leakage equivalent circuit of the rotor in Fig. 10.4c is expressed in terms of the reduced rotor values. The equivalent circuits shown in Fig. 10.4 can be used to construct the equivalent circuit of the machine. The construction of the equivalent circuit of the machine, in this case, is reduced to connecting the input terminals c–d of the leakage equivalent circuit of the rotor (Fig. 10.4c) with the output terminals c–d of the magnetizing equivalent circuit of the air gap (Fig. 10.4a), and also to connecting the input terminals a–b of the magnetizing equivalent circuit of the air gap (Fig. 10.4a) with the input terminals a–b of the leakage equivalent circuit of the stator (Fig. 10.4b). As noted above, such a connection is known as the cascade connection. As a result, the equivalent circuit of the machine takes the form shown in Fig. 10.2b. The equivalent circuit in Fig. 10.2b through the voltages of the stator and rotor windings can be reduced to the form in Fig. 10.3. Using the equivalent circuit in Fig. 10.3, the voltage equations can be written out. These voltage equations can correspond to the "no-load and under-load" concept or to the "inductively coupled circuits" concept. The voltage equations obtained in this way have found a wide practical application.

In the case of application of the "no-load and load" concept, the equivalent circuit of the machine (Fig. 10.3) contains the leakage reactance of the rotor winding $x'_{2\sigma}$ reduced to the stator winding. When using the "inductively coupled circuits" concept, in the equivalent circuit of the machine (Fig. 6.5), the leakage reactances of the stator winding $x_{1\sigma}$ and the rotor winding $x_{2\sigma}$ are used, respectively. To determine the leakage reactances $x_{1\sigma}$ and $x_{2\sigma}$, it is necessary to find the values of the leakage emf of the stator winding $\dot{Э}_{1\sigma}$ and the leakage emf of the rotor winding $\dot{Э}_{2\sigma}$. The leakage emfs $\dot{Э}_{1\sigma}$ and $\dot{Э}_{2\sigma}$ are caused by the leakage flux linkages of the stator $\Psi_{1\sigma}$ and the rotor $\Psi_{2\sigma}$ calculated in accordance with the "inductively coupled circuits" concept by the equations given in (17.8) in the form $\Psi_{1\sigma} = \Psi_1 - \Psi_{21}$ and $\Psi_{2\sigma} = \Psi_2 - \Psi_{12}$. The separation of the leakage flux linkages from the full self-induction flux linkages of the machine windings as a result of the difference of the very close values makes it difficult to ensure the required level of reliability in determining the leakage reactances of the machine windings.

In the technical literature, a different approach to determining the leakage reactances of the machine windings has become widespread. This approach is based on the "no-load and under-load" concept and on the fact that in the load mode the leakage magnetic fields occur in the machine. In this case, the leakage magnetic fields are caused by the currents flowing in the stator and rotor windings in the load mode of the machine. Therefore, in this case, the determination of the leakage reactances of the machine windings is carried out on the basis of the leakage flux linkages caused by the volumes of the currents in the stator and rotor slots.

There is a total leakage magnetic field in the machine, since this leakage magnetic field is created by the currents flowing in the stator and rotor windings in the load mode. For convenience, let us single out the leakage magnetic field of the stator winding and the leakage magnetic field of the rotor winding. In Appendix B (Fig. B.4), it is shown that in the load mode the radial component of the magnetic flux density becomes zero at the surface 0–0 which passes through the middle of the radial length of the air gap of the machine. With respect to this air gap surface, the total leakage magnetic field can be divided into the leakage magnetic field of the stator winding and the leakage magnetic field of the rotor winding. Now, to find the leakage reactances of the stator winding $x_{1\sigma}$ and the rotor winding $x_{2\sigma}$, it is necessary to set with respect to the surface 0–0 the leakage magnetic fields caused by the volume of the currents in the stator and rotor slots, and also calculate the leakage emfs $\dot{Э}_{1\sigma}$ and $\dot{Э}_{2\sigma}$ induced in the machine windings by these leakage magnetic fields. Then, for the leakage reactances $x_{1\sigma}$ and $x_{2\sigma}$, we can use the relations in the form $x_{1\sigma} = \dot{Э}_{1\sigma}/\dot{I}_{1н}$ and $x_{2\sigma} = \dot{Э}_{2\sigma}/\dot{I}_2$. Here, the second equation can be presented in relation to the reduced rotor values. Then, for the reactance $x'_{2\sigma}$, we can have that $x'_{2\sigma} = \dot{Э}'_{2\sigma}/\dot{I}'_2$. Thus, the determination of the leakage reactances $x_{1\sigma}$ and $x_{2\sigma}$ can be realized separately for each of the stator and the rotor windings of the machine, which simplifies the computational process.

17.8 Equivalent Circuits for Fragments of emf Equations: Specific System of Units

The equivalent circuits shown in Fig. 10.4 will be represented in the specific system of units. For this, we use equations of the form

$$\dot{Э} = \oint \dot{E}dl \qquad \dot{F} = \oint \dot{H}dl \tag{17.12}$$

where $\dot{F} = (mwk_w/\pi p)\sqrt{2}\dot{I}$.

With regard to the stator and rotor windings, the equations shown in (17.12) take the form

$$\begin{aligned} \sqrt{2}\dot{Э}_1 &= k_{1Э}\dot{E}_1 \qquad & \sqrt{2}\dot{Э}_2 &= k_{2Э}\dot{E}_2 \\ \sqrt{2}\dot{I}_1 &= k_{1I}\dot{H}_1 \qquad & \sqrt{2}\dot{I}_2 &= k_{2I}\dot{H}_2 \end{aligned} \tag{17.13}$$

where $k_{1I} = \dfrac{\tau_0 p}{m_1 w_1 k_{w1}}; k_{2I} = \dfrac{\tau_0 p}{m_2 w_2 k_{w2}}; k_{1Э} = 2w_1 k_{w1} l_0; k_{2Э} = 2w_2 k_{w2} l_0.$

In the equivalent circuits obtained in Fig. 10.4, the reduced rotor values are used $(\dot{Э}_2', \dot{I}_2')$. Through the quantities $\dot{Э}_2'$ and $\dot{I}_2'$, the conditions (17.13) take the form

$$\begin{aligned} \sqrt{2}\dot{Э}_1 &= k_{1Э}\dot{E}_1 \qquad & \sqrt{2}\dot{Э}_2' &= k_{1Э}\dot{E}_2 \\ \sqrt{2}\dot{I}_1 &= k_{1I}\dot{H}_1 \qquad & \sqrt{2}\dot{I}_2' &= k_{1I}\dot{H}_2 \end{aligned} \tag{17.14}$$

Let us use the conditions (17.14) to represent the equivalent circuits in Fig. 10.4 in the specific system of units. To convert the reactances used in the equivalent circuits in Fig. 10.4, it is necessary to divide them by the dimensionless coefficient K_A defined from (17.13) and (17.14) as

$$K_A = \frac{2m_1 (w_1 k_{w1})^2 l_0}{\tau_0 p} \tag{17.15}$$

Now, taking into account the relations obtained in (17.14) and (17.15), the equivalent circuits in Fig. 10.4 can be represented in the form in Fig. 12.1.

17.9 Machine Equivalent Circuit: Specific System of Units

We proceed from the equivalent circuits shown in Fig. 12.1. The equivalent circuit shown in Fig. 12.1a is the converted magnetizing equivalent circuit of the machine, since in it the magnetizing reactance x_m^* takes into account the sections of the stator and rotor magnetic circuits. The leakage equivalent circuit of the stator is shown in Fig. 12.1b. The leakage equivalent circuit of the rotor is given in Fig. 12.1c. The equivalent circuits shown in Fig. 12.1 can be used to construct the equivalent circuit of the machine. The construction of the equivalent circuit of the machine, in this case, is reduced to connecting the input terminals c–d of the rotor leakage equivalent circuit (Fig. 12.1c) with the output terminals c–d of the air gap magnetizing equivalent circuit (Fig. 12.1a), and also to connect the input terminals a–b of the air gap magnetizing equivalent circuit (Fig. 12.1a) with input terminals a–b of the stator leakage equivalent circuit (Fig. 12.1b). As a result, the equivalent circuit of the machine takes the form shown in Fig. 12.2a. The equivalent circuit in Fig. 12.2a through the voltages of the stator and rotor windings can be reduced to the form in Fig. 12.2b.

With the help of the equivalent circuit in Fig. 12.2b, voltage equations can be written out. These voltage equations can correspond to the "no-load and under-load" concept or the "inductively coupled circuits" concept. The resulting voltage equations and the parameters used in them are represented in the specific system of units.

17.10 Machine Equivalent Circuit Based on Resulting Magnetic Field

The equivalent circuit of the machine built using the specific system of units and the "no-load and under-load" concept is shown in Fig. 12.2b. In this equivalent circuit, the magnetizing field of the machine is taken into account using the lumped magnetizing reactance x_m^*, and the leakage magnetic fields of the stator and rotor are represented using the lumped reactances $x_{1\sigma}^*$ and $x_{2\sigma}^*$. The method of using the lumped reactances adopted in the technical literature does not provide the sufficiently strict consideration of the peculiarities of the spatial distribution of the magnetizing field of the machine and the leakage magnetic fields of the stator and rotor and, consequently, the distribution of the resulting magnetic field of the machine. Below, we will consider the construction of the equivalent circuit of the machine in terms of taking into account the spatial distribution of the two-dimensional magnetizing field of the machine and the one-dimensional leakage magnetic fields of the stator and rotor. In this case, the specific system of units $(\dot{E}, \dot{H})$ is used.

"Unfolded" magnetizing equivalent circuits of the stator and rotor. When using the specific system of units, the construction of the "unfolded" magnetizing equivalent circuits of the stator and rotor is based on the representation of the magnetizing circuit of the machine consisting of the flat layers (Fig. 13.1). The layers of the model of the magnetizing circuit of the machine are represented by the L-shaped equivalent circuits (with the exception of the stator yoke layer and the rotor yoke layer, which are replaced by the equivalent circuit of the two pole) [1]. Therefore, the "unfolded" magnetizing equivalent circuits of the stator and rotor result from the cascade connection of the L-shaped equivalent circuits reflecting the corresponding layers of the model of the magnetizing circuit of the machine (Fig. 13.1). As follows from Appendix B, the magnetizing equivalent circuits of the stator and rotor can be set with respect to the surface 0–0 which runs in the middle of the radial length of the air gap of the machine. Under these conditions, the stator magnetizing circuit includes the air gap sublayer adjacent to the stator bore surface, the layer of the slot-wedge area, the layer of the winding location area, and the layer of the stator bore yoke area (Fig. 13.1). The rotor magnetizing circuit contains the air gap sublayer adjacent to the rotor surface, the layer of the slot-wedge area, the layer of the winding location area, and the layer of the rotor yoke area (Fig. 13.1). Then, the "unfolded" equivalent circuit of the stator magnetizing circuit takes the form in Fig. 13.3a, and the "unfolded" equivalent circuit of the rotor magnetizing circuit takes the form shown in Fig. 13.4a. The expressions for the reactances used in these equivalent circuits are given in (13.3)–(13.6). In the future, we will use the "unfolded"

equivalent circuits obtained for the magnetizing circuits of the stator and rotor in Figs. 13.3a and 13.4a and reflecting more strictly the real distribution of the two-dimensional magnetizing field of the machine.

"Unfolded" leakage equivalent circuits of the stator and rotor. In practice, the leakage magnetic field calculations are performed for the single slot. Therefore, with respect to the transverse leakage magnetic flux, the single slot of the stator or rotor can be considered as a system consisting of the conditional layers that propagate in the tangential direction over the length of the tooth division. This representation of the stator and rotor slots is given in Appendix F and shown in Fig. F.5b. In what follows, we will proceed from the models of the stator and rotor magnetic leakage circuits shown in Fig. F.5.

"Unfolded" leakage equivalent circuit of the stator. The "unfolded" leakage equivalent circuit of the stator corresponding to the model in Fig. F.5b is shown in Fig. 13.3b. For the reactances used in the equivalent circuit in Fig. 13.3b, we can use the expressions obtained in (13.10).

"Unfolded" leakage equivalent circuit of the rotor. The "unfolded" leakage equivalent circuit of the rotor following from the model in Fig. F.5b is shown in Fig. 13.4b. For the reactances used in the equivalent circuit in Fig. 13.4b, we can use the expressions received in (13.12).

"Unfolded" equivalent circuit with the mixed connection of the elements of the stator. We consider the "unfolded" magnetizing equivalent circuit of the stator (Fig. 13.3a) and the "unfolded" leakage equivalent circuit of the stator (Fig. 13.3b). "Unfolded" magnetizing equivalent circuit of the stator shown in Fig. 13.3a corresponds to the no-load mode and the "unfolded" leakage equivalent circuit of the stator shown in Fig. 13.3b reflects the load mode of the machine. In accordance with the "no-load and under-load" concept, these equivalent circuits can be superimposed, as a result of which the stator equivalent circuit in Fig. 13.5 can be obtained. The equivalent circuit in Fig. 13.5 reflects the resulting magnetic field of the machine.

The "unfolded" equivalent circuit obtained in Fig. 13.5 represents the result of the cascade connection of the L-shaped equivalent circuits reflecting the corresponding layers of the model of the magnetizing circuit of the stator (with the exception of the layer of the stator yoke region). The equivalent circuit in Fig. 13.5 reflects the spatial distribution of the two-dimensional magnetizing field of the machine and the one-dimensional leakage magnetic field of the stator, as a result of which it has a mixed connection of the elements. All this leads to the complication of the structure of the "unfolded" equivalent circuit of the stator of the machine. In order to give the form that is convenient for practical use, the "unfolded" equivalent circuit in Fig. 13.5 is subjected to the transformation for which the multiple transfer technique can be used [1].

Reducing the "unfolded" stator equivalent circuit with the mixed connection of the elements to the equivalent circuit with the parallel connection of the elements. In order to obtain the stator equivalent circuit with the parallel connection of the elements, we will use the "unfolded" stator equivalent circuit with the mixed connection of the elements (Fig. 13.5). In the equivalent circuit in Fig. 13.5, its parallel branches k–k, z–z, 1–1, and a–a, we transfer to the terminals of the transverse branch 0 – 0. The process of reducing the parallel branches (k–k, z–z, 1–1, and a–a) can be implemented using the multiple transfer technique [1]. A feature of this technique is the introduction of the transfer coefficients (Appendix G). Thus, the "unfolded" stator equivalent circuit with the mixed connection of the elements (Fig. 13.5) can be reduced to the form of the stator equivalent circuit with the parallel connection of the elements (Fig. 13.7).

Magnetizing circuit loop and magnetizing reactance of the stator. In the equivalent circuit in Fig. 13.7, the parallel branches (0–0, k–k, z–z, and a–a) represent the magnetizing circuit loops of the layers of the stator model (Fig. 13.1). In this regard, they can be combined into the single common contour, which will represent the stator magnetizing circuit loop. In this case, the reactance x_{m1}^* used in the common circuit loop (magnetizing circuit loop) reflects the stator magnetizing reactance. On the bases of the equivalent circuit in Fig. 13.7, the expression obtained for the stator magnetizing reactance x_{m1}^* is given in (13.20).

Leakage circuit loop and leakage reactance of the stator. In the equivalent circuit in Fig. 13.7, the parallel branch 1–1 represents the stator leakage circuit loop. The impedance $\overset{\circ}{Z}_1$ used in the stator leakage circuit loop is determined by the Eq. (13. 18a). Based on relation (13.18a), the ohmic resistance of the stator winding r_1^* determined by (13.23) can be introduced, and the expression for the leakage reactance of the stator winding $x_{1\sigma}^*$ given in (13.28) is also obtained. The expression for the reactance $x_{1\sigma}^*$ is presented in the traditional form, which facilitates its practical use.

"Folded" equivalent circuit of the stator. Taking into account the relation for the reactance x_{m1}^* obtained in (13.20), as well as the expressions for the ohmic resistance r_1^* and the leakage reactance $x_{1\sigma}^*$ given in (13.23) and (13.28), respectively, the equivalent circuit in Fig. 13.7 can be represented as in Fig. 13.8. In this case, Fig. 13.8 represents the "folded" stator equivalent circuit which follows from the "unfolded" stator equivalent circuit with the parallel connection of the elements shown in Fig. 13.7. As follows from the equivalent circuit in Fig. 13.8, with respect to $\dot{E}_m$, the stator of the machine is represented in the form of two independent parallel circuits loops: the magnetizing circuit loop containing the reactance x_{m1}^*, and the stator leakage circuit loop having the reactance $x_{1\sigma}^*$ and the ohmic resistance r_1^*. The stator equivalent circuit in Fig. 13.8 has a simple structure and is therefore convenient for practical use.

"Unfolded" equivalent circuit with the mixed connection of the elements of the rotor. Let us consider the "unfolded" magnetizing equivalent circuit of the rotor (Fig. 13.4a) and the "unfolded" leakage equivalent circuit of the rotor (Fig. 13.4b). The "unfolded" magnetizing equivalent circuit of the rotor (Fig. 13.4a) corresponds to the no-load mode and the "unfolded" leakage equivalent circuit of the rotor (Fig. 13.4b) reflects the load mode of the machine. In accordance with the "no-load and under-load" concept, these equivalent circuits can be superimposed as a result of which the rotor equivalent circuit in Fig. 13.6 can be obtained. The equivalent circuit in Fig. 13.6 reflects the resulting magnetic field of the machine.

The "unfolded" equivalent circuit obtained in Fig. 13.6 represents the result of the cascade connection of the L-shaped equivalent circuits, reflecting the corresponding layers of the model of the rotor magnetic circuit (with the exception of the layer of the rotor yoke area). The equivalent circuit in Fig. 13.6 reflects the spatial distribution of the two-dimensional magnetizing field of the machine and the one-dimensional leakage magnetic field of the rotor as a result of which it has the mixed connection of the elements. All this leads to the complication of the structure of the "unfolded" equivalent circuit of the rotor of the machine. In order to give the form that is convenient for practical use, the "unfolded" equivalent circuit in Fig. 13.6 can be subjected to the transformation for which the multiple transfer technique is used [1].

Reducing the "unfolded" rotor equivalent circuit with the mixed connection of the elements to the equivalent circuit with the parallel connection of the elements. In order to obtain the rotor equivalent circuit with the parallel connection of the elements, we will use the "unfolded" rotor equivalent circuit with the mixed connection of the elements (Fig. 13.6). In the equivalent circuit in Fig. 13.6, the parallel branches (k–k, z–z, 2–2, and a–a) are transferred to the terminals of the transverse branch 0–0. The process of transferring the parallel branches (k–k, z–z, 2–2, and a–a) can be implemented using the multiple transfer technique [1]. A feature of this technique is the introduction of the transfer coefficients (Appendix F). Thus, the "unfolded" rotor equivalent circuit with the mixed connection of the elements (Fig. 13.6) can be reduced to the form of the rotor equivalent circuit with the parallel connection of the elements (Fig. 13.9).

Magnetizing circuit loop and magnetizing reactance of the rotor. In the equivalent circuit in Fig. 13.9, the parallel branches (0–0, k–k, z–z, and a–a) represent the magnetizing circuit loops of the layers of the rotor model (Fig. 13.1). In this regard, they can be combined into a single common circuit loop which will represent the rotor magnetizing circuit loop. In this case, the reactance x_{m2}^* used in the common circuit loop (magnetizing circuit loop) reflects the magnetizing reactance of the rotor. The expression obtained for the stator magnetizing reactance x_{m2}^* based on the equivalent circuit in Fig. 13.9 is given in (13.36).

Leakage circuit loop and leakage reactance of the rotor. In the equivalent circuit in Fig. 13.9, the parallel branch 2–2 represents the rotor leakage circuit loop. The impedance $\overset{\circ}{Z}_2$ used in the leakage circuit loop of the rotor is determined by the relation (13.38). Based on the relation (13.38), the ohmic resistance of the rotor winding r_2^*/s determined by (13.39) can be introduced, and the expression for the leakage reactance of the rotor winding $x_{2\sigma}^*$ given in (13.44) is also obtained. The expression for the reactance $x_{2\sigma}^*$ is presented in the traditional form, which facilitates its practical use.

"Folded" equivalent circuit of the rotor. Considering the relation for the reactance x_{m2}^* obtained in (13.36), as well as the expressions for the ohmic resistance r_2^*/s and the leakage reactance $x_{2\sigma}^*$ given in (13.39) and (13.44), respectively, the equivalent circuit in Fig. 13.9 can be shown in the form in Fig. 13.10. In this case, Fig. 13.10 represents the "folded" rotor equivalent circuit which follows from the

"unfolded" rotor equivalent circuit with the parallel connection of the elements, shown in Fig. 13.9. As follows from the equivalent circuit in Fig. 13.10, in relation to $\dot{E}_m$, the rotor of the machine is represented in the form of the two independent parallel circuits loops: the magnetizing circuit loop containing the reactance x^*_{m2}, the rotor leakage circuit loop having the reactance $x^*_{2\sigma}$, and the ohmic resistance r^*_2/s. The rotor equivalent circuit in Fig. 13.10 has a simple structure and is therefore convenient for practical use.

Equivalent circuit of the machine. Based on the "folded" equivalent circuits obtained for the stator and rotor in Figs. 13.8 and 13.10, respectively, we can build the machine equivalent circuit. To do this, the terminals 0–0 of the equivalent circuit in Fig. 13.8 should be connected to the terminals 0–0 of the equivalent circuit in Fig. 13.10 and the magnetizing reactance of the machine, defined as $x^*_m = x^*_{m1}x^*_{m2}/\left(x^*_{m1} + x^*_{m2}\right)$ should be interred. Then, the equivalent circuits in Figs. 13.8 and 13.10 can be combined into the single common equivalent circuit as a result of which the equivalent circuit of the machine follows in the form in Fig. 12.2b. This means that for the asynchronous machine of the normal design, the equivalent circuit obtained on the basis of the consideration of the resulting magnetic field, and the equivalent circuit constructed on the basis of the decomposition of the resulting magnetic field into the "partial" magnetic fields takes the same form (Fig. 12.2b).

17.11 Main Stages of Constructing Machine Equivalent Circuit

The description of the method for constructing the equivalent circuit of the machine is given above. This method is based on the use of the resulting magnetic field which is determined using the "no-load and under-load" concept. Below we will trace the main stages and fix them by formulating the provisions that establish the logical sequence of the actions aimed at constructing the machine equivalent circuit. In doing so, these provisions are formulated for the two cases when the resulting magnetic field of the machine is determined using the "no-load and under-load" concept and is established applying the "field" concept.

17.11.1 Applying "No-Load and Under-Load" Concept

Below, we present the main stages of constructing the equivalent circuit of the machine, when its two-dimensional resulting magnetic field is determined based on the application of the "no-load and under-load" concept:

- establishment of the magnetizing circuit of the machine and the leakage magnetic circuits of the stator and rotor,

- decomposition of the magnetizing circuit of the machine into the layers taking into account that within each layer the magnetic field intensity is constant,
- obtaining the magnetizing circuit model of the machine,
- decomposition of the magnetizing circuit model of the machine into the magnetizing circuit model of the stator and the magnetizing circuit model of the rotor,
- representation of the layers of the stator and rotor magnetizing circuit models in the form of the corresponding L-shaped equivalent circuits (with the exception of the stator yoke layer and the rotor yoke layer),
- construction of the “unfolded” magnetizing equivalent circuits of the stator and the rotor as a result of the cascade way of connecting the L-shaped equivalent circuits of the layers into which the magnetizing circuit of the machine is subdivided,
- determination of the reactances used in the “unfolded” magnetizing equivalent circuits of the stator and the rotor,
- decomposition of the leakage magnetic circuits of the machine into the layers taking into account that within each layer the magnetic field intensity is constant (with the exception of the layers of the stator and rotor winding location areas),
- -obtaining the leakage magnetic circuit model of the machine
- decomposition of the leakage magnetic circuit model of the machine into the leakage magnetic circuit model of the stator and the leakage magnetic circuit model of the rotor,
- construction of the “unfolded” leakage equivalent circuit of the stator and the “unfolded” leakage equivalent circuit of the rotor,
- determination of the reactances used in the “unfolded” leakage equivalent circuits of the stator and the rotor,
- superposition of the “unfolded” magnetizing equivalent circuit of the stator and the “unfolded” leakage equivalent circuit of the stator,
- obtaining the “unfolded” stator equivalent circuit with the mixed connection of the elements,
- superposition of the “unfolded” magnetizing equivalent circuit of the rotor and the “unfolded” leakage equivalent circuit of the rotor,
- obtaining the “unfolded” rotor equivalent circuit with the mixed connection of the elements,
- transformation of the stator and the rotor “unfolded” equivalent circuits with the mixed connection of the elements using the multiple transfer technique,
- reducing the stator and the rotor “unfolded” equivalent circuits with the mixed connection of the elements to the equivalent circuits with the parallel connection of the elements,
- introduction of the magnetizing circuit loop of the stator and the determination of the magnetizing reactance of the stator,
- introduction of the magnetizing circuit loop of the rotor and the determination of the magnetizing reactance of the rotor,
- introduction of the leakage circuit loop of the stator and the determination leakage reactance of the stator,

- introduction of the leakage circuit loop of the rotor and the determination leakage reactance of the rotor,
- construction of the "folded" equivalent circuit of the stator and the "folded" equivalent circuit of the rotor,
- combination of the "folded" equivalent circuits of the stator and rotor, the introduction of the magnetizing reactance of the machine, and the construction of the machine equivalent circuit,
- drawing up the voltage equations for the stator and rotor windings for what the machine equivalent circuit is used.

The provisions described above correspond to the case when the two-dimensional resulting magnetic field of the machine is determined as a result of the superposition of the two-dimensional magnetizing field of the machine and the one-dimensional leakage magnetic fields of the stator and rotor windings. These provisions can be considered as the initial basis to construct the equivalent circuit of the machine when its resulting magnetic field is determined using the field equations ("field" concept).

17.11.2 Applying "Field" Concept

The above are the provisions arising from the construction of the equivalent circuit of the machine based on the consideration of its resulting magnetic field. In this case, the resulting magnetic field of the machine was visualized using the stator and rotor "unfolded" equivalent circuits with the mixed connection of the elements. These "unfolded" equivalent circuits with the mixed connection of the elements are obtained by superimposing the "unfolded" magnetizing equivalent circuits and the "unfolded" leakage equivalent circuits of the stator and rotor. Let us generalize the provisions described above for the case when the construction of the machine equivalent circuit is implemented using the resulting magnetic field determined to be Maxwell's equations. Maxwell's equations usually deal with the resulting magnetic field of the machine. In this regard, based on the provisions formulated above, the construction of the machine equivalent circuit can be reduced, in this case, to the following:

- construction of the layered model of the electric machine. In this case, it is usually assumed that there is a two-dimensional magnetic field in the cross section of the machine and there is a one-dimensional electric field in the axial direction of the machine. In this regard, we can talk about the construction of the layered model of the machine. In such a model, the areas of the stator and rotor tooth–slot zones are represented in the form of the homogeneous layers, for which the average values of the radial and tangential components of the magnetic permeability can be obtained. The layered model of the machine can include the isotropic, anisotropic, conductive, nonconductive, passive, and active layers

- decomposition of the layered model of the machine into the layered model of the stator and the layered model of the rotor in relation to the surface 0–0 passing in the middle of the radial length of the air gap
- introduction of the air gap sublayer adjacent to the stator bore surface to the model of the stator
- introduction of the air gap sublayer adjacent to the rotor surface to the model of the rotor
- solving the field equations using layered models of the stator and rotor and obtaining the equations for the tangential components of the electric and magnetic field intensities
- representation of the field equations in terms of the tangential components of the electric and magnetic field intensities given on the outer and inner surfaces of the ith layer
- reducing the field equations for the components of the electric and magnetic field intensities given on the outer and inner surfaces of the ith layer to the surface 0–0 passing in the middle of the radial length of the air gap of the machine
- representation of the field equations of the ith layer of the stator and the rotor in the form of the equations for the four-port network (with the exception of the layers of the stator and the rotor yoke areas)
- representation of the ith layer of the stator and the rotor in the form of the L-shaped equivalent circuit (with the exception of the layers of the stator and the rotor yoke areas which are replaced by the two-pole equivalent circuits) and determination of the parameters used in it
- introduction of the "own" circuit loop and the "connection" circuit loop of the ith layer. The "connection" circuit loop connects the "own" circuit loops of the adjacent layers of the models
- representation of the conducting layer "own" circuit loop consisting of the two circuit loops: the "own" magnetizing circuit loop and the "own" circuit loop of the eddy currents induced in the conducting layer
- construction of the stator and rotor equivalent circuits by the cascade connection of the L-shaped equivalent circuits reflecting the corresponding layers of the stator and rotor models
- obtaining the "unfolded" equivalent circuits with the mixed connection of the elements for the stator and the rotor
- transformation of the "unfolded" equivalent circuits with the mixed connection of the elements for the stator and the rotor using the multiple transfer technique,
- reducing the "unfolded" equivalent circuits with the mixed connection of the elements for the stator and rotor to the equivalent circuits with the parallel connection of the elements
- introduction of the magnetizing circuit loop of the stator and the determination of the magnetizing reactance of the stator
- introduction of the magnetizing circuit loop of the rotor and the determination of the magnetizing reactance of the rotor
- introduction of the leakage circuit loop of the stator and the determination of the leakage reactance of the stator,

- introduction of the leakage circuit loop of the rotor and the determination of the leakage reactance of the rotor
- construction of the "folded" equivalent circuit of the stator and the "folded" equivalent circuit of the rotor
- combination of the "folded" equivalent circuits of the stator and rotor, the introduction of the magnetizing reactance of the machine, and the construction of the machine equivalent circuit
- drawing up the voltage equations for the stator and rotor windings for what the machine equivalent circuit is used.

The equivalent circuit of the machine obtained in this way can be reduced to the form in Fig. 12.2b and it can be considered containing the three circuit loops: the magnetizing circuit loop of the machine, the leakage circuit loop of the stator, and the leakage circuit loop of the rotor. Therefore, the introduction of the machine equivalent circuit, in this case, means that the resulting magnetic field determined using Maxwell's equations can be represented as decomposed into the magnetizing field of the machine and the leakage magnetic fields of the stator and rotor which is usually done in the classical theory of the electric machines.

The provisions described above are used to construct the equivalent circuit of the asynchronous machine when its resulting magnetic field is determined by applying Maxwell's equations. The results obtained in this way are presented in Chaps. 15 and 16 of this work. For the multi-circuit electric machines, these provisions are used in [1].

17.12 Summary

1. The field equations (Maxwell's equations) usually describe the resulting magnetic field of the machine. The modern theory of the electric machine is based on the use of the "partial" magnetic fields resulting from the "no-load and under-load" concept. Consequently, it follows from this that in order to successfully solve the problem associated with the practical application of the field equations, it is necessary to adapt the theory of calculating the electric machine used in engineering practice to the conditions that arise when using the directly resulting magnetic field of the machine.
2. According to the "no-load and under-load" concept, the resulting magnetic field of the electric machine can be obtained as a result of the superposition of the magnetizing field of the machine and the leakage magnetic fields of the stator and rotor windings. The magnetizing field of the machine can be visualized using the magnetizing equivalent circuit of the machine and the leakage magnetic fields of the stator and rotor are represented as the leakage equivalent circuits of the stator and rotor. When considering the two-dimensional magnetizing field of the machine, the magnetizing equivalent circuit of the machine follows as a result of the cascade connection of the L-shaped equivalent circuits representing the

corresponding layers of the magnetizing circuit model of the machine. If the magnetizing equivalent circuit of the machine is decomposed into the magnetizing equivalent circuits of the stator and the rotor, then, for the magnetizing equivalent circuit of the stator and the leakage equivalent circuit of the stator as well as the magnetizing equivalent circuit of the rotor and the leakage equivalent circuit of the rotor, the superposition principle can be applied. As a result, the "unfolded" equivalent circuits with the mixed connection of the elements can be obtained for the stator and rotor of the machine. The "unfolded" equivalent circuits with the mixed connection of the elements correspond to the resulting magnetic field of the machine and represent a result of the cascade connection of the L-shaped equivalent circuits reflecting the layers of the magnetic circuit models of the machine.

3. In this work, the multiple transfer technique is developed. This technique arises from the application of the "no-load and under-load" concept. As a result of using this technique, the "unfolded" equivalent circuits with the mixed connection of the elements obtained for the stator and rotor are reduced to the equivalent circuits with the parallel connection of the elements. Further, the equivalent circuits with the parallel connection of the elements are converted into the "folded" equivalent circuits of the stator and the rotor for which the magnetizing circuit loops and the leakage circuit loops of the stator and the rotor are introduced. Now the magnetizing reactance of the machine is determined and the "folded" equivalent circuits of the stator and rotor are combined into the equivalent circuit of the machine. Thus, the construction of the equivalent circuit of the machine is achieved directly from the consideration of its resulting magnetic field.
4. Provisions are formulated that fix the logical sequence of the actions associated with the construction of the machine equivalent circuit based on the use of the resulting magnetic field arising from the application of the "no-load and under-load" concept.
5. In this work, these provisions are applied in the construction of the machine equivalent circuit arising from the analysis of the resulting magnetic field determined using Maxwell's equations. Therefore, the calculation methods obtained using the field equations will be similar to the calculation methods used in practice which opens up new opportunities for improving the engineering methods for calculating the electric machines.

Reference

1. Asanbayev, V.: Alternating Current Multi-Circuit Electric Machines: A New Approach to Steady-State Parameter Determination, Springer Charm Heidelberg, New York (2015)

Appendix A: Some Features of Asynchronous Machine Calculation Method

First Calculation Methods

The first calculation methods had been assumed the use of the equivalent circuits and the vector diagrams of the electric machine. Therefore, the corresponding quantitative data had been obtained from the construction of the vector diagrams. The basis of such constructions was represented by the vectors reflecting the mmf and the magnetic flux density waves created by the symmetric multiphase system of the currents of the stator and rotor windings. In this case, the following vector diagrams had been built; the vector diagram of the mmf of the stator and rotor windings; the vector diagram representing the voltage equation for the stator winding; the vector diagram representing the voltage equation for the rotor winding. These vector diagrams were the basis for compiling the voltage equations for the stator and rotor windings. The machine equivalent circuit is built using the system of the relative units. This led to the fact that the equivalent circuits have become quite widespread in the design practice. It was also very important to introduce the complex quantities that had been opened the way to the symbolic method. Together with the symbolic method, the vector diagrams and the equivalent circuits were the main tools for analyzing the operating modes of the electric machine. Here, the voltage equations had not been actively used in the design practice of the electric machine.

Modern Calculation Methods

The use of the voltage equations (Kirchhoff's equations) compiled for the stator and rotor windings can be considered as the basis for the modern method of calculation of the electric machine. Therefore, the modern calculation method of the electric machine is based on the solution of the voltage equations using the various techniques and procedures, as well as using modern computing facilities [1–6]. The

V. Asanbayev, *Asynchronous Machines*, https://doi.org/10.1007/978-3-030-92284-9

voltage equations contain the parameters that need to be determined. In the theory of the electric machines, the issues related to the determination of the parameters are the independent area of the research, for example [7, 8]. In this regard, the theory of the electric machines can be considered consisting of the techniques and procedures focused on the compilation of the voltage equations for the machine windings and the determination of the parameters used in these equations.

In the modern theory of electrical machines, the voltage equations are considered as the main tools for analyzing the electromagnetic processes taking place in the electric machine. At present, the vector diagrams and the equivalent circuits have been moved away to the background. The voltage equations written using the matrices and the tensors, as well as adapted to the use of the modern computing facilities, allow us to cope with the most complex and very laborious tasks of the calculation and analysis of the various operating modes of the electric machines [6, 9, 10]. However, until now, the task of providing the required level of a reliability of the calculation and analysis results continues to be a serious problem. Basically, this problem is due to the fact that there are still no reliable methods for determining the parameters used in the voltage equations of modern electric machines. Especially, this problem is appeared very sharply in the calculation and analysis of the electromagnetic processes in the multi-winding electric machines (powerful turbine generators, large high-speed synchronous motors, machines with solid rotor, etc.).

Principles, Concepts and Main Provisions of Implementation

The calculation methods applied in the theory of the asynchronous machine are based on the use of the various principles, concepts, and simplifying provisions. Below, we will briefly review the main ones.

Superposition principle. In order to simplify and achieve clarity of an analysis, the electromagnetic processes taking place in the electric machine are considered with the use of the superposition principle. Superposition principle can be used for the linear system. Electromagnetic processes associated with the formation of the magnetic fluxes and with the emf induction in the windings, etc., are considered in the linear approximation (in the absence of saturation of the stator and rotor iron). The currents of the windings create distributed in the space the mmf, which causes the magnetic fluxes in the air gap of the electric machine. In accordance with the electromagnetic induction law (Faraday's law), the magnetic fluxes induce the emf in the windings. Moreover, the magnetic flux of each winding induces the emf in its own winding and also in the other windings. Therefore, following the superposition principle, the resulting emf of any winding can be considered as a result of the superposition of the emf induced in the given winding by the magnetic flux from the own current as well as by the magnetic fluxes created by the currents of the other windings. Now, following to the second Kirchhoff's law according to which the sum of the voltage applied to the terminals of the winding and the emf induced in it is equal to the voltage drop across the ohmic resistance of this winding, we can write the voltage equations for the machine windings. Hence, the voltage equations used for the analysis of the electromagnetic processes taking place in the electric machine directly arise from applying the superposition principle and the second Kirchhoff's law.

Space vectors. In order to simplify, the voltage equations of the machine windings can be expressed through the calculated quantities known as the space vectors of the current, voltage, emf, flux linkage, etc., which are the functions of the actual phase currents, voltages, emf, flux linkages, etc. In this case, the actual currents, voltages, emf, flux linkages, etc., of the windings as the function of a time can be determined from the known values of the space vectors of the current, voltage, emf, flux linkage, etc.

In the symmetrical electric machine and with a symmetrical power supply of its windings, as well as with the sinusoidal currents and voltages, the sinusoidal space waves of the rotating magnetic field and the mmf occur in the air gap of the electric machine. Like any sinusoidal function, the sinusoidal space waves of the rotating magnetic field and the mmf can be represented by the vectors coinciding in a space with their amplitude quantities both in the value and the direction. The magnetic field in the machine and the mmf of the windings are determined by the joint action of the instantaneous values of the phase currents, and in the idealized machine, the magnetic field and the mmf are proportional to the currents. Therefore, we can note the existence in the electric machine of some spatial vectors of both magnetic field, mmf and the currents, emf, voltages, flux linkages, etc. In this case, the absolute values of the spatial vectors of the currents, emf, voltages, flux linkages, etc., are determined in such a way that they coincide with the amplitudes of the instantaneous values of the phase currents, emf, voltages, flux linkages, etc. In other words, the instantaneous values of the phase currents, emf, voltages, flux linkages, etc., follow, in this case, as a result of the projection of these vectors on the axis of the corresponding phases. Consequently, the phenomena in the symmetrical machine can be represented as caused by the rotating space vectors in such a way that the processes in all phases will be similar but will proceed with a time shift by the phase angle 120°. Hence, when using the space vectors for the analysis of the processes in the electric machine, one can restrict oneself to considering the voltage equations for the single phase of the stator and single phase of the rotor of the machine which greatly simplifies the calculations.

Coordinate transformation. In the voltage equations compiled using of the space vectors, the magnitudes of the current, voltage, emf, flux linkage, etc., are the space–time vectors depicting, in addition, to the time complexes, and the rotating waves of the mmf, magnetic and electric fields in the electric machine. The asynchronous machine consists of the fixed stator and the rotating rotor separated by the uniform air gap. Therefore, the waves of the mmf and the magnetic field created by the currents flowing in the stator and rotor windings are rotated at different speeds. Hence, the currents flowing in the stator and rotor windings have different frequencies. This situation makes the calculations extremely difficult. In this regard, the need to switch from the stator system to the rotor one and vice versa arises (coordinate transformation). This procedure is implemented by the transformation of the space vectors used in the voltage equations of the stator and rotor windings. When analyzing the steady-state processes, the space vectors are usually transferred from the rotating rotor system to the stator one. In this case, the asynchronous machine can be considered as the static transformer that simplifies an analysis. As a result of this transformation, the waves of the mmf and the magnetic fields of the stator and rotor are rotated at the same speed. Therefore, the currents flowing in the stator and rotor windings have the same frequency equal to the frequency of the power supply network.

"Inductively coupled circuits" concept. This concept can be considered as a result of the physical interpretation of the superposition principle. Below, we will show it. In this work, the symmetrical three-phase electric machine having the same phase resistances and supplied with the sinusoidal currents and voltages is considered. Only the fundamental field harmonics generated by the currents of the stator and rotor windings are taken into account. With the symmetrical load and the symmetrical design of the machine, the electromagnetic processes proceed in all its phases in the same way and, as a consequence of this, the currents and voltages in each phase are also the same. In this case, the phase currents and the voltages are shifted by the angle 120°. As stated previously, the operating modes of such machines can be described, in this case, on the basis of the consideration of the single phase. In the operating modes, the currents flow in the stator and rotor windings of the electric machine. These currents create the total self-induction magnetic fluxes of these windings. In this case, a part of the total self-induction magnetic flux of the stator winding is coupled with the rotor winding. Thus, the mutual induction magnetic flux of the rotor winding is formed. A part of the total self-induction magnetic flux of the rotor winding is coupled with the stator winding. Thus, the mutual induction magnetic flux of the stator winding is created. Now, the resulting magnetic fluxes of the stator and rotor windings can be represented as a result of the superposition of the total self-induction magnetic flux of the considered winding and the mutual induction magnetic flux caused by the current of another winding.

In electrical machines, multi-turn windings are usually used. In this regard, the magnetic fluxes noted previously create corresponding flux linkages: the total self-induction flux linkages and the mutual induction flux linkages of the machine windings. When the flux linkages of the total self-induction are changed in a time, the emf of the total self-induction is induced in the stator and rotor windings. When the flux linkages of the mutual induction are changed in a time, the emf of the mutual induction is induced in the stator and rotor windings. Thus, the resulting magnitudes of the emf induced in the stator and rotor windings can be considered as a result of the superposition of the total self-induction emf and mutual induction emf. The windings in which the emf of the mutual induction is induced are the inductively coupled windings. Hence, the stator and rotor windings can be represented as the inductively coupled system. The voltage equations for the inductively coupled stator and rotor windings can be obtained by the second Kirchhoff's law. In this case, the stator winding is usually considered as a receiver, and the rotor winding is represented as the source of an electrical energy.

The reactances representing in the voltage equations for the inductively coupled windings are difficult to the precise calculation, for example [1]. In this regard, an analysis of the electromagnetic processes by Kirchhoff's equations compiled using the "inductively coupled circuits" concept is associated with the relatively large error. In practice, this problem is solved by using the referred rotor quantities in the voltage equations. Below, we will use this approach.

"No-load and under-load" concept. The simplification of the description of the operating modes of the electric machine can be achieved under the following conditions of the interpretation of the formation of the machine resulting magnetic field. When the rotor winding is open (no-load mode), the stator winding current caused by the action of the power supply network voltage creates in the machine the rotating magnetic field that, crossing the stator and rotor windings, induces the emf in them. If the rotor winding is closed to the load impedance (under-load mode), then under the action of the emf induced in it in the no-load mode, the load current arises in the rotor winding. The load current of the rotor winding causes the rotor magnetic flux. One part of this rotor magnetic flux is closed around the rotor winding creating the rotor leakage magnetic flux. Another part of the rotor magnetic flux crosses the stator winding and induces in it the emf under of the action of which the load component of the current arises in the stator winding. The load component of the stator winding current causes the stator magnetic flux. One part of this magnetic flux is closed around the stator winding creating the stator leakage magnetic flux. Another part of the stator magnetic flux is directed opposite to the rotor magnetic flux and completely compensates for it. Thus, in the under-load mode, the stator and rotor winding currents create in the machine only the leakage magnetic fluxes. Hence, in accordance with the "no-load and load" concept, the machine resulting magnetic flux can be represented as a result of the superposition of the magnetic flux taking place in the machine under the no-load mode and the leakage magnetic fluxes of the stator and rotor windings (under the load mode). In this case, the stator current is determined as the sum of the magnetizing current (no-load mode) and the current corresponding to the under-load mode. Electromagnetic torque acting on the rotor of the machine can be represented as a result of the interaction of the rotor load current with the resulting magnetic flux of the electric machine. Thus, the machine operating mode can be considered as a result of the superposition of the no-load mode (characterized by the appearance in the stator winding of the magnetizing current and the induction in the stator and rotor windings of the emf) and the under-load mode (characterized by the appearance in the closed rotor winding of the load current and in the stator winding of the load component of the current). In practice, the superposition of the machine no-load mode and under-load mode is implemented by the use of the referred quantities of the emf and the current of the rotor winding, for example [1–10].

Reduction technique. In the technical literature, the reduction technique is used for the implementation of the "no-load and load" concept. When using the "no-load and load" concept (consequently, when using the technique for reducing the windings of the machine), the voltage equations represent the windings of the machine as the conditionally electrically connected circuit loops. In the conditions of reducing the rotor winding to the stator winding, the magnetizing field induces in the stator winding and in the referred rotor winding the emf of the same magnitude. This provision is the basis for the conditional electrical connection of the stator and rotor windings and for the use in the machine equivalent circuit of the common branch for the currents of the stator and rotor windings. As a result, the stator and rotor windings are represented as the conditionally electrically connected circuit loops. The voltages equations compiled on the basis of the conditionally electrically connected circuit loops do not have those computational difficulties which the voltage equations obtained with the use of the "inductively coupled circuits" concept have [1].

The voltage equations obtained on the basis of the conditionally electrically connected circuit loops have a fairly simple form, they are easily amenable to the physical interpretation, and also they can be easily visualized as the vector diagram and the equivalent circuit. All this greatly facilitates the computational analysis of the processes taking place in the electric machine.

Magnetizing flux. In the no-load mode which is usually realized in the absence of the current in the rotor winding, there is in the machine the magnetizing flux that passes through the main magnetic circuit. The value of this flux is determined by the fundamental harmonic of the distribution curve of the magnetic flux density in the machine air gap. The magnetizing flux axis represents the radius passing through the point with the amplitude value of the magnetic flux density distributed in the air gap. The magnetizing flux induces the emf in the stator and rotor windings. In this case, for the fixed magnetizing flux, the emf is induced of the same magnitude in all machine windings (in the stator winding and the referred rotor winding). Consequently, in this case, it is possible to imagine the presence in the machine of the magnetizing flux common to all of its windings. This flux induces in the windings of the electric machine the emf of the same value. In this sense, the concept of the magnetizing flux of the electric machine should be understood.

In the presence of the magnetizing flux common for all machine windings, to determine the emf induced in the rotor winding, it is sufficient to use the single magnetizing curve defined, for example, in relation to the stator winding terminals. This provision greatly facilitates the procedure for determining the emf induced in the rotor windings.

Leakage flux. In the under-load mode, the load currents flowing in the stator and rotor windings do not create the common flux of the mutual induction and, as noted previously, these currents create only the leakage fluxes of the stator and rotor windings. The magnetic flux coupled to only single particular winding represents the leakage flux of this winding. Consequently, the leakage fluxes occur between the two windings, in other words, the leakage flux of the winding can only be established with respect to the other winding. The leakage fluxes arise in the electric machine when load currents flow around the stator and rotor windings (under-load mode). In this regard, the common picture of the leakage flux takes place in the electric machine. The separation of the leakage flux of the stator winding and the leakage flux of the rotor winding from the common picture of the machine leakage flux represents, in the general case, the undefined task. In practice, it can be assumed that the leakage flux of the stator winding and the leakage flux of the rotor winding which are spread to the surface $0 - 0$ passing through the middle of the radial length of the air gap. In this work, this provision is used.

In the case of a large number of windings, each of the windings has different leakage fluxes in relation to all other windings. Therefore, the determination of the leakage fluxes of a large number of the windings is associated with great difficulties. In practice, the problem of determining the leakage fluxes of the multi-winding machine is solved based on the adoption of the simplifying provision that all machine windings are interlocked with the same mutual induction magnetic flux (magnetizing flux). In this case, each machine winding in relation to the winding causing the magnetizing flux has only single leakage flux that simplified the calculations. With the existence of the magnetizing flux common to all machine windings, the problem of determining the leakage fluxes becomes definite, since the leakage fluxes are obtained in relation to the winding which creates the magnetizing flux (in practice, it is usually the stator winding). In other words, the introduction of the machine magnetizing flux is associated with fixing the winding the currents of which create the magnetizing flux of the machine. Then, in relation to this fixed winding, the leakage fluxes of the other windings are determined.

The provisions considered previously can be used when the calculation elements of the electric machines are determined by applying the Maxwell's equations. In this work, in this way, the direct connection between the theory of engineering calculations of the electric machines and the theory of the electromagnetic field is achieved.

Summary

1. The theory for calculating the asynchronous machine is based on the use of the "no-load and under-load" concept. According to this concept, the operating mode of the electric machine is represented as the superposition of the no-load mode and the under-load mode. In this work, the no-load mode of the electric machine is achieved in the absence of the current in the rotor winding. The under-load mode of the machine is ensured when there is an under-load current in the rotor winding. The use of the "partial" modes the superposition of which provides the operating mode of the electric machine significantly simplifies an analysis.
2. When using the "no-load and under-load" concept, the magnetic field created in the machine by the currents of the stator and rotor windings can be considered as a result of the superposition of the machine magnetizing field and the slot leakage magnetic fields of the stator and rotor windings. The machine magnetizing field and the slot leakage magnetic fields in comparison with the resulting magnetic field taking place in the machine in the operating mode have a simple structure. Consequently, the application of the machine magnetizing field and the slot leakage magnetic fields greatly simplifies the calculations.
3. The introduction of the machine magnetizing field and the slot leakage magnetic fields of the stator and rotor windings make it possible to establish the machine magnetizing circuit loop and the slot leakage circuit loops of the stator and rotor

windings, determine the parameters used in them, construct the machine equivalent circuit, and compile the voltage equations for the machine windings. Consequently, the determination of the calculation elements of the electric machine is based on the decomposition of the resulting magnetic field of the machine into the magnetizing field and the slot leakage magnetic fields of the stator and rotor windings. Hence, it follows that when using Maxwell's equations, the determination of the calculation elements of the machine is reduced to the decomposition of the resulting magnetic field obtained applying these equations into the "partial" magnetic fields: the magnetizing field and the slot leakage magnetic fields of the stator and rotor windings.

References

1. Voldek. A.: Electric Machines, Energiya, Leningrad (1974)
2. Ivanov-Smolensky, A.: Electric Machines, Energiya, Moscow (1980)
3. Vazhnov, A.: Electric Machines, Energiya, Leningrad (1968)
4. Alger, P.: Induction Machines. Their Behavior and Uses, Gordon and Beach, New York (1970)
5. Livshits, A.: Electric Machines, vol. 1-3, United Scientific and Technical Publishing, Moscow (1936)
6. White, D., Woodson, H.: Electromechanical Energy Conversion, John Wiley and Sons, Inc., New York (1064)
7. Kazovsky, E.: Transient Processes in AC Electric Machines, Publishing of Academy of Science USSR, Moscow-Leningrad (1962)
8. Danilevich, Y., Dombrovsky, V., Kazovsky, E.: Parameters of AC Electric Machines, Publishing Nauka, Moscow (1965)
9. Sipailov, G., Kononenko, E., Kharkov, K.: Electric Machines. Special Course, Vysshaya Shkola, Moscow (1987)
10. Treshchev, I.: Electromechanical Processes in AC Electric Machines, Energiya, Leningrad (1980)

Appendix B: Magnetic Field in Flat Air Gap of Machine

Magnetic Flux Density Components

We assume that the traveling magnetic field sinusoidally distributed in a space exists in the air gap of the electric machine. Such magnetic field is created by the sinusoidally distributed in space surface currents of the stator and rotor windings, for example [1–15]. The stator and rotor surfaces are smooth, and for the stator and rotor steel, the condition of the form $\mu_c = \infty$ is true. We proceed from the fact that the electric machine is relatively long in the axial direction, which is basis for neglecting the end effects. In these conditions, the air gap magnetic field has the radial and tangential components of the field intensities $\dot{H}_y$ and $\dot{H}_x$, since the axial component of the magnetic field intensity $\dot{H}_z$ is equal to zero, i.e., $\dot{H}_z = 0$. The values of $\dot{H}_y$ and $\dot{H}_x$ do not depend on the coordinate z. The flat air gap model corresponding to these conditions is shown in Fig. B.1a.

As noted previously, we consider the traveling magnetic field sinusoidally distributed in the air gap. Therefore, the stator and rotor currents can be represented as the traveling sinusoidal waves of the linear densities of the surface currents. The linear density of the surface current of the stator winding can be expressed in the complex form as (when $y = \delta/2$) (Fig. B.1a)

$$\dot{\varkappa}_1(x) = \dot{\varkappa}_1 e^{-j\alpha x} \tag{B.1}$$

The linear density of the surface current of the rotor winding in the complex form (when $y = -\delta/2$) (Fig. B.1a)

$$\dot{\varkappa}_2(x) = \dot{\varkappa}_2 e^{-j\alpha x} \tag{B.2}$$

In (B.1) and (B.2), the values of the $\dot{\varkappa}_1$ and $\dot{\varkappa}_2$ represent the complex amplitudes of the linear densities of the stator and rotor surface currents, respectively and $\alpha = \pi/\tau_0$. Here, τ_0 is the pole pitch on the surface $0 - 0$ passing through the middle of the radial length of the air gap.

V. Asanbayev, *Asynchronous Machines*, https://doi.org/10.1007/978-3-030-92284-9

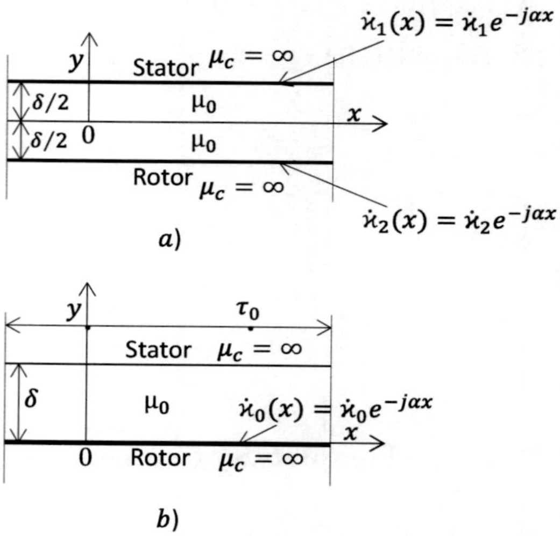

Fig. B.1 To calculation of the air gap magnetic field: (**a**) air gap flat model (operating mode); (**b**) air gap flat model (no-load mode)

The traveling sinusoidal waves of the stator and rotor winding currents determined by (B.1) and (B.2) create the traveling sinusoidal waves of the magnetic field intensities for the components of which, we can obtain

$$\begin{aligned} \dot{H}_x(x) &= \dot{\mathrm{H}}_x e^{-j\alpha x} \\ \dot{H}_y(x) &= \dot{\mathrm{H}}_y e^{-j\alpha x} \end{aligned} \tag{B.3}$$

where $\dot{\mathrm{H}}_x$ and $\dot{\mathrm{H}}_y$ are the complex amplitudes of the magnetic field intensity components.

In accordance, for example, with [12, 14, 16, 17], for $\dot{\mathrm{H}}_x$ and $\dot{\mathrm{H}}_y$, we have

$$\begin{aligned} \dot{H}_x &= C_1 \sinh \alpha y + C_2 \cosh \alpha y \\ \dot{H}_y &= C_3 \sinh \alpha y + C_4 \cosh \alpha y \end{aligned} \tag{B.4}$$

where C_1, C_2, C_3, and C_4 are the integration factors.

To determine the integration factors used in (B.4), it is necessary to establish the border conditions for $\dot{H}_x$ and $\dot{H}_y$ existing on the stator and rotor surfaces. According to Fig. B.1a, as such surfaces are considered the surfaces $y = \pm \delta/2$. The border conditions for $\dot{H}_x$ can be defined proceed from Fig. B.1a. On the basis of the first equation of the system (B.3), for the tangential components of the magnetic field intensities on the stator and rotor surfaces, the conditions of the form are implemented

$$\begin{aligned}\left[\dot{\mathrm{H}}_x(x)\right]_{y=\delta/2} &= \dot{\mathrm{H}}_{x1}e^{-j\alpha x}\\ \left[\dot{\mathrm{H}}_x(x)\right]_{y=-\delta/2} &= \dot{\mathrm{H}}_{x2}e^{-j\alpha x}\end{aligned} \tag{B.5}$$

On the stator and rotor surfaces, the contours of the infinite small tangential length can be established. For these contours, we can use the total current law $\int Hdl = \sum i$, for example, [1–15]. Using the conditions (B.2) and (B.3), the components of the magnetic field intensities tangential to the stator and rotor border surfaces can be expressed as

$$\begin{aligned}\left[\dot{\mathrm{H}}_x(x)\right]_{y=\delta/2} &= -\dot{\varkappa}_1e^{-j\alpha x}\\ \left[\dot{\mathrm{H}}_x(x)\right]_{y=-\delta/2} &= \dot{\varkappa}_2e^{-j\alpha x}\end{aligned} \tag{B.6}$$

Now, on the basis of the conditions (B.5) and (B.6), we have

$$\begin{aligned}\left[\dot{\mathrm{H}}_{x1}\right]_{y=\delta/2} &= -\dot{\varkappa}_1 = -\frac{\pi\dot{F}_1}{\tau_0}\\ \left[\dot{\mathrm{H}}_{x2}\right]_{y=-\delta/2} &= \dot{\varkappa}_2 = \frac{\pi\dot{F}_2}{\tau_0}\end{aligned} \tag{B.7}$$

where $\dot{\varkappa}_1 = \pi\dot{F}_1/\tau_0$; $\dot{\varkappa}_2 = \pi\dot{F}_2/\tau_0$; $\dot{F}_1$ and $\dot{F}_2$ are the mmf of the stator and rotor windings.

In the first equation of the system (B.4), we substitute the coordinates $y = \delta/2$ and $y = -\delta/2$, and also substitute from (B.7) the values $\dot{\mathrm{H}}_x$ corresponding to these coordinates. As a result, we obtain the following conditions

$$\begin{aligned}C_1\sinh\frac{\pi\delta}{2\tau_0} + C_2\cosh\frac{\pi\delta}{2\tau_0} &= -\frac{\pi\dot{F}_1}{\tau_0}\\ -C_1\sinh\frac{\pi\delta}{2\tau_0} + C_2\cosh\frac{\pi\delta}{2\tau_0} &= \frac{\pi\dot{F}_2}{\tau_0}\end{aligned}$$

From here, the integration factors C_1 and C_2 acquire the form

$$C_1 = -\frac{\pi\left(\dot{F}_2 + \dot{F}_1\right)}{2\tau_0\sinh\frac{\pi\delta}{2\tau_0}} \qquad C_2 = \frac{\pi\left(\dot{F}_2 - \dot{F}_1\right)}{2\tau_0\cosh\frac{\pi\delta}{2\tau_0}}$$

The integration factors C_1 and C_2, we substitute in the first equation of the system (B.4). Using the conditions of the form $\sinh(b \pm d) = \sinh b \cdot \cosh d \pm \cosh b \cdot \sinh d$, and also $\dot{B}_x = \mu_0\dot{H}_x$, we can receive

$$\dot{B}_x = -\frac{\mu_0 \pi}{\tau_0 \sinh \frac{\pi\delta}{\tau_0}} \left[\dot{F}_1 \sinh \frac{\pi}{\tau_0}\left(y + \frac{\delta}{2}\right) + \dot{F}_2 \sinh \frac{\pi}{\tau_0}\left(y - \frac{\delta}{2}\right)\right] \tag{B.8}$$

This expression can be used to analyze the distribution of the tangential component of the magnetic flux density in the air gap of the electric machine.

Now, we define the border conditions for $\dot{H}_y$ proceed from the condition of kind $divH = 0$ for which, in our case, arises

$$\frac{\partial \dot{H}_x}{\partial x} + \frac{\partial \dot{H}_y}{\partial y} = 0$$

As the border condition for $\dot{H}_y$ can be taken the value of $\frac{\partial \dot{H}_y}{\partial y}$ for which we have that $\frac{\partial \dot{H}_y}{\partial y} = -\frac{\partial \dot{H}_x}{\partial x}$. Now, with consideration of (B.5), it is followed

$$\begin{aligned} \left[\frac{\partial \dot{H}_y(x)}{\partial y}\right]_{y=\delta/2} &= -ja\dot{\varkappa}_1 e^{-jax} \\ \left[\frac{\partial \dot{H}_y(x)}{\partial y}\right]_{y=-\delta/2} &= ja\dot{\varkappa}_2 e^{-jax} \end{aligned} \tag{B.9}$$

We differentiate the second equation of the system (B.3) with respect to y and determine the results of the differentiation for the $y = \delta/2$ and $y = -\delta/2$. On the basis of these data and the conditions given (B.9), the border conditions for $\frac{d\dot{H}_y}{dy}$ acquire the form

$$\begin{aligned} \left[\frac{d\dot{H}_y}{dy}\right]_{y=\delta/2} &= -ja\dot{\varkappa}_1 = -ja\frac{\pi \dot{F}_1}{\tau_0} \\ \left[\frac{d\dot{H}_y}{dy}\right]_{y=-\delta/2} &= ja\dot{\varkappa}_2 = ja\frac{\pi \dot{F}_1}{\tau_0} \end{aligned} \tag{B.10}$$

We differentiate the second equation of the system (B.4) with respect to y and determine the results of the differentiation for the $y = \delta/2$ and $y = -\delta/2$. Now, we compare the equations obtained in this way with the conditions given in (B.10). Then, the new equations of the form arise

$$\begin{aligned} C_3 \cosh \frac{\pi\delta}{2\tau_0} + C_4 \sinh \frac{\pi\delta}{2\tau_0} &= -j\frac{\pi \dot{F}_1}{\tau_0} \\ -C_3 \cosh \frac{\pi\delta}{2\tau_0} - C_4 \sinh \frac{\pi\delta}{2\tau_0} &= j\frac{\pi \dot{F}_2}{\tau_0} \end{aligned}$$

From here, for the integration factors C_3 and C_4, we can have

$$C_3 = \frac{j\pi\left(\dot{F}_2 - \dot{F}_1\right)}{2\tau_0 \cosh \frac{\pi\delta}{2\tau_0}} \qquad C_4 = -\frac{j\pi\left(\dot{F}_2 + \dot{F}_1\right)}{2\tau_0 \sinh \frac{\pi\delta}{2\tau_0}}$$

The integration factors C_3 and C_4 can be substituted into the second equation of the system (B.4). Using the conditions of the form $\cosh(b \pm d) = \cosh b \cdot \cosh d \pm \sinh b \cdot \sinh d$, and also $\dot{B}_y = \mu_0 \dot{H}_y$, the second equation of the system (B.4) acquires

$$\dot{B}_y = -\frac{j\mu_0\pi}{\tau_0 \sinh \frac{\pi\delta}{\tau_0}} \left[\dot{F}_1 \cosh \frac{\pi}{\tau_0}\left(y + \frac{\delta}{2}\right) + \dot{F}_2 \cosh \frac{\pi}{\tau_0}\left(y - \frac{\delta}{2}\right)\right] \qquad \text{(B.11)}$$

The equations (B.8) and (B.11) can be used to analyze the magnetic flux densities in the machine air gap. Subsequently, we will consider the two special cases representing the machine no-load mode and the under-load mode.

No-Load Mode

In the machine no-load mode, the magnetic field can be created in the air gap by the stator winding current (in the absence of the rotor winding current) or by the rotor winding current (in the absence of the stator winding current). When considering the flat air gap (Fig. B.1a), it does not matter which winding (stator or rotor winding) current creates the magnetic field in the air gap. We can proceed from the fact that in accordance with Fig. B.1b, the magnetic field corresponding to the machine no-load mode is created in the air gap by the rotor winding current. In this case, the mmf of the stator winding F_1 is equal to zero, i.e., $F_1 = 0$. In addition, the horizontal axis of the coordinate system is aligned with the rotor surface. This means that in equations (B.8) and (B.11), the value of y should be replaced by the $(y - \delta/2)$. In the machine no-load mode, in equations (B.8) and (B.11) instead of the rotor winding mmf F_2, the no-load mmf F_0 should be used, i.e., $F_2 \to F_0$. With consideration of these conditions, equations (B.8) and (B.11) acquire the form

$$\begin{aligned} \dot{B}_x &= -\frac{\mu_0\pi}{\tau_0 \sinh \dfrac{\pi\delta}{\tau_0}} \dot{F}_0 \sinh \frac{\pi}{\tau_0}(y - \delta) \\ \dot{B}_y &= -\frac{j\mu_0\pi}{\tau_0 \sinh \dfrac{\pi\delta}{\tau_0}} \dot{F}_0 \cosh \frac{\pi}{\tau_0}(y - \delta) \end{aligned} \qquad \text{(B.12)}$$

The equation (B.12) can be used to analyze the distribution in the air gap of the components of the magnetic flux densities in the machine no-load mode.

When $y = \delta$ (stator surface), we have from (B.12)

$$\dot{B}_x = 0 \;\; \dot{B}_y = -\frac{j\mu_0\pi}{\tau_0 \sinh\frac{\pi\delta}{\tau_0}}\dot{F}_0 \tag{B.13}$$

When $y = 0$ (rotor surface)

$$\dot{B}_x = \frac{\mu_0\pi}{\tau_0}\dot{F}_0 = \mu_0\dot{\varkappa}_0 \quad \dot{B}_y = -\frac{j\mu_0\pi}{\tau_0 \sinh\frac{\pi\delta}{\tau_0}}\dot{F}_0 \cosh\frac{\pi\delta}{\tau_0} \tag{B.14}$$

To analyze the distribution in the air gap of the radial component of the magnetic flux density, we introduce the dimensionless factor k_p determined as

$$k_p = \frac{\left[\dot{B}_y\right]_{y=0}}{\left[\dot{B}_y\right]_{y=\delta}} = \cosh\frac{\pi\delta}{\tau_0} \tag{B.15}$$

The expression (B.15) sets the level of the difference of the radial component of the magnetic flux density on the stator surface from its value on the rotor surface in the machine no-load mode.

For the electric machine of the normal design, the largest value of the parameter $\frac{\pi\delta}{\tau_0}$ can be approximately taken equal to $\pi\delta/\tau_0 \approx 0.05$. For this case, we define the value of the dimensionless factor k_p by the expression (B.15). Then, we have that $k_p = 1.0013$. Consequently, it is followed that $k_p \cong 1.0$. This means that when $\pi\delta/\tau_0 = 0.05$, the radial component of the magnetic flux density calculated on the stator and rotor surfaces has almost the same value. Therefore, in the machine no-load mode, the change along the length of the air gap of the radial component of the magnetic flux density can be ignored, which, in practice, it is usually taken for the electric machine of the normal design.

In Fig. B.2, the distribution along the radial length of the air gap of the tangential and radial components of the magnetic flux densities is shown. When $\pi\delta/\tau_0 \leq 0.05$, how it is followed from Fig. B.2, the radial component in the machine no-load mode does not, practically, change along the radial length of the air gap. In the machine no-load mode, the tangential component of the magnetic flux density gets the greatest value on the rotor surface (Fig. B.2) and from this value it decreases almost linearly to zero on the stator surface. From (B.13), it is followed that on the rotor surface the tangential component of the magnetic flux density is determined by the

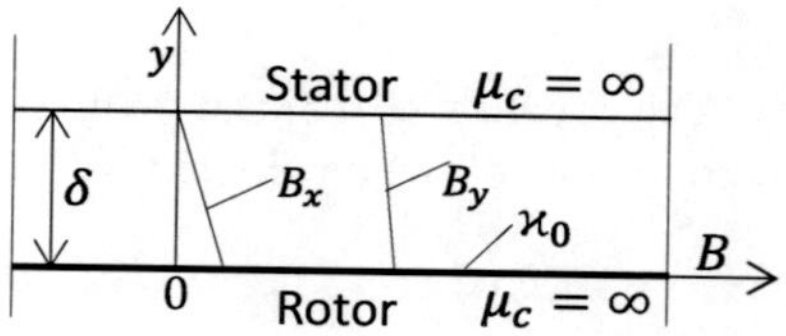

Fig. B.2 Distribution of the tangential and radial components of the magnetic flux densities along the radial length of the air gap (no-load mode)

rotor winding mmf corresponding to the machine no-load mode. The fact that the radial component of the magnetic flux density does not almost change along the radial length of the air gap indicates the relatively small value of the tangential component of the magnetic flux density in the air gap in the machine no-load mode. Therefore, the presence in the air gap of the tangential component of the magnetic flux density does not any significant effect, in many cases, on the calculation results, for example, [1–15].

Under-Load Mode

As it was shown previously, in the machine under-load mode, the condition of the form $\dot{F}_1 = -\dot{F}_2$ is implemented. For this condition, the flat air gap model is shown in Fig. B.3. We use the condition $\dot{F}_1 = -\dot{F}_2$ in the expressions (B.8) and (B.11) obtained for the tangential and radial components of the magnetic flux densities in the machine air gap. Then, we have that

$$\begin{aligned} \dot{B}_x &= \frac{\mu_0 \pi \dot{F}_2}{\tau_0 \sinh \dfrac{\pi\delta}{\tau_0}} \left[\sinh \frac{\pi}{\tau_0}\left(y + \frac{\delta}{2}\right) - \sinh \frac{\pi}{\tau_0}\left(y - \frac{\delta}{2}\right) \right] \\ \dot{B}_y &= \frac{j\mu_0 \pi \dot{F}_2}{\tau_0 \sinh \dfrac{\pi\delta}{\tau_0}} \left[\cosh \frac{\pi}{\tau_0}\left(y + \frac{\delta}{2}\right) - \cosh \frac{\pi}{\tau_0}\left(y - \frac{\delta}{2}\right) \right] \end{aligned} \tag{B.16}$$

Using the known conditions for the hyperbolic functions of the since and cosine, the equations (B.16) can be given as

$$\dot{B}_x = \frac{\mu_0 \pi \dot{F}_2}{\tau_0 \cosh \dfrac{\pi\delta}{2\tau_0}} \cosh \frac{\pi}{\tau_0} y \quad \dot{B}_y = \frac{j\mu_0 \pi \dot{F}_2}{\tau_0 \cosh \dfrac{\pi\delta}{2\tau_0}} \sinh \frac{\pi}{\tau_0} y \tag{B.17}$$

The expressions obtained in (B.17) can be used to analyze the magnetic flux density components in the machine air gap in the under-load mode.

By expression (B.17), the calculations of the distribution in the air gap of the magnetic flux density components $\dot{B}_x$ and $\dot{B}_y$ are implemented. The calculation results obtained for $\pi\delta/\tau_0 = 0.05$ are represented in Fig. B.4. As it is followed from

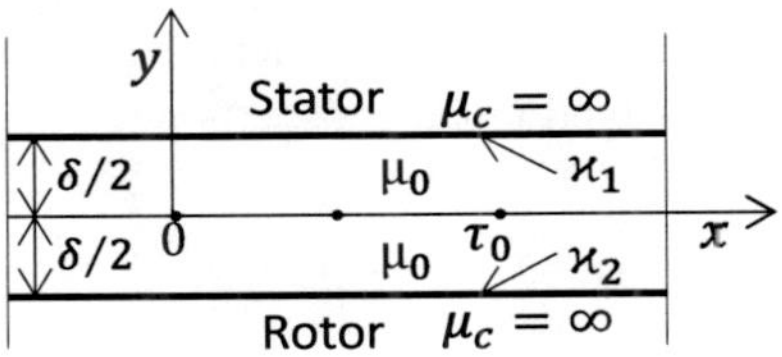

Fig. B.3 Air gap flat model (under-load mode)

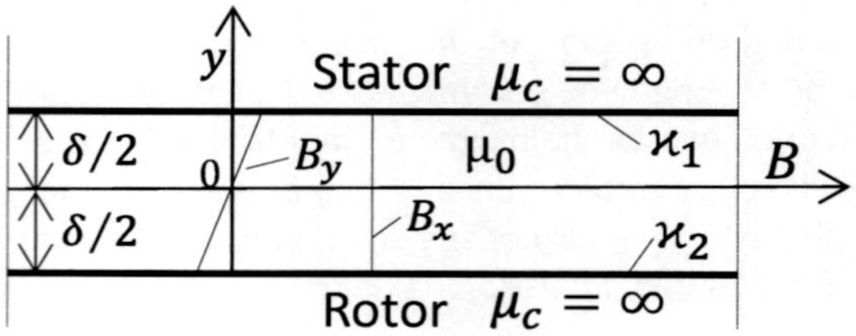

Fig. B.4 Distribution of the tangential and radial components of the magnetic flux densities along the radial length of the air gap (under-load mode)

this figure, the value of the tangential component of the magnetic flux density $\dot{B}_x$ defined on the middle of the radial length of the air gap is slightly less than on the stator and rotor surfaces. However, this value of $\dot{B}_x$ differs little from the values of $\dot{B}_x$ taken place on the stator and rotor surfaces. Therefore, in practice, such change of the tangential component of the magnetic flux density $\dot{B}_x$ is neglected and it is assumed that $\dot{B}_x$ has the constant value along the radial length of the air gap in the machine under-load mode.

In Fig. B.4, the distribution along the radial length of the air gap of the radial component of the magnetic flux density $\dot{B}_y$ is shown in the machine under-load mode. In the machine under-load mode, the mutual induction magnetic flux created in the air gap by the rotor winding current is compensated by the mutual induction magnetic flux caused by the stator winding current, i.e., $\Phi_{11} = -\Phi_{22}$. It can be indicated from Fig. B.4 that the condition $\Phi_{11} = -\Phi_{22}$ is implemented on the surface $0 - 0$ passing through the middle of the radial length of the air gap, since on this surface the radial component of the magnetic flux density $\dot{B}_y$ takes the value equal to zero, i.e., $\dot{B}_y = 0$. The distribution along the radial length of the air gap of the magnetic flux density $\dot{B}_y$ occurs in such way that in relation to the surface $0 - 0$ the flux density $\dot{B}_y$ changes the sign. In this case, the value of $\dot{B}_y$ is relatively small. Indeed, by the expressions shown in (B.17), for the ratio of the radial component of the magnetic flux density $\dot{B}_y$ to the tangential component of the magnetic flux density $\dot{B}_x$ determined on the stator surface, we can have that $k_y = \frac{[\dot{B}_y]_{y=\delta/2}}{[\dot{B}_x]_{y=\delta/2}} = \tanh \frac{\pi\delta}{2\tau_0}$. When $\pi\delta/\tau_0 \approx 0.05$, the value of the dimensionless factor k_y is $k_y = \tanh 0.025 \approx 0.025$. It can be shown that the dimensionless factor k_y takes a similar value (with the opposite sign) on the rotor surface. Consequently, in our case, we have that $k_y = \pm 0.025$. This means that when $\pi\delta/\tau_0 = 0.05$, the radial component of the magnetic flux density corresponding to the air gap takes a relatively small value in the machine under-load mode. This is evidenced by the fact that the tangential component of the magnetic flux density does not almost change, as it is followed along the radial length of the air gap (Fig. B.4). Therefore, in practice, this small value of the radial component of the magnetic flux density $\dot{B}_y$ taken place in the air gap in the machine under-load mode is usually neglected.

Summary

1. As a result of the analysis of the magnetic field in the air gap in no-load mode, we can conclude that in the gap of the machine of the normal design, the tangential component of the magnetic flux density is relatively small, and therefore, it can be neglected. In this case, there is only the radial component of the magnetic flux density in the machine air gap. We can note that its value does not change along the radial length of the air gap.
2. In the under-load mode, the radial component of the magnetic flux density takes the value equal to zero on the surface $0-0$ passing through the middle of the radial length of the air gap. The distribution along the length of the air gap of the radial component of the magnetic flux density occurs in such way that in relation to the surface $0-0$ it changes the sign. Shown that the radial component of the magnetic flux density taking place in the air gap in the under-load mode has a relatively small value and it is usually neglected. In this regard, the tangential component of the magnetic flux density does not almost change along the radial length of the air gap. Therefore, in the under-load mode, the tangential component of the magnetic flux density exists in the air gap of the electric machine.

References

1. Kostenko, M., Piotrovsky, L.: Electric Machines, Energiya, Leningrad (1973)
2. Piotrovskij, L.: Electric Machines, Publishing Energiya, Leningrad (1975)
3. Voldek. A.: Electric Machines, Energiya, Leningrad (1974)
4. McPherson, G., Laramore, R.: Electric Machines and Transformers, 2nd ed., John Wiley and Sons, Inc., New York (1990)
5. Say, M.: Alternating Current Machines, John Wiley and Sons, New York (1902)
6. Fitzgerald, A., Kingsley, C., Umans, S.: Electric Machinery, McGraw-Hill, Inc., 5th ed., New York (1990)
7. Alger, P.: Induction Machines, 2nd ed., Gordon and Beach (1970)
8. Del Toro, V.: Electric Machines and Power Systems, Prentice-Hall, Inc., Englewood Cliffs, New Jersey (1985)
9. Krause, P.: Analysis of Electric Machinery, McGraw-Hill, New York (1986)
10. Chapman, S.: Electric Machinery Fundamentals, 4th ed., McGraw-Hill, Inc., New York (2005)
11. Vazhnov, A.: Electric Machines, Energiya, Leningrad (1968)
12. Asanbayev, V.: Alternating Current Multi-Circuit Electric Machines. A New Approach to Steady-State Parameter Determination, Springer Cham Heidelberg, New York, Dordrecht London (2015)
13. Seely, S.: Electromechanical Energy Conversion, McGraw Book Company Inc., New York (1962)
14. Voldek, A.: Induction Magneto-Hydrodynamic Machines with Liquid-Metal Working Substance, Publishing Energiya, Leningrad (1970)
15. Ivanov-Smolensky, A.: Electric Machines, Publishing Energiya, Moscow (1980)
16. Shimoni, K.: Theoretical Electrical Engineering (Translation from German), Publishing Mir, Moscow (1973)
17. Dombrovsky, V., Zaichik, V.: Asynchronous Machines: Theory, Calculation, Elements of Design, Publishing Energoatomizdat, Leningrad (1990)

Appendix C: EMF Equations of Air Gap in Form of A: Transformation

Air Gap emf Equations in Form of *A*

The transformation techniques considered in this chapter are developed on the basis of applying the air gap emf equations represented in (6.19) in the form of *Z*

$$\begin{aligned}(-\dot{Э}_1) &= z_{11}\dot{I}_1 + z_{21}\dot{I}_2 \\ (-\dot{Э}_2) &= z_{12}\dot{I}_1 + z_{22}\dot{I}_2\end{aligned} \tag{C.1}$$

where $z_{11} = jx_{11}$; $z_{21} = jx_{21}$; $z_{12} = jx_{12}$ and $z_{22} = jx_{22}$.

The air gap emf equations can be represented in the form of *A*. Then, the equation (C.1) in relation to the stator winding emf $\dot{Э}_1$ and the current $\dot{I}_1$ acquires the form

$$\begin{aligned}(-\dot{Э}_1) &= a_{11}(-\dot{Э}_2) + a_{21}\dot{I}_2 \\ \dot{I}_1 &= a_{12}(-\dot{Э}_2) + a_{22}\dot{I}_2\end{aligned} \tag{C.2}$$

The coefficients of the system of the equations (C.2) are expressed through the coefficients of the equations (C.1) as

$$\begin{aligned}a_{11} &= \frac{z_{11}}{z_{12}} \quad a_{21} = \frac{\Delta_z}{z_{12}} \\ a_{12} &= \frac{1}{z_{12}} \quad a_{22} = \frac{z_{22}}{z_{12}}\end{aligned} \tag{C.3}$$

where $\Delta_z = z_{11}z_{22} - z_{21}z_{12}$.

On the basis of the expressions obtained in (7.21), for the coefficients of the system of the equations (C.1), we have

V. Asanbayev, *Asynchronous Machines*, https://doi.org/10.1007/978-3-030-92284-9

$$\begin{aligned} x_{11} &= \frac{\omega_1\mu_0\tau_0^2}{\pi^2\delta'}k_1 \quad x_{21} = \frac{\omega_1\mu_0\tau_0^2}{\pi^2\delta'}k_{21} \\ x_{12} &= \frac{\omega_1\mu_0\tau_0^2}{\pi^2\delta'}k_{12} \quad x_{22} = \frac{\omega_1\mu_0\tau_0^2}{\pi^2\delta'}k_2 \end{aligned} \tag{C.4}$$

Here, the coefficients k_1, k_{21}, k_{12}, and k_2 are determined by the expressions given in (7.21) as

$$\begin{aligned} k_1 &= \frac{2m_1(w_1k_{w1})^2 l_0}{p\tau_0} \quad k_{21} = \frac{2m_2(w_1k_{w1}w_2k_{w2})l_0}{p\tau_0} \\ k_{12} &= \frac{2m_1(w_1k_{w1}w_2k_{w2})l_0}{p\tau_0} \quad k_2 = \frac{2m_2(w_2k_{w2})^2 l_0}{p\tau_0} \end{aligned} \tag{C.5}$$

The coefficients of the system of the equations (C.2) from (C.3) and (C.4) take the form

$$\begin{aligned} a_{11} &= \frac{z_{11}}{z_{12}} = \frac{w_1k_{w1}}{w_2k_{w2}} \quad a_{21} = \frac{\Delta_z}{z_{12}} = 0 \\ a_{12} &= \frac{1}{z_{12}} = \frac{1}{\frac{\omega_1\mu_0\tau_0^2}{\pi^2\delta'}k_{12}} \quad a_{22} = \frac{z_{22}}{z_{12}} = \frac{1}{\frac{m_1}{m_2}\frac{w_1k_{w1}}{w_2k_{w2}}} \end{aligned} \tag{C.6}$$

where $\Delta_z = z_{11}z_{22} - z_{21}z_{12} = 0$.

With considering the (C.6), the system of the equations (C.2) acquires the form

$$\begin{aligned} (-\dot{Э}_1) &= a_{11}(-\dot{Э}_2) + 0 \\ \dot{I}_1 &= a_{12}(-\dot{Э}_2) + a_{22}\dot{I}_2 \end{aligned} \tag{C.7}$$

since, in our case, in accordance with (C.6) for the coefficient a_{21} of the system of the equation (C.2), we have that $a_{21} = 0$.

Approach to Transformation of emf Equations of Air Gap

To the system of the equations (C.1) and (C.2), the equivalent circuit having the mutual induction reactances shown in Fig. 6.4a corresponds. The transformation of the equivalent circuit in Fig. 6.4a is usually done in order to simplify its structure by eliminating in it the mutual induction reactances. Such transformations of the equivalent circuit in Fig.6.4a are implemented previously. Here, the transformations made previously, we will consider on the basis of the equations obtained in (C.7). In this case, we will assume that in accordance with the “no-load and under-load” concept, the stator winding current is determined by the equation of the form

$$\dot{I}_1 = \dot{I}_0 + (-\dot{I}'_2) \tag{C.8}$$

Here, $\dot{I}_0$ represents the current in stator winding in the machine no-load mode and $(-\dot{I}'_2)$ reflects the current taken place in the stator winding in the machine under-load mode. Above, the "no-load and under-load" concept was used to convert the air gap emf equations shown in (C.1). Now, this concept is used to convert the air gap emf equations of the form given in (C.7).

In accordance with the equation (C.7), it follows that the transformation of the system of the equation (C.7) is reduced to the determination of the referred values of the rotor winding emf $\dot{Э}'_2$ and the current $\dot{I}'_2$ and also to the representation of the equations (C.7) through the referred values of the rotor winding emf $\dot{Э}'_2$ and the current $\dot{I}'_2$. As a result, equation (C.7) acquires the transformed form. Below, we will show this by the example of implementing in equation (C.7) the transformation techniques described previously (stator winding current decomposition technique, transfer technique, and reduction technique).

Stator Winding Current Decomposition Technique

We proceed from the equations obtained in (C.7). The currents $\dot{I}_0$ and $(-\dot{I}'_2)$ can be determined by the equations (C.7). The current $\dot{I}_0$ follows from the equation (C.7) when for the current $\dot{I}_2$, we have that $\dot{I}_2 = 0$. For the current $(-\dot{I}'_2)$ from (C.8), we can obtain

$$(-\dot{I}'_2) = \dot{I}_1 - \dot{I}_0 \tag{C.9}$$

No-load mode. On the basis of the equations (C.7), for the emf $\dot{Э}_1$ and the current $\dot{I}_1$, we receive when $\dot{I}_2 = 0$

$$(-\dot{Э}_1) = a_{11}(-\dot{Э}_2) \quad \dot{I}_0 = a_{12}(-\dot{Э}_2) \tag{C.10}$$

Here, the first equation can be reduced to the form

$$(-\dot{Э}_1) = a_{11}(-\dot{Э}_2) = (-\dot{Э}'_2)$$

From here, it is followed

$$\dot{Э}'_2 = a_{11}\dot{Э}_2 = k_Э\dot{Э}_2 \tag{C.11}$$

where $k_Э = a_{11}$ is the reduction factor of the rotor winding emf; $\dot{Э}'_2$ is the referred to the stator winding value of the rotor winding emf.

From (C.6) and (C.11), the emf reduction factor $k_Э$ is determined as

$$k_Э = a_{11} = \frac{w_1 k_{w1}}{w_2 k_{w2}} \tag{C.12}$$

From (C.10), the current $\dot{I}_0$ follows

$$\dot{I}_0 = a_{12}(-\dot{Э}_2) = \frac{a_{12}}{a_{11}}(-\dot{Э}'_2) = \frac{(-\dot{Э}_1)}{jx_0} \tag{C.13}$$

where $x_0 = a_{11}/a_{12}$.

Here, x_0 represents the air gap magnetizing reactance for which from (C.6) and (C.13), we receive

$$x_0 = \frac{a_{11}}{a_{12}} = \frac{\omega_1 \mu_0 \tau_0^2}{\pi^2 \delta'} k_1 \tag{C.14}$$

where $k_1 = \dfrac{2m_1 (w_1 k_{w1})^2 l_0}{p\tau_0}$.

Under-load mode. In this case, the current $(-\dot{I}'_2)$ is determined from the equation (C.9) in the form $(-\dot{I}'_2) = \dot{I}_1 - \dot{I}_0$. In accordance with this equation, to find the current $(-\dot{I}'_2)$ it is necessary to have the currents $\dot{I}_0$ and $\dot{I}_1$. The value of the current $\dot{I}_0$ arises from the expression (C.13). The value of the current $\dot{I}_1$ can be established by the system of the equations (C.7). From the second equation of the system (C.7), we have that $\dot{I}_1 = a_{12}(-\dot{Э}_2) - a_{22}\dot{I}_2$. From the first equation of the system (C.7), the emf $\dot{Э}_2$ can be defined as $(-\dot{Э}_2) = (-\dot{Э}_1)/a_{11}$. Then, for the current $\dot{I}_1$ it is followed that $\dot{I}_1 = \frac{a_{12}}{a_{11}}(-\dot{Э}_1) - a_{22}\dot{I}_2$. Now, taking into account from (C.13) that $\dot{I}_0 = \frac{a_{12}}{a_{11}} \times (-\dot{Э}_1)$, for the current $(-\dot{I}'_2)$, from (C.9), we have

$$(-\dot{I}'_2) = \dot{I}_1 - \dot{I}_0 = -a_{22}\dot{I}_2$$

From here

$$\dot{I}'_2 = a_{22}\dot{I}_2 = \frac{\dot{I}_2}{k_I} \tag{C.15}$$

where $k_I = 1/a_{22}$ is the reduction factor of the rotor winding current; $\dot{I}'_2$ is the referred to the stator winding value of the rotor winding current.

From (C.6) and (C.15), the current reduction factor of the rotor winding acquires the form

$$k_I = \frac{1}{a_{22}} = \frac{m_1 w_1 k_{w1}}{m_2 w_2 k_{w2}} \tag{C.16}$$

Operating mode. The air gap emf equations corresponding to the operating mode can be obtained by the system of the equations given in (C.7). For this purpose, equation (C.7), it is necessary to express through the referred values of the rotor winding emf $\dot{Э}'_2$ and the current $\dot{I}'_2$ determined from (C.11) and (C.15). Then, we can obtain

$$\begin{aligned} \dot{Э}_1 &= \dot{Э}'_2 \\ \dot{I}_1 &= \frac{(-\dot{Э}_1)}{jx_0} - \dot{I}'_2 \end{aligned} \tag{C.17}$$

To the system of the equations (C.17) represented in the form of *A*, the air gap equivalent circuit shown in Fig. 9.2 corresponds. It can be shown that the air gap emf equations expressed in the form of *Z* obtain from (C.17) the known form

$$\begin{aligned} (-\dot{Э}_1) &= jx_0\dot{I}_0 \\ (-\dot{Э}'_2) &= jx_0\dot{I}_0 \\ \dot{I}_0 &= \dot{I}_1 + \dot{I}'_2 \end{aligned} \tag{C.18}$$

We can note that to the equations (C.17) and (C.18) represented in the different forms, the air gap equivalent circuit given in Fig. 8.3b is true.

Transfer Technique

The transfer technique is implemented by the equivalent circuit obtained on the basis of the equations (C.7). In this case, it can be shown that the air gap equivalent circuit following the equations (C.7) can be constructed in the form of Fig. 9.1. Above, the equivalent circuit in Fig. 9.1 was obtained by the equations of the form (C.1) represented in the form of *Z* and it is transformed to the form of Fig. 8.3b. Therefore, the realization of the transfer technique on the basis of the equivalent circuit constructed by equation (C.7) can be implemented by the transformation techniques described previously.

Reduction Technique

As it was shown above, the reduction technique is implemented when using the conditions of the form $\dot{Э}_1 = 0$ and $\dot{I}_2 = 0$. We will realize these conditions in equation (C.7).

No-load mode. When $\dot{I}_2 = 0$ from (C.7) it follows

$$(-\dot{Э}_1) = a_{11}\dot{Э}_2 \quad \dot{I}_0 = a_{12}(-\dot{Э}_2) \tag{C.19}$$

From here, for the emf $\dot{Э}'_2$ and the current $\dot{I}_0$, we have

$$(-\dot{Э}'_2) = a_{11}(-\dot{Э}_2) = k_Э(-\dot{Э}_2) \quad \dot{I}_0 = \frac{(-\dot{Э}'_2)}{jx_0} = \frac{(-\dot{Э}_1)}{jx_0} \tag{C.20}$$

where $\dot{Э}_1 = \dot{Э}'_2; k_Э = a_{11}; x_0 = a_{11}/a_{12}$.

Under-load mode. In the under-load mode $(\dot{Э}_1 = 0)$, from the second equation of the system (C.7), we can obtain that $\dot{I}_1 = -a_{22}\dot{I}_2 = -\dot{I}'_2$. From here, the current $\dot{I}'_2$ acquires

$$\dot{I}'_2 = a_{22}\dot{I}_2 = (1/k_I)\dot{I}_2 \tag{C.21}$$

where $k_I = 1/a_{22}$.

Now, the referred values of the rotor winding emf $\dot{Э}'_2$ and the current $\dot{I}'_2$ determined from (C.20) and (C.21) can be used in equation (C.7), and then, to reduce these equations to the form of the air gap emf equations shown in (C.17).

Thus, the air gap equivalent circuit and the air gap emf equations suitable for practical use can be obtained by the system of equations (C.1) and (C.2) represented in the form of *Z* and in the form of *A*, respectively.

Summary

The procedures and the techniques used to transform the air gap emf equations shown in the form of *Z* can be applied when transforming the air gap emf equations represented in the form of *A*. The air gap equivalent circuit resulting from the application of these two ways of transforming has the same structure.

Reference

1. Asanbayev, V.: Alternating Current Multi-Circuit Electric Machines: A New Approach to Steady-State Parameter Determination, Springer Cham Heidelberg, New-York, Dordrecht London (2015).

Appendix D: Practical Technique for Constructing Equivalent Circuits of Machine

Equivalent Circuit of Machine with Single-Winding Rotor

We consider the asynchronous machine with the single-winding rotor. The machine equivalent circuit containing the inductively coupled elements is given in Fig. 6.5. In order to obtain the equivalent circuit convenient for practical use, the equivalent circuit in Fig. 6.5 can be transformed. However, this goal can be achieved by transforming the voltage equations (6.15) and (6.16). The transformation of the voltage equation (6.15) can be implemented by the use of the "no-load and under-load" concept applied, for example, in [1–6].

Machine equivalent circuit: No-load mode. We consider the equations (6.15) and (6.16) in the absence of the current in the rotor winding $\left(\dot{I}_2 = 0\right)$. When $\dot{I}_2 = 0$, the first equation of the system (6.15) acquires the form

$$\dot{U}_1 = r_1 \dot{I}_m + jx_{1\sigma}\dot{I}_m + (-\dot{Э}_1) \tag{D.1}$$

where $\dot{I}_m$ is the no-load current in the stator winding.

Here, the emf $(-\dot{Э}_1)$ is induced in the stator winding by the magnetic flux Φ_m which induces in the rotor winding the emf $(-\dot{Э}_2)$. When $\dot{I}_2 = 0$, the system of the equations (6.16) can be reduced to the form

$$(-\dot{Э}_1) = jx_m\dot{I}_m \qquad (-\dot{Э}_2') = jx_m\dot{I}_m \tag{D.2}$$

where $x_m = x_{11}; x_m = k_Э x_{12}; \dot{Э}_2' = k_Э \dot{Э}_2$.

As it follows from these equations, the emf induced in the rotor winding takes the referred to the stator winding value $\dot{Э}_2'$. With considering (D.2), the equation (D.1) acquires

V. Asanbayev, *Asynchronous Machines*, https://doi.org/10.1007/978-3-030-92284-9

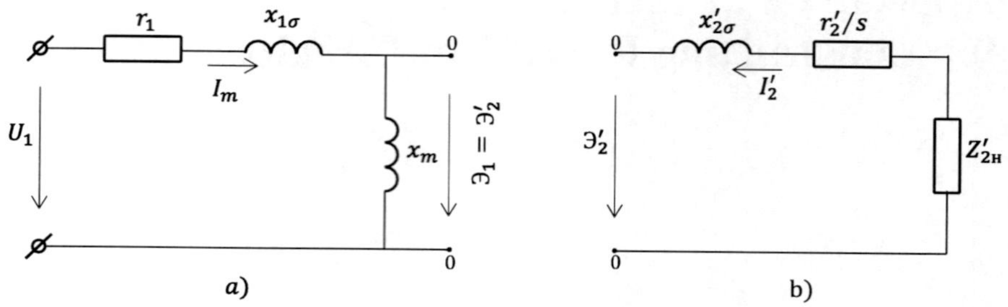

Fig. D.1 To the implementation of the "no-load and under-load" concept as applied to the electric machine with the single-winding rotor: (**a**) stator equivalent circuit for the no-load mode; (**b**) rotor equivalent circuit for the under-load mode

$$\begin{aligned}\dot{U}_1 &= r_1\dot{I}_m + jx_{1\sigma}\dot{I}_m + jx_m\dot{I}_m \\ (-\dot{Э}_1) &= (-\dot{Э}_2^{'}) = jx_m\dot{I}_m\end{aligned} \tag{D.3}$$

Now, on the basis of equation (D.3), the stator equivalent circuit obtains the form in Fig. D.1a. It is shown in the equivalent circuit in Fig. D.1a, the emf $\dot{Э}_2^{'}$ induced in the rotor winding has the referred to the stator winding value and it is determined by the expression given in (8.2).

Rotor equivalent circuit: Under-load mode. In the under-load mode, the rotor winding is closed to the load impedance $Z_{2н}$ (impedance $Z_{2н}$ has the referred to the frequency of the stator winding current value). Then, the voltage drop across the impedance $Z_{2н}$ can be replaced by the voltage on the rotor winding terminals, i.e., $\dot{U}_2/s = \dot{I}_2 Z_{2н}$. In these conditions, we can accept that the rotor winding current $\dot{I}_2$ arises under the influence of the emf $\dot{Э}_2$ induced in the rotor winding in the no-load mode. Then, for the rotor winding voltage equation, we have from (6.15)

$$\dot{Э}_2 = [(r_2/s) + jx_{2\sigma} + Z_{2н}]\dot{I}_2 \tag{D.4}$$

In the no-load mode, the terminals 0 – 0 of the stator equivalent circuit (Fig. D.1a) are under the influence of the referred to the stator winding value of the rotor winding emf $\dot{Э}_2^{'}$. Therefore, we can proceed from the fact that in the under-load mode the rotor winding current arises under the influence of the referred to the stator winding value of the rotor winding emf $\dot{Э}_2^{'}$. Since in this case $m_2\dot{Э}_2\dot{I}_2 = m_1\dot{Э}_2^{'}\dot{I}_2^{'}$, we have that $\dot{I}_2 = k_Э\frac{m_1}{m_2}\dot{I}_2^{'} = k_I\dot{I}_2^{'}$, where $k_I = k_Э\frac{m_1}{m_2}$. From here, it is followed that $\dot{I}_2^{'} = \dot{I}_2/k_I$. Consequently, the current $\dot{I}_2^{'}$ having the referred to the stator winding value is flowed in the rotor winding. Then, the rotor winding voltage equation given in (D.4) obtains through the referred values of the emf $\dot{Э}_2^{'}$ and the current $\dot{I}_2^{'}$ the form

$$\dot{Э}_2' = \left[\left(r_2'/s\right) + jx_{2\sigma}' + Z_{2\text{H}}'\right]\dot{I}_2' \tag{D.5}$$

where $x_{2\sigma}' = k_2 x_{2\sigma}$; $r_2' = k_2 r_2, Z_{2\text{H}}' = k_2 Z_{2\text{H}}$ and $k_2 = k_Э k_I$.

On the basis of this equation, the rotor equivalent circuit corresponding to the machine under-load mode can be represented in Fig. D.1b.

Machine equivalent circuit: Operating mode. In accordance with (8.11), the "no-load and under-load" concept is realized when the stator winding current is determined as

$$\dot{I}_1 = \dot{I}_m + \left(-\dot{I}_2'\right) \tag{D.6}$$

The stator winding current $\dot{I}_m$ corresponding to the no-load mode is flowed in the equivalent circuit in Fig. D.1b. The current $\dot{I}_2'$ reflecting the under-load mode is flowed in the equivalent circuit in Fig. D.1b. In this regard, to construct the machine equivalent circuit satisfying the condition (D.6) for the stator winding current, the input terminals $0 - 0$ of the rotor equivalent circuit corresponding to the under-load mode (Fig. D.1b) it is necessary to connect with the output terminals $0 - 0$ of the stator equivalent circuit following from the no-load mode (Fig. D.1a). Such connection is feasible since the terminals $0 - 0$ of the equivalent circuits in Fig. D.1 are under the influence of the same emf $\dot{Э}_2'$. As a result, the machine equivalent circuit convenient for the description of the operating mode acquires the form in Fig. 10.6. When receiving the equivalent circuit in Fig. 10.6, the condition of the form $\dot{I}_2' Z_{2H}' = \dot{I}_2' Z_{2H} k_2 = \left(\dot{U}_2 k_Э / s\right) = \dot{U}_2'/s$, where $\dot{U}_2' = \dot{U}_2 k_Э$ and $k_2 = k_Э k_I$, is used. On the basis of the equivalent circuits given in Figs. 6.4 and 10.6, it can be noted that as a result of applying the "no-load and under-load" concept, the equivalent circuit having the inductively coupled elements (Fig. 6.4) is transformed into the equivalent circuit in which the stator and rotor windings are represented as the conditionally electrically connected circuit loops (Fig. 10.6). The machine equivalent circuit shown in Fig. 10.6 is widely used in practice.

Equivalent Circuit of Machine with Three-Winding Rotor

In the machine with the three-winding rotor, there are the mutual induction magnetic fluxes of the two types: the mutual induction magnetic flux closing on the main magnetic circuit and embracing the stator and rotor windings and also the mutual induction magnetic fluxes closing along the rotor leakage paths and embracing the rotor windings. This circumstance significantly complicates the process of constructing the equivalent circuit of the machine with the three-winding rotor. Here, to construct the equivalent circuit of the machine with the three-winding rotor, we use the approximate technique. In this case, the "no-load and under-load" concept used above can be applied to construct the equivalent circuit of the machine with the three-winding rotor.

The asynchronous machine having the on the rotor the distributed and symmetrical three-phase windings of the same design is considered. We proceed from the fact that the rotor windings are closed and they are not powered from the external source (from the exciter).

Machine equivalent circuit: No-load mode. Suppose that the rotor is stopped and its windings are open. The currents in the rotor windings are absent. In the stator winding, the no-load current $\dot{I}_m$ is flowed. Then, for the stator winding, the voltage equation takes the form

$$\dot{U}_1 = r_1\dot{I}_m + jx_{1\sigma}\dot{I}_m + (-\dot{\text{Э}}_1) \tag{D.7}$$

In the no-load mode, the emf $\dot{\text{Э}}_2$, $\dot{\text{Э}}_3$, and $\dot{\text{Э}}_4$ are induced in the open rotor windings. For convenience, in the case of using on the rotor the windings of the same design, for the emf $\dot{\text{Э}}_2$, $\dot{\text{Э}}_3$ and $\dot{\text{Э}}_4$, the conditions of the form $\dot{\text{Э}}_2 = \dot{\text{Э}}_3 = \dot{\text{Э}}_4$ are implemented. The emf $\dot{\text{Э}}_2$, $\dot{\text{Э}}_3$ and $\dot{\text{Э}}_4$ should be reduced to the stator winding emf $\dot{\text{Э}}_1$. The referred values of the emf $\dot{\text{Э}}'_2$, $\dot{\text{Э}}'_3$ and $\dot{\text{Э}}'_4$ take

$$\dot{\text{Э}}'_2 = k_{\text{Э}}\dot{\text{Э}}_2 \quad \dot{\text{Э}}'_3 = k_{\text{Э}}\dot{\text{Э}}_3 \quad \dot{\text{Э}}'_4 = k_{\text{Э}}\text{Э}_4 \tag{D.8}$$

For the emf reduction factor $k_{\text{Э}}$, we have

$$k_{\text{Э}} = \frac{w_1 k_{w1}}{w_i k_{wi}} \tag{D.9}$$

where $i = 2, 3, 4$.

In (D.7), the value of the emf $(-\dot{\text{Э}}_1)$ can be represented as $(-\dot{\text{Э}}_1) = jx_m\dot{I}_m$. Then, it follows

$$\dot{U}_1 = (r_1 + jx_{1\sigma})\dot{I}_m + jx_m\dot{I}_m \tag{D.10}$$

In Fig. D.2, the distribution of the stator winding leakage fluxes and the magnetizing flux of the machine with the three-winding rotor is represented. From Fig. D.2 and the equation (D.10), the stator equivalent circuit acquires the form in Fig. D.3a.

Rotor equivalent circuits: Under-load mode. In the considered case, the three identical windings are located on the rotor. The under-load mode can be implemented for each winding separately.

Equivalent circuit of the rotor with the upper winding: Under-load mode. Suppose that the upper rotor winding is closed. In the upper rotor winding, the current I'_2 is flowed under the influence of the referred to the stator winding value of the emf $\text{Э}'_2$ induced in the upper rotor winding in the no-load mode. The referred value of the upper rotor winding current I'_2 is determined as

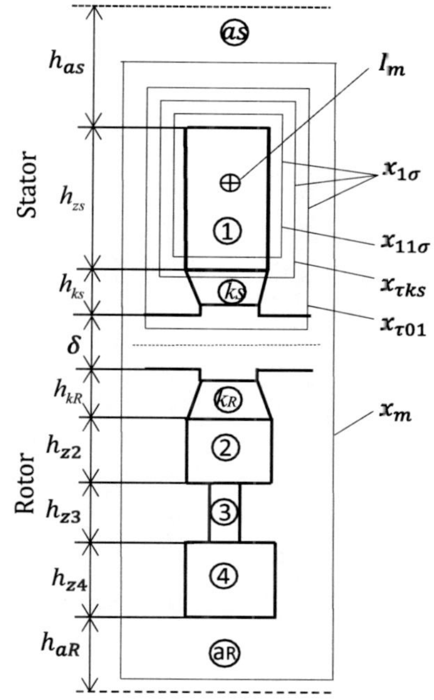

Fig. D.2 Picture of the conditional lines of the machine magnetizing flux and the stator slot leakage magnetic fluxes: no-load mode

$$\dot{I}'_2 = \frac{\dot{I}_2}{k_I} \tag{D.11}$$

where $k_I = \frac{m_1}{m_2}\frac{w_1 k_{w1}}{w_2 k_{w2}}$.

The voltage equation for the upper rotor winding follows in the form

$$\dot{Э}'_2 = \left(\frac{r'_2}{s} + jx'_{2\sigma}\right)\dot{I}'_2 \tag{D.12}$$

Here, the referred to the stator winding values of the ohmic resistance and the slot leakage reactance of the upper rotor winding are used. For which, we have

$$r'_2 = k_2 r_2 \quad x'_{2\sigma} = k_2 x_{2\sigma} \tag{D.13}$$

where $k_2 = k_Э k_I$ and $k_2 = \frac{m_1}{m_2}\frac{(w_1 k_{w1})^2}{(w_2 k_{w2})^2}$.

The referred to the stator winding value of the slot leakage reactance of the upper rotor winding $x'_{2\sigma}$ can be represented as on the basis of Fig. D.4.

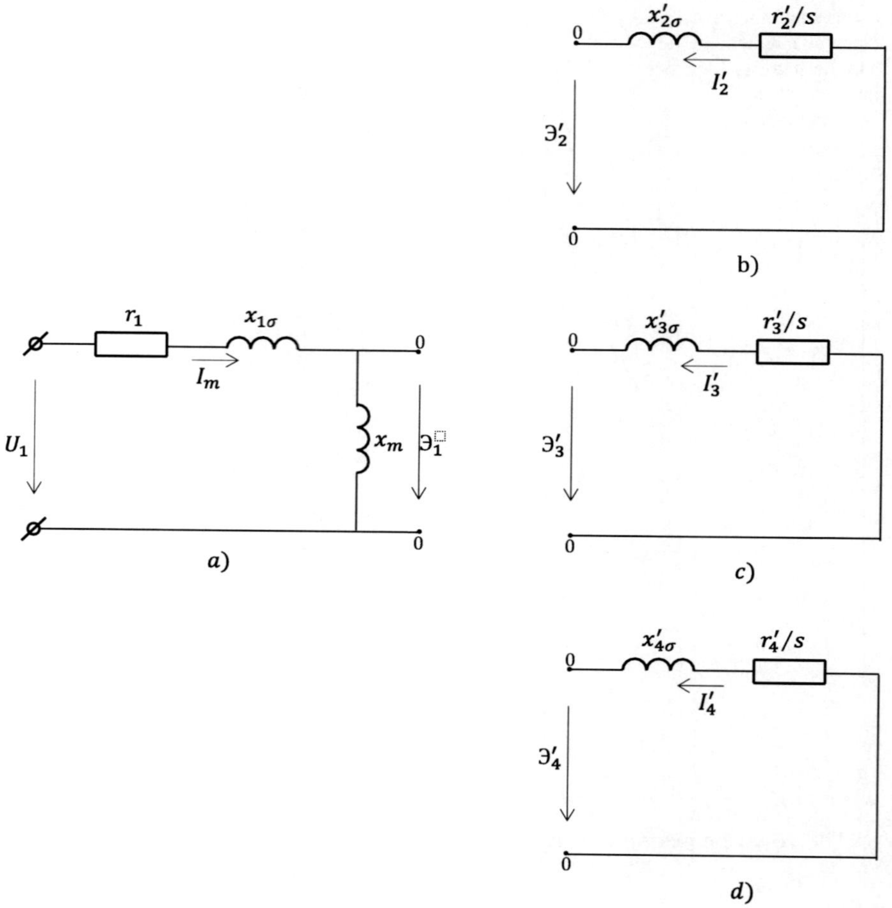

Fig. D.3 To the implementation of the "no-load and under-load" concept as applied to the electric machine with the three-winding rotor: (**a**) stator equivalent circuit for the no-load mode; (**b**) rotor upper winding equivalent circuit for the under-load mode; (**c**) rotor middle winding equivalent circuit for the under-load mode; (**d**) rotor lower winding equivalent circuit for the under-load mode

$$x'_{2\sigma} = x'_{22\sigma} + x'_{\tau kR} + x'_{\tau 02} \tag{D.14}$$

In accordance with the equation (D.12), the upper rotor winding equivalent circuit is obtained in the form represented in Fig. D.3b.

Equivalent circuit of the rotor with the middle winding: Under-load mode. Now, we proceed from the fact that the middle rotor winding is closed. In the middle rotor winding, the current I'_3 is flowed under the influence of the referred to the stator winding value of the emf $Э'_3$ induced in the middle rotor winding in the no-load mode. The referred value of the middle rotor winding current I'_3 is established as

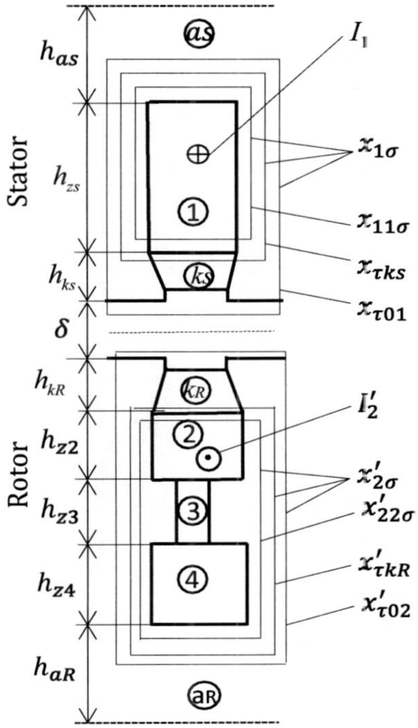

Fig. D.4 Picture of the conditional lines of the stator winding leakage magnetic fluxes and the rotor upper winding leakage magnetic fluxes

$$\dot{I}_3' = \frac{\dot{I}_3}{k_I} \tag{D.15}$$

Here, the current reduction factor k_I is determined by the expression given in (D.11).

The voltage equation for the middle rotor winding obtains

$$\dot{Э}_3' = \left(\frac{r_3'}{s} + jx_{3\sigma}'\right)\dot{I}_3' \tag{D.16}$$

In (D.16), the referred to the stator winding values of the ohmic resistance and the slot leakage reactance of the middle rotor winding are used. Therefore, we have

$$r_3' = k_2 r_3 \quad x_{3\sigma}' = k_2 x_{3\sigma} \tag{D.17}$$

The factor k_2 used in (D.17) is determined by the expression shown in (D.13).

Fig. D.5 Picture of the conditional lines of the stator winding leakage magnetic fluxes and the rotor middle winding leakage magnetic fluxes

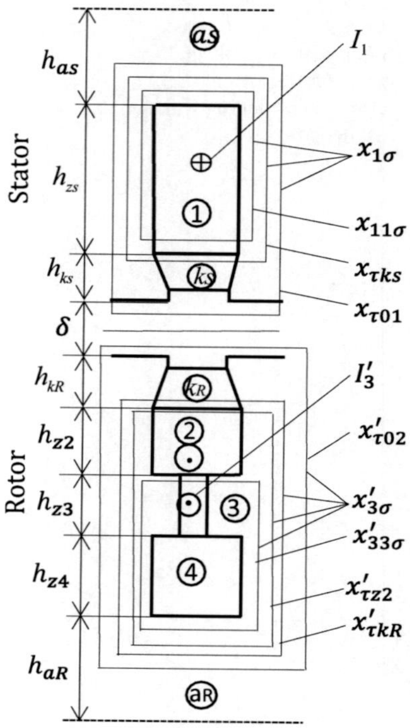

The referred stator winding value of the slot leakage reactance of the middle rotor winding $x'_{3\sigma}$ can be represented as on the basis of Fig. D.5.

$$x'_{3\sigma} = x'_{33\sigma} + x'_{\tau z2} + x'_{\tau kR} + x'_{\tau 02} \tag{D.18}$$

On the basis of the equation (D.16), the middle rotor winding equivalent circuit is obtained in the form represented in Fig. D.3c.

Equivalent circuit of the rotor with the lower winding: Under-load mode. We consider the case when the lower rotor winding is closed. In the lower rotor winding, the current I'_4 is flowed under the influence of the referred to the stator winding value of the emf $Э'_4$ induced in the lower rotor winding in the no-load mode. The referred value of the lower rotor winding current I'_4 is determined as

$$\dot{I}'_4 = \frac{\dot{I}_4}{k_I} \tag{D.19}$$

In (D.19), the current reduction factor k_I is defined by the expression given in (D.11).

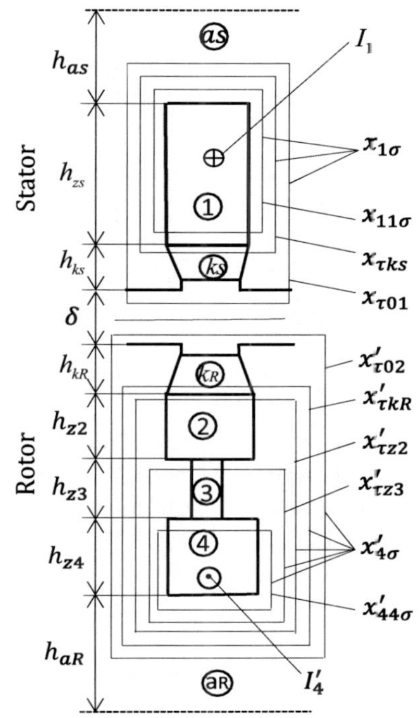

Fig. D.6 Picture of the conditional lines of the stator winding leakage magnetic fluxes and the rotor lower winding leakage magnetic fluxes

The voltage equation for the lower rotor winding acquires

$$\dot{Э}'_4 = \left(\frac{r'_4}{s} + jx'_{4\sigma}\right)\dot{I}'_4 \tag{D.20}$$

In (D.20), the referred to the stator winding values of the ohmic resistance and the slot leakage reactance of the lower rotor winding are applied. Consequently, we have

$$r'_4 = k_2 r_4 \quad x'_{4\sigma} = k_2 x_{4\sigma} \tag{D.21}$$

In (D.21), the factor k_2 is determined by the expression obtained in (D.13).

The referred stator winding value of the leakage reactance of the lower rotor winding $x'_{4\sigma}$ takes the form on the basis of Fig. D.6.

$$x'_{4\sigma} = x'_{44\sigma} + x'_{\tau z3} + x'_{\tau z2} + x'_{\tau kR} + x'_{\tau 02} \tag{D.22}$$

By equation (D.20), the equivalent circuit for the lower rotor winding can be reduced to the form of Fig. D.3d.

Machine equivalent circuit: Operating mode. The machine equivalent circuit corresponding to the operating mode can be obtained by the stator and rotor equivalent circuits given in Fig. D.3. In these equivalent circuits, for the emf

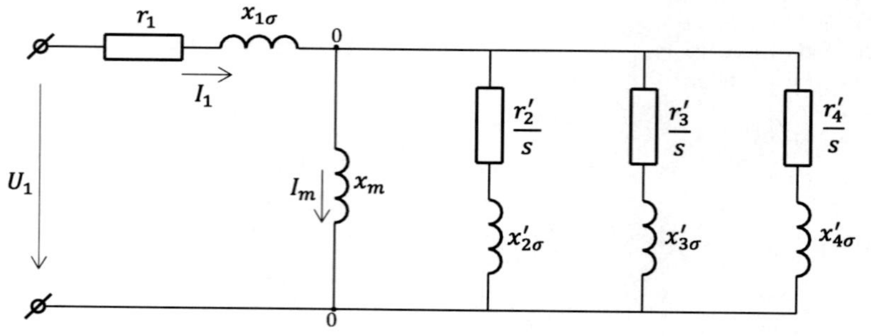

Fig. D.7 Equivalent circuit of the electric machine with the three-winding rotor

$\dot{Э}_1, \dot{Э}_2', \dot{Э}_3'$, and $\dot{Э}_4'$, the condition of the form $\dot{Э}_1 = \dot{Э}_2' = \dot{Э}_3' = \dot{Э}_4'$ is true. In this regard, the equivalent circuits in Fig. D.3 can be connected in such a way that for the stator winding current, the following condition is provided in relation to the terminals 0 – 0

$$\dot{I}_1 = \dot{I}_m - (\dot{I}_2' + \dot{I}_3' + \dot{I}_4') \tag{D.23}$$

Now, with considering the equation shown in (D.23) for the stator winding current and the equivalent circuits obtained in Fig. D.3d, the equivalent circuit for the asynchronous machine with the three-winding rotor acquires the form in Fig. D.7. In this equivalent circuit, the leakage reactances are determined by the expressions given in (D.14), (D.18), and (D.22).

In [1], the equivalent circuit of the asynchronous machine with the three-winding rotor is recommended. This equivalent circuit is obtained using the more rigorous technique. It follows from the comparison of the equivalent circuits obtained in [1] and given in this work that the equivalent circuit in Fig. D.7 can be applied when the reduction factor used in the equivalent circuit proposed in [1] can be taken equal to the one. In this regard, the practical application of the equivalent circuit in Fig. D.7 is limited by these conditions.

Summary

1. The equivalent circuit of the asynchronous machine can be constructed using the equation for the current of the stator winding having the form $\dot{I}_1 = \dot{I}_m + (-\dot{I}_2')$ and representing the "no-load and under-load" concept. In accordance with this equation, to obtain the machine equivalent circuit, it is necessary to construct the machine magnetizing equivalent circuit (when $\dot{I}_2 = 0$) and the rotor equivalent circuit (when $\dot{I}_m = 0$) and then to combine them together. The practical technique for constructing machine equivalent circuit has a simple implementation scheme.

2. The practical technique for constructing machine equivalent circuit can be applied to obtain the equivalent circuit for the machine with the multi-winding rotor. It should be noted here that this technique can be effectively used in the area of the weak skin effect. In the area of the strong skin effect, the application of this technique is significantly complicated.
3. This practical technique greatly simplifies the procedure for constructing the equivalent circuits for the electric machines. Therefore, this method has become widespread in practice. However, the considered practical technique used to construct the machine equivalent circuit is not compatible with the technique arising from the electromagnetic field theory. This is especially true for the region of the strong skin effect.

References

1. Asanbayev, V.: Alternating Current Multi-Circuit Electric Machines: A New Approach to Steady-State Parameter Determination, Springer Cham Heidelberg, New York, Dordrecht London (2015)
2. Kostenko, M., Piotrovsky, L .: Electric Machines, Energiya Publishing House, Leningrad Branch, Leningrad (1973)
3. Say, M.: Alternating Current Machines, John Wiley and Sons, New York (1902)
4. Livshits, A.: Electric Machines, vol. 1-3, United Scientific and Technical Publishing, Moscow (1936)
5. Voldek. A.: Electric Machines, Energiya Publishing House, Leningrad Branch, Leningrad (1974)
6. Ivanov-Smolensky, A.: Electric Machines, Energiya Publishing House, Moscow (1980)

Appendix E: "Unfolded" Equivalent Circuits of Stator and Rotor: Phase System of Units

"Unfolded" Magnetizing Equivalent Circuits of Stator and Rotor

When calculating the machine magnetic circuit, it is assumed that the loop L of the passage of the magnetizing flux (Fig. 10.7) is divided into the sections with the constant values of the magnetic field intensity (within the specific section), i.e., $\dot{H}_n = const$, for example [1–4]. In these conditions, the equation for the total current law can be represented as

$$0.5 \oint \dot{H} dl = 0.5 \sum \dot{H}_n h_n = \frac{m_1 w_1 k_{w1}}{p\pi} \sqrt{2} \dot{I}_m \tag{E.1}$$

where $\dot{H}_n$ and h_n are the magnetic field intensity and the thickness of the nth section of the machine magnetic circuit.

In accordance with Fig. 10.7, the machine magnetic circuit is represented consisted of the seven sections ($m = 7$), where $m = as$, zs, ks, 0, kR, zR, aR. Now, for the magnetic field intensity of the nth section, we have that $\dot{H}_n = \dot{\Phi}_m / \mu_n S_n$. Here, $\dot{\Phi}_n, \mu_n$ and S_n are the magnetizing flux, the magnetic permeability, and the cross section of the nth region of the magnetic circuit. We note that $n = zs$, ks, 0, kR, and zR. The values of the magnetic field intensities in the stator and rotor yoke sections are calculated as $\dot{H}_s = \dot{\Phi}_m / 2\mu_s S_s$, where $s = as$ and aR. The value of S_s is the radial cross section of the stator and rotor yoke regions. In this case, the value of h_n used in (E.1) is determined from the condition of kind $h_n \to b_s$, where $s = as$ and aR. The value of b_s represents the conditional length of the magnetic flux density line in the stator and rotor yoke areas. Now, taking into account that $\dot{\Phi}_m = \sqrt{2} \dot{\mathrm{Э}}_m / \omega_1 w_1 k_{w1}$, the expression (E.1) can be reduced to the form

V. Asanbayev, *Asynchronous Machines*, https://doi.org/10.1007/978-3-030-92284-9

$$\frac{\dot{Э}_m}{\frac{2m_1 w_1^2 k_{w1}^2 l_0}{p\tau_0}} \left(\sum \frac{1}{\frac{\omega_1 \mu_n S_n \tau_0}{2\pi h_n l_0}} + \sum \frac{1}{\frac{2\omega_1 \mu_s S_s \tau_0}{\pi b_s l_0}} \right) = \dot{I}_m \tag{E.2}$$

In this expression, we can use the following reactances

$$\begin{aligned} x_n &= \frac{\omega_1 \mu_n S_n \tau_0}{2\pi h_n l_0} k_1 \\ x_s &= \frac{2\omega_1 \mu_s S_s \tau_0}{\pi b_s l_0} k_1 \end{aligned} \tag{E.3}$$

where $k_1 = \frac{2m_1 w_1^2 k_{w1}^2 l_0}{p\tau_0}; S_n = \frac{2}{\pi}\tau_n l_n; b_s = \frac{2}{\pi}\tau_s; S_s = h_s l_s$

$n = zs, ks, 0, kR, zR; s = as, aR.$

With considering (E.3), the expression (E.2) acquires

$$\dot{Э}_m \left(\sum \frac{1}{x_n} + \sum \frac{1}{x_s} \right) = \dot{I}_m \tag{E.4}$$

Or assuming that $n = zs$, ks, 0, kR, zR, and $s = as$, aR, we can have

$$\dot{I}_m = \frac{\dot{Э}_m}{x_m} = \dot{Э}_m \left(\frac{1}{x_{as}} + \frac{1}{x_{zs}} + \frac{1}{x_{ks}} + \frac{1}{x_0} + \frac{1}{x_{kR}} + \frac{1}{x_{zR}} + \frac{1}{x_{aR}} \right)$$

This equation corresponds to the machine magnetizing equivalent circuit shown in Fig. E.1.

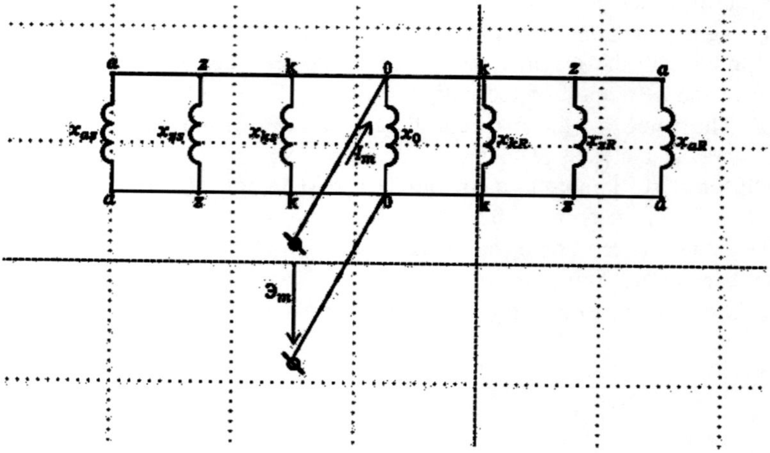

Fig. E.1 "Unfolded" magnetizing equivalent circuit of machine

In Fig. E.1, for the reactance x_0, the condition of kind $x_0 = x_{01}x_{02}/(x_{01} + x_{02})$ or $1/x_0 = 1/x_{01} + 1/x_{02}$ can be used. Then, the equation for the current $\dot{I}_m$ can be represented as

$$\dot{I}_m = \dot{Э}_m\left(\frac{1}{x_{as}}+\frac{1}{x_{zs}}+\frac{1}{x_{ks}}+\frac{1}{x_{01}}+\frac{1}{x_{02}}+\frac{1}{x_{kR}}+\frac{1}{x_{zR}}+\frac{1}{x_{aR}}\right)$$
$$\dot{Э}_m\left(\frac{1}{x_{as}}+\frac{1}{x_{zs}}+\frac{1}{x_{ks}}+\frac{1}{x_{01}}\right)+\dot{Э}_m\left(\frac{1}{x_{02}}+\frac{1}{x_{kR}}+\frac{1}{x_{zR}}+\frac{1}{x_{aR}}\right)=\frac{\dot{Э}_m}{x_{m1}}+\frac{\dot{Э}_m}{x_{m2}} \qquad \text{(E.5)}$$

where $x_{m1} = \dfrac{1}{\frac{1}{x_{01}}+\frac{1}{x_{ks}}+\frac{1}{x_{zs}}+\frac{1}{x_{as}}}$; $x_{m2} = \dfrac{1}{\frac{1}{x_{02}}+\frac{1}{x_{kR}}+\frac{1}{x_{zR}}+\frac{1}{x_{aR}}}$.

From (E.5), the reactance x_{m1} takes the form

$$x_{m1} = \frac{x_{01}}{1 + x_{01}\left(\frac{1}{x_{ks}}+\frac{1}{x_{zs}}+\frac{1}{x_{as}}\right)} = \frac{x_{01}}{k_{\text{H}1}} \qquad \text{(E.6a)}$$

From (E.5), the reactance x_{m2} can be obtained as

$$x_{m2} = \frac{x_{02}}{1 + x_{02}\left(\frac{1}{x_{kR}}+\frac{1}{x_{zR}}+\frac{1}{x_{aR}}\right)} = \frac{x_{02}}{k_{\text{H}2}} \qquad \text{(E.6b)}$$

Here, the values of $k_{н1}$ and $k_{н2}$ represent the stator and rotor saturation coefficients and they are determined by the magnetizing reactances expressed in the phase system of units.

On the basis of the expressions (E.6), the stator and rotor magnetizing equivalent circuits take the form shown in Fig. E.2. The equivalent circuit in Fig. E.2a represents the "unfolded" stator magnetizing equivalent circuit. The equivalent circuit obtained in Fig. E.2b shows the "unfolded" rotor magnetizing equivalent circuit.

We define the reactances used in the equivalent circuits shown in Fig. E.2. For this purpose, the expressions given in (E.1) can be used. These expressions contain the magnetic permeabilities which need to be determined for the sections representing the stator and rotor slot-teeth zones. For definiteness, the magnetic permeability of the section zR reflecting the rotor winding location area can be considered. We proceed from the condition of the equality of the magnetic flux corresponding to the one tooth pitch and the magnetic flux passing through the rotor tooth, i.e. $B_m t_{zR} = B_{zR} b_{zR}$, where t_{zR} is the tooth pitch of the zR section and b_{zR} is the width of the rotor tooth. From here, we have that $B_m = B_{zR}(b_{zR}/t_{zR})$. Since $B_m = \mu_n H_{zR} = \mu_{yzR} H_{zR}$ and $B_z = \mu_{zR} H_{zR}$, then, the magnetic permeability of the zR section is determined as $\mu_n = \mu_{yzR} = \mu_{zR}(b_{zR}/t_{zR})$. Here, the value of μ_{zR} represents the magnetic permeability of the tooth section reflecting the rotor winding location

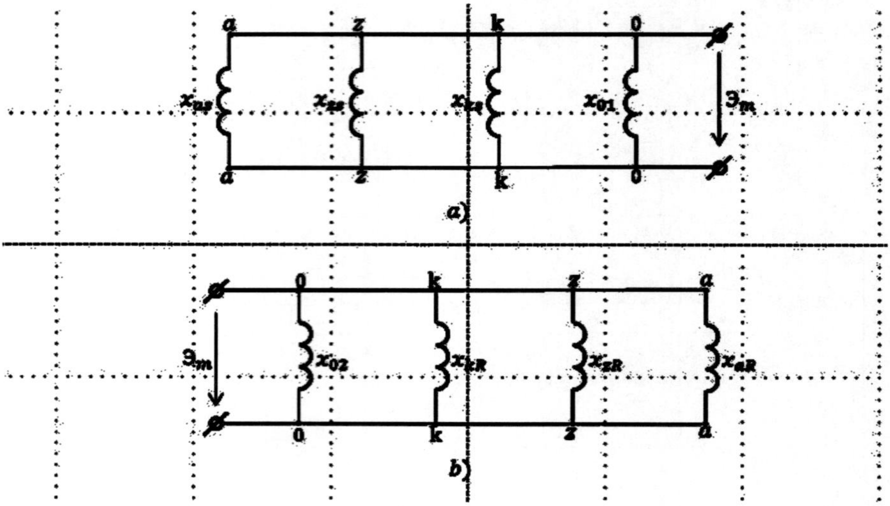

Fig. E.2 "Unfolded" magnetizing equivalent circuits of stator and rotor. (**a**) "Unfolded" stator magnetizing equivalent circuit; (**b**) "unfolded" rotor magnetizing equivalent circuit

area. The similar expressions can be obtained for the magnetic permeabilities of the other sections of the stator and rotor slot-tooth areas. Keeping in mind that on the basis of the expression (E.3), the reactances used in the stator magnetizing equivalent circuit (Fig. E.2a) can be represented as

$$\begin{aligned} x_{01} &= \frac{\omega_1 \mu_0 \tau_0^2}{\pi^2 \delta_1} k_1 \\ x_{ks} &= \frac{\omega_1 \mu_{ks} \tau_{ks}^2}{\pi^2 h_{ks}} \frac{b_{ks}}{t_{ks}} \frac{\xi_{Lks}}{\xi_{\tau ks}} k_1 \\ x_{zs} &= \frac{\omega_1 \mu_{zs} \tau_{zs}^2}{\pi^2 h_{zs}} \frac{b_{zs}}{t_{zs}} \frac{\xi_{Lzs}}{\xi_{\tau zs}} k_1 \end{aligned} \tag{E.7}$$

For the reactances applied in the rotor magnetizing equivalent circuit (Fig. E.2b), we can have

$$\begin{aligned} x_{02} &= \frac{\omega_1 \mu_0 \tau_0^2}{\pi^2 \delta_2} k_1 \\ x_{kR} &= \frac{\omega_1 \mu_{kR} \tau_{kR}^2}{\pi^2 h_{kR}} \frac{b_{kR}}{t_{kR}} \frac{\xi_{LkR}}{\xi_{\tau kR}} k_1 \\ x_{zR} &= \frac{\omega_1 \mu_{zR} \tau_{zR}^2}{\pi^2 h_{zR}} \frac{b_{zR}}{t_{zR}} \frac{\xi_{LzR}}{\xi_{\tau zR}} k_1 \end{aligned} \tag{E.8}$$

The reactance x_{as} acquires

$$x_{as} = \omega_1 \mu_{as} t_0 \frac{h_{as}}{t_{as}} \xi_{Las} k_1 \tag{E.9}$$

where $t_0 = 2p\tau_0/Z_1$; $t_{as} = 2p\tau_{as}/Z_1$.

The reactance x_{aR} can be expressed as

$$x_{aR} = \omega_1 \mu_{aR} t_0 \frac{h_{aR}}{t_{aR}} k_R \xi_{LaR} k_1 \tag{E.10}$$

where $t_{aR} = 2p\tau_{aR}/Z_2$; $k_R = Z_1/Z_2$.

The equivalent circuits obtained in Fig. E.2 and the reactances used in them correspond to the case, when the machine magnetizing field is represented as the one-dimensional magnetic field spreading in the machine cross-section area in the radial direction.

"Unfolded" Slot Leakage Equivalent Circuits of Stator and Rotor

We consider the single slot of the simplest configuration (Fig. E.3). The one coil side is laid in this slot, for example [1–12]. In Fig. E.3 shown for the stator and rotor, the slot-teeth areas are divided into the three sections along which the slot leakage magnetic fields are closed. The slot sections of the height equal to h_{zn}, where $n = s, R$, contain the stator and rotor windings. The slot sections of the height equal to h_{kn} represent the stator and rotor slot-wedge areas. The sections of the height equal to $\delta/2$ reflect the air gap sublayers adjoining to the stator or to the rotor surfaces. In Fig. E.3, the diagrams of the distribution of the slot leakage magnetic

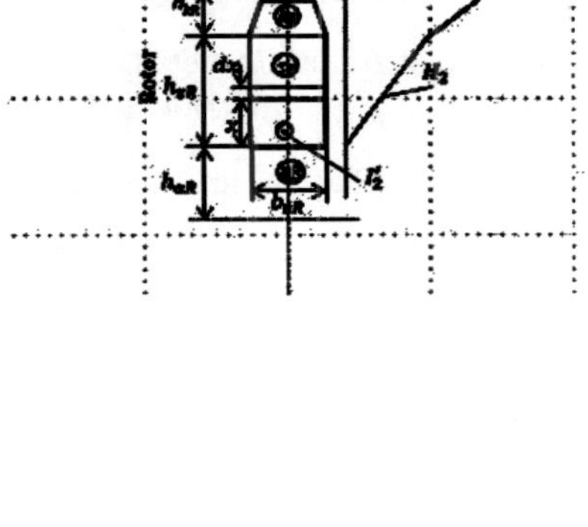

Fig. E.3 To the calculation of the components of the slot leakage reactances of stator and rotor

field intensities for the case of the uniform distribution of the current density in the stator and rotor slot areas are given.

The flux linkage in the rotor winding location area. We consider the slot section occupied by the rotor winding. In this slot section, the elementary magnetic flux tube of the width dx and the length b_{nR} and located at the height x from the slot base (Fig. E.3) is selected. Let this elementary magnetic flux tube embraces the S_x conductors of the one side of the winding coil. We note that the S_n conductors of the one side of the winding coil are laid in the slot. The magnetic flux of the elementary tube interlinked with the S_x conductors of the slot coil is directly proportional to the mmf and inversely proportional to the reluctance of the flux tube, i.e.,

$$\Phi_{dx} = \frac{S_x\sqrt{2}\dot{I}_2}{R_{dx}} = \mu_0\sqrt{2}\dot{I}_2 S_x l_{zR}\frac{dx}{b_{nR}} \tag{E.11}$$

where $R_{dx} = \dfrac{b_{пR}}{\mu_0 dx l_{zR}}$.

Here R_{dx} represents the reluctance of the elementary magnetic flux tube (Fig. E.3).

The flux Φ_{dx} forms the flux linkage Ψ_{dx} with the conductors S_x located at the height x from the slot base. Consequently, we can have

$$\Psi_{dx} = S_x\Phi_{dx} = \mu_0\sqrt{2}\dot{I}_2 S_x^2 l_{zR}\frac{dx}{b_{пR}} \tag{E.12}$$

Summing the flux linkages of all elementary magnetic flux tubes, for the total flux linkage with the conductors of the coil side, we can obtain

$$\Psi_{п2} = \sum S_x\Phi_{dx} = \mu_0\sqrt{2}\dot{I}_2\sum S_x^2 l_{zR}\frac{dx}{b_{пR}} \tag{E.13}$$

This expression can be expressed through the total number of the conductors S_n located in the slot. Then, it is followed

$$\Psi_{п2} = \mu_0\sqrt{2}\dot{I}_2 S_{п}^2 l_{zR}\sum\left(\frac{S_x}{S_{п}}\right)^2\frac{dx}{b_{пR}} \tag{E.14}$$

The value of $\sum\left(\frac{S_x}{S_n}\right)^2\frac{dx}{b_{nR}}$ used in (E.14) can be determined for the rectangular slot. We consider the rectangular slot filled evenly with the coil conductors. The slot height is h_{zR} and the slot width is b_{nR}. The conductors S_x located at the height x from the slot base is determined, in this case, as $S_x = \frac{x}{h_{zR}}S_n$. Now, for $\sum_0^{h_{zR}}\left(\frac{S_x}{S_n}\right)^2\frac{dx}{b_{nR}}$, we can have

$$\sum_{0}^{h_{zR}}\left(\frac{S_x}{S_n}\right)^2\frac{dx}{b_{nR}}=\int_{0}^{h_{zR}}\frac{x^2}{h_{zR}^2 b_{nR}}dx=\frac{h_{zR}}{3b_{nR}}$$

This is the well-known expression for the leakage conductance of the rectangular slot uniformly filled with the coil conductors.

Taking into account the number of the slots filled with the single phase is $Z_2 = 2pq_2$ and the number of the turns in the phase is equal to $w_2 = S_n pq_2$, the expression (E.14) for the Ψ_{n2} can be represented as

$$\Psi_{22}=2pq_2\Psi_{\text{п}2}=2\mu_0\frac{h_{zR}}{3b_{\text{п}R}}\frac{w_2^2}{pq_2}l_{zR}\sqrt{2}\dot{I}_2 \tag{E.15}$$

The number of the rotor teeth is determined by the expression of kind $Z_2 = 2pq_2m_2$, and the tooth pitch on the surface on which the rotor winding is leaned arises from the condition of the form $t_{zR}=\frac{2p\tau_{zR}}{Z_2}$. Then, the condition (E.15) acquires

$$\Psi_{22}=\mu_0 t_{zR}\frac{h_{zR}}{3b_{\text{п}R}}\frac{2m_2w_2^2l_{zR}}{p\tau_{zR}}\sqrt{2}\dot{I}_2 \tag{E.16}$$

The expression (E.16) can be represented through the values of τ_0 and l_0. In this case, we obtain

$$\Psi_{22}=\mu_0 t_0\frac{h_{zR}}{3b_{\text{п}R}}k_R\xi_{LzR}\frac{2m_2w_2^2l_0}{p\tau_0}\sqrt{2}\dot{I}_2 \tag{E.17}$$

where $k_R=\frac{Z_1}{Z_2}$; $t_{zR}=\frac{2p\tau_{zR}}{Z_2}$; $t_0=\frac{2p\tau_0}{Z_1}$; $\xi_{LzR}=\frac{l_{zR}}{l_0}$; $\xi_{\tau zR}=\frac{\tau_{zR}}{\tau_0}$.

The expression (E.17) is considered as the flux linkage for the section representing the rotor winding location area.

The flux linkage in the rotor slot-wedge area. The magnetic flux in the section reflecting the rotor slot-wedge area

$$\Phi_{\tau kR}=B_{xk}h_{kR}l_{kR}$$

where $B_{xk}=\frac{\mu_0}{b_k}S_n\sqrt{2}\dot{I}_2$; $b_{kR}=(b_{ш}+b_n)/2$ is the average width of the section of the rotor slot-wedge area (Fig. E.3).

The magnetic flux $\Phi_{\tau kR}$ creates the flux linkage equal to

$$\Psi_{\tau\text{п}}=S_{\text{п}}\Phi_{\tau kR}=\mu_0\frac{h_{kR}}{b_{kR}}S_{\text{п}}^2 l_{kR}\sqrt{2}\dot{I}_2 \tag{E.18}$$

The condition (E.18) can be represented in the form of the expression shown in (E.13). Taking into account that $S_n = \frac{w_2}{pq_2} = \frac{2m_2w_2}{Z_2}$ and $pq_2 = \frac{Z_2}{2m_2}$, as well as $t_{kR} = \frac{2p\tau_{kR}}{Z_2}$, where τ_{kR} is the pole pitch on the rotor surface, the expression (E.18) acquires

$$\Psi_{\tau kR} = \mu_0 t_{kR} \frac{h_{kR}}{b_{kR}} \frac{2m_2 w_2^2}{p\tau_{zR}} l_{kR} \sqrt{2} \dot{I}_2 \tag{E.19}$$

The expression (E.19) can be represented through the values of τ_0 and l_0. Then, we can have

$$\Psi_{\tau kR} = \mu_0 t_0 \frac{h_{kR}}{b_{kR}} k_R \xi_{LkR} \frac{2m_2 w_2^2 l_0}{p\tau_0} \sqrt{2} \dot{I}_2 \tag{E.20}$$

where $k_R = \frac{Z_1}{Z_2}$; $t_{kR} = \frac{2p\tau_{kR}}{Z_2}$; $t_0 = \frac{2p\tau_0}{Z_1}$; $\xi_{LkR} = \frac{l_{kR}}{l_0}$; $\xi_{\tau kR} = \frac{\tau_{kR}}{\tau_0}$.

The expression (E.20) represents the flux linkage of the section showing the rotor slot-wedge area.

Flux linkage in the air gap sublayer adjoining to the rotor surface. The magnetic flux in the air gap sublayer adjoining to the rotor surface and having the thickness equal to $\delta/2$ is calculated by the expression of kind

$$\Phi_{\tau 02} = B_{x0} \frac{\delta}{2} l_0$$

where $B_{x0} = \frac{\mu_0}{t_{02}} S_\text{п} \sqrt{2} \dot{I}_2$; $t_{02} = \frac{2p\tau_0}{Z_2}$.

The expression shown above for the magnetic flux $\Phi_{\tau 02}$ is obtained under the condition of the straightness of the magnetic flux density lines in the machine air gap. The distortion of the magnetic flux density lines taken place in a reality can be taken into account in the final expressions using the certain coefficient $k_{2\sigma}$ determined, for example, in [3, 5].

The magnetic flux $\Phi_{\tau 02}$ causes the flux linkage

$$\Psi_{\tau 02} = S_\text{п} \Phi_{\tau 02} = \mu_0 \frac{\delta}{2t_{02}} k_{2\sigma} S_\text{п}^2 l_0 \sqrt{2} \dot{I}_2 \tag{E.21}$$

The condition (E.21) can be represented in the form of the expression obtained in (E.13). Assuming that $S_n = \frac{w_2}{pq_2} = \frac{2m_2w_2}{Z_2}$ and $pq_2 = \frac{Z_2}{2m_2}$ as well as $t_{02} = \frac{2p\tau_0}{Z_2}$, the expression (E.21) takes

$$\Psi_{\tau 02} = \mu_0 t_0 \frac{\delta}{2t_{02}} k_{2\sigma} k_R \frac{2m_2 w_2^2 l_0}{p\tau_0} \sqrt{2} \dot{I}_2 \tag{E.22}$$

where $k_R = \frac{t_{02}}{t_0} = \frac{Z_1}{Z_2}$; $t_{02} = \frac{2p\tau_0}{Z_2}$; $t_0 = \frac{2p\tau_0}{Z_1}$.

Slot leakage reactance of the rotor winding. From (E.17), (E.20), and (E.22), the total flux linkage of the rotor winding is determined as

$$\Psi_{2\sigma} = \Psi_{\tau 02} + \Psi_{\tau kR} + \Psi_{22}$$

Or in another form

$$\Psi_{2\sigma} = \mu_0 t_0 \left(\frac{\delta}{2t_{02}} k_{2\sigma} + \frac{h_{kR}}{b_{kR}} \xi_{LkR} + \frac{h_{zR}}{3b_{\text{п}R}} \xi_{LzR} \right) k_R \frac{2m_2 w_2^2 l_0}{p\tau_0} \sqrt{2} \dot{I}_2 \quad \text{(E.23)}$$

Taking into account that $\sqrt{2}\dot{Э}_{2\sigma} = \omega_1 \Psi_{2\sigma}$, the slot leakage emf of the rotor winding acquires from (E.23) the form

$$\dot{Э}_{2\sigma} = \omega_1 \mu_0 t_0 \left(\frac{\delta}{2t_{02}} k_{2\sigma} + \frac{h_{kR}}{b_k} \xi_{LkR} + \frac{h_{zR}}{3b_{\text{п}R}} \xi_{LzR} \right) k_R \frac{2m_2 w_2^2 l_0}{p\tau_0} \dot{I}_2 \quad \text{(E.24)}$$

In accordance with (E.24) the value of the leakage emf $\dot{Э}_{2\sigma}$ is determined by the rotor phase winding having the number of the turns equal to w_2. Meanwhile, the current in the rotor winding $\dot{I}_2$ arises under the influence of the emf $\dot{Э}_2$ induced in the rotor winding in the no-load mode. When calculating the value of the emf $\dot{Э}_2$, the calculated number of the rotor winding turns equal to $w_2 k_{w2}$, where k_{w2} is the winding factor, is used. In this regard, the values of the leakage emf $\dot{Э}_{2\sigma}$ and the current $\dot{I}_2$ used in (E.24) should be reduced to the calculated number of the rotor winding turns $w_2 k_{w2}$. This reduction is reduced to the fact that the right side of the expression (E.24) is multiplied by the factor k_{w2}^2. Then, the expression (E.24) acquires the following form

$$\dot{Э}_{2\sigma} = \omega_1 \mu_0 t_0 \left(\frac{\delta}{2t_{02}} k_{2\sigma} + \frac{h_{kR}}{b_k} \xi_{LkR} + \frac{h_{zR}}{3b_{\text{п}R}} \xi_{LzR} \right) k_R \frac{2m_2 w_2^2 k_{w2}^2 l_0}{p\tau_0} \dot{I}_2 \quad \text{(E.25)}$$

In practice, the referred values of the rotor winding emf and current determined as $\dot{Э}'_{2\sigma} = \dot{Э}_{2\sigma} k_Э$ and $\dot{I}'_2 = \dot{I}_2 / k_I$, where $k_Э = \frac{w_1 k_{w1}}{w_2 k_{w2}}$ and $k_I = \frac{m_1 w_1 k_{w1}}{m_2 w_2 k_{w2}}$, are applied. When using the referred values of the rotor winding emf and the current, the expression (E.25) can be shown as

$$\dot{Э}'_{2\sigma} = \omega_1 \mu_0 t_0 \left(\frac{\delta}{2t_{02}} k_{2\sigma} + \frac{h_{kR}}{b_k} \xi_{LkR} + \frac{h_{zR}}{3b_{\text{п}R}} \xi_{LzR} \right) k_R k_1 \dot{I}'_2 \quad \text{(E.26)}$$

where $k_1 = \dfrac{2m_1 w_1^2 k_{w1}^2 l_0}{p\tau_0}$.

Now, the expression (E.26) obtains the form

$$\begin{aligned} \dot{Э}'_{2\sigma} &= \omega_1\mu_0 t_0\left(\frac{\delta}{2t_{02}}k_{2\sigma} + \frac{h_{kR}}{b_k}\xi_{LkR} + \frac{h_{zR}}{3b_{пR}}\xi_{LzR}\right)k_R k_1 \dot{I}'_2 \\ &= \left(x'_{\tau02} + x'_{\tau kR} + x'_{22\sigma}\right)\dot{I}'_2 = x'_{2\sigma}\dot{I}'_2 \end{aligned} \tag{E.27}$$

where $x'_{2\sigma} = x'_{\tau02} + x'_{\tau kR} + x'_{22\sigma}$.

According to (E.27), the slot leakage reactance of the rotor winding $x'_{2\sigma}$ represents the sum of the three reactances: the leakage reactance $x'_{\tau02}$ of the air gap sublayer adjoining to the rotor surface, the leakage reactance $x'_{\tau kR}$ of the section reflecting the rotor slot-wedge area, and the leakage reactance $x'_{22\sigma}$ of the section representing the rotor winding location area . The reactances $x'_{\tau02}, x'_{\tau kR}$, and $x'_{22\sigma}$ can be considered as the components of the slot leakage reactance of the rotor winding $x'_{2\sigma}$. On the basis of the expression (E.27), for these reactances, we have

$$\begin{aligned} x'_{\tau02} &= \omega_1\mu_0 t_0\frac{\delta}{2t_{02}}k_{2\delta}k_R k_1 \\ x'_{\tau kR} &= \omega_1\mu_0 t_0\frac{h_{kR}}{b_{kR}}k_R\xi_{LkR}k_1 \\ x'_{22\sigma} &= \omega_1\mu_0 t_0\frac{h_{zR}}{3b_{пR}}k_R\xi_{LzR}k_1 \end{aligned} \tag{E.28}$$

Here, in this expression for the reactance $x'_{\tau02}$, the coefficient $k_{2\delta}$ is used that can be set, for example, from [3]. In the air gap, the real picture of the leakage magnetic field distribution is extremely complex and difficult to describe. In the technical literature, a large number of works are devoted to the problem of determining the air gap leakage reactance, for example [3, 5]. However, until now, there are no sufficiently reliable expressions for determining the air gap leakage reactance.

By the expressions (E.27) and (E.28), the similar expression for the slot leakage reactance of the stator winding $x_{1\sigma}$ can be received. It can be shown that the slot leakage reactance of the stator winding $x_{1\sigma}$ consists of similar leakage reactances $x_{\tau01}$, $x_{\tau ks}$, and $x_{11\sigma}$. Therefore, we can have

$$x_{1\sigma} = x_{\tau01} + x_{\tau ks} + x_{11\sigma} \tag{E.29}$$

According to (E.29), the slot leakage reactance of the stator winding $x_{1\sigma}$ represents the sum of the three reactances: the leakage reactance $x_{\tau01}$ of the air gap sublayer adjoining to the stator surface, the leakage reactance $x_{\tau ks}$ of the section reflecting the stator slot-wedge area, and the leakage reactance $x_{11\sigma}$ of the section representing the stator winding location area .

The reactances $x_{\tau01}$, $x_{\tau ks}$ and $x_{11\sigma}$ can be considered as the components of the slot leakage reactance of the stator winding $x_{1\sigma}$. The expressions for these components of the reactance $x_{1\sigma}$ can be obtained by (E.28). In this case, for the coefficient k_R used in (E.28), the condition of kind $k_R = 1.0$ should be applied, since the tooth pitch t_{02} arising from (E.22) is determined in relation to the number of the stator teeth. Because of this, for the coefficient $\xi_{\tau i}$, the condition of kind $\xi_{\tau i} = \tau_i/\tau_0 = t_i/t_0$ is implemented. With considering these conditions in (E.28), we can obtain

$$\begin{aligned} x_{\tau 01} &= \omega_1 \mu_0 t_0 \frac{\delta}{2t_0} k_{1\sigma} k_1 \\ x_{\tau ks} &= \omega_1 \mu_0 t_0 \frac{h_{ks}}{b_{ks}} \xi_{Lks} k_1 \\ x_{11\sigma} &= \omega_1 \mu_0 t_0 \frac{h_{zs}}{3b_{\text{пs}}} \xi_{Lzs} k_1 \end{aligned} \tag{E.30}$$

Here, the coefficient $k_{1\sigma}$ takes into account the distortion of the slot leakage magnetic flux density lines in the air gap sublayer adjoining to the stator surface, for example, [3, 5].

From (8.20) and (E.29), the equation for the emf induced in the stator winding by the slot leakage magnetic field is derived as

$$\dot{Э}_m = [j(x_{\tau 01} + x_{\tau ks} + x_{11\sigma}) + r_1 + Z_c]\dot{I}_1 \tag{E.31}$$

Here, the condition of kind $\dot{Э}_1 = \dot{Э}_m$ is implemented. The equation (E.31) describes the stator slot leakage circuit loop. In Fig. E.4a, the stator slot leakage equivalent circuit is shown.

From (8.21) and (E.27), the equation for the emf induced in the rotor winding by the slot leakage magnetic field, we have

$$\dot{Э}_m = \left[j\left(x'_{\tau 02} - x'_{\tau kR} + x'_{22\sigma}\right) + r'_2/s + Z'_{2\text{н}}\right]\dot{I}'_2 \tag{E.32}$$

Here, the condition of kind $\dot{Э}'_2 = \dot{Э}_m$ is true. The equation (E.32) describes the rotor slot leakage circuit loop. In Fig. E.4b, the rotor slot leakage equivalent circuit is reflected.

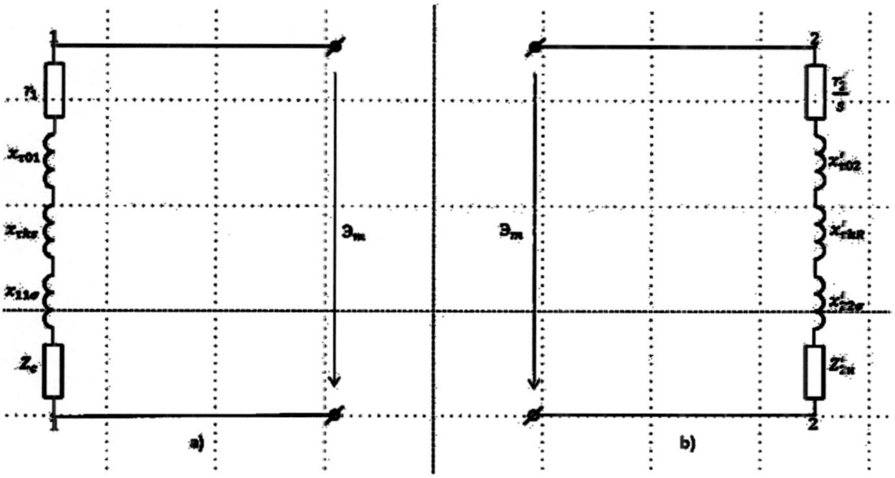

Fig. E.4 "Unfolded" slot leakage equivalent circuits of stator and the rotor. (**a**) "Unfolded" stator slot leakage equivalent circuit; (**b**) "unfolded" rotor slot leakage equivalent circuit

Summary

1. The stator and rotor magnetizing circuits and the stator and rotor slot leakage magnetic circuits are subdivided into sections in such a way that in the sections obtained in this way the values of the magnetic permeabilities do not change. In the sections of the stator and rotor magnetizing circuits and the stator and rotor slot leakage magnetic circuits, the one-dimensional magnetic field is considered. All this greatly simplifies the calculations of the magnetic circuits of the electric machine.
2. The stator and rotor magnetizing equivalent circuits are constructed, as well as the stator and rotor slot leakage equivalent circuits are built. For the construction of these equivalent circuits, the analysis of the stator and rotor magnetic circuits as well as the stator and rotor slot leakage magnetic circuits have been carried out. In this case, the phase system of units is used.
3. The stator and rotor magnetizing equivalent circuits correspond to the no-load mode of the machine. The stator and rotor slot leakage equivalent circuits reflect the machine under-load mode. In accordance with the "no-load and under-load" concept, these equivalent circuits can be used to construct the equivalent circuits that reflect the resulting magnetic fields taking place in the stator and rotor of the electric machine. For this purpose, the superposition principle can be applied.

References

1. Asanbayev, V.: Alternating Current Multi-Circuit Electric Machines: A New Approach to Steady-State Parameter Determination, Springer Cham Heidelberg, New York, Dordrecht London (2015)
2. Say, M.: Alternating Current Machines, John Wiley and Sons, New York (1902)
3. Liwschitz-Garik, M., Whipple, C.: Alternating Current Machines, D. Van Nastran Company, Inc., New York (1961)
4. Kostenko, M, Piotrovskij, L.: Electric Machines, Publishing Energiya, Leningrad (1973)

Appendix F: "Unfolded" Equivalent Circuits of Stator and Rotor: Specific System of Units

"Unfolded" Magnetizing Equivalent Circuits of Stator and Rotor

When calculating the machine magnetic circuit, it is represented subdivided into the corresponding sections. The concept of the "magnetic circuit section" is used when the phase system of units is applied ($\dot{Э}, \dot{I}$). In the specific system of units, the tangent components of the electric and magnetic field intensities $\left(\dot{E}, \dot{H}\right)$ are used. The tangent component of the magnetic field intensity $\dot{H}$ is proportional to the sum of the winding currents attributed per unit length of the stator circumference. The axial component of the electric field intensity $\dot{E}$ is proportional to the emf induced in the winding and calculated per unit length of the machine active length and determined by the average value of the radial component of the magnetic flux density taking place in the considered section of the machine magnetic circuit. Due to this nature of the introduction of the components of the electric and magnetic field intensities, the surfaces of the magnetic circuit sections can be considered as the conditionally "smooth" and the sections can be represented as the homogeneous layers, for example [1–5]. Then, in relation to the components of the electric and magnetic field intensities determined in these ways, the machine magnetic circuit can be considered as the system of the layers with the "smooth" surfaces.

In Fig. F.1, the flat model of the machine magnetizing circuit is given. In this model, the sections of the stator and rotor slot-teeth areas are represented as the homogeneous layers having the "smooth" surfaces. For these stator and rotor sections, the calculated values of the magnetic permeabilities are used [1–5]. We will demonstrate this provision by the example of considering the layer of the slot-tooth area corresponding to the rotor winding location area. The tooth pitch of the considered layer is t_{zR}. The width, height, and axial length of the single tooth are designated as b_{zR}, h_{zR} and l_{zR}. The magnetic permeability of the single tooth is μ_{zR}.

V. Asanbayev, *Asynchronous Machines*, https://doi.org/10.1007/978-3-030-92284-9

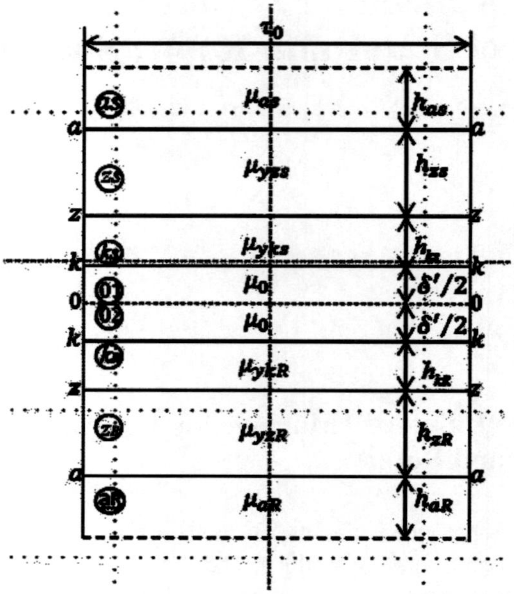

Fig. F.1 Flat model of the magnetizing circuit of the machine

Let us assume that the radial magnetic flux determined by the one tooth pitch passes through the one tooth which can be taken when $B_{zi} \leq 1.8\ T$. In this case, the reluctance of the single tooth R_{zR} is determined by the geometric dimensions and the magnetic permeability μ_{zR} of the tooth. In this regard, for the reluctance R_{zR}, we can obtain that $R_{zR} = \frac{h_{zR}}{\mu_{zR} b_{zR} l_{zR}}$. Now, in relation to the radial magnetic flux, the slot-tooth area located within the one tooth pitch can be represented as the homogenous layer with the average value of the magnetic permeability μ_{yzR}. Then, the reluctance R_{tR} of the homogenous layer representing the slot-tooth area within the one tooth pitch can be determined as $R_{tR} = \frac{h_{zR}}{\mu_{yzR} t_{zR} l_{zR}}$. We proceed from the fact that the reluctance of the single tooth R_{zR} and the reluctance of the homogenous layer R_{tR} have the same value, i.e., $R_{zR} = R_{tR}$. This condition can be used to determine the magnetic permeability of the homogenous layer μ_{yzR}. Taking into account that $R_{zR} = \frac{h_{zR}}{\mu_{zR} b_{zR} l_{zR}}$ and $R_{tR} = \frac{h_{zR}}{\mu_{yzR} t_{zR} l_{zR}}$ and also $R_{zR} = R_{tR}$, for the magnetic permeability of the homogenous layer μ_{yzR}, we can have that $\mu_{yzR} = \mu_{zR}(b_{zR}/t_{zR})$. In relation to μ_{yzR}, the rotor slot-tooth area is represented as the homogenous layer. Acting in the similar way, for the rotor slot-wedge area considered as the homogenous layer, the condition of kind $\mu_{ykR} = \mu_{kR}(b_{kR}/t_{kR})$ follows. Now, the magnetic permeabilities of the layers reflecting the rotor slot-wedge and the rotor slot-tooth areas can be expressed as

$$\mu_{ykn} = \mu_{kn}\frac{b_{kn}}{t_{kn}} \quad \mu_{yzn} = \mu_{zn}\frac{b_{zn}}{t_{zn}} \tag{F.1}$$

where $n = s, R$.

Proceeding from (F.1), we can note that the slot-tooth areas of the stator and rotor can be replaced by the corresponding layers. Now, it is shown in Fig. F.1 that the machine magnetizing circuit is represented consisted of the seven homogenous layers. In Fig. F.1, the layers have the same pole pitch τ_0 and the axial length equal to l_0. As shown in [1], this condition is a result of using in the layered model in Fig. F.1 the calculated values for the electric and magnetic field intensities $\dot{E}_n$ and $\dot{H}_n$ determined as $\dot{E}_n = \dot{E}_{zn}\xi_{Ln}$ and $\dot{H}_n = \dot{H}_{xn}\xi_{\tau n}$, where $\xi_{Ln} = l_n/l_0$, $\xi_{\tau n} = \tau_n/\tau_0$. Here, $\dot{E}_{zn}$ and $\dot{H}_{xn}$ represent the initial values of the electric and magnetic field intensities of the nth layer.

We proceed from the flat layered model in Fig. F.1. We assume that the one-dimensional magnetic field takes place in this model. In this case, the total current law can be represented as

$$\frac{1}{2}\oint \dot{H}dl = \frac{1}{2}\sum \dot{H}_n h_n = \dot{\varkappa}_m \frac{\tau_0}{\pi} \tag{F.2}$$

where $\dot{H}_n$ and h_n are the magnetic field intensity and the thickness of the nth layer, $\dot{\varkappa}_m$ is the linear density of the stator winding surface current, τ_0 is the pole pitch on the surface passing through the middle of the radial length of the air gap.

According to Fig. F.1, the machine magnetizing circuit model consists of the seven layers ($m = 7$), where $m = as, zs, ks, 0, kR, zR, aR$. Now, for the magnetic field intensity of the nth layer, where $n = zs, ks, 0, kR$, and zR, we have that $\dot{H}_n = \dot{B}_n/\mu_n$. Here, $\dot{B}_n$ is the magnetic flux density of the nth layer, μ_n is the magnetic permeability of the nth layer. The values of the magnetic field intensities in the layers representing the stator and rotor yoke areas are calculated as $\dot{H}_s = (\dot{B}_s\tau_s)/\mu_s\pi h_s$, where $s = as$ and aR. With considering the conditions for $\dot{H}_n$ and $\dot{H}_s$, the expression (F.2) can be reduced to the form

$$\dot{B}_n\sum\frac{1}{\frac{\mu_n}{h_n}\frac{\tau_0}{\pi}} + \dot{B}_s\sum\frac{1}{\frac{2\mu_s h_s\tau_0}{l_s\tau_s}} = \dot{\varkappa}_m \tag{F.3a}$$

In (F.3a), the value of h_n is used two times.

In (F.3a), the values of $\dot{B}_n$ and $\dot{B}_s$ can be expressed through the magnetic flux density $\dot{B}_m$ determined on the air gap surface $0-0$ for that the following conditions $\dot{B}_n = (\tau_0/\tau_n)(\dot{B}_m/\xi_{Ln})$, where $\xi_{Ln} = l_n/l_0$), and $\dot{B}_s = (\tau_0/\tau_s)(\dot{B}_m/\xi_{Ls})$, where $\xi_{Ls} = l_s/l_0$), are used. Now, the expression (F.3a) acquires

$$\dot{B}_m\left(\sum\frac{1}{\frac{\mu_n\tau_n\xi_{Ln}}{\pi h_n}}+\sum\frac{1}{\frac{2\mu_s h_s\xi_{Ls}}{l_s}}\right)=\dot{\varkappa}_m \tag{F.3b}$$

Keeping in mind that $\dot{E}_m=\upsilon_1\dot{B}_m=\frac{\omega_1\tau_0}{\pi}\dot{B}_m$, where $\upsilon=\frac{\omega_1\tau_0}{\pi}$, the equation (F.3b) can be reduced to the form

$$\dot{E}_m\left(\sum\frac{1}{\frac{\omega_1\mu_n\tau_n^2}{\pi^2 h_n}\frac{\xi_{Ln}}{\xi_{\tau n}}}+\sum\frac{1}{\frac{2\omega_1\mu_s h_s\tau_s}{\pi l_s}\frac{\xi_{Ls}}{\xi_{\tau s}}}\right)=\dot{\varkappa}_m \tag{F.4}$$

where $\xi_{Ls}=l_s/l_0$, $\xi_{\tau s}=\tau_s/\tau_0$.

In this equation, the following reactances can be used. As a result, we can have

$$\begin{aligned}x_n^*&=\frac{\omega_1\mu_n\tau_n^2}{\pi^2 h_n}\frac{\xi_{Ln}}{\xi_{\tau n}}\\ x_s^*&=\frac{2\omega_1\mu_s h_s\tau_s}{\pi l_s}\frac{\xi_{Ls}}{\xi_{\tau s}}\end{aligned} \tag{F.5a}$$

where $n=zs$, ks, 0, kR, zR and $s=as$, aR.

Here, the value of l_s representing the conditional length of the magnetic flux density line in the layers reflecting the stator and rotor yoke areas is defined in [1] as $l_s\approx(2/\pi)\tau_s$. Now, from (F.5a), the reactance x_s^* follows

$$x_s^*=\omega_1\mu_s h_s\frac{\xi_{Ls}}{\xi_{\tau s}} \tag{F.5b}$$

With considering the conditions shown in (F.5), the equation (F.4) can be given as

$$\dot{E}_m\left(\sum\frac{1}{x_n^*}+\sum\frac{1}{x_s^*}\right)=\dot{\varkappa}_m \tag{F.6}$$

Or assuming that $n=zs$, ks, 0, kR, zR, and $s=as$, aR and also keeping in mind that $\dot{\varkappa}_m=\dot{H}_m$, we can obtain

$$\dot{H}_m=\dot{E}_m\left(\frac{1}{x_{as}^*}+\frac{1}{x_{zs}^*}+\frac{1}{x_{ks}^*}+\frac{1}{x_0^*}+\frac{1}{x_{kR}^*}+\frac{1}{x_{zR}^*}+\frac{1}{x_{aR}^*}\right)=\frac{\dot{E}_m}{x_m^*} \tag{F.7}$$

Here, the value of x_m^* represents the machine magnetizing reactance. In (F.7), the condition of kind $\frac{1}{x_0^*}=\frac{1}{x_{01}^*}+\frac{1}{x_{02}^*}$ can be used. Then, the equation (F.7) acquires the form

$$\dot{H}_m = \dot{E}_m\left(\frac{1}{x^*_{as}} + \frac{1}{x^*_{zs}} + \frac{1}{x^*_{ks}} - \frac{1}{x^*_{01}}\right) + \dot{E}_m\left(\frac{1}{x^*_{02}} + \frac{1}{x^*_{kR}} + \frac{1}{x^*_{zR}} + \frac{1}{x^*_{aR}}\right) \\ = \frac{\dot{E}_m}{x^*_{m1}} + \frac{\dot{E}_m}{x^*_{m2}} \tag{F.8a}$$

where $x^*_{m1} = \dfrac{1}{\dfrac{1}{x^*_{01}} + \dfrac{1}{x^*_{ks}} + \dfrac{1}{x^*_{zs}} + \dfrac{1}{x^*_{as}}}$; $x^*_{m2} = \dfrac{1}{\dfrac{1}{x^*_{02}} + \dfrac{1}{x^*_{kR}} + \dfrac{1}{x^*_{zR}} + \dfrac{1}{x^*_{aR}}}$.

From (F.8a), the stator magnetizing reactance x^*_{m1} can be expressed as

$$x^*_{m1} = x^*_{01}\frac{1}{1 - x^*_{01}\left(\frac{1}{x^*_{ks}} + \frac{1}{x^*_{zs}} + \frac{1}{x^*_{as}}\right)} = \frac{x^*_{01}}{k_{\text{н}1}} \tag{F.8b}$$

where $k_{\text{н}1} = 1 + x^*_{01}\left(\frac{1}{x^*_{ks}} + \frac{1}{x^*_{zs}} + \frac{1}{x^*_{as}}\right)$.

From (F.8a), the rotor magnetizing reactance x^*_{m2} can be represented in the form

$$x^*_{m2} = x^*_{02}\frac{1}{1 + x^*_{02}\left(\frac{1}{x^*_{kR}} + \frac{1}{x^*_{zR}} + \frac{1}{x^*_{aR}}\right)} = \frac{x^*_{02}}{k_{\text{н}2}} \tag{F.8c}$$

where $k_{\text{н}2} = 1 + x^*_{02}\left(\frac{1}{x^*_{kR}} + \frac{1}{x^*_{zR}} + \frac{1}{x^*_{aR}}\right)$.

The dimensionless values of $k_{н1}$ and $k_{н2}$ used in (F.8b) and (F.8c) are considered as the stator and rotor saturation factors.

On the basis of (F.7) and (F.8), we can obtain the following conditions

$$\frac{1}{x^*_m} = \frac{1}{x^*_{m1}} + \frac{1}{x^*_{m2}} \quad x^*_m = \frac{x^*_{m1}x^*_{m2}}{x^*_{m1} + x^*_{m2}}$$

On the basis of (F.8), the magnetizing equivalent circuits of the stator and rotor can be obtained in Fig. F.2. The equivalent circuit shown in Fig. F.2a corresponds to the "unfolded" stator magnetizing equivalent circuit. The equivalent circuit received in Fig. F.2b reflects the "unfolded" rotor magnetizing equivalent circuit. From (F.5) it follows that the reactances used in the "unfolded" stator magnetizing equivalent circuit can be expressed as

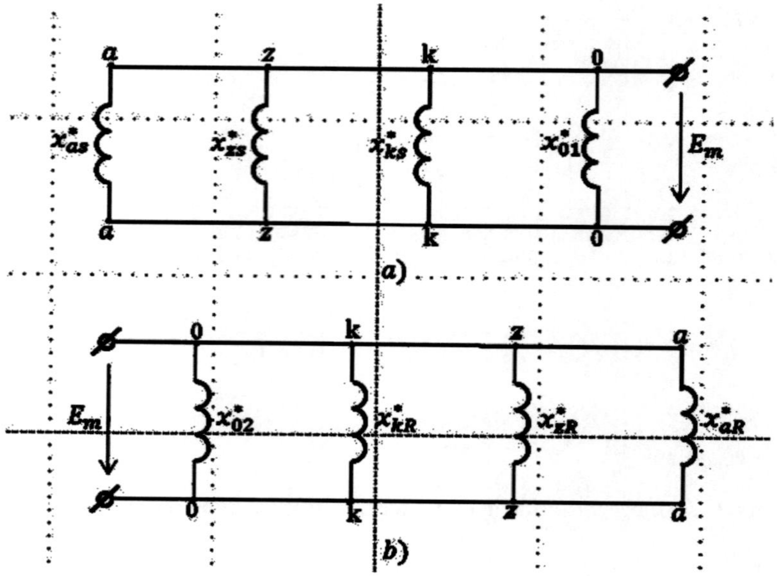

Fig. F.2 "Unfolded" magnetizing equivalent circuits of stator and rotor: (**a**) "Unfolded" stator magnetizing equivalent circuit; (**b**) "unfolded" rotor magnetizing equivalent circuit

$$x^*_{01} = \frac{\omega_1 \mu_0 \tau_0^2}{\pi^2 \delta_1}$$
$$x^*_{ks} = \frac{\omega_1 \mu_{yks} \tau_{ks}^2}{\pi^2 h_{ks}} \frac{\xi_{Lks}}{\xi_{\tau ks}} \tag{F.9}$$
$$x^*_{zs} = \frac{\omega_1 \mu_{yzs} \tau_{zs}^2}{\pi^2 h_{zs}} \frac{\xi_{Lzs}}{\xi_{\tau zs}}$$

The reactance used in the transverse magnetizing circuit loop reflecting the stator yoke layer is determined as

$$x^*_{as} = \omega_1 \mu_{as} t_0 \frac{h_{as}}{t_{as}} \xi_{Las} \tag{F.10}$$

where $t_0 = 2p\tau_0/Z_1$; $t_{as} = 2p\tau_{as}/Z_1$.

Here, t_{as} represents the conditional tooth pitch determined on the inner stator yoke layer surface in relation to the number of the stator teeth.

The reactances used in the longitudinal magnetizing circuit loops representing the air gap sublayer adjoining to the rotor surface and reflecting the layers of the slot-wedge and the rotor winding location areas are defined as

$$x_{02}^{*} = \frac{\omega_1 \mu_0 \tau_0^2}{\pi^2 \delta_2}$$
$$x_{kR}^{*} = \frac{\omega_1 \mu_{ykR} \tau_{kR}^2}{\pi^2 h_{kR}} \frac{\xi_{LkR}}{\xi_{\tau kR}} \quad \text{(F.11)}$$
$$x_{zR}^{*} = \frac{\omega_1 \mu_{yzR} \tau_{zR}^2}{\pi^2 h_{zR}} \frac{\xi_{LzR}}{\xi_{\tau zR}}$$

The reactance used in the transverse magnetizing circuit loop representing the rotor yoke layer is expressed as

$$x_{aR}^{*} = \omega_1 \mu_{aR} t_0 \frac{h_{aR}}{t_{aR}} k_R \xi_{LaR} \quad \text{(F.12)}$$

where $t_{aR} = 2p\tau_{aR}/Z_2$; $k_R = Z_1/Z_2$.

Here, t_{aR} represents the conditional tooth pitch determined on the outer rotor yoke layer surface in relation to the number of the rotor teeth.

The "unfolded" equivalent circuits obtained in Fig. F.2 and the reactances used in them correspond to the case, when the machine magnetizing field is represented as the one-dimensional magnetic field spreading in the machine cross-section area in the radial direction.

"Unfolded" Slot Leakage Equivalent Circuits of Stator and Rotor

In the under-load mode, the total leakage magnetic flux formed by the combined action of the stator and rotor winding currents is existed in the electric machine. However, for the convenience of an analysis, the total leakage magnetic flux should be subdivided into the stator winding leakage magnetic flux that closes only around this winding and into the rotor winding leakage magnetic flux that embraces only the rotor winding. Such, the separation of the total leakage magnetic flux can be carried out in relation to the air gap surface $0 - 0$ passing through the middle of the radial length of the air gap. In Appendix B, it is shown that in the under-load mode the radial component of the magnetic flux density is equal to zero on the air gap surface $0 - 0$ and the tangential component of the magnetic flux density is not practically changed along the radial length of the air gap. Consequently, in relation to the air gap surface $0 - 0$, the currents flowing in the stator and rotor windings create the stator leakage magnetic flux and the rotor leakage magnetic flux embracing the stator and rotor windings individually. For the convenience of an analysis, the stator and rotor leakage magnetic fluxes are subdivided into several parts the main ones of which are the slot leakage (the leakage flux around the coils laid in the slot), the flank, and the differential leakages. These leakage fluxes are taken into account by the corresponding reactances of the analytical expressions for which are usually given in the same form. In this work, the flank and the differential leakages of the stator and rotor windings are not considered. In this regard, we will consider only the slot leakage magnetic fields of the stator and rotor windings. This means that the expressions for the slot leakage reactances are obtained on the basis of the analysis of the slot leakage magnetic fields created by the currents flowing in the stator and rotor windings.

In Fig. F.3a, it is given the picture of the conditional lines of the slot leakage magnetic flux densities for the windings of the stator and rotor having each the three

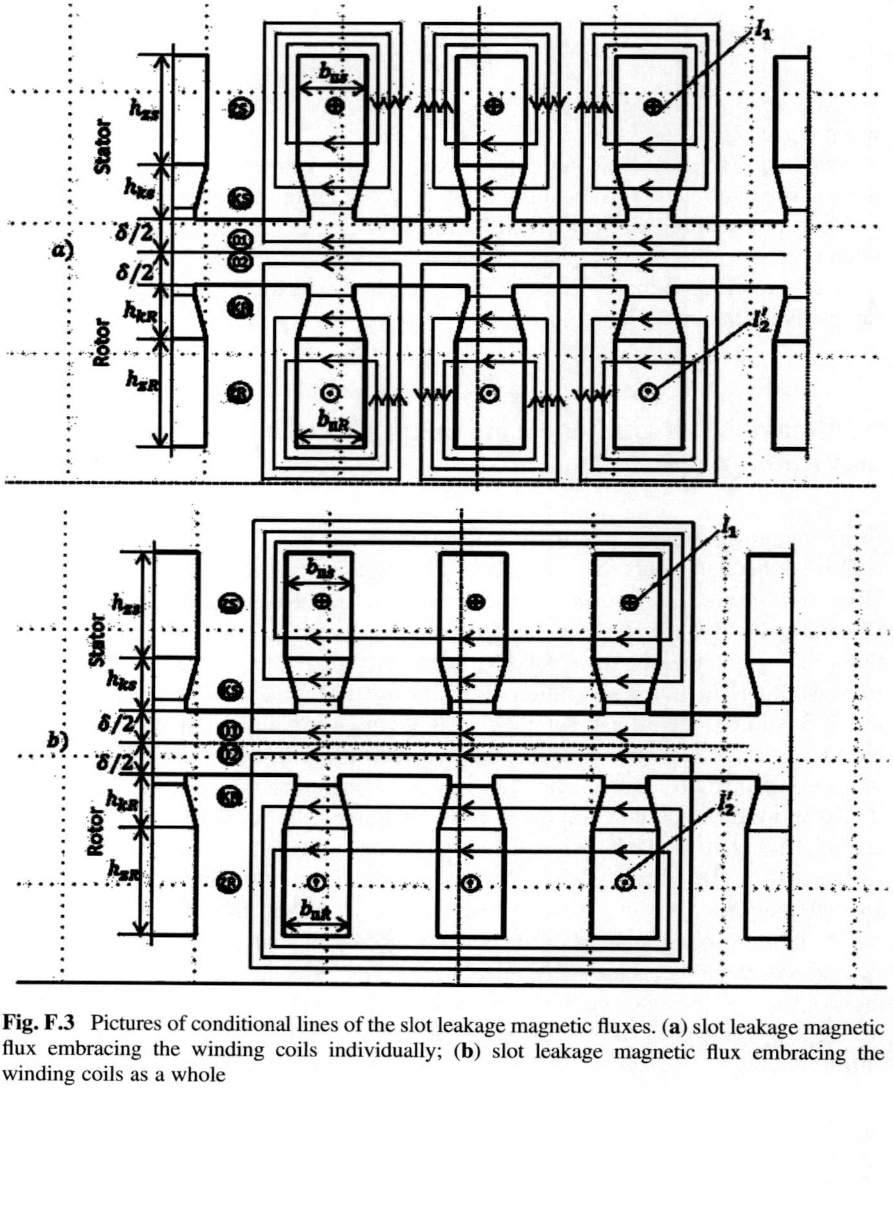

Fig. F.3 Pictures of conditional lines of the slot leakage magnetic fluxes. (**a**) slot leakage magnetic flux embracing the winding coils individually; (**b**) slot leakage magnetic flux embracing the winding coils as a whole

slots, for example [6]. One can imagine that the leakage magnetic fields embrace each coil of the winding and they are spreading within the single tooth pitch. As shown in Fig. F.3a, in the teeth, the radial components of the leakage magnetic fields are mutually compensated, and as a result, the total leakage magnetic fields intersecting the slots in the transverse directions are formed (Fig. F.3b). According to Fig. F.3a, the total slot leakage magnetic field is closed along the three transverse magnetic circuits. The first transverse magnetic circuit represents the slot area occupied by the winding. The second transverse magnetic circuit passes the slot-wedge area. The third transverse magnetic circuit reflects the air gap sublayers having the thickness equal to $\delta/2$ and adjoining to the stator or rotor surfaces. Thus, the slot leakage magnetic field can be considered to be closed in the three main transverse stator and rotor magnetic circuits representing the slot area occupied by the winding, the slot-wedge area, and the areas of the air gap sublayers adjoining to the stator or rotor surfaces. Depending on the configuration of the slots, the leakage magnetic field can be represented by the different number of the transverse magnetic circuits.

For the slot leakage magnetic fluxes, the magnetic circuits shown in Fig. F.3b correspond to the case when the stator and rotor magnetic permeabilities are taken equal to $\mu_c \approx \infty$. In this regard, the leakage magnetic fields can be taken existing in the stator and rotor slots and in the air gap sublayers. The picture of the stator and rotor slot leakage magnetic fields corresponding to this case is given in Fig. F.4a. The further idealization of the slot leakage magnetic field pattern can be based on the representation of the slot sections as the system of the layers with the "smooth" surfaces (Fig. F.4b). As shown previously, for this, it is necessary to introduce the average values of the magnetic permeabilities μ_{xn}, where $n = ks,\ zs,\ kR,\ zR$, for the stator and rotor slot-tooth layers. The definition of μ_{xn} is described subsequently.

In practice, the slot leakage calculations are performed for the single slot. In this case, the single slot can be considered as the single layer spreading in the tangential direction by the length of the single tooth pitch. For the example, we consider the rotor slot section occupied by the winding and represent this slot section as the layer with a height equal to h_{nR} and with the magnetic permeability equal to μ_{xzR} (Fig. F.5b). This slot layer has the width of the single tooth pitch equal to t_{zR}. We will define the value of μ_{xzR}. For this purpose, the reluctance of the single slot having the width b_{nR} and the height h_{zR} (Fig. F.5b) can be given, as well as, the reluctance of the single slot layer having the width t_{zR} and the height h_{zR} (Fig. F.5b) can be considered. We will note that the leakage magnetic field intensity created by the current of the winding laid in the slot increases linearly from the base of the slot. Here, the tooth pitch t_{zR} is determined on the upper surface of the layer reflecting the rotor winding location area. The slot leakage magnetic flux can be obtained as $\Phi_{nR} = \mu_0 \frac{h_{zR} l_{zR}}{2b_{nR}} S_n \sqrt{2} \dot{I}_2$. From here, the reluctance of the single slot is expressed in the form $R_n = \frac{2b_{nR}}{\mu_0 h_{zR} l_{zR}}$. The leakage magnetic flux of the slot layer can be obtained as $\Phi_t = \mu_{xzR} \frac{h_{zR} l_{zR}}{2t_{zR}} S_n \sqrt{2} \dot{I}_2$. The reluctance of the slot layer is represented from this expression in the form $R_t = \frac{2t_{zR}}{\mu_{xzR} h_{zR} l_{zR}}$. We proceed from the fact that $R_n = R_t$. Using this condition and keeping in mind that $R_n = \frac{2b_{nR}}{\mu_0 h_{zR} l_{zR}}$ and $R_t = \frac{2t_{zR}}{\mu_{xzR} h_{zR} l_{zR}}$, the expression

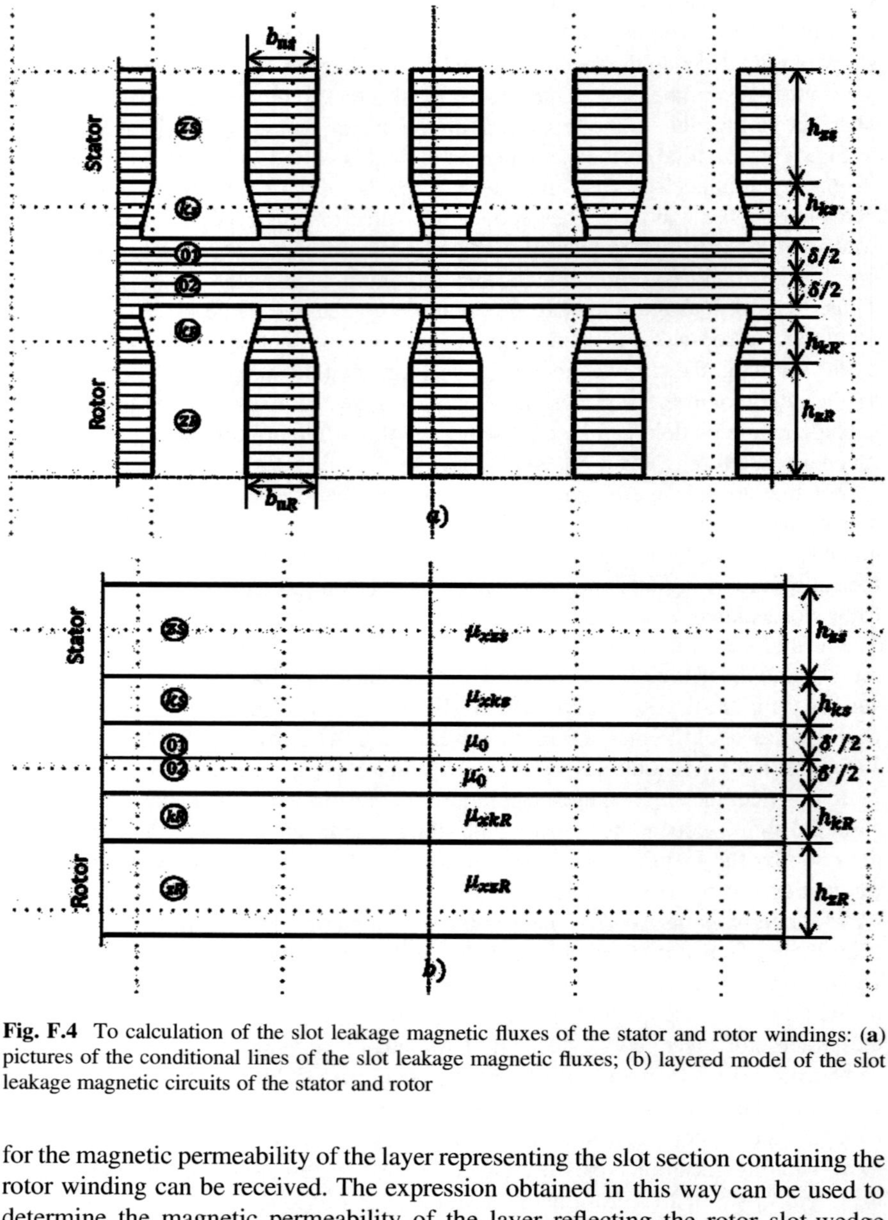

Fig. F.4 To calculation of the slot leakage magnetic fluxes of the stator and rotor windings: (**a**) pictures of the conditional lines of the slot leakage magnetic fluxes; (b) layered model of the slot leakage magnetic circuits of the stator and rotor

for the magnetic permeability of the layer representing the slot section containing the rotor winding can be received. The expression obtained in this way can be used to determine the magnetic permeability of the layer reflecting the rotor slot-wedge section. As a result, we can have

$$\mu_{xkR} = \mu_0 \frac{t_{kR}}{b_{kR}} \quad \mu_{xzR} = \mu_0 \frac{t_{zR}}{b_{\text{п}R}} \tag{F.13}$$

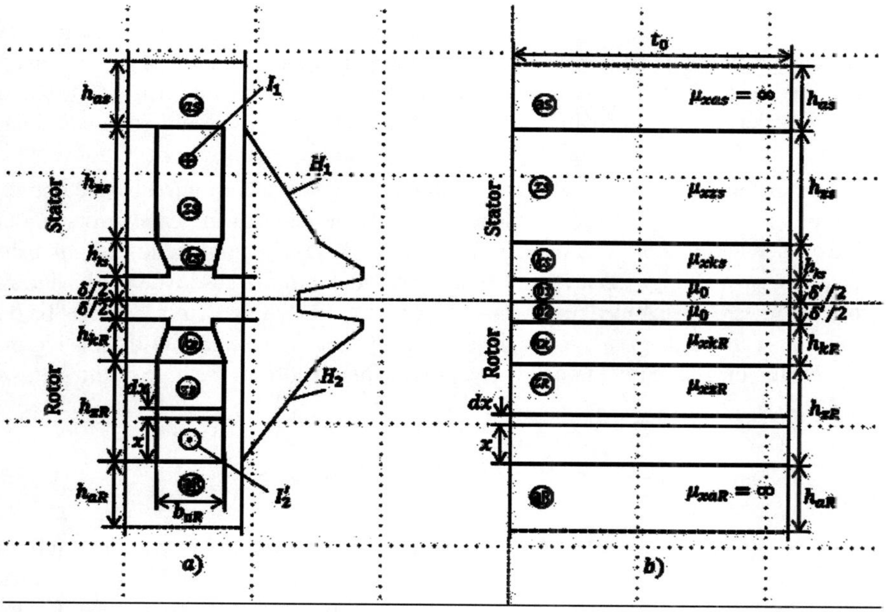

Fig. F.5 To calculation of the single slot leakage magnetic fluxes of the stator and rotor: (**a**) Leakage magnetic flux distribution plot of the single slot of the stator and rotor; (**b**) layered model of the single slot leakage magnetic circuits of the stator and rotor

Acting in the similar way, the magnetic permeabilities of the slot layers representing stator winding and slot-wedge areas can be determined (Fig. F.5b). Now, we can obtain

$$\mu_{xks} = \mu_0 \frac{t_{ks}}{b_{ks}} \quad \mu_{xzs} = \mu_0 \frac{t_{zs}}{b_{ns}} \tag{F.14}$$

Thus, in relation to the slot leakage transverse magnetic flux, the sections reflecting the stator and rotor slot-tooth areas can be represented as the multilayer system. In Figs. F.4b and F.5b, the multilayer models for the stator and rotor slot leakage magnetic circuits are shown. In Fig. F.4b, the slot leakage multilayer model corresponds to the stator and rotor phase zones. For the single tooth pitch, the slot leakage multilayer model obtained in Fig. F.5b is given. In this work, the multilayer model corresponding to the single tooth pitch is used (Fig. F.5b).

We will consider the single slot of the simplest configuration in which the one coil side is laid (Fig. F.5a). The layered model of the slot is given in Fig. F.5b. In this model, the stator winding in the form of the infinitely thin current filaments is uniformly distributed in the layer having the thickness h_{zs} and width t_{zs} and the rotor winding in the form of the infinitely thin current filaments is uniformly distributed in the layer having the thickness h_{zR} and width t_{zR} (Fig. F.5b). In this regard, in these stator and rotor layers (Fig. F.5b), the diagram of the distribution of the leakage magnetic field intensity has the configuration similar to the case when the winding conductors in the stator and rotor slots (Fig. F.5a) are uniformly distributed. We consider the slot layer uniformly filled with the rotor winding conductors and having the magnetic permeability equal to $\mu_{xzR} = \mu_0 \frac{t_{zR}}{b_{nR}}$ (Fig. F.5b). On the upper surface of this layer, the magnetic field intensity H_{zR1} is determined by the expression of kind $H_{zR1} = \sqrt{2}I_2S_n/t_{zR}$, where S_n is the number of the conductors of one side of the winding coil. The picture of the distribution of the magnetic field intensity H_{zR1} is shown in Fig. F.5a. Figure F.5a corresponds to the case when the winding conductors are distributed uniformly in the slot. As shown from Fig. F.5b, the diagram of the leakage magnetic field intensity H_{zR1} is straight. With considering this, for the magnetic field intensity H_{zR1} on the height x from the base of the slot, we can obtain

$$H_{zRx} = H_{zR1}\frac{x}{h_{zR}} \quad B_{zRx} = \mu_{xzR}H_{zR1}\frac{x}{h_{zR}} \tag{F.15}$$

In this slot layer, we select the elementary magnetic flux tube with the width dx and the length t_{zR} and located at the height x from the base of the slot (Fig. F.5b). The axial length of the elementary magnetic flux tube is taken equal to the one. On the height dx the leakage magnetic flux density B_{zRx} creates the magnetic flux $d\Phi_{zR}$ determined as

$$d\Phi_{zR} = B_{zRx}dx = \mu_{xzR}H_{zR1}\frac{x}{h_{zR}}dx$$

The magnetic flux linkage $d\Psi_{zR}$ forming by the magnetic flux $d\Phi_{zR}$ determined on the slot height equal to x/h_{zR} can be expressed as

$$d\Psi_{zR} = \mu_{xzR}H_{zR1}\frac{x^2}{h_{zR}^2}dx \tag{F.16}$$

The total flux linkage

$$\Psi_{zR} = \int_0^{h_{zR}} d\Psi_{zR} = \mu_{xzR}\frac{h_{zR}}{3}H_{zR1} \tag{F.17}$$

On the basis of (F.17), the electric field intensity of the emf induced in the rotor winding can be represented as

$$E_{zR} = \omega_1\Psi_{zR} = \omega_1\mu_{xzR}\frac{h_{zR}}{3}H_{zR1} \tag{F.18}$$

In relation to the values $E'_{zR} = E_{zR}\xi_{LzR}$ and $H'_{zR1} = H_{zR1}\xi_{\tau zR}$, where $\xi_{LzR} = l_{zR}/l_0$ and $\xi_{\tau zR} = \tau_{zR}/\tau_0$, reduced to the air gap surface 0 – 0, the leakage reactance representing the layer of the rotor winding location area is multiplied by the coefficient $(\xi_{LzR}/\xi_{\tau zR})$ determined as $E'_{zR}/H'_{zR1} = (E_{zR}/H_{zR1})(\xi_{LzR}/\xi_{\tau zR}) = x^*_{22\sigma}(\xi_{LzR}/\xi_{\tau zR})$, where $x^*_{22\sigma} = E_{zR}/H_{zR1}$. With considering this condition, we have from (F.18)

$$x^*_{22\sigma} = \omega_1\mu_{xzR}\frac{h_{zR}}{3}\frac{\xi_{LzR}}{\xi_{\tau zR}} \tag{F.19}$$

In (F.19), we consider the coefficient $\xi_{\tau zR}$ determined as $\xi_{\tau zR} = \tau_{zR}/\tau_0$. Taking into account that $t_{zR} = 2p\tau_{zR}/Z_2$ and $t_0 = 2p\tau_0/Z_1$, for τ_{zR}/τ_0, we have that $\tau_{zR}/\tau_0 = (1/k_R)(t_{zR}/t_0)$, where $k_R = (Z_1/Z_2)$. With considering the condition $\tau_{zR}/\tau_0 = (1/k_R)(t_{zR}/t_0)$, the expression (F.19) can be represented as

$$x^*_{22\sigma} = \omega_1\mu_{xzR}t_0\frac{h_{zR}}{3t_{zR}}k_R\xi_{LzR} \tag{F.20}$$

where $k_R = Z_1/Z_2$.

The expression obtained in (F.20) corresponds to the case when the rotor winding location area is represented as the layer having the thickness h_{zR} and magnetic permeability equal to $\mu_{xzR} = \mu_0\frac{t_{zR}}{b_{nR}}$. Using this condition for μ_{xzR}, one can show that the known expression obtaining on the basis of considering the leakage magnetic field in the single rotor slot can be derived from (F.20). Acting in the similar way and taking into account that the leakage magnetic field intensity does not change along the layer height, the leakage reactance $x^*_{\tau kR}$ of the layer representing the rotor slot-wedge area can be reduced to the form shown in (F.19). Now, from (F.19) the leakage reactance $x^*_{\tau kR}$ can be expressed in the form

$$x^*_{\tau kR} = \omega_1\mu_{xkR}h_{kR}\frac{\xi_{LkR}}{\xi_{\tau kR}} \tag{F.21}$$

In (F.21), we consider the coefficient $\xi_{\tau kR}$ determined as $\xi_{\tau kR} = \tau_{kR}/\tau_0$. Taking into account that $t_{kR} = 2p\tau_{kR}/Z_2$ and $t_0 = 2p\tau_0/Z_1$, for τ_{kR}/τ_0, it is followed that $\tau_{kR}/\tau_0 = (1/k_R)(t_{kR}/t_0)$, where $k_R = (Z_1/Z_2)$. Since $\tau_{kR}/\tau_0 = (1/k_R)(t_{kR}/t_0)$, the expression (F.21), can be reduced to the form

$$x^*_{\tau kR} = \omega_1\mu_{xkR}t_0\frac{h_{kR}}{t_{kR}}k_R\xi_{LkR} \tag{F.22}$$

From (F.21), for the leakage reactance of the air gap sublayer adjoining to the rotor surface, we have

$$x^*_{\tau 02} = \omega_1 \mu_0 t_0 \frac{\delta}{2t_0} \tag{F.23}$$

This formula can be expressed in relation to the value of t_{02} calculated as $t_{02} = 2p\tau_0/Z_2$. Then, it is followed

$$x^*_{\tau 02} = \omega_1 \mu_0 t_0 \frac{\delta}{2t_{02}} k_R \tag{F.24}$$

where $k_R = t_{02}/t_0 = Z_1/Z_2$.

Now, on the basis of the expressions (F.19), (F.22), (F.24) the slot leakage reactance of the rotor winding $x^*_{2\sigma}$ can be represented as the sum of the components $x^*_{\tau 02}, x^*_{\tau kR}$ and $x^*_{22\sigma}$. Then, we can obtain

$$x^*_{2\sigma} = x^*_{\tau 02} + x^*_{\tau kR} + x^*_{22\sigma} \tag{F.25}$$

Here, the reactances $x^*_{\tau 02}, x^*_{\tau kR}$ and $x^*_{22\sigma}$ acquire

$$\begin{aligned} x^*_{\tau 02} &= \omega_1 \mu_0 t_0 \frac{\delta}{2t_{02}} k_{2\sigma} k_R \\ x^*_{\tau kR} &= \omega_1 \mu_{xkR} t_0 \frac{h_{kR}}{t_{kR}} k_R \xi_{LkR} \\ x^*_{22\sigma} &= \omega_1 \mu_{xzR} t_0 \frac{h_{zR}}{3t_{zR}} k_R \xi_{LzR} \end{aligned} \tag{F.26}$$

Acting in the similar way, the leakage reactance of the stator winding $x^*_{1\sigma}$ can be expressed as

$$x^*_{1\sigma} = x^*_{\tau 01} + x^*_{\tau ks} + x^*_{11\sigma} \tag{F.27}$$

By analogy with (F.26), the reactances $x^*_{\tau 01}, x^*_{\tau ks}$ and $x^*_{11\sigma}$ can be reduced to the form

$$\begin{aligned} x^*_{\tau 01} &= \omega_1 \mu_0 t_0 \frac{\delta}{2t_0} k_{1\sigma} \\ x^*_{\tau ks} &= \omega_1 \mu_{xks} t_0 \frac{h_{ks}}{t_{ks}} \xi_{Lks} \\ x^*_{11\sigma} &= \omega_1 \mu_{xzs} t_0 \frac{h_{zs}}{3t_{zs}} \xi_{Lzs} \end{aligned} \tag{F.28}$$

The equation for $\dot{E}_m$ of the emf induced in the stator winding from (12.42b) and (F.27) can be represented as

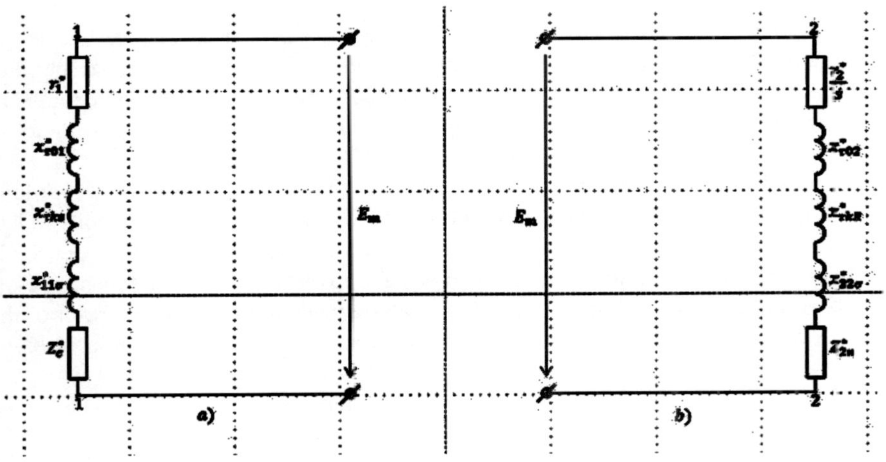

Fig. F.6 "Unfolded" slot leakage equivalent circuits of the stator and the rotor. (**a**) "unfolded" stator slot leakage equivalent circuit; (**b**) "unfolded" rotor slot leakage equivalent circuit

$$\dot{E}_m = \left[j\left(x^*_{\tau 01} + x^*_{\tau ks} + x^*_{11\sigma}\right) + r^*_1 + Z^*_c \right] \dot{H}_1 \tag{F.29}$$

The equation (F.29), represents the stator slot leakage circuit loop.

The equation for $\dot{E}_m$ of the emf induced in the rotor winding from (12.46b) and (F.25) can be expressed as

$$\dot{E}_m = \left[j\left(x^*_{\tau 02} + x^*_{\tau kR} + x^*_{22\sigma}\right) + r^*_2/s + Z^*_{2\text{H}} \right] \dot{H}_2 \tag{F.30}$$

The equation (F.30) describes the rotor slot leakage circuit loop.

Here, the equations (F.29) and (F.30) are represented in the specific system of units. On the basis of the equations (F.29) and (F.30), the "unfolded" slot leakage equivalent circuits of the stator and rotor can be obtained. The "unfolded" stator slot leakage equivalent circuit is given in Fig. F.6a, and the "unfolded" rotor slot leakage equivalent circuit is shown in Fig. F.6b.

Summary

1. For the stator and rotor magnetizing circuits and the stator and rotor slot leakage magnetic circuits, the multilayer models are proposed. For this purpose, the average values of the magnetic permeabilities are applied for the layers representing the stator and rotor slot-teeth areas. The use of the average values of the magnetic permeabilities simplifies an analysis.

2. The stator and rotor magnetizing equivalent circuits are constructed, as well as the stator and rotor slot leakage equivalent circuits are built. For the construction of these equivalent circuits, the analysis of the stator and rotor magnetic circuits, as well as the stator and rotor slot leakage magnetic circuits, has been carried out. In this case, the specific system of units is used.
3. The stator and rotor magnetizing equivalent circuits correspond to the no-load mode of the machine. The stator and rotor slot leakage equivalent circuits reflect the machine under-load mode. In accordance with the "no-load and under-load" concept, these equivalent circuits can be used to construct the equivalent circuits that reflect the resulting magnetic fields taking place in the stator and rotor of the electric machine. For this purpose, the superposition principle can be applied.

References

1. Asanbayev, V.: Alternating Current Multi-Circuit Electric Machines: A New Approach to Steady-State Parameter Determination, Springer Cham Heidelberg, New York, Dordrecht London (2015)
2. Boldea, I., Bobescu, M.: Multilayer Approach to the Analysis of Single-Sided Linear Induction Motor, Pross.IEE,vol.125, N 4, 1978, pp. 220-287
3. Willianson, S.: The Anisotropic Layer Theory of Induction Machines and Devices, Inst. Math. And Appl., 1976, vol.17, N 1, pp.69-84
4. Inkin, A., Litvinov, B.: Synthesis of Cascade Equivalent Circuits for Induction Electric Machines, Elektrotechnika, 1977, n.1, pp.29-34
5. Veselovsky, O., Konyaev, A., Sarapulov, F.: Linear Asynchronous Motor, Publishing Energoatomizdat, Moscow (1991)
6. Костенко, М., Пиотровский, А.: Электрические Машины Переменного Тока, Изд-во Энергия, Ленинградское Отделение, Ленинград (1973)

Appendix G: Reducing "Unfolded" Equivalent Circuits of Stator and Rotor

Reducing "Unfolded" Stator Equivalent Circuit to Equivalent Circuit with Parallel Connection of Elements

We will proceed from the equivalent circuit shown in Fig. G.1. In Fig. G.1, the "unfolded" stator equivalent circuit with the mixed connection of the elements is shown. This equivalent circuit needs to be transformed. For this purpose, the multiple transfer technique can be used [1]. It should be noted that when implementing the procedure of transferring, the equivalent circuit in Fig. G.1 can be reduced to the form of the equivalent circuit in Fig.14.4a in relation to the parallel branch transferred to its terminals $0 - 0$. Then, the process of transferring is simplified and becomes clearer.

We will transform the equivalent circuit shown in Fig. G.1. This process can be started from transferring the parallel branch $k - k$ to its terminals $0 - 0$. In relation to the parallel branch $k - k$, the left part of the equivalent circuit in Fig. G.1 can be represented as the impedance Z^*_{kza} determined by the equivalent circuit in Fig. G.2. Then, with considering the equivalent circuit in Fig. G.2, the equivalent circuit (Fig. G.1) acquires the form shown in Fig. G.3. In this equivalent circuit, the parallel branch $k - k$ can be transferred to the terminals $0 - 0$. When transferring the parallel branch $k - k$ of the equivalent circuit in Fig. G.3 to its terminals $0 - 0$, the longitudinal reactance $x^*_{\tau 01}$ goes into the parallel branch $k - k$ and adds up with its reactance x^*_{ks}, i.e. $x^*_{ks} + x^*_{\tau 01} = x^*_{ks}\left(1 + x^*_{\tau 01}/x^*_{ks}\right) = x^*_{ks}c_{\tau ks}$, where $c_{\tau ks} = 1 + x^*_{\tau 01}/x^*_{ks}$. Now, in the converted equivalent circuit (Fig. G.4), the reactance of the parallel branch $k - k$ becomes equal to $x^*_{ks}c_{\tau ks}$. Here, the value of $x^*_{ks}c_{\tau ks}$ represents the reactance of the magnetizing circuit loop of the layer reflecting the stator slot-wedge area. So, we can have

V. Asanbayev, *Asynchronous Machines*, https://doi.org/10.1007/978-3-030-92284-9

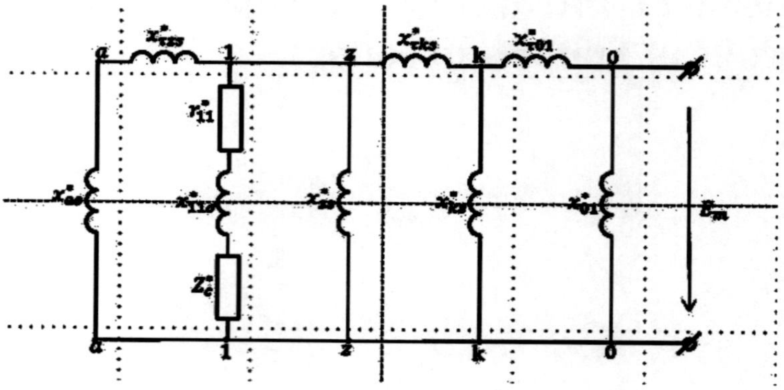

Fig. G.1 "Unfolded" stator equivalent circuit with the mixed connection of the elements

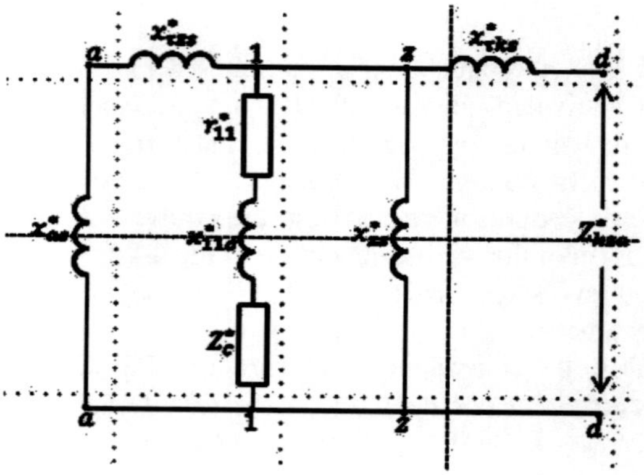

Fig. G.2

$$\overset{\circ}{x}_{ks} = x^*_{ks} c_{\tau ks} = x^*_{ks} + x^*_{\tau 01} \tag{G.1}$$

For the transfer factor $c_{\tau ks}$, we can obtain

$$c_{\tau ks} = 1 + \frac{x^*_{\tau 01}}{x^*_{ks}} \tag{G.2}$$

In (G.2), the value of $c_{\tau ks}$ represents the transfer factor of the parallel branch $k - k$ of the equivalent circuit in Fig. G.3.

Further, in the equivalent circuit in Fig. G.3, when transferring its parallel branch $k - k$, the impedance of the parallel branch $d - d$ is multiplied by the factor $c^2_{\tau ks}$ and acquires the value equal to $Z^*_{kza} c^2_{\tau ks}$. In addition, multiplied by the factor $c_{\tau ks}$ the

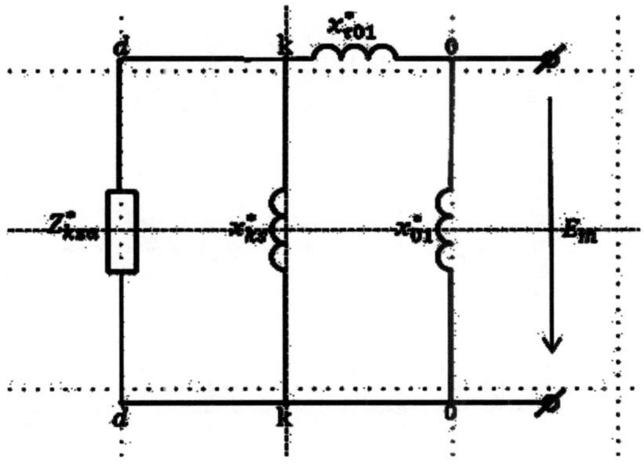

Fig. G.3

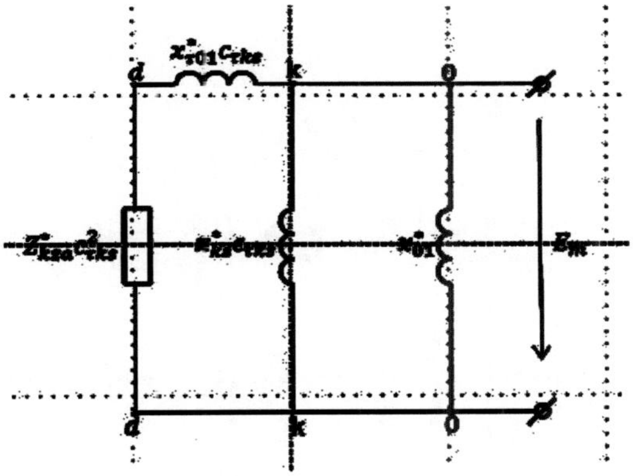

Fig. G.4

longitudinal reactance $x^*_{\tau 01}$, i.e., $x^*_{\tau 01}c_{\tau ks}$, is added up to the corrected impedance of the parallel branch $d - d$ determined as $Z^*_{kza}c^2_{\tau ks}$. As a result, in the converted equivalent circuit (Fig. G.4), we have the impedance $Z^*_{kza}c^2_{\tau ks} + x^*_{\tau 01}c_{\tau ks}$ instead of the impedance Z^*_{kza} (Fig. G.3). Thus, the equivalent circuit received in Fig. G.3 acquires the form in Fig. G.4.

Now, the equivalent circuits in Figs. G.2 and G.4 can be considered. We can obtain the stator equivalent circuit for what the terminals of the parallel branch $d - d$ of the equivalent circuit in Fig. G.4 can be connected to the input terminals $d - d$ of the equivalent circuit in Fig. G.2. As follows from the equivalent circuit in Fig. G.4, the impedance of its parallel branch $d - d$ is multiplied by the factor $c^2_{\tau ks}$. Thus, when connecting the terminals of the parallel branch $d - d$ of the equivalent circuit in Fig. G.4 to the input terminals $d - d$ of the equivalent circuit in Fig. G.2, the impedances used in Fig. G.2 are multiplied by the factor $c^2_{\tau ks}$. Now, the stator equivalent circuit acquires the form in Fig. G.5.

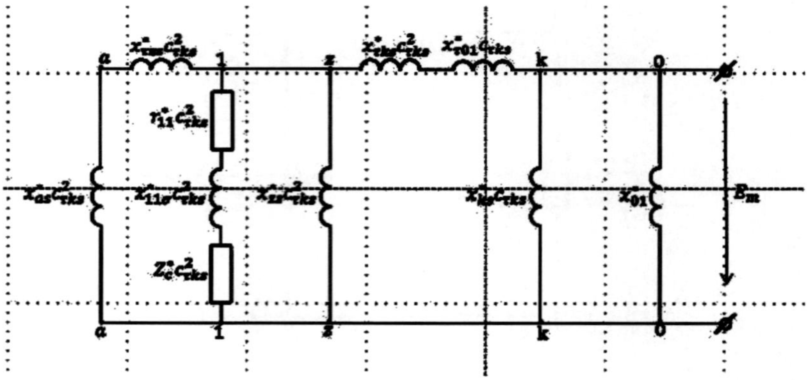

Fig. G.5

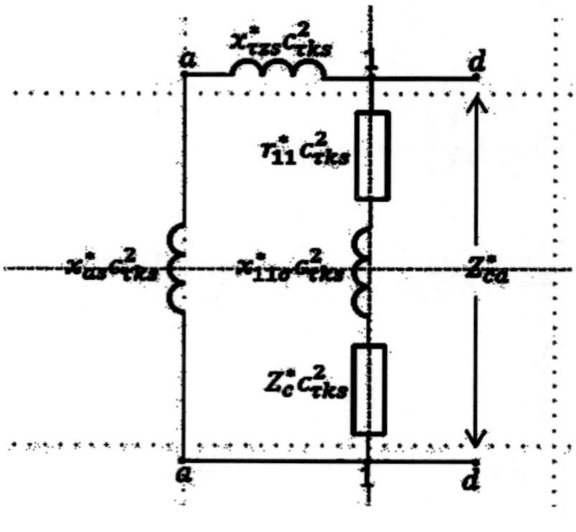

Fig. G.6

The parallel branch $z - z$ used in the equivalent circuit in Fig. G.5 can be transferred to its terminals $k - k$. For this purpose, in relation to the parallel branch $z - z$, the left part of the equivalent circuit in Fig. G.5 can be represented as the impedance Z^*_{ca} determined by the equivalent circuit in Fig. G.6. Then, the equivalent circuit in Fig. G.5 acquires the form shown in Fig. G.7. In the equivalent circuit in Fig. G.7, the parallel branch $z - z$ can be transferred to the terminals $k - k$. When transferring the parallel branch $z - z$ of the equivalent circuit in Fig. G.7 to its terminals $k - k$, the series-connected longitudinal reactances $\left(x^*_{\tau 01} c_{\tau ks} + x^*_{\tau ks} c^2_{\tau ks}\right)$ are going into the parallel branch $z - z$ and are added up with its reactance $x^*_{zs} c^2_{\tau ks}$, i.e., $x^*_{zs} c^2_{\tau ks} + x^*_{\tau 01} c_{\tau ks} + x^*_{\tau ks} c^2_{\tau ks} = x^*_{zs} c^2_{\tau ks}\left(1 + \frac{x^*_{\tau 01} c_{\tau ks} + x^*_{\tau ks} c^2_{\tau ks}}{x^*_{zs} c^2_{\tau ks}}\right) = x^*_{zs} c_{\tau zs} c^2_{\tau ks}$,

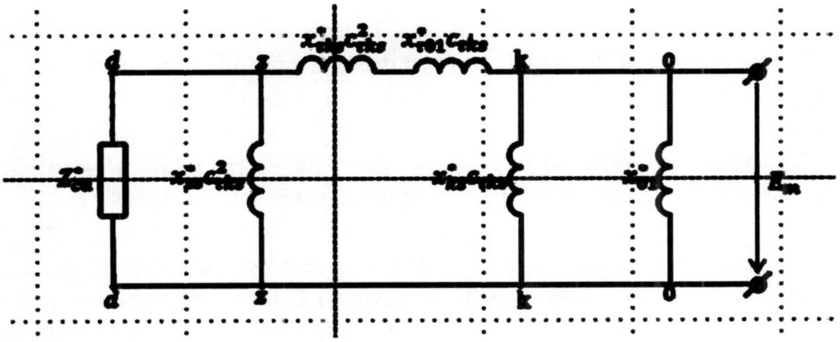

Fig. G.7

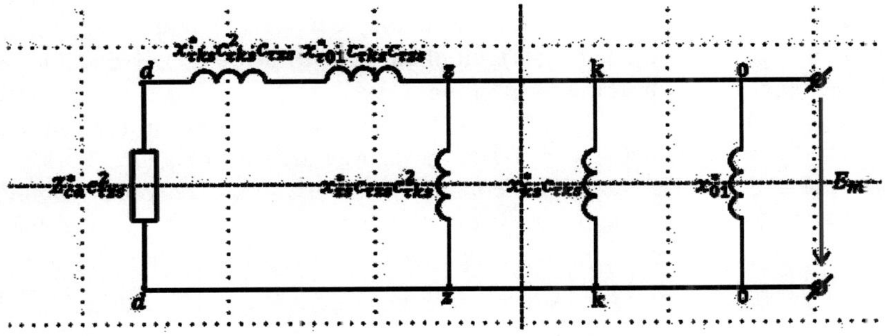

Fig. G.8

where $c_{\tau zs} = 1 + \frac{x^*_{\tau 01} c_{\tau ks} + x^*_{\tau ks} c^2_{\tau ks}}{x^*_{zs} c^2_{\tau ks}}$. Consequently, in the converted equivalent circuit (Fig. G.8), the reactance of the parallel branch $z - z$ becomes equal to $x^*_{zs} c_{\tau zs} c^2_{\tau ks}$. Here, the value of $x^*_{zs} c_{\tau zs} c^2_{\tau ks}$ represents the reactance of the magnetizing circuit loop of the layer reflecting the stator winding location area. For this reactance, we can have

$$\overset{\circ}{x}_{zs} = x^*_{zs} c_{\tau zs} c^2_{\tau ks} = x^*_{zs} c^2_{\tau ks} + x^*_{\tau ks} c^2_{\tau ks} + x^*_{\tau 01} c_{\tau ks} \tag{G.3}$$

The factor $c_{\tau zs}$ used in (G.3) determined as

$$c_{\tau zs} = 1 + \frac{x^*_{\tau 01} c_{\tau ks} + x^*_{\tau ks} c^2_{\tau ks}}{x^*_{zs} c^2_{\tau ks}} \tag{G.4}$$

In (G.4), the value of $c_{\tau zs}$ reflects the transfer factor of the parallel branch $z - z$ of the equivalent circuit in Fig. G.8.

Further, in the equivalent circuit in Fig. G.7, when transferring its parallel branch $z - z$, the reactance of the parallel branch $d - d$ is multiplied by the factor $c^2_{\tau zs}$ and acquires the value equal to $Z^*_{ca}c^2_{\tau zs}$. In addition, multiplied by the factor $c_{\tau zs}$ the series connected longitudinal reactances $\left(x^*_{\tau 01}c_{\tau ks} + x^*_{\tau ks}c^2_{\tau ks}\right)$, i.e., $\left(x^*_{\tau 01}c_{\tau ks} + x^*_{\tau ks}c^2_{\tau ks}\right)c_{\tau zs}$, are added up to the corrected reactance of the parallel branch $d - d$ determined as $Z^*_{ca}c^2_{\tau zs}$. As a result, in the converted equivalent circuit (Fig. G.8), we have the impedance $Z^*_{ca}c^2_{\tau zs} + x^*_{\tau 01}c_{\tau ks}c_{\tau zs} + x^*_{\tau ks}c^2_{\tau ks}c_{\tau zs}$ instead the impedance Z^*_{ca} (Fig. G.7). Thus, the equivalent circuit obtained in Fig. G.7 acquires the form in Fig. G.8.

Now, the equivalent circuits in Figs. G.6 and G.8 can be considered. We can obtain the stator equivalent circuit for what the terminals of the parallel branch $d - d$ of the equivalent circuit in Fig. G.8 can be connected to the input terminals $d - d$ of the equivalent circuit in Fig. G.6. As follows from the equivalent circuit in Fig. G.8, the impedance of its parallel branch $d - d$ is multiplied by the factor $c^2_{\tau zs}$. Thus, when connecting the terminals of the parallel branch $d - d$ of the equivalent circuit in Fig. G.8 to the input terminals $d - d$ of the equivalent circuit in Fig. G.6, the impedance used in Fig. G.6 are multiplied by the factor $c^2_{\tau zs}$. Now, the stator equivalent circuit acquires the form in Fig. G.9.

The parallel branch 1 – 1 used in the equivalent circuit shown in Fig. G.9 can be transferred to the terminals $z - z$. When transferring the parallel branch 1 – 1 of the equivalent circuit in Fig. G.9 to its terminals $z - z$, the series-connected longitudinal reactances $\left(x^*_{\tau 01}c_{\tau ks}c_{\tau zs} + x^*_{\tau ks}c^2_{\tau ks}c_{\tau zs}\right)$ go into the parallel branch 1 – 1, and they are added up with its impedances $\left(r^*_{11} + jx^*_{11\sigma} + Z^*_c\right)c^2_{\tau ks}c^2_{\tau zs}$, i.e. $\left(r^*_{11} + jx^*_{11\sigma} + Z^*_c\right)c^2_{\tau ks}c^2_{\tau zs} + x^*_{\tau 01}c_{\tau ks}c_{\tau zs} + x^*_{\tau ks}c^2_{\tau ks}c_{\tau zs} = \left(r^*_{11} + jx^*_{11\sigma} + Z^*_c\right)c^2_{\tau ks}c^2_{\tau zs}\left(1 + \frac{x^*_{\tau 01}c_{\tau ks}c_{\tau zs} + x^*_{\tau ks}c^2_{\tau ks}c_{\tau zs}}{\left(r^*_{11} + jx^*_{11\sigma} + Z^*_c\right)c^2_{\tau ks}c^2_{\tau zs}}\right) = \left(r^*_{11} + jx^*_{11\sigma} + Z^*_c\right)c^2_{\tau ks}c^2_{\tau zs}c_{\tau 1}$, where $c_{\tau 1} = 1 + \frac{x^*_{\tau 01}c_{\tau ks}c_{\tau zs} + x^*_{\tau ks}c^2_{\tau ks}c_{\tau zs}}{\left(r^*_{11} + jx^*_{11\sigma} + Z^*_c\right)c^2_{\tau ks}c^2_{\tau zs}}$. Consequently, in the converted equivalent circuit (Fig. G.10), the impedance of the parallel branch 1 – 1 becomes equal to $\left(r^*_{11} + jx^*_{11\sigma} + Z^*_c\right)c^2_{\tau ks}c^2_{\tau zs}c_{\tau 1}$.

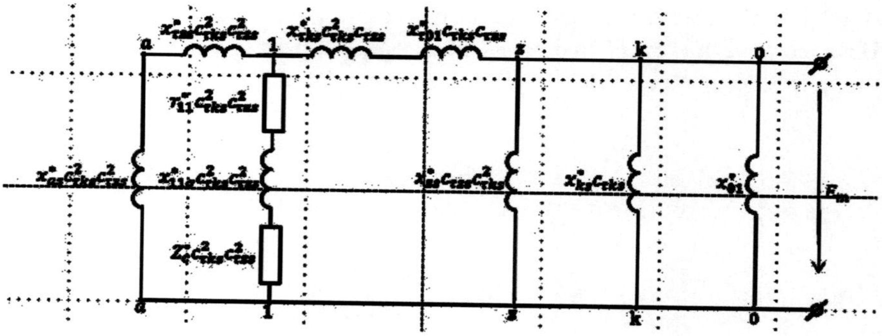

Fig. G.9

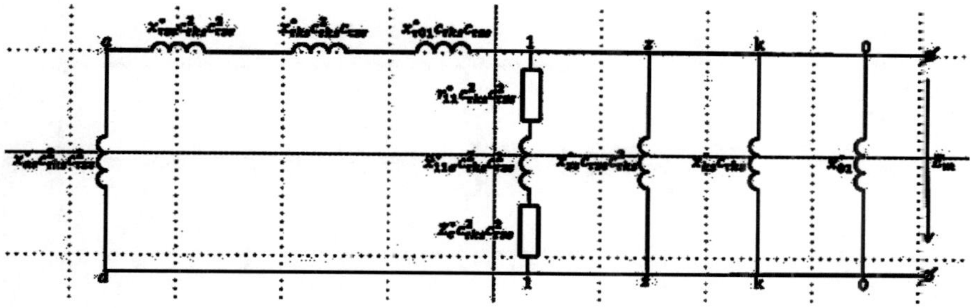

Fig. G.10

Here, the value of $\left(r_{11}^{*}+jx_{11\sigma}^{*}+Z_{c}^{*}\right)c_{\tau ks}^{2}c_{\tau zs}^{2}c_{\tau 1}$ represents the impedance of the slot leakage circuit loop of the stator winding. For this impedance, we can have

$$\begin{aligned} Z_1^{\circ} &= r_{11}^{*}c_{\tau ks}^{2}c_{\tau zs}^{2}+j\left(x_{11\sigma}^{*}c_{\tau ks}^{2}c_{\tau zs}^{2}+x_{\tau ks}^{*}c_{\tau ks}^{2}c_{\tau zs}+x_{\tau 01}^{*}c_{\tau ks}c_{\tau zs}\right)+Z_{c}^{*}c_{\tau ks}^{2}c_{\tau zs}^{2} \\ &= r_{11}^{*}c_{\tau ks}^{2}c_{\tau zs}^{2}+jx_{1\sigma}^{*}+Z_{c}^{*}c_{\tau ks}^{2}c_{\tau zs}^{2} \end{aligned} \tag{G.5a}$$

In another form

$$\begin{aligned} Z_1^{\circ} &= r_{11}^{*}c_{\tau ks}^{2}c_{\tau zs}^{2} \\ &+j\left[x_{11\sigma}^{*}+x_{\tau ks}^{*}(1/c_{\tau zs})+x_{\tau 01}^{*}(1/c_{\tau ks}c_{\tau zs})\right]c_{\tau ks}^{2}c_{\tau zs}^{2}+Z_{c}^{*}c_{\tau ks}^{2}c_{\tau zs}^{2} \\ &= \left(r_{11}^{*}+j\widetilde{x}_{11\sigma}^{*}+Z_{c}^{*}\right)c_{\tau ks}^{2}c_{\tau zs}^{2} \end{aligned} \tag{G.5b}$$

where $\widetilde{x}_{11\sigma}^{*}=x_{11\sigma}^{*}+x_{\tau ks}^{*}(1/c_{\tau zs})+x_{\tau 01}^{*}(1/c_{\tau ks}c_{\tau zs}); x_{1\sigma}^{*}=\widetilde{x}_{11\sigma}^{*}c_{\tau ks}^{2}c_{\tau zs}^{2}$.

For the transfer factor $c_{\tau 1}$, we have

$$c_{\tau 1}=1+\frac{x_{\tau 01}^{*}c_{\tau ks}c_{\tau zs}+x_{\tau ks}^{*}c_{\tau ks}^{2}c_{\tau zs}}{\left(r_{11}^{*}+jx_{11\sigma}^{*}+Z_{c}^{*}\right)c_{\tau ks}^{2}c_{\tau zs}^{2}} \tag{G.6}$$

In (G.6), the value of $c_{\tau 1}$ represents the transfer factor of the parallel branch $1-1$ of the equivalent circuit in Fig. G.9.

Further, in the equivalent circuit in Fig. G.9, when transferring its parallel branch $1-1$, the series-connected reactances $(x_{\tau zs}+x_{as})c_{\tau ks}^{2}c_{\tau zs}^{2}$ are multiplied by the factor $c_{\tau 1}^{2}$ and acquire the value equal to $(x_{\tau zs}+x_{as})c_{\tau ks}^{2}c_{\tau zs}^{2}c_{\tau 1}^{2}$. In addition, multiplied by the factor $c_{\tau 1}$ the series-connected longitudinal reactances $\left(x_{\tau 01}^{*}c_{\tau ks}c_{\tau zs}+x_{\tau ks}^{*}c_{\tau ks}^{2}c_{\tau zs}\right)$, i.e., $\left(x_{\tau 01}^{*}c_{\tau ks}c_{\tau zs}+x_{\tau ks}^{*}c_{\tau ks}^{2}c_{\tau zs}\right)c_{\tau 1}$, are added up to the corrected reactance of the parallel branch $1-a-a$ determined as $(x_{\tau zs}+x_{as})c_{\tau ks}^{2}c_{\tau zs}^{2}c_{\tau 1}^{2}$. As a result, instead of the reactances $[(x_{\tau zs}+x_{as})c_{\tau ks}^{2}c_{\tau zs}^{2}]$ used in Fig. G.9, we have the reactances of kind

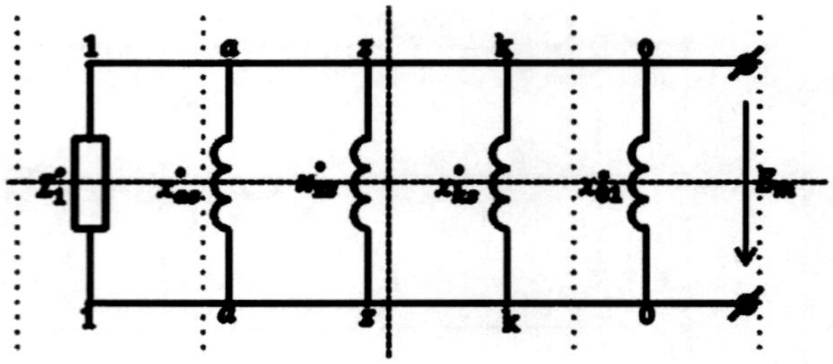

Fig. G.11 "Unfolded" stator equivalent circuit with the parallel connection of the elements

$\left[\left(x^*_{\tau 01}c_{\tau ks} + x^*_{\tau ks}c^2_{\tau ks}\right)c_{\tau zs}c_{\tau 1} + (x_{\tau zs} + x_{as})c^2_{\tau ks}c^2_{\tau zs}c^2_{\tau 1}\right]$ in the converted equivalent circuit (Fig. G.10). Now, for the reactance $\overset{\circ}{x}_{as}$ used in the equivalent circuit (Fig. G.11), we have

$$\overset{\circ}{x}_{as} = \left(x^*_{\tau 01}c_{\tau ks} + x^*_{\tau ks}c^2_{\tau ks}\right)c_{\tau zs}c_{\tau 1} + (x_{\tau zs} + x_{as})c^2_{\tau ks}c^2_{\tau zs}c^2_{\tau 1} \tag{G.7}$$

In (G.7), the value of $\overset{\circ}{x}_{as}$ represents the reactance of the magnetizing circuit loop of the stator yoke layer.

The electric machines used in the power systems are designed with a relatively high efficiency. This means that the impedance of the power supply network is much higher than the impedance of the electric machine winding, i.e., $Z^*_c > \left|r^*_{11} + jx^*_{11\sigma}\right|$. Therefore, for the value of the transfer factor $c_{\tau 1}$ determined by (G.6), the condition of kind $c_{\tau 1} \approx 1.0$ is fulfilled. Consequently, the expression (G.7) for the reactance $\overset{\circ}{x}_{as}$ can be written without taking into account the transfer factor $c_{\tau 1}$. Then, we can obtain

$$\overset{\circ}{x}_{as} = x_{as}c^2_{\tau ks}c^2_{\tau zs} + x_{\tau zs}c^2_{\tau ks}c^2_{\tau zs} + x^*_{\tau ks}c^2_{\tau ks}c_{\tau zs} + x^*_{\tau 01}c_{\tau ks}c_{\tau zs} \tag{G.8}$$

For the reactances used in the equivalent circuit in Fig. G.10, the expressions obtained in (G.1), (G.3), (G.5b) and (G.8) can be applied. Then, the equivalent circuit received in Fig. G.10 can be reduced to the form in Fig. G.11. In Fig. G.11, the "unfolded" stator equivalent circuit with the parallel connection of the elements is shown.

We can note that the stator magnetizing circuit loop and the stator slot leakage circuit loop can be introduced and the reactances applying in them can be determined using the equivalent circuit given in Fig. G.11. As a result of this, the "folded" stator equivalent circuit arises.

Reducing "Unfolded" Rotor Equivalent Circuit to Equivalent Circuit with Parallel Connection of Elements

We will consider the equivalent circuit shown in Fig. G.12. In Fig. G.12, the "unfolded" rotor equivalent circuit with the mixed connection of the elements is given. This equivalent circuit needs to be transformed. For this purpose, the multiple transfer technique can be used. It should be noted that when implementing the procedures of transferring, the equivalent circuit in Fig. G.12 can be reduced to the form of the equivalent circuit in Fig.14.4a in relation to the parallel branch transferred to the terminals 0 – 0. Then, the process of transferring is simplified and becomes clearer.

The transformation of the equivalent circuit obtained in Fig. G.12 can be started from transferring the parallel branch $k - k$ to its terminals 0 – 0. In relation to the parallel branch $k - k$, the right part of the equivalent circuit in Fig. G.12 can be represented as the impedance Z^*_{kza} determined by the equivalent circuit in Fig. G.13. Then, with considering the equivalent circuit in Fig. G.13 the equivalent circuit in Fig. G.12 acquires the form shown in Fig. G.14. In this equivalent circuit, the parallel branch $k - k$ can be transferred to the terminals 0 – 0. When transferring the parallel branch $k - k$ of the equivalent circuit in Fig. G.14 to the terminals 0 – 0, the longitudinal reactance $x^*_{\tau 02}$ goes into the parallel branch $k - k$, and it is added up with its reactance x^*_{kR}, i.e. $x^*_{kR} + x^*_{\tau 02} = x^*_{kR}\left(1 + x^*_{\tau 02}/x^*_{kR}\right) = x^*_{kR} c_{\tau kR}$, where $c_{\tau kR} = 1 + x^*_{\tau 02}/x^*_{kR}$. Now, in the converted equivalent circuit (Fig. G.15), the reactance of the parallel branch $k - k$ becomes equal to $x^*_{kR} c_{\tau kR}$. Here, the value of $x^*_{kR} c_{\tau kR}$ represents the reactance of the magnetizing circuit loop of the layer reflecting the rotor slot-wedge area. So, we can have

$$\overset{\circ}{x}_{kR} = x^*_{kR} c_{\tau kR} = x^*_{kR} + x^*_{\tau 02} \tag{G.9}$$

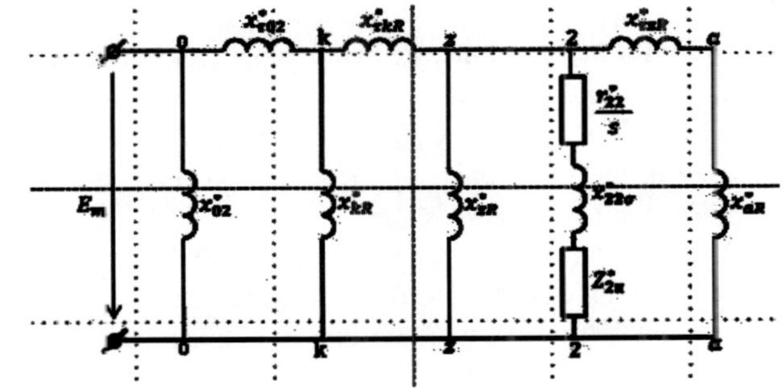

Fig. G.12 "Unfolded" rotor equivalent circuit with the mixed connection of the elements

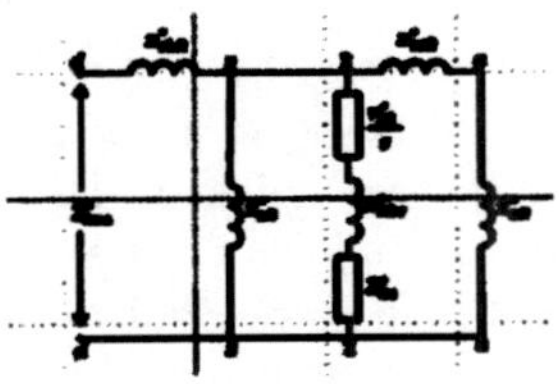

Fig. G.13

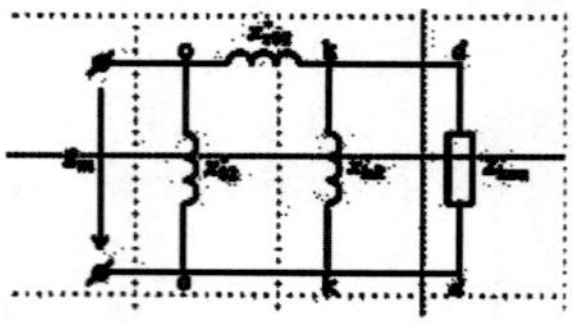

Fig. G.14

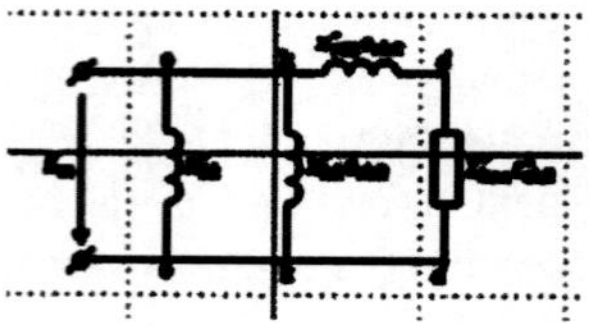

Fig. G.15

The factor $c_{\tau kR}$ is determined as

$$c_{\tau kR} = 1 + \frac{x^*_{\tau 02}}{x^*_{kR}} \tag{G.10}$$

In (G.10), the value of $c_{\tau kR}$ represents the transfer factor for the parallel branch $k - k$ of the equivalent circuit in Fig. G.14.

Further, in the equivalent circuit in Fig. G.14, when transferring its parallel branch $k - k$, the impedance of the parallel branch $d - d$ is multiplied by the factor $c^2_{\tau kR}$ and acquires the value equal to $Z^*_{kza}c^2_{\tau kR}$. In addition, multiplied by the factor $c_{\tau kR}$ the longitudinal reactance $x^*_{\tau 02}$, i.e., $x^*_{\tau 02}c_{\tau kR}$, is added up to the corrected impedance of the parallel branch $d - d$ determined as $Z^*_{kza}c^2_{\tau kR}$. As a result, in the converted equivalent circuit (Fig. G.15), we have the impedance of kind $Z^*_{kza}c^2_{\tau kR} + x^*_{\tau 02}c_{\tau kR}$ instead of the impedance Z^*_{kza} (Fig. G.14). Thus, the equivalent circuit (Fig. G.14) acquires the form in Fig. G.15.

We can obtain the rotor equivalent circuit using the equivalent circuits shown in Figs. G.13 and G.15. For this, the terminals of the parallel branch $d - d$ of the equivalent circuit in Fig. G.15 can be connected to the input terminals $d - d$ of the equivalent circuit in Fig. G.13. As follows from the equivalent circuit shown in Fig. G.15, the impedance of its parallel branch $d - d$ is multiplied by the factor $c^2_{\tau kR}$. Thus, when connecting the terminals of the parallel branch $d - d$ (Fig. G.15) to the terminals $d - d$ (Fig. G.13) the impedances used in Fig. G.13 are multiplied by the factor $c^2_{\tau kR}$. Now, the rotor equivalent circuit acquires the form in Fig. G.16.

The parallel branch $z - z$ corresponding to the equivalent circuit obtained in Fig. G.16 can be transferred to its terminals $k - k$. For this purpose in relation to the parallel branch $z - z$, the right part of the equivalent circuit in Fig. G.16 can be represented as the impedance Z^*_{ca} determined by the equivalent circuit in Fig. G.17. Then, the equivalent circuit obtained in Fig. G.16 acquires the form shown in Fig. G.18. In the equivalent circuit in Fig. G.18, the parallel branch $z - z$ can be

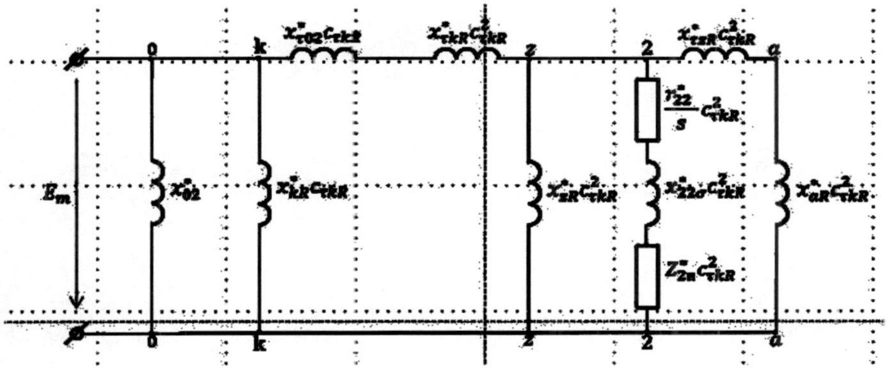

Fig. G.16

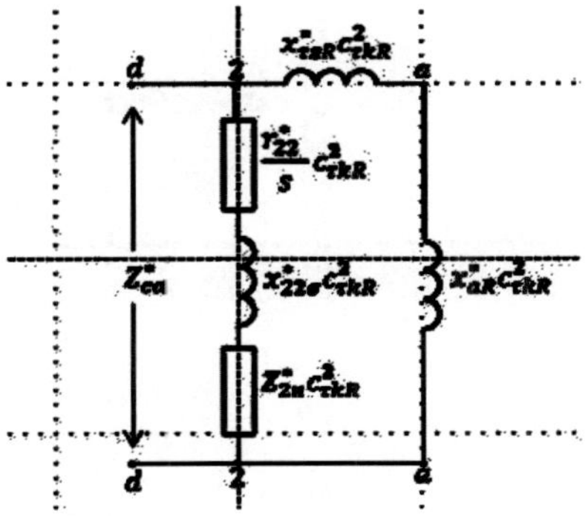

Fig. G.17

transferred to the terminals $k - k$. When transferring the parallel branch $z - z$ of the equivalent circuit in Fig. G.18 to its terminals $k - k$, the series-connected longitudinal reactances $\left(x^*_{\tau 02} c_{\tau kR} + x^*_{\tau kR} c^2_{\tau kR}\right)$ go into the parallel branch $z - z$, and they are added up with its reactance $x^*_{zR} c^2_{\tau kR}$, i.e. $x^*_{zR} c^2_{\tau kR} + x^*_{\tau 02} c_{\tau kR} + x^*_{\tau kR} c^2_{\tau kR} = x^*_{zR} c^2_{\tau kR} \left(1 + \frac{x^*_{\tau 02} c_{\tau kR} + x^*_{\tau kR} c^2_{\tau kR}}{x^*_{zs} c^2_{\tau ks}}\right) = x^*_{zR} c^2_{\tau kR} c_{\tau zR}$, where $c_{\tau zR} = 1 + \frac{x^*_{\tau 02} c_{\tau kR} + x^*_{\tau kR} c^2_{\tau kR}}{x^*_{zR} c^2_{\tau kR}}$. Consequently, in the converted equivalent circuit (Fig. G.19), the reactance of the parallel branch $z - z$ becomes equal to $x^*_{zR} c^2_{\tau kR} c_{\tau zR}$. Here, the value of $x^*_{zR} c^2_{\tau kR} c_{\tau zR}$ represents the reactance of the magnetizing circuit loop of the layer reflecting the rotor winding location area. For this reactance, we can have

$$\overset{\circ}{x}_{zR} = x^*_{zR} c^2_{\tau kR} c_{\tau zR} = x^*_{zR} c^2_{\tau kR} + x^*_{\tau 02} c_{\tau kR} + x^*_{\tau kR} c^2_{\tau kR} \tag{G.11}$$

The factor $c_{\tau zR}$ used in the converted equivalent circuit (Fig. G.19) is determined as

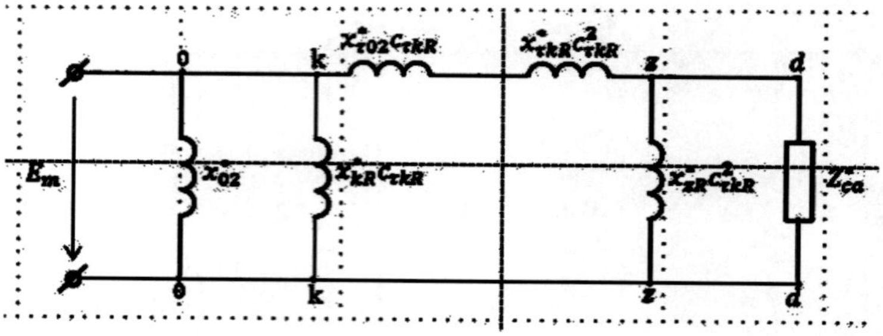

Fig. G.18

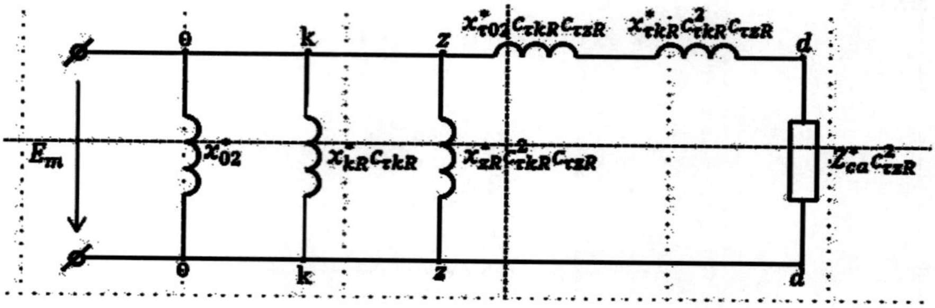

Fig. G.19

$$c_{\tau zR} = 1 + \frac{x^{*}_{\tau 02} c_{\tau kR} + x^{*}_{\tau kR} c^{2}_{\tau kR}}{x^{*}_{zR} c^{2}_{\tau kR}} \tag{G.12}$$

In (G.12), the value of $c_{\tau zR}$ reflects the transfer factor for the parallel branch $z - z$ of the equivalent circuit in Fig. G.18.

Further, in the equivalent circuit in Fig. G.18, when transferring its parallel branch $z - z$, the impedance of the parallel branch $d - d$ is multiplied by the factor $c^{2}_{\tau zR}$ and acquires the value equal to $Z^{*}_{ca} c^{2}_{\tau zR}$. In addition, multiplied by the factor $c_{\tau zR}$ the series-connected longitudinal reactances $\left(x^{*}_{\tau 02} c_{\tau kR} + x^{*}_{\tau kR} c^{2}_{\tau kR}\right)$ т.е. $\left(x^{*}_{\tau 02} c_{\tau kR} + x^{*}_{\tau kR} c^{2}_{\tau kR}\right) c_{\tau zR}$ are added up to the corrected impedance of the parallel branch $d - d$ determined as $Z^{*}_{ca} c^{2}_{\tau zR}$. Following this, we can note that instead of the impedance Z^{*}_{ca} used in Fig. G.18, the converted impedance of kind $Z^{*}_{ca} c^{2}_{\tau zR} + x^{*}_{\tau 02} c_{\tau kR} c_{\tau zR} + x^{*}_{\tau kR} c^{2}_{\tau kR} c_{\tau zR}$ follows in the converted equivalent circuit obtained in Fig. G.19. Thus, the equivalent circuit in Fig. G.18 acquires the form in Fig. G.19.

Now, by the equivalent circuits (Figs. G.17 and G.19), we can obtain the rotor equivalent circuit for what the terminals of the parallel branch $d - d$ of the equivalent circuit in Fig. G.19 can be connected to the input terminals $d - d$ of the equivalent circuit in Fig. G.17. As follows from the equivalent circuit (Fig. G.19), the impedance of its parallel branch $d - d$ is multiplied by the factor $c^{2}_{\tau zR}$. Thus, when

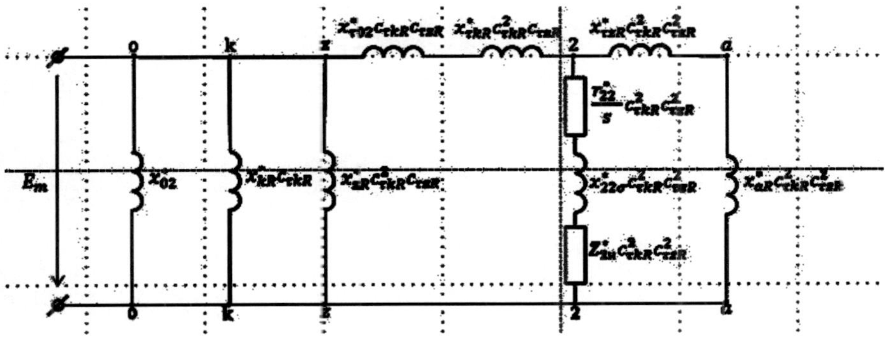

Fig. G.20

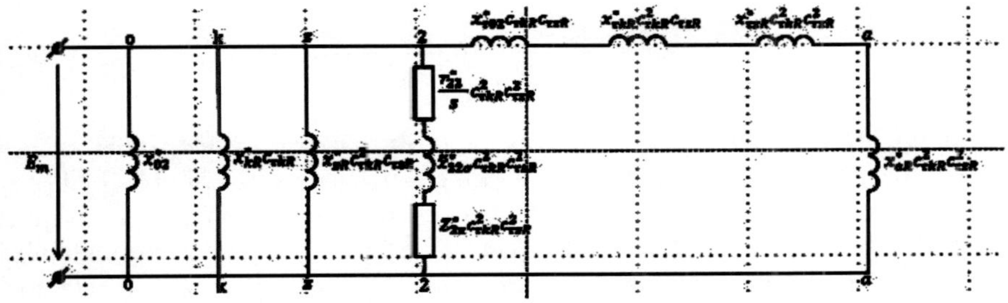

Fig. G.21

connecting the terminals of the parallel branch $d - d$ of the equivalent circuit (Fig. G.19) to the terminals $d - d$ of the equivalent circuit (Fig. G.17) the impedances used in Fig. G.17 are multiplied by the factor $c^2_{\tau zR}$. Now, the rotor equivalent circuit acquires the form in Fig. G.20.

In the equivalent circuit (Fig. G.20), the parallel branch $2 - 2$ can be transferred to the terminals $z - z$. In the equivalent circuit (Fig. G.20), when transferring the parallel branch $2 - 2$ to its terminals $z - z$, the series-connected longitudinal reactances $(x^*_{\tau 02}c_{\tau kR}c_{\tau zR} + x^*_{\tau kR}c^2_{\tau kR}c_{\tau zR})$ go into the parallel branch $2 - 2$, and they are added up with its impedances $\left(r^*_{22}/s + jx^*_{22\sigma} + Z^*_{2H}\right)c^2_{\tau kR}c^2_{\tau zR}$, i.e. $\left(r^*_{22}/s + jx^*_{22\sigma} + Z^*_{2H}\right)c^2_{\tau kR}c^2_{\tau zR} + x^*_{\tau 02}c_{\tau kR}c_{\tau zR} + x^*_{\tau kR}c^2_{\tau kR}c_{\tau zR} = \left(r^*_{22}/s + jx^*_{22\sigma} + Z^*_{2H}\right)c^2_{\tau kR}c^2_{\tau zR}\Big(1 + \frac{x^*_{\tau 02}c_{\tau kR}c_{\tau zR} + x^*_{\tau kR}c^2_{\tau kR}c_{\tau zR}}{\left(r^*_{22}/s + jx^*_{22\sigma} + Z^*_{2H}\right)c^2_{\tau kR}c^2_{\tau zR}}\Big) = \left(r^*_{22}/s + jx^*_{22\sigma} + Z^*_{2H}\right)c^2_{\tau kR}c^2_{\tau zR}c_{\tau 2}$, where $c_{\tau 2} = 1 + \frac{x^*_{\tau 02}c_{\tau kR}c_{\tau zR} + x^*_{\tau kR}c^2_{\tau kR}c_{\tau zR}}{\left(r^*_{22}/s + jx^*_{22\sigma} + Z^*_{2H}\right)c^2_{\tau kR}c^2_{\tau zR}}$. Consequently, in the converted equivalent circuit (Fig. G.21), the impedance of the parallel branch $2 - 2$ becomes equal to $\left(r^*_{22}/s + jx^*_{22\sigma} + Z^*_{2H}\right)c^2_{\tau kR}c^2_{\tau zR}c_{\tau 2}$. The value of $\left(r^*_{22}/s + jx^*_{22\sigma} + Z^*_{2H}\right)c^2_{\tau kR}c^2_{\tau zR}c_{\tau 2}$ represents the impedance of the slot leakage circuit loop of the rotor winding. For this impedance, we can have

$$\begin{aligned} \overset{\circ}{Z}_2 &= (r_{22}^*/s)c_{\tau kR}^2 c_{\tau zR}^2 \\ &+j(x_{22\sigma}^* c_{\tau kR}^2 c_{\tau zR}^2 + x_{\tau kR}^* c_{\tau kR}^2 c_{\tau zR} + x_{\tau 02}^* c_{\tau kR} c_{\tau zR}) \\ &+Z_{2\text{H}}^* c_{\tau kR}^2 c_{\tau zR}^2 \end{aligned} \tag{G.13a}$$

In the another form

$$\begin{aligned} \overset{\circ}{Z}_2 &= (r_{22}^*/s)c_{\tau kR}^2 c_{\tau zR}^2 \\ &+j\left[x_{22\sigma}^* + x_{\tau kR}^*(1/c_{\tau zR}) + x_{\tau 02}^*(1/c_{\tau kR} c_{\tau zR})\right] c_{\tau kR}^2 c_{\tau zR}^2 + Z_{2\text{H}}^* c_{\tau kR}^2 c_{\tau zR}^2 \\ &= (r_{22}^*/s + j\widetilde{x}_{22\sigma}^* + Z_{2\text{H}}^*) c_{\tau kR}^2 c_{\tau zR}^2 \end{aligned} \tag{G.13b}$$

where $\widetilde{x}_{22\sigma}^* = x_{22\sigma}^* + x_{\tau kR}^*(1/c_{\tau zR}) + x_{\tau 02}^*(1/c_{\tau kR} c_{\tau zR}); x_{2\sigma}^* = \widetilde{x}_{22\sigma}^* c_{\tau kR}^2 c_{\tau zR}^2$.

We have for the factor $c_{\tau 2}$

$$c_{\tau 2} = 1 + \frac{x_{\tau 02}^* c_{\tau kR} c_{\tau zR} + x_{\tau kR}^* c_{\tau kR}^2 c_{\tau zR}}{(r_{22}^*/s + jx_{22\sigma}^* + Z_{2\text{H}}^*) c_{\tau kR}^2 c_{\tau zR}^2} \tag{G.14}$$

In (G.14), the value of $c_{\tau 2}$ represents the transfer factor for the parallel branch $2 - 2$ transferring to the terminals $z - z$ of the equivalent circuit in Fig. G.20.

Further, in the equivalent circuit in Fig. G.20, when transferring its parallel branch $2 - 2$, the series-connected reactances $(x_{\tau zR}^* + x_{aR}^*) c_{\tau kR}^2 c_{\tau zR}^2$ are multiplied by the factor $c_{\tau 2}^2$ and acquire the value equal to $(x_{\tau zR}^* + x_{aR}^*) c_{\tau kR}^2 c_{\tau zR}^2 c_{\tau 2}^2$. In addition, multiplied by the factor $c_{\tau 2}$ the series-connected longitudinal reactances $(x_{\tau 02}^* c_{\tau kR} c_{\tau zR} + x_{\tau kR}^* c_{\tau kR}^2 c_{\tau zR})$, i.e., $(x_{\tau 02}^* c_{\tau kR} c_{\tau zR} + x_{\tau kR}^* c_{\tau kR}^2 c_{\tau zR}) c_{\tau 2}$ are added up to the corrected reactances of the parallel branch $2 - a - a$ determined as $(x_{\tau zR}^* + x_{aR}^*) c_{\tau kR}^2 c_{\tau zR}^2 c_{\tau 2}^2$. As a result, we have in the converted equivalent circuit (Fig. G.21) instead of the reactances $(x_{\tau zR}^* + x_{aR}^*) c_{\tau kR}^2 c_{\tau zR}^2$ used in Fig. G.20

$$\overset{\circ}{x}_{aR} = (x_{\tau 02}^* c_{\tau kR} + x_{\tau kR}^* c_{\tau kR}^2) c_{\tau zR} c_{\tau 2} + (x_{\tau zR}^* + x_{aR}^*) c_{\tau kR}^2 c_{\tau zR}^2 c_{\tau 2}^2 \tag{G.15a}$$

In (G.15a), the value of $\overset{\circ}{x}_{aR}$ represents the reactance of the magnetizing circuitloop of the rotor yoke layer.

In this work, we assume that the impedance of the outside network Z_{2H}^* is much higher than the impedance of the rotor winding, i.e., $Z_{2H}^* > |r_{22}^*/s + jx_{22\sigma}^*|$. Therefore, for the value of the transfer factor $c_{\tau 2}$ determined by (G.14), the condition of kind $c_{\tau 2} \approx 1.0$ is fulfilled. Consequently, the expression (G.15) for the reactance $\overset{\circ}{x}_{aR}$ can be written without taking into account the transfer factor $c_{\tau 2}$. Then, we can obtain

$$\overset{\circ}{x}_{aR} = x_{aR}^* c_{\tau kR}^2 c_{\tau zR}^2 + x_{\tau zR}^* c_{\tau kR}^2 c_{\tau zR}^2 + x_{\tau kR}^* c_{\tau kR}^2 c_{\tau zR} + x_{\tau 02}^* c_{\tau kR} c_{\tau zR} \tag{G.15b}$$

For the reactances used in the equivalent circuit in Fig. G.21, the expressions obtained in (G.9), (G.11), (G.13b), and (G.15b) can be applied. Then, the equivalent

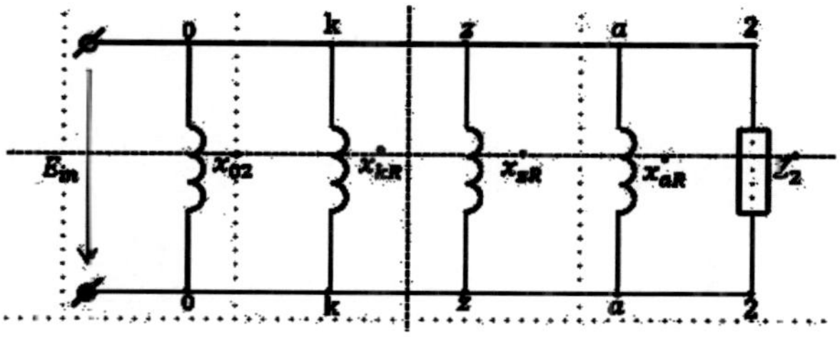

Fig. G.22 "Unfolded" rotor equivalent circuit with the parallel connection of the elements

circuit shown in Fig. G.21 can be reduced to the form in Fig. G.22. In Fig. G.22, the "unfolded" rotor equivalent circuit with the parallel connection of the elements is shown.

The rotor magnetizing circuit loop and the rotor slot leakage circuit loop can be introduced and the reactances applying in them can be determined using the equivalent circuit given in Fig. G.22. As a result of this, the "folded" rotor equivalent circuit follows.

Summary

1. In this section, the application of the multiple transfer technique is shown. This technique is used to transform the "unfolded" stator and rotor equivalent circuits with the mixed connection of the elements into the "unfolded" stator and rotor equivalent circuits with the parallel connection of the elements. The multiple transfer technique is based on the ideas of the "no-load and under-load" concept.
2. The "unfolded" stator and rotor equivalent circuits with the parallel connection of the elements
3. are the bases for introducing the stator and rotor magnetizing circuit loops and the stator and rotor slot leakage circuit loops as well as for determining the parameters included in them. As a result of this, the "folded" stator and rotor equivalent circuits can be obtained. Now, the "folded" stator and rotor equivalent circuits can be used to construct the equivalent circuit of the electric machine.

Reference

1. Asanbayev, V.: Alternating Current Multi-Circuit Electric Machines: A New Approach to Steady-State Parameter Determination, Springer Charm Heidelberg, New-York, Dordrecht London (2015)

Appendix H: Calculation Elements of Conducting Layer: "Field" Concept

Generalized Equations of Conducting Layer

The generalized equations for the ith layer can be obtained based on the description of the electromagnetic field in the layer using Maxwell's equations. In this regard, below we will briefly consider the approximate field equations associated with a neglect of the displacement currents.

The system of the equations for the electromagnetic field. Based on the equations (2.8) and (2.17), the system of the equations for the electromagnetic field can be represented in the form

$$\begin{aligned} rotH &= \sigma E + \delta_{\text{CT}} \\ rotE &= -\frac{\partial B}{\partial t} \end{aligned} \tag{H.1}$$

In accordance with the equation (H.1), the electromagnetic field is described, in our case (when $\rho = const$, $\mu = const$ and the displacement currents are neglected) by the three vectors E, H, and B. Using the condition $B = \mu H$, the equations (H.1) can be represented only through the vectors E and B or only through the vectors E and H. However, to determine these vectors, the equations (H.1) are not sufficient, since these vectors are not uniquely determined for the given rotor. To ensure the unambiguous determination of the vectors E, H, and B, it is necessary to specify the divergence of the vector B (the condition for the solenoidality of the magnetic field) in the form

$$\text{div } B = 0 \tag{H.2}$$

Then, the systems of the equations of the electromagnetic field for a medium with the constant electromagnetic parameters ($\rho = const$ and $\mu = const$) can be represented with respect to the vectors E and B and also to the vectors E and H

V. Asanbayev, *Asynchronous Machines*, https://doi.org/10.1007/978-3-030-92284-9

$$
\begin{aligned}
rotB &= \mu(\sigma E + \delta_{cm}) & rotH &= \sigma E + \delta_{cm} \\
rotE &= -\frac{\partial B}{\partial t} & rotE &= -\mu\frac{\partial H}{\partial t} \\
divB &= 0 & divH &= 0
\end{aligned}
\tag{H.3}
$$

These systems of the equations are valid for describing the electromagnetic field in a stationary medium associated with the stationary coordinate system (in our case, the stator coordinate system), and they represent the system of Maxwell's equations in the differential form. These equations are often used in solving the various problems of the electrodynamics, and they assume the determination of the vectors *E*, *H*, and *B*. The operations rot and div used in (H.3) represent the first partial derivatives of the projections of the vectors *E*, *H*, and *B*. Therefore, Maxwell's equations reflect the system partial differential equations of the first order with respect to the projections of the field vectors, for example, *E* and *H*, i.e., E_x, E_y, E_z and H_x, H_y, H_z. Solving the system of Maxwell's equations for the general case is not an easy task. However, when analyzing specific practical problems, it turns out to be sufficient to determine the solution taking into account certain simplifying assumptions. As practice shows, this provides the rather serious simplification of the field equations, and at the same time, the quite acceptable reliability of the results is achieved.

The systems of the equations (H.3) contain only the vectors *E* and *B* or only the vectors *E* and *H*. The vectors *E* and *B* are considered as the main vectors of the electromagnetic field, since the interaction forces characterizing the electromagnetic field are determined by these vectors. In this work, we are mainly interested in the electromagnetic processes, considered in relation to the voltages and currents given at the terminals of the stator and rotor windings of the electric machine. In the theory of the electric machines, for this purpose, Kirchhoff's equations compiled for the circuit loops of the machine windings are used. In the general case, to obtain Kirchhoff's equations, it does not matter which the vectors (*E* and *B* or *E* and *H*) are used. It should be noted, however, that Kirchhoff's equations characterize the electric machine from the energy point of a view. It is convenient to describe the energy processes using the Poynting vector. This vector is expressed in terms of the field vectors *E* and *H*. In addition, the vectors *E* and *H* are more preferable when considering the processes in the electric machine with the implicitly expressed windings (e.g., the machine with the solid rotor, etc.), when it becomes necessary to resolve the issue associated with the formation of the circuit loops of the eddy currents induced, for example, in the solid rotor of the machine. In this regard, in what follows, we will proceed from the system of the equations (H.3), represented through the vectors *E* and *H*. When solving the system of the equations (H.3), the value δ_{cm} can be taken as given. In addition, in these equations, the values ρ and μ are given. Therefore, the solution of equations (H.3) can be reduced to finding the unknown values of the vectors *E* and *H*. In this case, the vector of magnetic flux density B can be found using the simple mathematical transformation of the relation of the form $B = \mu H$. However, in our case, we can do without implementing this operation, since when compiling the Kirchhoff's equations for the circuit loops of the machine windings, it is sufficient to use the vectors E and H.

Complex form of the electromagnetic field equations. We consider the case when the values of the vectors E, H, B, δ, and δ_{cm} used in the equations for the electromagnetic field, change sinusoidally in a space and in a time. The sine wave of some quantity can be represented by the space vector directed along the axis passing through the point where this quantity has the amplitude value. On the complex plane, this quantity is represented as the complex number, similar to the complex numbers used to determine the values of the voltage $\dot{U}$ and the current $\dot{I}$. In this case, the magnitudes of the voltage $\dot{U}$ and the current $\dot{I}$ are usually in practice taken equal to their effective values and it is desirable to take the values of the space vectors equal to the amplitude values.

We consider some vector A which represents the vectors of the electromagnetic field as E, H, B, δ and δ_{cm}. Let the vector A changes in a time and space according to the cosine law and forms the traveling wave along the coordinate x (in the tangential direction). Then, we can have

$$A = A_m \cos(\omega t - \alpha x + \varphi) \tag{H.4}$$

where A_m is the amplitude value of the vector A at the considered point of the medium; ω is the angular frequency; t is a time; $\alpha = \pi/\tau$; φ is the initial phase at $x = 0$. Considering that $\cos(\omega t - \alpha x + \varphi) = Re\,[\cos(\omega t - \alpha x + \varphi) + j\sin(\omega t - \alpha x + \varphi)] = Re\, e^{j(\omega t - \alpha x + \varphi)}$, we can have instead of (H.4),

$$A = A_m\, Re\, e^{j(\omega t - \alpha x + \varphi)} \tag{H.5}$$

In (H.5), we introduce the new quantity defined as

$$\dot{A} = A_m e^{j\varphi} \tag{H.6}$$

In (H.6), the quantity $\dot{A}$ represents the complex amplitude of the vector A.

From (H.5) and (H.6), we receive

$$A = Re\, \dot{A} e^{-j\alpha x} e^{j\omega t} = Re\, \dot{A}(x) e^{j\omega t} \tag{H.7}$$

where $\dot{A}(x) = \dot{A} e^{-j\alpha x}$.

Similarly, the field vectors E, H, B, δ and δ_{cm} can be represented through the complex amplitudes of these vectors. Then, we can get

$$\begin{aligned} E &= Re\, \dot{E}(x) e^{j\omega t} \quad & H &= Re\, \dot{H}(x) e^{j\omega t} \\ B &= Re\, \dot{B}(x) e^{j\omega t} \quad & \delta &= Re\, \dot{\delta}(x) e^{j\omega t} \end{aligned} \tag{H.8}$$

where $\dot{E}(x) = \dot{E} e^{-j\alpha x}$; $\dot{H}(x) = \dot{H} e^{-j\alpha x}$; $\dot{B}(x) = \dot{B} e^{-j\alpha x}$; $\dot{\delta}(x) = \dot{\delta} e^{-j\alpha x}$.

With the help of the expressions (H.7) and (H.8), the equations for the electromagnetic field can be represented in the complex form, if all terms of the equations can be canceled by $e^{j\omega t}$. For stationary design zones of the electric machine, the field equations in the complex form take

$$\begin{array}{ll} rot\dot{B} = \mu\left(\sigma\dot{E} + \dot{\delta}_{cm}\right) & rot\dot{H} = \sigma\dot{E} + \dot{\delta}_{cm} \\ rot\dot{E} = -j\omega\dot{B} & rot\dot{E} = -j\omega\mu\dot{H} \\ div\dot{B} = 0 & div\dot{H} = 0 \end{array} \tag{H.9}$$

These systems of equations must be supplemented with the material equations expressed in the complex form

$$\dot{B} = \mu\dot{H} \quad \dot{\delta} = \sigma\left(\dot{E} + \dot{E}_{\text{CT}}\right) \tag{H.10}$$

For the rotating design zones of the electric machine, the field equations in the complex form (H.9) are supplemented by the relations of the form

$$\dot{B} = \mu\dot{H} \quad \dot{\delta} = \sigma\left\{\dot{E} + \dot{E}_{\text{CT}} + \left[\upsilon\dot{B}\right]\right\} \tag{H.11}$$

The region of the existence of the electromagnetic field in the electric machine is conveniently represented as consisting of the separate fragments reflecting the design features of the machine and having the different resistivity ρ and magnetic permeability μ. Therefore, to calculate the electromagnetic field in the machine, the equations given above should be supplemented by the equations for the surfaces, represented, in the general case, as

$$f_{nm}(x, y, z) = 0 \tag{H.12}$$

The equations of the form (H.12) can be used to separate the fragments of the machine design in such a way that the separated fragments have different but constant values of the electromagnetic parameters ρ and μ for the considered fragment. On the surfaces of the adjacent fragments (s and m fragments), the boundary conditions are established for the tangential components of the electric and magnetic field intensities. For the complex amplitudes of the field intensities, these conditions take the form

$$\begin{array}{c} \dot{E}_{st} = \dot{E}_{mt} \qquad \dot{H}_{st} = \dot{H}_{mt} \\ \dot{B}_{sn} = \mu_{rs}\dot{H}_{sn} = \mu_{rm}\dot{H}_{mn} = \dot{B}_{mn} \end{array} \tag{H.13}$$

where $\dot{E}_{st}, \dot{E}_{mt}$ and $\dot{H}_{st}, \dot{H}_{mt}$ are the tangential components of the complex amplitudes of the electric and magnetic field intensities; $\dot{B}_{sn}, \dot{B}_{mn}$ are the normal components of the complex amplitudes of the magnetic flux densities at the boundaries of the s and m fragments.

The solution of the system of the equations (H.9) given in general form is a very difficult task. As shown, for example, in [1–3], the general equations for the complex amplitudes of the vectors $\dot{E}, \dot{H}$, and $\dot{\delta}$ obtained from the system (H.9) are rather cumbersome, and therefore, they are of no practical interest. In this regard, it is of practical interest to consider the special cases arising from such simplification of the problem, which does not lead to the significant distortion of the real picture of the field distribution in the electric machine. Then, for the specific case, the solution of the field equations becomes much easier. In the future, we will adhere to this approach to solving the field equations in the electric machine.

As a result of solving the field equations presented in the complex form, we can obtain the complex amplitudes for the field vectors $\dot{E}, \dot{H}, \dot{\delta}$, and $\dot{B}$. The transition to the real physical quantities as the functions of the coordinates and time is carried out by multiplying by $e^{j\omega t}$ these quantities $\dot{E}, \dot{H}, \dot{\delta}$, and $\dot{B}$ found after solving the field equations, and also by taking the real parts of the results of these multiplications. However, this procedure of the reverse transition is not always necessary, since when calculating the most important quantities associated with the electromagnetic processes occurring in the electric machine, one can directly operate with the complex amplitudes of the quantities $\dot{E}, \dot{H}, \dot{\delta}, \dot{\delta}_{cm}$, and $\dot{B}$.

Solution of the field equations using the complex amplitudes for the vectors $\dot{E}$ and $\dot{H}$. The accepted above assumptions that simplify the picture of the electromagnetic field in the electric machine can be represented as $\dot{E}_x = \dot{E}_y = 0; \dot{E} = \dot{E}_z$; $\dot{H}_z = 0$; $\dot{\delta}_{cm} = 0$; $\partial\dot{E}/\partial z = \partial\dot{E}^2/\partial z^2 = 0$; $\partial\dot{H}_x/\partial z = 0$; $\partial\dot{H}_y/\partial z = 0$. Here, the x coordinate is in the tangential direction, the y coordinate is oriented in the radial direction, and the z coordinate defines the axial direction of the machine. We consider the field equations given in (H.9). Taking into account the assumptions made above, the equations (H.9) can be represented as

$$\begin{aligned} rot_z\dot{H} &= \frac{d\dot{H}_y}{dx} - \frac{d\dot{H}_x}{dy} = \frac{\dot{E}}{\rho} \\ rot_z\dot{E} &= \frac{d\dot{E}}{dy} = -j\omega_1 s\mu_x\dot{H}_x \\ rot_y\dot{E} &= \frac{d\dot{E}}{dx} = j\omega_1 s\mu_y\dot{H}_y \end{aligned} \tag{H.14}$$

In (H.14), we differentiate the second and third equations with respect to the variables y and x and substitute the obtained results into the first equation. As a result, we get

$$\frac{d^2\dot{E}}{dy^2} - k_\mu^2\frac{d^2\dot{E}}{dx^2} = j\omega_1 s\mu_x\frac{\dot{E}}{\rho} \tag{H.15}$$

where $k_\mu = \sqrt{\mu_x/\mu_y}$ is the coefficient of the magnetic anisotropy.

In accordance with the harmonic law, the electromagnetic field existing in the machine changes in a time and in the direction of the x coordinate that is determined as the function $e^{j(\omega t - \alpha x + \varphi)}$, where $\alpha = \pi/\tau$. Taking this condition into account, the equation (H.15) acquires

$$\frac{d^2\dot{E}}{dy^2} = \left[k_\mu^2\left(\frac{\pi}{\tau}\right)^2 + \frac{j\omega_1 s\mu_x}{\rho}\right]\dot{E} = \beta^2\dot{E} \tag{H.16}$$

where $\beta^2 = k_\mu^2\left(\frac{\pi}{\tau}\right)^2 + \frac{j\omega_1 s\mu_x}{\rho}$; ρ and μ_x are the resistivity and magnetic permeability of the considered medium.

Using the equations (H.14) and (H.16), for the axial component of the electric field intensity $\dot{E}_i$ and for the tangential component of the magnetic field intensity $\dot{H}_{xi}$, we have in the ith layer of the rotor model

$$\begin{aligned} \dot{E}_i &= Ce^{\beta_i y} + De^{-\beta_i y} \\ \dot{H}_{xi} &= -\frac{\beta_i}{j\omega_1 s\mu_{xi}}\left(Ce^{\beta_i y} - De^{-\beta_i y}\right) \end{aligned} \tag{H.17}$$

where $\beta_i = \sqrt{k_{\mu i}^2\left(\frac{\pi}{\tau_i}\right)^2 + \frac{j\omega_1 s\mu_{xi}}{\rho_i}}$; $k_{\mu i} = \sqrt{\mu_{xi}/\mu_{yi}}$; C and D are the integration constants.

Generalized layer equations in the form of Z. In (H.17), the integration constants C and D should be expressed through the values of the electric and magnetic field intensities given on the upper and lower surfaces of the ith layer (Fig. 14.2). In this case, in (H.17), instead of $\dot{E}_i$ and $\dot{H}_{xi}$, the calculated values $\dot{E}_{ik}$ and $\dot{H}_{ik}$, where $k = 1$, 2, defined as $\dot{E}_{ik} = \dot{E}_i\xi_{Li}$ and $\dot{H}_{ik} = \dot{H}_{xi}\xi_{\tau i}$, where $\xi_{Li} = l_i/l_0$; $\xi_{\tau i} = \tau_i/\tau_0$, should be used. Now, based on the equations (H.17), we have at $y = 0$ (the upper surface of the layer)

$$\begin{aligned} \dot{E}_{i1} &= \xi_{Li}(C + D) \\ \dot{H}_{i1} &= -\frac{\beta_i\xi_{\tau i}}{j\omega_1 s\mu_{xi}}(C - D) \end{aligned}$$

For $y = h_i$ (the lower surface of the layer), we obtain

$$\begin{aligned} \dot{E}_{i2} &= \xi_{Li}\left(Ce^{\beta_i h_i} + De^{-\beta_i h_i}\right) \\ \dot{H}_{i2} &= -\frac{\beta_i\xi_{\tau i}}{j\omega_1 s\mu_{xi}}\left(Ce^{\beta_i h_i} - De^{-\beta_i h_i}\right) \end{aligned}$$

The integration constants C and D based on the second equations of these systems get the form

$$C = \frac{j\omega_1 s\mu_{xi}}{\beta_i \xi_{\tau i}} \frac{\dot{H}_{i1} e^{-\beta_i h_i} - \dot{H}_{i2}}{e^{\beta_i h_i} - e^{-\beta_i h_i}} \quad D = \frac{j\omega_1 s\mu_{xi}}{\beta_i \xi_{\tau i}} \frac{\dot{H}_{i1} e^{\beta_i h_i} - \dot{H}_{i2}}{e^{\beta_i h_i} - e^{-\beta_i h_i}} \tag{H.18}$$

Using the relations (H.18) for the integration constants C and D, the first equations of the systems shown above can be reduced to the form

$$\begin{aligned}\dot{E}_{i1} &= \frac{j\omega_1 \mu_{xi}}{\beta_i \tanh \beta_i h_i} \frac{\xi_{Li}}{\xi_{\tau i}} \dot{H}_{i1} + \frac{j\omega_1 \mu_{xi}}{\beta_i \sinh \beta_i h_i} \frac{\xi_{Li}}{\xi_{\tau i}} \dot{H}_{i2} = Z_{11}^i \dot{H}_{i1} + Z_{21}^i \dot{H}_{i2} \\ \dot{E}_{i2} &= \frac{j\omega_1 \mu_{xi}}{\beta_i \sinh \beta_i h_i} \frac{\xi_{Li}}{\xi_{\tau i}} \dot{H}_{i1} + \frac{j\omega_1 \mu_{xi}}{\beta_i \tanh \beta_i h_i} \frac{\xi_{Li}}{\xi_{\tau i}} \dot{H}_{i2} = Z_{12}^i \dot{H}_{i1} + Z_{22}^i \dot{H}_{i2}\end{aligned} \tag{H.19}$$

where
$$\begin{aligned}Z_{11}^i &= \frac{j\omega_1 \mu_{xi}}{\beta_i \tanh \beta_i h_i} \frac{\xi_{Li}}{\xi_{\tau i}}; Z_{21}^i = \frac{j\omega_1 \mu_{xi}}{\beta_i \sinh \beta_i h_i} \frac{\xi_{Li}}{\xi_{\tau i}} \\ Z_{12}^i &= \frac{j\omega_1 \mu_{xi}}{\beta_i \sinh \beta_i h_i} \frac{\xi_{Li}}{\xi_{\tau i}}; Z_{22}^i = \frac{j\omega_1 \mu_{xi}}{\beta_i \tanh \beta_i h_i} \frac{\xi_{Li}}{\xi_{\tau i}}\end{aligned}$$

The system shown in (H.19) represents the generalized equations for the ith layer, expressed in the form of Z.

Generalized layer equations in the form of A. When considering the problems associated with the conversion of an energy, it is convenient to represent the equations for the ith layer in the form of the system of the equations expressed in the form of A. In this regard, the equations (H.19) are transformed to the equations presented in the form of A. For this purpose, we write equations (H.19) in relation to the quantities $\dot{E}_{i1}$ and $\dot{H}_{i1}$ belonging to the upper surface of the ith layer (Fig. 14.2). Then, we get

$$\begin{aligned}\dot{E}_{i1} &= \frac{Z_{11}^i}{Z_{12}^i} \dot{E}_{i2} + \frac{Z_{11}^i Z_{22}^i - Z_{12}^i Z_{21}^i}{Z_{12}^i} \dot{H}_{i2} = a_{11}^i \dot{E}_{i2} + a_{21}^i \dot{H}_{i2} \\ \dot{H}_{i1} &= \frac{1}{Z_{12}^i} \dot{E}_{i2} + \frac{Z_{22}^i}{Z_{12}^i} \dot{H}_{i2} = a_{12}^i \dot{E}_{i2} + a_{22}^i \dot{H}_{i2}\end{aligned} \tag{H.20}$$

where $a_{11}^i = \frac{Z_{11}^i}{Z_{12}^i}; a_{21}^i = \frac{Z_{11}^i Z_{22}^i - Z_{12}^i Z_{21}^i}{Z_{12}^i}; a_{12}^i = \frac{1}{Z_{12}^i}; a_{22}^i = \frac{Z_{22}^i}{Z_{12}^i}$.

Using the relations (H.19) and (H.20), we can obtain for the coefficients $a_{11}^i, a_{21}^i, a_{12}^i$ and a_{22}^i

$$\begin{aligned}a_{11}^i &= \cosh \beta_i h_i & a_{21}^i &= \frac{j\omega_1 \mu_{xi}}{\beta_i} \frac{\xi_{Li}}{\xi_{\tau i}} \sinh \beta_i h_i \\ a_{12}^i &= \frac{\beta_i}{j\omega_1 \mu_{xi}} \frac{\xi_{\tau i}}{\xi_{Li}} \sinh \beta_i h_i & a_{22}^i &= \cosh \beta_i h_i\end{aligned} \tag{H.21}$$

Equations (H.19) and (H.20) can be used to construct the L-shaped equivalent circuit for the ith layer. The L-shaped equivalent circuit of the ith layer obtained in this way is shown in Fig. 14.5. In this work, the L-shaped equivalent circuits are used in the construction for the electric machine of the "unfolded" equivalent circuits with the mixed connection of the elements.

L-shaped Equivalent Circuit for Layer: Analysis of Parameters

We proceed from the equivalent circuit shown in Fig. 14.5. Fig. 14.5 shows the L-shaped equivalent circuit built for the ith layer using the equations obtained in (H.19). For convenience, the L-shaped equivalent circuit for the ith layer is shown in Fig. H.1. This equivalent circuit for the ith layer contains the impedances Z_i^* and $Z_{\tau i}^*$ the expressions for which are given in (14.41). We consider the impedances of the L-shaped equivalent circuit for the ith layer in the areas of the weak and strong skin effect [4].

Weak Skin Effect. In accordance with (14.41), the impedance Z_i^* is determined by the expression of the form

$$Z_i^* = \frac{j\omega_1 \mu_{xi}}{\beta_i \tanh \beta_i h_i} \frac{\xi_{Li}}{\xi_{\tau i}} \tag{H.22}$$

where $\xi_{\tau i} = \tau_i/\tau_0$; $\xi_{Li} = l_i/l_0$; h_i is the height of the ith layer.

For the coefficient β_i used in (H.22), we have

$$\beta_i = \sqrt{\left(k_{\mu i}\frac{\pi}{\tau_i}\right)^2 + \frac{j\omega_1 s \mu_{xi}}{\rho_i}} \tag{H.23}$$

Here, $k_{\mu i} = \sqrt{\mu_{xi}/\mu_{yi}}$; τ_i is the pole pitch of the ith layer.

In (H.22), for the hyperbolic tangent function, we use the approximate condition of the form

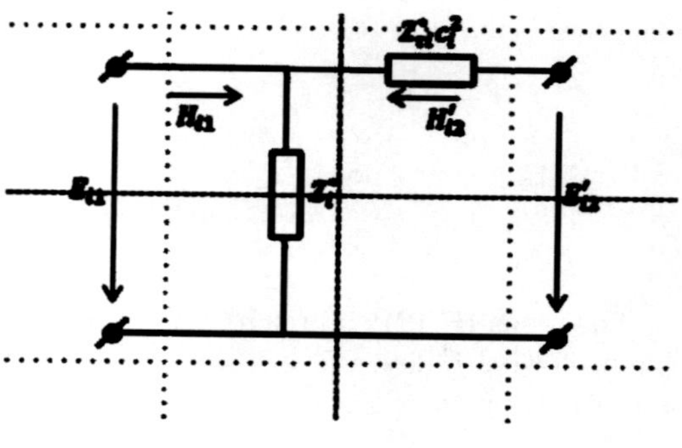

Fig. H.1 L-shaped equivalent circuit of the ith layer

$$\tanh \beta_i h_i \approx \beta_i h_i - \frac{(\beta_i h_i)^3}{3} = \beta_i h_i \left(1 - \frac{(\beta_i h_i)^2}{3}\right)$$

For the coefficient β_i^2 applied in the second term of this expression, we use the approximate condition defined as $\beta_i^2 \approx \frac{j\omega_1 s\mu_{xi}}{\rho_i}$. Then, for the function of the hyperbolic tangent, we can have

$$\tanh \beta_i h_i \approx \sqrt{\left(k_{\mu i}\frac{\pi}{\tau_i}\right)^2 + \frac{j\omega_1 s\mu_{xi}}{\rho_i}} \cdot h_i \left(1 - \frac{j\omega_1 s\mu_{xi}}{\rho_i}\frac{h_i^2}{3}\right) \tag{H.24}$$

Taking into account (H.24), the expression (H.22) can be reduced to the form

$$\begin{aligned} Z_i^* &\approx \frac{j\omega_1 \mu_{xi}}{\left[\left(k_{\mu i}\frac{\pi}{\tau_i}\right)^2 h_i + \frac{j\omega_1 s\mu_{xi} h_i}{\rho_i}\right]\left(1 - \frac{j\omega_1 s\mu_{xi}}{\rho_i}\frac{h_i^2}{3}\right)}\frac{\xi_{Li}}{\xi_{\tau i}} \\ &\approx \frac{j\omega_1 \mu_{xi}}{\left[\left(k_{\mu i}\frac{\pi}{\tau_i}\right)^2 h_i + \frac{j\omega_1 s\mu_{xi} h_i}{\rho_i}\right]\frac{1}{\left(1 + \frac{j\omega_1 s\mu_{xi}}{\rho_i}\frac{h_i^2}{3}\right)}}\frac{\xi_{Li}}{\xi_{\tau i}} \end{aligned} \tag{H.25}$$

The expression (H.25) can be reduced to the form

$$\begin{aligned} Z_i^* &\approx \frac{j\omega_1 \mu_{xi}}{\left(k_{\mu i}\frac{\pi}{\tau_i}\right)^2 h_i + \frac{j\omega_1 s\mu_{xi} h_i}{\rho_i}\frac{1}{\left(1 + \frac{j\omega_1 s\mu_{xi}}{\rho_i}\frac{h_i^2}{3}\right)}}\frac{\xi_{Li}}{\xi_{\tau i}} \\ &= \frac{1}{\frac{1}{\frac{j\omega_1 \mu_{xi}\tau_i^2}{k_{\mu i}^2\pi^2 h_i}\frac{\xi_{Li}}{\xi_{\tau i}}} + \frac{1}{\frac{\rho_i}{h_i s}\left(1 + \frac{j\omega_1 s\mu_{xi}}{\rho_i}\frac{h_i^2}{3}\right)\frac{\xi_{Li}}{\xi_{\tau i}}}} \\ &= \frac{1}{\frac{1}{\frac{j\omega_1 \mu_{xi}\tau_i^2}{k_{\mu i}^2\pi^2 h_i}\frac{\xi_{Li}}{\xi_{\tau i}}} + \frac{1}{\frac{\rho_i}{h_i s}\frac{\xi_{Li}}{\xi_{\tau i}} + j\omega_1\mu_{xi}\frac{h_i}{3}\frac{\xi_{Li}}{\xi_{\tau i}}}} \end{aligned} \tag{H.26}$$

Here, we introduce the magnetizing reactance of the ith layer, defined as

$$x_i^* = \frac{j\omega_1 \mu_{xi}\tau_i^2}{k_{\mu i}^2\pi^2 h_i}\frac{\xi_{Li}}{\xi_{\tau i}} = \frac{j\omega_1 \mu_{yi}\tau_i^2}{\pi^2 h_i}\frac{\xi_{Li}}{\xi_{\tau i}} \tag{H.27}$$

where $k_{\mu i} = \sqrt{\mu_{xi}/\mu_{yi}}$.

The ohmic resistance of the ith layer

$$\frac{r_{ii}^*}{s} = \frac{\rho_i}{h_i s} \frac{\xi_{Li}}{\xi_{\tau i}} \tag{H.28}$$

The leakage reactance of the eddy currents induced in the ith layer

$$x_{ii\sigma}^* = \omega_1 \mu_{xi} \frac{h_i}{3} \frac{\xi_{Li}}{\xi_{\tau i}} = \omega_1 \mu_{xi} t_0 \frac{h_i}{3t_i} k_R \xi_{Li} \tag{H.29}$$

where $\xi_{\tau i} = \frac{\tau_i}{\tau_0} = \frac{t_i}{t_0} \frac{Z_2}{Z_1} = \frac{t_i}{t_0} \frac{1}{k_R}; k_R = \frac{Z_1}{Z_2}$.

In the case of placing the winding in the slots, the reactance $x_{ii\sigma}^*$ will represent the own slot leakage circuit loop of the winding.

Now, taking into account the relations (H.27)–(H.29), the expression (H.26) for the impedance Z_i^* can be reduced to the form

$$Z_i^* = \frac{1}{\frac{1}{x_i^*} + \frac{1}{r_{ii}^*/s + jx_{ii\sigma}^*}} \tag{H.30}$$

In accordance with (H.30), the ith layer is represented in the form of the two independent circuit loops: the own magnetizing circuit loop of the ith layer and the own leakage circuit loop of the eddy currents induced in the ith layer. Here, the own magnetizing circuit loop of the ith layer contains the reactance x_i^* and the own leakage circuit loop of the eddy currents induced in the ith layer is represented by the impedance equal to $Z_{i\sigma}^* = r_{ii}^*/s + jx_{ii\sigma}^*$. The expression (H.30) can be used in the area of the weak skin effect to determine the impedance representing the transverse branch $a - b$ of the equivalent circuit shown in Fig. H.1.

The longitudinal impedance $Z_{\tau i}^*$ used in the equivalent circuit in Fig. H.1 can be represented from (14.41) as

$$Z_{\tau i}^* = j\omega_1 \mu_{xi} h_i \frac{\tanh \beta_i h_i}{\beta_i h_i} \frac{\xi_{Li}}{\xi_{\tau i}} = j\omega_1 \mu_{xi} h_i \frac{\xi_{Li}}{\xi_{\tau i}} k_i \tag{H.31}$$

where $k_i = \frac{\tanh \beta_i h_i}{\beta_i h_i}$.

This expression taking into account the first term in the series expansion of the hyperbolic tangent function when the condition of kind $k_i = 1.0$ is satisfied obtains the form

$$x_{\tau i}^* \approx \omega_1 \mu_{xi} h_i \frac{\xi_{Li}}{\xi_{\tau i}} = \omega_1 \mu_{xi} t_0 \frac{h_i}{t_i} k_R \xi_{Li} \tag{H.32}$$

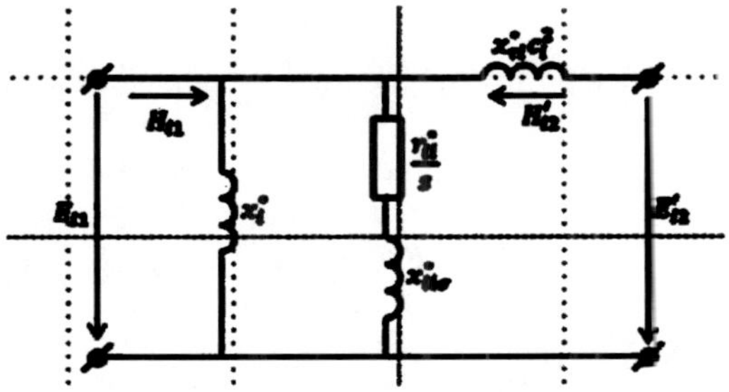

Fig. H.2 L-shaped equivalent circuit of the ith layer for the weak skin effect

On the bases of the equivalent circuit shown in Fig. H.1 and the expressions obtained in (H.30) and (H.32), the equivalent circuit of the ith layer which is valid for the weak skin effect takes the form in Fig. H.2.

Strong Skin Effect. We consider the impedance Z_i^* determined from (H.22) as

$$Z_i^* = \frac{j\omega_1 \mu_{xi}}{\beta_i \tanh \beta_i h_i} \frac{\xi_{Li}}{\xi_{\tau i}} \tag{H.33}$$

where $\xi_{\tau i} = \tau_i/\tau_0$; $\xi_{Li} = l_i/l_0$; h_i is the height of the ith layer.

In the considered strong skin effect, for the coefficient β_i used in (H.33), we have the approximate condition of the form

$$\beta_i \approx \sqrt{\frac{j\omega_1 s\mu_i}{\rho_i}} = \frac{1+j}{\Delta_i} \tag{H.34}$$

where $\Delta_i = \sqrt{2\rho_i/\omega_1 s\mu_i}$.

Taking into account (H.34), the expression (H.33) for the impedance Z_i^* can be represented in the form

$$Z_i^* = \frac{j\omega_1 \mu_{xi}}{\beta_i \tanh \beta_i h_i} \frac{\xi_{Li}}{\xi_{\tau i}} = \frac{\rho_i}{h_i s} \frac{\beta_i h_i}{\tanh \beta_i h_i} \frac{\xi_{Li}}{\xi_{\tau i}} \tag{H.35}$$

Here, the complex quantity $\beta_i h_i$ is expressed through the real and imaginary components. Then, we get

$$\beta_i h_i = (1+j)\frac{h_i}{\Delta_i} = (1+)\varepsilon_i \tag{H.36}$$

where $\varepsilon_i = h_i/\Delta_i$.

Now, the function of the hyperbolic tangent ($\tanh \beta_i h_i$) used in (H.35), can be represented as

$$\tanh \beta_i h_i = \tanh(1 + \mathrm{j})\varepsilon_i = \frac{\sinh 2\varepsilon_i + j \sin 2\varepsilon_i}{\cosh 2\varepsilon_i + \cos 2\varepsilon_i}$$

Using this condition, the impedance Z_i^* from (H.35) is converted to the form

$$Z_i^* = \frac{\rho_i}{h_i s} \frac{\beta_i h_i}{\tanh \beta_i h_i} \frac{\xi_{Li}}{\xi_{\tau i}} = \frac{\rho_i}{h_i s} \frac{\xi_{Li}}{\xi_{\tau i}} \frac{\varepsilon_i (1 + j)(\cosh 2\varepsilon_i + \cos 2\varepsilon_i)}{\sinh 2\varepsilon_i + j \sin 2\varepsilon_i} \quad \text{(H.37)}$$

By this expression, we select the real and imaginary components of the impedance Z_i^*, and as a result, the impedance Z_i^* is represented as

$$Z_i^* = r_{ii}^*/s + jx_{ii\sigma}^* \quad \text{(H.38)}$$

Here, the active resistance r_{ii}^*/s and the leakage reactance $x_{ii\sigma}^*$ are caused by the eddy currents induced in the ith layer. The active resistance r_{ii}^*/s from (H.37) can be expressed as

$$\frac{r_{ii}^*}{s} = \frac{\rho_i}{h_i s} \frac{\xi_{Li}}{\xi_{\tau i}} k_r \quad \text{(H.39)}$$

where $k_r = \varepsilon_i \dfrac{\sinh 2\varepsilon_i + \sin 2\varepsilon_i}{\cosh 2\varepsilon_i - \cos 2\varepsilon_i}$.

From (H.37), the leakage reactance of the eddy currents $x_{ii\sigma}^*$ takes the form

$$x_{ii\sigma}^* = \omega_1 \mu_{xi} \frac{h_i}{3} \frac{\xi_{Li}}{\xi_{\tau i}} k_x = \omega_1 \mu_{xi} t_0 \frac{h_i}{3t_i} k_R \xi_{Li} k_x \quad \text{(H.40)}$$

where $k_x = \dfrac{3}{2\varepsilon_i} \dfrac{\sinh 2\varepsilon_i - \sin 2\varepsilon_i}{\cosh 2\varepsilon_i - \cos 2\varepsilon_i}; k_R = \dfrac{Z_1}{Z_2}$.

We consider the impedance $Z_{\tau i}^*$ following from (14.41) in the form

$$Z_{\tau i}^* = \frac{j\omega_1 \mu_{xi}}{\beta_i} \frac{\xi_{Li}}{\xi_{\tau i}} \tanh \beta_i h_i = \frac{j\omega_1 \mu_{xi}}{\beta_i} \frac{\xi_{Li}}{\xi_{\tau i}} \frac{\sinh 2\varepsilon_i + j \sin 2\varepsilon_i}{\cosh 2\varepsilon_i + \cos 2\varepsilon_i} \quad \text{(H.41)}$$

From (H.41), we represent the impedance $Z_{\tau i}^*$ as the sum of the real and imaginary components, i.e.,

$$Z_{\tau i}^* = r_{\tau i}^*/s + jx_{\tau i}^* \quad \text{(H.42)}$$

Based on (H.41) and (H.42), we can get for the active resistance $r_{\tau i}^*/s$

$$\frac{r_{\tau i}^*}{s} = \frac{\rho_i}{h_i s} \frac{\xi_{Li}}{\xi_{\tau i}} k_{\tau ir} \quad \text{(H.43)}$$

where $k_{\tau ir} = \varepsilon_i \dfrac{\sinh 2\varepsilon_i - \sin 2\varepsilon_i}{\cosh 2\varepsilon_i + \cos 2\varepsilon_i}$.

The reactance $x_{\tau i}^*$

$$x_{\tau i}^* = \omega_1 \mu_{xi} h_i \frac{\xi_{Li}}{\xi_{\tau i}} k_{\tau ix} = \omega_1 \mu_{xi} t_0 \frac{h_i}{t_i} k_R \xi_{Li} k_{\tau ix} \tag{H.44}$$

where $k_{\tau ix} = \dfrac{1}{2\varepsilon_i} \dfrac{\sinh 2\varepsilon_i + \sin 2\varepsilon_i}{\cosh 2\varepsilon_i + \cos 2\varepsilon_i}$.

The impedance $Z_{\tau i}^*$ used in the L-shaped equivalent circuit in Fig. H.1 is multiplied, in the general case, by the complex coefficient c_i^2. Through the real and imaginary components, this coefficient can be represented as

$$c_i^2 = k_{ir} + jk_{ix} \tag{H.45}$$

Here, for the coefficients k_{ir} and k_{ix}, we have from [4]

$$\begin{aligned} k_{ir} &= 0.5(1 + \cosh 2\varepsilon_i \cdot \cos 2\varepsilon_i) \\ k_{ix} &= 0.5 \sinh 2\varepsilon_i \cdot \sin 2\varepsilon_i \end{aligned} \tag{H.46}$$

Using the relations (H.42) and (H.45), for the impedance $Z_{\tau i}^* c_i^2$ used in the equivalent circuit in Fig. H.1, we can obtain

$$\begin{aligned} Z_{\tau i}^* c_i^2 &= \left(r_{\tau i}^*/s + jx_{\tau i}^*\right)\left(k_{ir} + jk_{ix}\right) \\ &= \frac{r_{\tau i}^*}{s} k_{ir} \left(1 - \frac{x_{\tau i}^* s}{r_{\tau i}^*} \frac{k_{ix}}{k_{ir}}\right) + jx_{\tau i}^* k_{ir} \left(1 + \frac{r_{\tau i}^*}{x_{\tau i}^* s} \frac{k_{ix}}{k_{ir}}\right) \\ &= \widetilde{r}_{\tau i}^*/s + j\widetilde{x}_{\tau i}^* \end{aligned} \tag{H.47}$$

where $\dfrac{\widetilde{r}_{\tau i}^*}{s} = \dfrac{r_{\tau i}^*}{s} k_{ir} \left(1 - \dfrac{x_{\tau i}^* s}{r_{\tau i}^*} \dfrac{k_{ix}}{k_{ir}}\right)$; $\widetilde{x}_{\tau i}^* = x_{\tau i}^* k_{ir} \left(1 + \dfrac{r_{\tau i}^*}{x_{\tau i}^* s} \dfrac{k_{ix}}{k_{ir}}\right)$.

Based on the expressions (H.38) and (H.47), the L-shaped equivalent circuit obtained for the layer in Fig. H.1 takes the form in Fig. H.3.

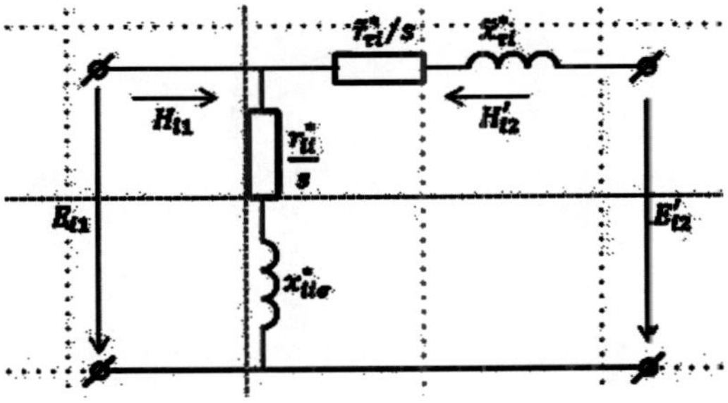

Fig. H.3 L-shaped equivalent circuit of the ith layer for the strong skin effect

Summary

1. The elements for calculating the conducting layer are given. The circuit loops, parameters, equivalent circuits, and voltage equations are considered as the elements for calculating the conductive layer. The elements for calculating the conducting layer follow from the solution of Maxwell's equations in this layer under consideration ("field" concept).
2. The elements for calculating the conducting layer are obtained on the basis of its L-shaped equivalent circuit. In the general case, the L-shaped equivalent circuit of the conductive layer consists of the two independent circuit loops: the "own" circuit loop of the layer and the "connecting" circuit loop that unites the "own" circuit loops of the adjacent layers. As a result, the L-shaped equivalent circuit of the layer takes the form convenient for practical use. Therefore, the L-shaped equivalent circuit is considered as the original circuit in determining the calculation elements of the electric machine.
3. The parameters of the L-shaped equivalent circuit of the conducting layer are analyzed in the area of the weak and strong skin effect. The expressions are obtained that are similar in the structure and physical meaning to the expressions used in the technical literature to calculate the parameters of the electric machine.

References

1. Shimoni, K.: Theoretical Electrical Engineering (Translation from German), Publishing Mir, Moscow (1973)
2. Voldek, A.: Induction Magneto-Hydrodynamic Machines with Liquid-Metal Working Substance, Publishing Energiya, Leningrad (1970)
3. Turovsky, Ya.: Technical Electrodynamics, (Translated from Polish), Energiya Publishing House, Moscow (1974)
4. Asanbayev, V.: Alternating Current Multi-Circuit Electric Machines: A New Approach to Steady-State Parameter Determination, Springer Cham Heidelberg, New York, Dordrecht London (2015)

Appendix I: Reducing “Unfolded” Equivalent Circuits of Stator and Rotor: “Field” Concept

Reducing “Unfolded” Stator Equivalent Circuit to Equivalent Circuit with Parallel Connection of Elements

We proceed from the stator equivalent circuit shown in Fig. 15.4. To reduce the “unfolded” stator equivalent circuit (Fig. 15.4) to the equivalent circuit with the parallel connection of the elements, the parallel branches $k - k$, $z - z$, $1 - 1$ and $a - a$ should be transferred to its terminals $0 - 0$. For this purpose, we will use the multiple transfer technique described above [1]. Note that Figs. G.1 and 15.4 show the “unfolded” equivalent circuits for the stator. These equivalent circuits have the same structure. In Fig. G.1, the “unfolded” stator equivalent circuit is derived using the superposition principle. In Fig. 15.4, the “unfolded” stator equivalent circuit is constructed using the “field” concept. In this regard, the reduction of the “unfolded” stator equivalent circuit (Fig. 15.4) to the equivalent circuit with the parallel connection of the elements can be carried out by an analogy with the reduction of the “unfolded” stator equivalent circuit (Fig. G.1) to the equivalent circuit with the parallel connection of the elements. Therefore, the description of the conversion process of the “unfolded” stator equivalent circuit (Fig. 15.4) is not shown. In Appendix G, the equivalent circuit in Fig. G.1 is reduced to the form given in Fig. G.11. Consequently, it can be shown that by the analogy with the equivalent circuit in Fig. G.11 the “unfolded” stator equivalent circuit with the mixed connection of the elements (Fig. 15.4) takes the form of the “unfolded” stator equivalent circuit with the parallel connection of the elements shown in Fig. I.1. This equivalent circuit corresponds to transferring the parallel branches $k - k$, $z - z$, $1 - 1$, and $a - a$ to the terminals $0 - 0$ of the equivalent circuit in Fig. 15.4. For the elements of the equivalent circuit in Fig. I.1 ($\overset{\circ}{x}_{ks}$, $\overset{\circ}{x}_{zs}$, $\overset{\circ}{Z}_1$, $\overset{\circ}{x}_{as}$), the following expressions can be obtained.

V. Asanbayev, *Asynchronous Machines*, https://doi.org/10.1007/978-3-030-92284-9

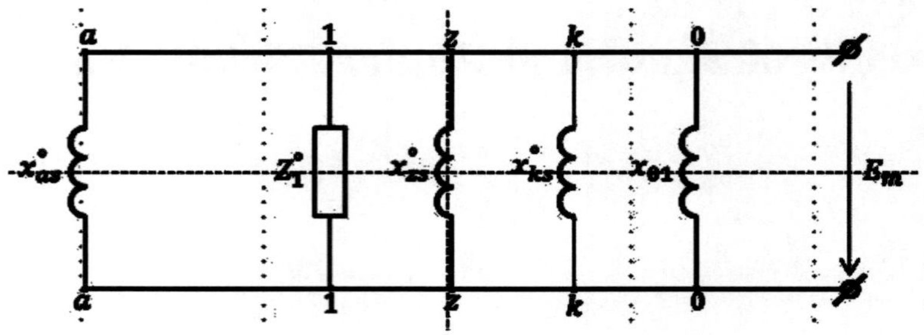

Fig. I.1 Stator equivalent circuit with the parallel connection of the elements

For the reactance $\overset{\circ}{x}_{ks}$

$$\overset{\circ}{x}_{ks} = x_{ks}c_{\tau ks} = x_{ks} + x_{\tau 01} \tag{I.1}$$

where $c_{\tau ks} = 1 + \dfrac{x_{\tau 01}}{x_{ks}}$.

We can receive for the reactance $\overset{\circ}{x}_{zs}$

$$\overset{\circ}{x}_{zs} = x_{zs}c^2_{\tau ks}c_{\tau zs} = x_{zs}c^2_{\tau ks} + x_{\tau ks}c^2_{\tau ks} + x_{\tau 01}c_{\tau ks} \tag{I.2}$$

where $c_{\tau zs} = 1 + \dfrac{x_{\tau ks}c^2_{\tau ks} + x_{\tau 01}c_{\tau ks}}{x_{zs}c^2_{\tau zs}}$.

The impedance $\overset{\circ}{Z}_1$ is obtained as

$$\begin{aligned}\overset{\circ}{Z}_1 &= r_{11}c^2_{\tau ks}c^2_{\tau zs} + j\left(x_{11\sigma}c^2_{\tau ks}c^2_{\tau zs} + x_{\tau ks}c^2_{\tau ks}c_{\tau zs} + x_{\tau 01}c_{\tau ks}c_{\tau zs}\right)\\ &+Z_c c^2_{\tau ks}c^2_{\tau zs} = r_{11}c^2_{\tau ks}c^2_{\tau zs} + jx_{1\sigma} + Z_c c^2_{\tau ks}c^2_{\tau zs}\end{aligned} \tag{I.3a}$$

where $x_{1\sigma} = x_{11\sigma}c^2_{\tau ks}c^2_{\tau zs} + x_{\tau ks}c^2_{\tau ks}c_{\tau zs} + x_{\tau 01}c_{\tau ks}c_{\tau zs}$.

In another form

$$\begin{aligned}\overset{\circ}{Z}_1 &= \{r_{11} + j[x_{11\sigma} + x_{\tau ks}(1/c_{\tau zs}) + x_{\tau 01}(1/c_{\tau ks}c_{\tau zs})] + Z_c\}c^2_{\tau ks}c^2_{\tau zs}\\ &= (r_{11} + j\widetilde{x}_{11\sigma} + Z_c)c^2_{\tau ks}c^2_{\tau zs}\end{aligned} \tag{I.3b}$$

where $\widetilde{x}_{11\sigma} = x_{11\sigma} + x_{\tau ks}(1/c_{\tau zs}) + x_{\tau 01}(1/c_{\tau ks}c_{\tau zs}); x_{1\sigma} = \widetilde{x}_{11\sigma}c^2_{\tau ks}c^2_{\tau zs}$.

The reactance $\overset{\circ}{x}_{as}$ is defined as

$$\overset{\circ}{x}_{as} = Z_{as}c^2_{\tau ks}c^2_{\tau zs} + x_{\tau zs}c^2_{\tau ks}c^2_{\tau zs} + x_{\tau ks}c^2_{\tau ks}c_{\tau zs} + x_{\tau 01}c_{\tau ks}c_{\tau zs} \tag{I.4}$$

Now, the "unfolded" stator equivalent circuit with the mixed connection of the elements (Fig. 15.4) takes the form of the "unfolded" stator equivalent circuit with the parallel connection of the elements shown in Fig. I.1. Note again that this equivalent circuit corresponds to transferring the parallel branches $k - k$, $z - z$, $1 - 1$, and $a - a$ to the terminals $0 - 0$ of the equivalent circuit in Fig. 15.4.

Reducing "Unfolded" Rotor Equivalent Circuit to Equivalent Circuit with Parallel Connection of Elements

We proceed from the equivalent circuit shown in Fig. 15.10. The Fig. 15.10 shows the single-cage rotor "unfolded" equivalent circuit with the mixed connection of the elements.

Below, the "unfolded" rotor equivalent circuit with the mixed connection of the elements is reduced to the "unfolded" equivalent circuit with the parallel connection of the elements. The reduction technique used in this work is associated with transferring the parallel branches $k - k$, $z - z$, $2 - 2$ and $a - a$ of the equivalent circuit in Fig. 15.10 to its terminals $0 - 0$. For this purpose, we will use the multiple transfer technique described above [1].

Note that Figs. G.12 and 15.4 show the "unfolded" equivalent circuits for the rotor. These equivalent circuits have the same structure, although the "unfolded" rotor equivalent circuit (Fig. G.12) is derived using the superposition principle, and the "unfolded" rotor equivalent circuit in Fig. 15.10 is constructed using the "field" concept. In this regard, the reduction of the "unfolded" rotor equivalent circuit (Fig. 15.10) to the equivalent circuit with the parallel connection of the elements can be carried out by an analogy with the reduction of the "unfolded" rotor equivalent circuit (Fig. G.12) to the equivalent circuit with the parallel connection of the elements. Therefore, the description of the conversion process of the "unfolded" rotor equivalent circuit (Fig. 15.10) is not given. In Appendix G, the equivalent circuit in Fig. G.12 is reduced to the form given in Fig. G.22. Consequently, it can be shown that by the analogy with the equivalent circuit in Fig. G.22 the "unfolded" rotor equivalent circuit with the mixed connection of the elements (Fig. 15.10) takes the form of the "unfolded" rotor equivalent circuit with the parallel connection of the elements shown in Fig. I.2. We note again that this equivalent circuit corresponds to transferring the parallel branches $k - k$, $z - z$, $2 - 2$, and $a - a$ to the terminals $0 - 0$ of the equivalent circuit in Fig. 15.10. The following expressions can be obtained for the elements $(\overset{\circ}{x}_{zR}, \overset{\circ}{x}_{zR}, \overset{\circ}{Z}_2, \overset{\circ}{x}_{aR})$ of the equivalent circuit in Fig. I.2.

For the reactance $\overset{\circ}{x}_{kR}$, it arises

$$\overset{\circ}{x}_{kR} = x_{kR} c_{\tau kR} = x_{kR} + x_{\tau 02} \tag{I.5}$$

where $c_{\tau kR} = 1 + \dfrac{x_{\tau 02}}{x_{kR}}$.

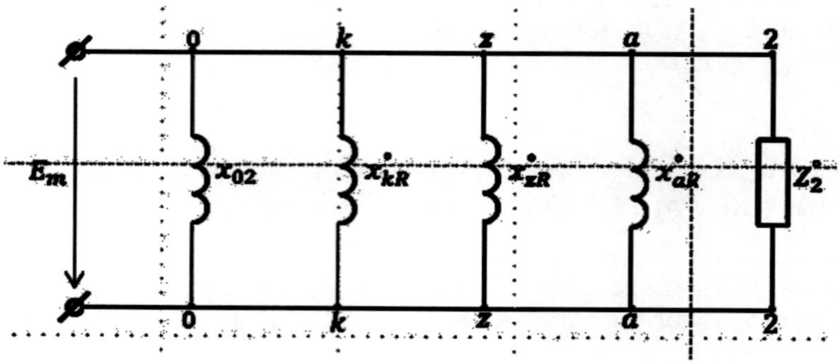

Fig. I.2 Single-cage rotor equivalent circuit with the parallel connection of the elements

The reactance $\mathring{x}_{zR}$ can be obtained

$$\mathring{x}_{zR} = x_{zR} c_{\tau zR} c_{\tau kR}^2 = x_{zR} c_{\tau kR}^2 + x_{\tau kR} c_{\tau kR}^2 + x_{\tau 02} c_{\tau kR} \tag{I.6}$$

where $c_{\tau zR} = 1 + \dfrac{x_{\tau 02} c_{\tau kR} + x_{\tau kR} c_{\tau kR}^2}{x_{zR} c_{\tau kR}^2}$.

We can receive the impedance $\mathring{Z}_2$

$$\begin{aligned} \mathring{Z}_2 &= (r_{22}/s) c_{\tau kR}^2 c_{\tau zR}^2 + j\left(x_{22\sigma} c_{\tau kR}^2 c_{\tau zR}^2 + x_{\tau kR} c_{\tau kR}^2 c_{\tau zR} + x_{\tau 02} c_{\tau kR} c_{\tau zR}\right) \\ &= (r_{22}/s) c_{\tau kR}^2 c_{\tau zR}^2 + j x_{2\sigma} \end{aligned} \tag{I.7a}$$

where $x_{2\sigma} = x_{22\sigma} c_{\tau kR}^2 c_{\tau zR}^2 + x_{\tau kR} c_{\tau kR}^2 c_{\tau zR} + x_{\tau 02} c_{\tau kR} c_{\tau zR}$.

Or

$$\begin{aligned} \mathring{Z}_2 &= \{r_{22}/s + j[x_{\tau 02}(1/c_{\tau kR} c_{\tau zR}) + x_{\tau kR}(1/c_{\tau zR}) + x_{22\sigma}]\} c_{\tau kR}^2 c_{\tau zR}^2 \\ &= (r_{22}/s + j\widetilde{x}_{22\sigma}) c_{\tau kR}^2 c_{\tau zR}^2 \end{aligned} \tag{I.7b}$$

where $\widetilde{x}_{22\sigma} = x_{\tau 02}(1/c_{\tau kR} c_{\tau zR}) + x_{\tau kR}(1/c_{\tau zR}) + x_{22\sigma}; x_{2\sigma} = \widetilde{x}_{22\sigma} c_{\tau kR}^2 c_{\tau zR}^2$.

The reactance $\mathring{x}_{aR}$ is determined as

$$\begin{aligned} \mathring{x}_{aR} &= x_{\tau 02} c_{\tau kR} c_{\tau zR} c_{\tau 2} + x_{\tau kR} c_{\tau kR}^2 c_{\tau zR} c_{\tau 2} \\ &+ x_{\tau zR} c_{\tau kR}^2 c_{\tau zR}^2 c_{\tau 2}^2 + x_{aR} c_{\tau kR}^2 c_{\tau zR}^2 c_{\tau 2}^2 \end{aligned} \tag{I.8}$$

Figure I.2 shows the "unfolded" single-cage rotor equivalent circuit with the parallel connection of the elements.

Summary

1. In this section, the "unfolded" stator and rotor equivalent circuits with the mixed connection of the elements are obtained on the basis of the "field" concept. The techniques used in Appendix G can be applied to transform the "unfolded" stator and rotor equivalent circuits with the mixed connection of the elements into the "unfolded" stator and rotor equivalent circuits with the parallel connection of the elements.
2. In Appendix G, the "unfolded" stator and rotor equivalent circuits with the mixed connection of the elements are constructed on the basis of the superposition principle. In this Appendix, the "unfolded" stator and rotor equivalent circuits with the mixed connection of the elements are received on the basis of the "field" concept. These "unfolded" equivalent circuits have the same structure. Therefore, in the case of the "field" concept, the reduction of the "unfolded" stator and rotor equivalent circuits with the mixed connection of the elements into the equivalent circuits with the parallel connection of the elements can be carried out in the same way as it was done in Appendix G. As a result, the "folded" stator and rotor equivalent circuits can be obtained from the use of the "field" concept. Now, the "folded" stator and rotor equivalent circuits can be used to construct the equivalent circuit of the electric machine.

Reference

1. Asanbayev, V.: Alternating Current Multi-Circuit Electric Machines: A New Approach to Steady-State Parameter Determination, Springer Cham Heidelberg, New York, Dordrecht London (2015)

Index

V. Asanbayev, *Asynchronous Machines*, https://doi.org/10.1007/978-3-030-92284-9

W